Gmelin Handbook of Inorganic and Organometallic Chemistry

8th Edition

Gmelin Handbook of Inorganic and Organometallic Chemistry

8th Edition

Gmelin Handbuch der Anorganischen Chemie

Achte, völlig neu bearbeitete Auflage

PREPARED AND ISSUED BY — Gmelin-Institut für Anorganische Chemie der Max-Planck-Gesellschaft zur Förderung der Wissenschaften
Director: Ekkehard Fluck

FOUNDED BY — Leopold Gmelin

8TH EDITION — 8th Edition begun under the auspices of the Deutsche Chemische Gesellschaft by R. J. Meyer

CONTINUED BY — E. H. E. Pietsch and A. Kotowski, and by Margot Becke-Goehring

Springer-Verlag
Berlin · Heidelberg · New York · London · Paris · Tokyo · Hong Kong · Barcelona · Budapest 1992

Organometallic Compounds in the Gmelin Handbook

The following listing indicates in which volumes these compounds are discussed or are referred to:

Ag	Silber B 5 (1975)
Au	Organogold Compounds (1980)
Be	Organoberyllium Compounds 1 (1987)
Bi	Bismut-Organische Verbindungen (1977)
Co	Kobalt-Organische Verbindungen 1, 2 (1973), Kobalt Erg.-Bd. A (1961), B 1 (1963), B 2 (1964)
Cr	Chrom-Organische Verbindungen (1971)
Cu	Organocopper Compounds 1 (1985), 2 (1983), 3 (1986), 4 (1987), Index (1987)
Fe	Eisen-Organische Verbindungen A 1 (1974), A 2 (1977), A 3 (1978), A 4 (1980), A 5 (1981), A 6 (1977), A 7 (1980), Organoiron Compounds A 8 (1986), A 9 (1989), A 10 (1991), Eisen-Organische Verbindungen B 1 (partly in English; 1976), Organoiron Compounds B 2 (1978), Eisen-Organische Verbindungen B 3 (partly in English; 1979), B 4, B 5 (1978), Organoiron Compounds B 6, B 7 (1981), B 8, B 9 (1985), B 10 (1986), B 11 (1983), B 12 (1984), B 13 (1988), B 14, B 15 (1989), B 16a, B 16b, B 17 (1990), B 18 (1991), B 19 (1992), Eisen-Organische Verbindungen C 1, C 2 (1979), Organoiron Compounds C 3 (1980), C 4, C 5 (1981), C 6a (1991), C 6b (1992), C 7 (1985), and Eisen B (1929 – 1932)
Ga	Organogallium Compounds 1 (1986)
Ge	Organogermanium Compounds 1 (1988), 2 (1989), 3 (1990)
Hf	Organohafnium Compounds (1973)
In	Organoindium Compounds 1 (1991)
Mo	Organomolybdenum Compounds 5 (1992), 6 (1990), 7 (1991), 8 (1992) **present volume**
Nb	Niob B 4 (1973)
Ni	Nickel-Organische Verbindungen 1 (1975), 2 (1974), Register (1975), Nickel B 3 (1966), and C 1 (1968), C 2 (1969)
Np, Pu	Transurane C (partly in English; 1972)
Os	Organoosmium Compounds A 1 (1992)
Pb	Organolead Compounds 1 (1987), 2 (1990), 3 (1992)
Po	Polonium Main Volume (1941)
Pt	Platin C (1939) and D (1957)
Re	Organorhenium 1, 2 (1989), 3 (1992)
Ru	Ruthenium Erg.-Bd. (1970)
Sb	Organoantimony Compounds 1, 2 (1981), 3 (1982), 4 (1986), 5 (1990)
Sc, Y, La to Lu	D 6 (1983)
Sn	Zinn-Organische Verbindungen 1, 2 (1975), 3, 4 (1976), 5 (1978), 6 (1979), Organotin Compounds 7 (1980), 8 (1981), 9 (1982), 10 (1983), 11 (1984), 12 (1985), 13 (1986), 14 (1987), 15, 16 (1988), 17 (1989), 18 (1990)
Ta	Tantal B 2 (1971)
Ti	Titan-Organische Verbindungen 1 (1977), 2 (1980), Organotitanium Compounds 3 (1984), 4 and Register (1984), 5 (1990)
U	Uranium Suppl. Vol. E 2 (1980)
V	Vanadium-Organische Verbindungen (1971), Vanadium B (1967)
Zr	Organozirconium Compounds (1973)

Gmelin Handbook of Inorganic and Organometallic Chemistry

8th Edition

Mo
Organomolybdenum Compounds

Part 8

With 55 illustrations

AUTHOR Hans Schumann (Bielefeld)

FORMULA INDEX Bernd Kalbskopf, Uwe Nohl, Hans-Jürgen Richter-Ditten, Edgar Rudolph

EDITOR Manfred Winter

CHIEF EDITOR Wolfgang Petz

Springer-Verlag
Berlin · Heidelberg · New York · London · Paris · Tokyo · Hong Kong · Barcelona · Budapest 1992

LITERATURE CLOSING DATE: END OF 1989
IN SOME CASES MORE RECENT DATA HAVE BEEN CONSIDERED

Library of Congress Catalog Card Number: Agr 25-1383

ISBN 3-540-93652-1 Springer-Verlag, Berlin · Heidelberg · New York · Tokyo
ISBN 0-387-93652-1 Springer-Verlag, New York · Heidelberg · Berlin · Tokyo

Typesetting, printing, and bookbinding: Wiesbadener Graphische Betriebe GmbH, Wiesbaden

Preface

This volume 8 is the fourth in a series dealing with organomolybdenum compounds. An Empirical Formula Index and a Ligand Formula Index provide ready access to the compounds covered.

Volume 5 describes mononuclear organomolybdenum compounds with isocyanide, carbene, carbyne, alkynyl, alkene, alkyne, ^3L, and ^4L ligands with and without additional CO groups. Volume 6 starts the description of mononuclear organomolybdenum compounds with one ^5L ligand, a ligand bonded to molybdenum by five carbon atoms. The compounds contain either zero or one CO group bonded to the molybdenum atom. Volume 7 continues the description of ^5L-molybdenum compounds containing two CO groups, but no additional ^nL ligands. **This volume** describes ^5L-molybdenum compounds with two CO groups and additional ^1L to ^4L ligands. Following the nomenclature used in this series of organomolybdenum compounds, ^nL is an organic ligand bonded by n C atoms to molybdenum, and ^mD is an electron donor ligand with m donor electrons. Thus ^2D denotes a ligand such as PR_3.

Many of the data, particular those in tables, are given in an abbreviated form without units; for explanations see p. X. Additional information, if necessary, is given before the individual table.

Frankfurt am Main
November 1992

Manfred Winter
Wolfgang Petz

Remarks on Abbreviations and Dimensions

Many compounds in this volume are presented in tables in which numerous abbreviations are used, the dimensions are omitted for the sake of conciseness. This necessitates the following clarifications.

Abbreviations used with **temperatures** are m.p. for melting point, dec. for decomposition, and b.p. for boiling point.

NMR represents **nuclear magnetic resonance**. Chemical shifts are given as δ values in ppm and to low field from the following reference substances: $Si(CH_3)_4$ for ^1H and ^{13}C, $CFCl_3$ for ^{19}F, and H_3PO_4 for ^{31}P. ^{17}O and ^{95}Mo NMR chemical shifts are given relative to H_2O and $Na_2[MoO_4]$ standards, respectively (downfield positive). Otherwise the reference substances and signs are given. Multiplicities of the signals are abbreviated as s, d, t, q, qui, sext, sept, and oct (singlet to octet), m (multiplet), and br (broad); terms like d of d are also used. Assignments referring to a labeled structural formula are given in the form C-4, H-3,5. Coupling constants J in Hz usually appear in parentheses after the δ value, along with the multiplicity and the assignment, and refer to the respective nucleus. If a more precise designation is necessary, it is given as, e.g., $^nJ(P, H)$ or J(H-1,3) referring to labeled formulas.

ESR represents **electron spin resonance**; the hyperfine interactions are given as $a(^{95}Mo)$ or a(P).

Optical spectra are labeled as IR (infrared) and UV (electronic spectrum including the visible region). IR and Raman bands are given in cm^{-1}. The assigned bands are usually labeled with the symbols ν for stretching, δ for deformation, ϱ for rocking, with the indexes sym and asym for symmetrical and asymmetrical. The UV absorption maxima, λ_{max}, are given in nm followed by the extinction coefficient ε (in $L \cdot cm^{-1} \cdot mol^{-1}$) or log ε in parentheses. The abbreviation sh is used for shoulder.

Solvents and **physical state** of the sample and the temperature (none is given if room temperature applies) are given in parentheses immediately after the spectral symbol, e.g., IR (solid), ^1H NMR (acetone-d_6, -30°C), or at the end of the data with their formula (C_6H_6 = benzene) or name (acetone-d_6), except THF, which represents tetrahydrofuran.

The abbreviation used with **electrochemical behavior** is SCE for saturated calomel electrode.

The bond distances in **Figures** are given in Å.

Further abbreviations:

D_{calc} for calculated density and D_{meas} for measured density.

Table of Contents

Organomolybdenum Compounds 8

1.5.1.3.2 Compounds with Additional ^{n}L Ligands

1.5.1.3.2.1 Compounds with Additional ^{1}L Ligands

1.5.1.3.2.1.1 Compounds of the Type $[^{5}LMo(CO)_2{}^{1}L]^-$ and $[^{5}LMo(CO)_2(^{1}L)X]^-$

This section contains anions of the type $[C_5H_5Mo(CO)_2{}^{1}L]^-$ with the two-electron donors ^{1}L = carbene or isonitrile (Nos. 1 to 7) and anions of formally Mo(II) of the type $[C_5H_5Mo(CO)_2(^{1}L)X]^-$ with ^{1}L = acyl or alkyl (Nos. 8 to 24). The approximately planar arrangement of ^{1}L, X, and the two CO groups gives rise to cis and trans isomers. Only compounds with $^{5}L = C_5H_5$ are described.

The compounds listed in Table 1 were prepared in most cases by the following methods:

Method I: $C_5H_5Mo(CO)_2(^{1}L)I$ (^{1}L = carbene) was reduced with two equivalents of sodium naphthalide in THF at −78°C. THF solutions of the compounds were obtained in good yields and used for further reactions without isolation [18].

Method II: $C_5H_5Mo(CO)_2(C{\equiv}NCH_3)Cl$ [4, 5, 8, 18] or trans-$(C_5H_5Mo(CO)_2C{\equiv}NCH_3)_2$ [20] was allowed to react in THF solution at room temperature with 1% Na/Hg for 30 to 45 min [5, 21]. A quantitative yield was assumed and the compound was used for further reactions without isolation.

Method III: $C_5H_5Mo(CO)_3{}^{1}L$ ($^{1}L = CH_3$, C_2H_5, $CH_2C_6H_5$, $(CH_2)_3CN$) reacted in refluxing H_2O/CH_3OH (1:3) for up to 30 min with 1.3 equivalents of KCN. After removal of the solvent the residue was extracted with THF and ether was added to the solution to precipitate the product. Coordinated THF could be removed by dissolving in methanol. Anion exchange with $[As(C_6H_5)_4]Cl$ in H_2O afforded the related arsonium salts [1 to 3]. No. 14 was prepared by reaction with $[N(C_2H_5)_4]GeCl_3$ in acetone for 23 h at 25°C followed by solvent evaporation and recrystallization [8]. Alternatively, $C_5H_5Mo(CO)_3E(C_6H_5)_3$ (E = Ge, Sn) was allowed to react in $CH_3OC_2H_4OCH_3$ with one equivalent of $LiCH_3$ followed by addition of aqueous $[N(CH_3)_4]Br$. The solids obtained were purified by recrystallization [10].

General Remarks. Method III produces at first exclusively the cis isomers (Nos. 12, 16, 18, and 20) which slowly isomerize in solution; equilibrium at a constant cis/trans ratio (about 1:4) was obtained after 0.5 to 2 h as monitored by ^{1}H NMR spectroscopy in CD_3OD solution at 40°C [1, 3]; the time-dependent spectra of Nos. 16 and 18 are depicted [3].

Alkylation of $[C_5H_5Mo(CO)_2(C(O)R)X]^-$ (X = $GeCl_3$, $Ge(C_6H_5)_3$, or $Sn(C_6H_5)_3$) with $[OR'_3]^+$ (R′ = CH_3, C_2H_5) gives the corresponding carbene compounds $C_5H_5Mo(CO)_2({=}C(OR')R)X$ [8, 10].

References on pp. 6/7

Table 1
Compounds of the Type $[^5LMo(CO)_2{}^1L]^-$ and $[^5LMo(CO)_2(^1L)X]^-$.
An asterisk indicates further information at the end of the table.
For explanations, abbreviations, and units see p. X.

No.	compound	method of preparation (yield) properties and remarks
compounds of the type $[^5LMo(CO)_2{}^1L]^-$ (^1L = carbene)		
1	$Na[C_5H_5Mo(CO)_2{=}C(CH_3)NHCH_3]$	I pale yellow solution in THF; no further information given [18]
*2	$Na[C_5H_5Mo(CO)_2{=}C(CH_2)_3NCH_3]$ ($=C(CH_2)_3NCH_3$ = [structure: N-methylpyrrolidin-2-ylidene])	I pale yellow solution in THF IR (THF): 1666, 1786 (ν(CO)); 1673, 1785 (ν(CO)) upon addition of 15-crown-5
3	$Na[C_5H_5Mo(CO)_2{=}C(CH_2)_4NCH_3]$ ($=C(CH_2)_4NCH_3$ = [structure: N-methylpiperidin-2-ylidene])	I pale yellow solution in THF; no further information given [18]
4	$Na[C_5H_5Mo(CO)_2{=}C(CH_2)_3O]$ ($=C(CH_2)_3O$ = [structure: oxolan-2-ylidene])	I reacts with excess $(C_6H_5)_3SnCl$ in THF to afford a > 50% yield of the carbene complex trans-$C_5H_5Mo(CO)_2(C(CH_2)_3O)Sn(C_6H_5)_3$ [18]
compounds of the type $[^5LMo(CO)_2{}^1L]^-$ (^1L = isonitrile)		
*5	$Na[C_5H_5Mo(CO)_2C{\equiv}NCH_3]$	II (not isolated) [4 to 6, 9, 13, 15, 19, 21] light yellow solution in THF [4, 5]; pale yellow-green solution in THF [21] ^1H NMR (THF-d_8): 2.90 (NCH_3), 4.95 (C_5H_5) [4, 5] IR (THF): 1710, 1765 (ν(CO)), 1875 (ν(CN)) [4, 5], given as figure in [5]; nearly identical data in [21]

References on pp. 6/7

Table 1 (continued)

No.	compound	method of preparation (yield) properties and remarks
6	$Na[C_5H_5Mo(CO)_2C{\equiv}NC_4H_9\text{-t}]$	II (not isolated) [21] pale yellow-green solution in THF [21] IR (THF): 1717, 1769 (ν(CO)), 1873 (ν(CN)) [21] reaction with $(C_6H_5)_3SnCl$ in THF affords 52% of $C_5H_5Mo(CO)_2(C{\equiv}NC_4H_9\text{-t})Sn(C_6H_5)_3$ [21]
7	$Na[C_5H_5Mo(CO)_2C{\equiv}NC_6H_5]$	II (not isolated) [7 to 11] reacts with one equivalent of CH_3I in THF at −78°C to afford a 78% yield of $C_5H_5Mo(CO)_2CCH_3{=}NC_6H_5$ [7 to 9, 11]
compounds of the type $[^5LMo(CO)_2(^1L)X]^-$		
*8	trans-$Li[C_5H_5Mo(CO)_2(C(O)CH_3)H]$	^1H NMR (THF, −25°C): −5.15 (s, MoH), 2.25 (s, CH_3), 4.90 (s, C_5H_5) [14, 17] ^{13}C NMR (THF, −50°C): 51.8 (CH_3), 92.1 (C_5H_5), 237.9 (CO), 306.9 (C=O) [14, 17] IR (THF, −20°C): 1816, 1906 (ν(CO)); given also as diagram [14, 17]
*9	trans-$Li[C_5H_5Mo(CO)_2(C(O)CH_3)D]$	^2H NMR (THF, −40°C): −4.9 (MoD) [17]
*10	trans-$Li[C_5H_5Mo(CO)_2(C(O)CD_3)H]$	^2H NMR (THF, −30°C): 2.3 (CD_3) [17]
*11	$Li[C_5H_5Mo(CO)_2(C(O)(CH_2)_3Br)H]$	^1H NMR (THF, low temperature): −6.5 (MoH) [22]
12	$K[C_5H_5Mo(CO)_2(C(O)CH_3)CN] \cdot n\ THF$	III (83%); THF-free after precipitation from THF with ether [1] pure trans isomer in the solid state, but cis/trans mixture in solution [1] ^1H NMR (D_2O): 2.65 (s, CH_3), 5.45 and 5.59 (s, C_5H_5, cis and trans isomer); (CD_3OD, 40°C): 2.53 (CH_3), 5.2 (C_5H_5, trans isomer), 5.4 (C_5H_5, cis isomer); for cis/trans isomerization, see General Remarks [1] IR (KBr): 1520 (ν(C=O)), 1832, 1930 (ν(CO)), 2080 (ν(CN)) [1] air-sensitive, hygroscopic, soluble in water and polar solvents but insoluble in benzene and aliphatic hydrocarbons [1] reaction with CH_3I in CH_3CN gives $C_5H_5Mo(CO)_2(C{\equiv}NCH_3)C(O)CH_3$ [1]
13	$[As(C_6H_5)_4][C_5H_5Mo(CO)_2(C(O)CH_3)CN]$	III [1] yellow needles (from THF/ether/pentane), cis/trans mixture [1] ^1H NMR (acetone-d_6): 2.3 (s, CH_3), 4.9 and 5.15 (s, C_5H_5), 7.92 (m, C_6H_5) [1]

References on pp. 6/7

Table 1 (continued)

No.	compound	method of preparation (yield) properties and remarks
13 (continued)		IR (KBr): 1570 (ν(C=O)), 1830, 1930 (ν(CO)), 2090 (ν(CN)) [1] soluble in most organic solvents but insoluble in water, more air-stable than No. 12 [1]
14	$[N(C_2H_5)_4][C_5H_5Mo(CO)_2(C(O)CH_3)GeCl_3]$	III (45%) [8] solid, m.p. 85 to 86°C (from acetone/ether) [8] ^{1}H NMR ($CDCl_3$): 1.42 (t, CH_3; J(H, H) = 7), 2.42 (s, $C(O)CH_3$), 3.52 (q, NCH_2; J(H, H) = 7), 5.13 (s, C_5H_5) [8] IR (acetone): 1857, 1935 (ν(CO)) [8]
15	trans-$[N(CH_3)_4][C_5H_5Mo(CO)_2(C(O)CH_3)Ge(C_6H_5)_3]$	III (20%) [10] yellow solid, m.p. 215 to 218°C (from acetone/ether) [10] ^{1}H NMR: 2.45 ($C(O)CH_3$), 4.83 (C_5H_5) [10] IR (THF): 1543 (ν(C=O)), 1811, 1890 (ν(CO)) [10]
16	$K[C_5H_5Mo(CO)_2(C(O)C_2H_5)CN]$	III (85%) [3, mentioned in 2] ^{1}H NMR (CD_3OD, 40°C): cis isomer: 0.98 (CH_3), 2.8 to 3.3 (ABX_3 system, CH_2), 5.48 (C_5H_5); trans isomer: 0.93 (CH_3), 3.05 (q, CH_2), 5.28 (C_5H_5); for cis/trans isomerization, see General Remarks [3] IR (KBr): 1562 (ν(C=O)), 1835, 1930 (ν(CO)), 2090 (ν(CN)) [3]
17	$[As(C_6H_5)_4][C_5H_5Mo(CO)_2(C(O)C_2H_5)CN]$	III [3] yellow needles (from THF/ether/pentane) [3]
18	$K[C_5H_5Mo(CO)_2(C(O)CH_2C_6H_5)CN]$ · n THF	III (80%); THF-free after precipitation from CH_3OH with ether [3] ^{1}H NMR (CD_3OD, 40°C): cis isomer: 4.34 and 4.49 (AB system, CH_2), 5.48 (C_5H_5), 7.3 (m, C_6H_5); trans isomer: 4.37 (s, CH_2), 5.27 (C_5H_5), 7.3 (m, C_6H_5); for cis/trans isomerization, see General Remarks [3] IR (KBr): 1560 (ν(C=O)), 1835, 1930 (ν(CO)), 2090 (ν(CN)) [3]
19	$[As(C_6H_5)_4][C_5H_5Mo(CO)_2(C(O)CH_2C_6H_5)CN]$	III [3] yellow needles (from THF/ether/pentane) [3]

References on pp. 6/7

Table 1 (continued)

No.	compound	method of preparation (yield) properties and remarks
20	$K[C_5H_5Mo(CO)_2(C(O)(CH_2)_3CN)CN]$ · n THF	III (73%); THF-free complex is obtained by precipitation from CH_3OH with ether [3] ^1H NMR (CD_3OD, 40°C): cis isomer: 1.85 (CH_2), 2.5 (CH_2CN), 3.14 and 3.26 (ABX_2 system, $C(=O)CH_2$), 5.5 (C_5H_5); trans isomer: 1.85 (CH_2), 2.5 (CH_2CN), 3.23 (t, $C(O)CH_2$), 5.34 (C_5H_5) [3] IR (KBr): 1564 (ν(C=O)), 1838, 1932 (ν(CO)), 2092 (ν(CN)), 2180 (ν(CN), nitrile) [3]
21	$[As(C_6H_5)_4][C_5H_5Mo(CO)_2(C(O)(CH_2)_3CN)CN]$	III [3] yellow needles (from THF/ether/pentane) [3]
22	trans-$[N(CH_3)_4][C_5H_5Mo(CO)_2(C(O)C_6H_5)Ge(C_6H_5)_3]$	III (45%) [10] yellow solid, m.p. 212 to 215°C (from acetone/ether) [10] ^1H NMR: 4.92 (C_5H_5) [10] IR (THF): 1543 (ν(C=O)), 1814, 1899 (ν(CO)) [10]
23	trans-$[N(CH_3)_4][C_5H_5Mo(CO)_2(C(O)C_6H_5)Sn(C_6H_5)_3]$	III (18%) [10] yellow solid, m.p. 216 to 217°C (from acetone/ether) [10] ^1H NMR: 5.02 (C_5H_5) [10] IR (THF): 1515 (ν(C=O)), 1811, 1892 (ν(CO)) [10]
24	$K[C_5H_5Mo(CO)_2(CH_2CH_2CH_2CN)CN]$	$C_5H_5Mo(CO)_3C_3H_6Br$ and two equivalents of KCN were refluxed in CH_3OH for 30 min; after removal of the solvent, the residue was extracted with THF (86%); THF-free product from methanol [2] golden-yellow solid [2] ^1H NMR (CD_3OD): 1.9 to 2.7 (m, α- and β-CH_2), 3.83 and 3.85 (t, γ-CH_2, trans and cis isomer), 5.17 and 5.42 (s, C_5H_5, trans and cis isomer) [2] IR (KBr): 1840, 1930 (ν(CO)), 2090 (ν(CN)), 2185 (ν(CN), nitrile) [2] mechanism of formation is discussed [2]

*Further information:

$Na[C_5H_5Mo(CO)_2{=}C(CH_2)_3NCH_3]$ (Table **1**, No. **2**). Addition of 15-crown-5 to a THF solution of the complex causes the ν(CO) absorptions to change which indicates that the anion exists in THF solution exclusively as tight ion pairs with the Na^+ bonded to the oxygen atom of a carbonyl group [18].

References on pp. 6/7

Reactions proceed with retention of the carbene ligand. Thus, protonation with acetic acid affords cis/trans-$C_5H_5Mo(CO)_2(=C(CH_2)_3NCH_3)H$ and X_2 (X = Br or I) gives the corresponding $C_5H_5Mo(CO)_2(=C(CH_2)_3NCH_3)X$. Excess of $(C_6H_5)_3SnCl$ produces trans-$C_5H_5Mo(CO)_2(=C(CH_2)_3$-$NCH_3)Sn(C_6H_5)_3$ in 66% yield [18].

$Na[C_5H_5Mo(CO)_2C{\equiv}NCH_3]$ (Table **1**, No. **5**; Na^+ is not specifically stated in all cases) reacts with a variety of agents in THF. The reactions are compiled in the following table:

reaction with (conditions)	product
Na/Hg	$(C_5H_5Mo(CO)_2C{\equiv}NCH_3)_2Hg$ [5]
$(CH_3)_3GeBr$ (−78 to +25°C)	cis/trans-$C_5H_5Mo(CO)_2(C{\equiv}NCH_3)Ge(CH_3)_3$ [4, 5]
$(CH_3)_3SnCl$ (−78 to +25°C)	cis/trans-$C_5H_5Mo(CO)_2(C{\equiv}NCH_3)Sn(CH_3)_3$ [4]
$(C_6H_5)_3SnCl$ (0.5 h, 25°C)	$C_5H_5Mo(CO)_2(C{\equiv}NCH_3)Sn(C_6H_5)_3$ [21]
$(C_6H_5)_3PbI$ (−78 to +25°C)	cis/trans-$C_5H_5Mo(CO)_2(C{\equiv}NCH_3)Pb(C_6H_5)_3$ [4, 5]
HgI_2 (−78 to +25°C)	$C_5H_5Mo(CO)_2(C{\equiv}NCH_3)HgI$ [5]
CH_3COOH (3 equivalents)	$C_5H_5Mo(CO)_2(C{\equiv}NCH_3)H$ [4, 5]
CH_3I (−78 to +25°C)	cis-$C_5H_5Mo(CO)_2C(CH_3)=NCH_3$ [9, 11, 19]
CH_3I (10 equivalents in refluxing THF for 38 h)	$C_5H_5Mo(CO)_2(C(CH_3)NHCH_3)I$ [9]
$I(CH_2)_nI$ (n = 3, 4)	cis-$C_5H_5Mo(CO)_2(=C(CH_2)_nNCH_3)I$ [13, 15, 19]

$Li[C_5H_5Mo(CO)_2(C(O)CH_3)X]$ (Table **1**, Nos. **8**, **9**, X = H or D) and **$Li[C_5H_5Mo(CO)_2(C(O)CD_3)H]$** (Table **1**, No. **10**) were spectroscopically identified as intermediates at −40°C by the reduction of $C_5H_5Mo(CO)_3R$ (R = CH_3, CD_3) with $Li[(C_2H_5)_3BX]$ (X = H, D) in THF at −78°C. In the first step, the corresponding hydride (or D^-) adducts at a CO ligand, $Li[C_5H_5Mo(CO)_2(C(O)X)R]$, were observed which rearrange to the title complexes to give the ultimate products $Li[C_5H_5Mo(CO)_2O{=}C(X)R\text{-}\eta^2]$ at ambient temperature [14, 17]; see also [16].

$Li[C_5H_5Mo(CO)_2(C(O)(CH_2)_3Br)H]$ (Table **1**, No. **11**) was formed as an intermediate like Nos. 8 and 9. Final results of further rearrangements at ambient temperature is $C_5H_5Mo(CO)_2CH(CH_2)_3O$ (No. 46, p. 126) (20%) probably via the intermediate anion $[C_5H_5Mo(CO)_2O{=}CH(CH_2)_3Br\text{-}\eta^2]^-$; small amounts of the allyl complex, $C_5H_5Mo(CO)_2CH_2CHCHCH_3\text{-}\eta^3$, were obtained as by-product; the mechanism of the reaction is discussed [22].

References:

[1] Kruck, T.; Höfler, M.; Liebig, L. (Chem. Ber. **105** [1972] 1174/83).
[2] Kruck, T.; Liebig, L. (Chem. Ber. **106** [1973] 1055/61).
[3] Kruck, T.; Liebig, L. (Chem. Ber. **106** [1973] 3588/94).
[4] Adams, R. D. (J. Organometal. Chem. **88** [1975] C 38/C 40).
[5] Adams, R. D. (Inorg. Chem. **15** [1976] 169/74).
[6] Adams, R. D.; Chodosh, D. F. (J. Am. Chem. Soc. **98** [1976] 5391/3).
[7] Adams, R. D.; Chodosh, D. F. (J. Organometal. Chem. **122** [1976] C 11/C 14).
[8] Dean, W. K.; Graham, W. A. G. (J. Organometal. Chem. **120** [1976] 73/86).
[9] Adams, R. D.; Chodosh, D. F. (J. Am. Chem. Soc. **99** [1977] 6544/50).
[10] Dean, W. K.; Graham, W. A. G. (Inorg. Chem. **16** [1977] 1061/7).

[11] Chodosh, D. F. (Diss. Univ. New York 1977).
[12] Adams, H.; Bailey, N. A.; Cahill, P.; Rogers, D.; Winter, M. J. (J. Chem. Soc. Chem. Commun. **1983** 831/3).

[13] Adams, H.; Bailey, N. A.; Osborne, V. A.; Winter, M. J. (J. Organometal. Chem. **284** [1984] C 1/C 4).
[14] Gauntlett, J. T.; Taylor, B. F.; Winter, M. J. (J. Chem. Soc. Chem. Commun. **1984** 420/1).
[15] Adams, H.; Bailey, N. A.; Osborn, V. A.; Winter, M. J. (Chem. Uses Molybdenum Proc. 5th Intern. Conf., Newcastle Upon Tyne, Engl., 1985, pp. 129/30; poster 1.17).
[16] Gauntlett, J. T.; Winter, M. J. (Chem. Uses Molybdenum Proc. 5th Intern. Conf., Newcastle Upon Tyne, Engl., 1985, pp. 95/6).
[17] Gauntlett, J. T.; Taylor, B. F.; Winter, M. J. (J. Chem. Soc. Dalton Trans. **1985** 1815/20).
[18] Osborn, V. A.; Winter, M. J. (J. Chem. Soc. Chem. Commun. **1985** 1744/5).
[19] Adams, H.; Bailey, N. A.; Osborn, V. A.; Winter, M. J. (J. Chem. Soc. Dalton Trans. **1986** 2127/35).
[20] Osborn, V. A.; Parker, C. A.; Winter, M. J. (J. Chem. Soc. Chem. Commun. **1986** 1185/6).
[21] Adams, H.; Bailey, N. A.; Bannister, C.; Faers, M. A.; Fedorko, P.; Osborn, V. A.; Winter, M. J. (J. Chem. Soc. Dalton Trans. **1987** 341/8).
[22] Adams, H.; Bailey, N. A.; Cahill, P.; Rogers, D.; Winter, M. J. (J. Chem. Soc. Dalton Trans. **1986** 2119/26).

1.5.1.3.2.1.2 Compounds of the Type $^{5}LMo(CO)_2{}^{1}L$ and Related Ions

All of the complexes or intermediates described in this section (16- and 17-electron species) have been observed only by spectroscopic methods due to their instability. Paramagnetic species of unknown nature were also detected by ESR spectroscopy during the photolysis of $C_5H_5Mo(CO)_3CH_3$ in $(t\text{-}C_4H_9)_2O_2$ and in the presence of $P(OC_4H_9\text{-}n)_3$ [10].

$C_5H_5Mo(CO)_2C(O)R$ (R = CH_3 [1, 2], C_2H_5, or $CH_2C_6H_5$ [2]). These 16-electron species were supposed to be intermediates during the reaction of $C_5H_5Mo(CO)_3CH_3$ with one equivalent of $P(C_6H_5)_3$ in THF at 35°C to give $C_5H_5Mo(CO)_2(P(C_6H_5)_3)C(O)CH_3$. The rate constants of formation, k_1 and k_{-2}, in the reaction sequence $C_5H_5Mo(CO)_3CH_3 \underset{k_{-1}}{\overset{k_1}{\rightleftharpoons}} C_5H_5Mo(CO)_2C(O)CH_3 \underset{k_{-2}}{\overset{k_2}{\rightleftharpoons}}$ $C_5H_5Mo(CO)_3(P(C_6H_5)_3)C(O)CH_3$ have been estimated; the presence of a solvent molecule was not excluded [1]. It is an intermediate during the reaction of $C_5H_5Mo(CO)_3R$ with $C_5H_5Mo(CO)_3H$ to give aldehydes [2].

$C_5H_5Mo(CO)_2CH_3$ was observed besides CO as a product of the UV irradiation of C_5H_5-$Mo(CO)_3CH_3$ in the range of 300 to 370 nm in a matrix at low temperatures [4, 5].

Infrared band positions (in cm^{-1}) observed in the ν(CO) region in various matrixes (PVC is polyvinyl chloride) at 12 K are given in the following table:

	CH_4 [4, 5]	Ar [5]	N_2 [5]	CO [5]	PVC [4; similar data in 7]
A′	1966.0	1972.0	1972.8	1962.4	1950
A″	1880.1	1886.8	1884.4	1876.8	1851

Similar values were also obtained in a paraffin matrix at 75 K; the spectrum is given as a figure in [6]. The energy-factored force and interaction constants (CH_4 matrix) were calculated: K = 1493.5 and k_i = 67.0 N/m [5]. The ν(CO) bands of ^{13}CO-enriched samples in a CH_4 matrix at 12 K were found for $C_5H_5Mo(^{12}CO)(^{13}CO)CH_3$ at 1852.7 and 1947.8 cm^{-1} and for $C_5H_5Mo(^{13}CO)_2CH_3$ at 1836.8 (A″) cm^{-1}; the A′ band (1921 cm^{-1} calculated) is obscured by $C_5H_5Mo(CO)_3CH_3$. A good agreement between calculated and observed CO stretching vibrations is found [5].

References on pp. 8/9

On warming the polyvinyl chloride matrix to 40 and 100 K the bands at 1851 and 1950 cm^{-1} decrease and the $C_5H_5Mo(CO)_3CH_3$ bands grow. Further warming to 293 K yields $C_5H_5Mo(CO)_3Cl$ [4, 7] and $C_5H_5Mo(CO)_3H$ [7].

$C_5H_5Mo(CO)_2C_2H_5$ was obtained either by the UV irradiation of $C_5H_5Mo(CO)_3C_2H_5$ in a matrix [6 to 8] or $C_5H_5Mo(CO)_3H$ in a CH_4 matrix with 5% ethylene [9] at low temperatures.

Infared bands (in cm^{-1}) observed in the ν(CO) region in various matrixes (PVC is polyvinyl chloride) at 12 K are given in the following table:

CH_4 [8]	CO [8]	CH_4 with 5% ethylene [9]	PVC [7]	paraffin (77 K) [6]
1957.8	1956.6	1955.0	1940	1958
1876.2	1876.3	1872.5	1852	1879

Warming the paraffin matrix up to 195 K affords trans-$C_5H_5Mo(CO)_2(H_2C{=}CH_2)H$ reversibly. Kinetic parameters for this conversion are: $k = 0.004\ s^{-1}$, $t_{1/2} = 170$ s; $\Delta G^{\ddagger} = 10$ kcal/mol, $\Delta H^{\ddagger} = 10$ kcal/mol, and $\Delta S^{\ddagger} = -1.18\ cal \cdot mol^{-1} \cdot K^{-1}$ (all ±25%) [6].

$[C_5H_5Mo(CO)_2CX_3]^{-\cdot}$ (X = H or D). The radical anion was detected by ESR spectroscopy during the UV photolysis of $C_5H_5Mo(CO)_3CX_3$ in a 2.3×10^{-3} M solution in toluene at room temperature; it was also detected by photolyzation of a mixture of $C_5H_5Fe(CO)_2CH_3$ and $(C_5H_5Mo(CO)_3)_2$ (1:1 mole ratio) under the same conditions [3]. The same radical product was obtained by the irradiation of $C_5H_5Mo(CO)_3CH_3$ in $(t\text{-}C_4H_9)_2O_2$; however, its nature as a radical anion is questionable [10].

For X = H, a signal (diagram given) was observed at g = 2.016 with the hyperfine coupling constants $a(^{95,97}Mo) = 29.4$ and $a(^1H) = 6.4$ G [3]; g = 2.016, $a(^{95,97}Mo) = 32.0$, $a(^1H) = 6.4$ G [10]. For X = D, a signal with an identical g value but without hydrogen hyperfine splitting is observed. The radical anions have lifetimes of about one hour and can also be detected by ESR after generation by photolysis outside the ESR cavity. A mechanism involving electron transfer from the radical $C_5H_5Mo(CO)_3^{\cdot}$ to the 16-electron species $C_5H_5Mo(CO)_2CH_3$ is suggested [3].

References:

[1] Mawby, R. J.; Rowson, C. A. (3rd Intern. Symp. Organometal. Chem., München 1967, pp. 322/3).

[2] Jones, W. D. (Diss. California Inst. Technol. 1979; Diss. Abstr. Intern. B **40** [1979] 1708).

[3] Samuel, E.; Rausch, M. D.; Gismondi, T. E.; Mintz, E. A.; Giannotti, C. (J. Organometal. Chem. **172** [1979] 309/15).

[4] Hitam, R. B.; Hooker, R. H.; Mahmound, K. A.; Narayanaswamy, R.; Rest, A. J. (J. Organometal. Chem. **222** [1981] C 9/C 13).

[5] Mahmoud, K. A.; Narayanaswamy, R.; Rest, A. J. (J. Chem. Soc. Dalton Trans. **1981** 2199/204).

[6] Kazlauskas, R. J.; Wrighton, M. S. (J. Am. Chem. Soc. **104** [1982] 6005/15).

[7] Hooker, R. H.; Rest, A. J. (J. Chem. Soc. Dalton Trans. **1984** 761/70).

[8] Mahmoud, K. A.; Rest, A. J.; Alt, H. G.; Eichner, M. E.; Jansen, B. M. (J. Chem. Soc. Dalton Trans. **1984** 175/86).

[9] Mahmoud, K. A.; Rest, A. J.; Alt, H. G. (J. Chem. Soc. Dalton Trans. **1984** 187/97).
[10] Solodovnikov, S. P.; Tumanskii, B. L.; Bubnov, N. N.; Kabachnik, M. I. (Izv. Akad. Nauk SSSR Ser. Khim. **1986** 2147/50; Bull. Acad. Sci. USSR Div. Chem. Sci. **1986** 1960/2).

1.5.1.3.2.1.3 Compounds of the Type $^5LMo(CO)_2(^1L)X$ (^1L = Isonitrile or Ylide)

This section covers compounds of the types $^5LMo(CO)_2(C{\equiv}NR)X$ and $^5LMo(CO)_2(CR_2{=}PR_3')X$ with $^5L = C_5H_5$ or $(CH_3)_5C_5$. The two CO groups can be coordinated cis or trans. Not described in this section are compounds in which the $C_5H_5Mo(CO)_2(X)C{\equiv}N$ fragments are bonded to a polymer like polystyrene or polyvinyl chloride [14, 16].

The compounds listed in Table 2 were prepared in the most cases by the following methods:

Method I: $Na[C_5H_5Mo(CO)_2C{\equiv}NR]$ generated from $C_5H_5Mo(CO)_2(C{\equiv}NR)Cl$ and sodium amalgam in THF, was allowed to react with CH_3COOH, HgI_2, or $R_3'EX$ ($R' = CH_3$, C_6H_5; E = Ge, Sn, or Pb; X = Cl, Br, or I) at −78 °C to room temperature for 15 to 30 min [6, 7, 18]. After removal of the solvent the residue was extracted with toluene [6, 7] or chromatographed with CH_2Cl_2 [18]. Further information is given in the table.

Method II: $C_5H_5Mo(CO)_3X$ was allowed to react with the isonitrile at room temperature to boiling point in THF, pentane, or C_6H_6 for 15 min to 20 h [4, 7, 12, 13, 15, 19]. Addition of a catalyst (see General Remarks) increased the yield and the rate [12, 13, 15]. Nos. 12 and 14 were also obtained by irradiation in C_6H_6 for 15 min [10].

General Remarks. Method II produces a mixture of cis trans isomers. The results of the thermal reaction with a slight excess of $t\text{-}C_4H_9N{\equiv}C$ strongly depend on the reaction conditions. Thus, the bromide No. 14 forms only in small amounts in C_6H_6 (6 h) and in about 50% in THF (17 h); main or by-product is $[C_5H_5Mo(C{\equiv}NC_4H_9\text{-}t)_4]Br$. No. 15 forms under similar conditions (C_6H_6, 12 h and THF, 20 h) in about 35% yield [4]. The use of the catalyst $(C_5H_5Mo(CO)_3)_2$ (2 mol%) and a 1:1 mole ratio of the reactants produces the appropriate compounds in boiling C_6H_6 under following conditions: No. 8, 30 min, 80%; No. 15, 30 min, 70%; No. 18, 15 min, 74% [13]. 95% yield of No. 15 (45 min) were obtained with equimolar amounts of the catalyst or by replacement of the Mo catalyst by $(C_5H_5Fe(CO)_2)_2$ whereas without a catalyst the yield is less than 5% [15]; a mechanism involving radicals is discussed [12, 13].

Table 2
Compounds of the Type $^5LMo(CO)_2(^1L)X$ (^1L = Isonitrile or Ylide).
An asterisk indicates further information at the end of the table.
For explanations, abbreviations, and units see p. X.

No.	compound	method of preparation (yield) properties and remarks
^1L = isonitrile		
1	$C_5H_5Mo(CO)_2(C{\equiv}NCH_3)H$	I, with a 2- to 3-fold excess of CH_3COOH at −78 °C [6, 7] yellow-brown crystals (from hexane at −78 °C), melts at room temperature [6, 7] ^1H NMR (C_6H_6): −5.05 (MoH), 2.17 (CH_3), 4.85 (C_5H_5) [6, 7]

References on p. 16

Table 2 (continued)

No.	compound	method of preparation (yield) properties and remarks
1 (continued)		IR (cyclohexane): 1900, 1963 (ν(CO)), 2136 (ν(CN)) [6, 7] decomposes at room temperature or in a vacuum at 50 to 60°C to afford a 3:1 mixture of $(C_5H_5)_2Mo_2(CO)_5C{\equiv}NCH_3$ and $(C_5H_5Mo(CO)_3)_2$ [7]
2	$C_5H_5Mo(CO)_2(C{\equiv}NCH_3)Ge(CH_3)_3$	I, with $(CH_3)_3GeBr$ at −78°C (39%) [6, 7] pink solid (from hexane), m.p. 137.5 to 139.5°C, trans to cis ratio 3.7 [6, 7] ^1H NMR ($CDCl_3$): cis isomer: 0.46 ($GeCH_3$), 3.44 (NCH_3), 5.04 (C_5H_5); trans isomer: 0.52 ($GeCH_3$), 3.49 (NCH_3), 4.98 (C_5H_5) [6, 7] IR (cyclohexane): 1890, 1898, 1937, 1950 (ν(CO)), 2142 (ν(CN)) [6, 7]
3	$C_5H_5Mo(CO)_2(C{\equiv}NCH_3)Sn(CH_3)_3$	I, with $(CH_3)_3SnCl$ at −78°C (46%) [6, 7] off-white solid (from hexane), m.p. 112 to 113.5°C, trans to cis ratio 4.6 [6, 7] ^1H NMR ($CDCl_3$): cis isomer: 0.27 ($SnCH_3$; J($^{117,\,119}Sn$, H) ≈ 45), 3.38 (NCH_3), 4.99 (C_5H_5); trans isomer: 0.32 ($SnCH_3$; J($^{117,\,119}Sn$, H) ≈ 45), 3.47 (NCH_3), 4.93 (C_5H_5) [6, 7] IR (cyclohexane): 1871, 1883, 1927, 1947 (ν(CO)), 2110, 2139 (ν(CN)) [6, 7]
4	$C_5H_5Mo(CO)_2(C{\equiv}NCH_3)Sn(C_6H_5)_3$	I, with $(C_6H_5)_3SnCl$ at room temperature (30%) [18] yellow solid (from CH_2Cl_2/light petroleum), m.p. 133 to 135°C [18] ^1H NMR ($CDCl_3$): 2.78 (s, CH_3, cis isomer), 3.53 (s, CH_3, trans isomer), 5.08 (s, C_5H_5, trans isomer), 5.21 (s, C_5H_5, cis isomer), 7.31, 7.55, and 7.62 (m, C_6H_5) [18] ^{13}C NMR ($CDCl_3$, −50°C): cis isomer: 31.1 (CH_3), 88.2 (C_5H_5), 125.7 (C_6H_5, C-4), 127.8 (C_6H_5, C-3), 136.6 (C_6H_5, C-2), 144.6 (C_6H_5, C-1), 164.5 (CN), 233.2, 238.8 (CO); trans isomer: 31.1 (CH_3), 88.2 (C_5H_5), 164.0 (CN), 230.0 (CO) [18] IR (light petroleum): 1861, 1928, 1944 (ν(CO)), 2145, 2164 (ν(CN)) [18]
5	$C_5H_5Mo(CO)_2(C{\equiv}NCH_3)Pb(C_6H_5)_3$	I, with $(C_6H_5)_3PbI$ at −78°C (71%) [6, 7] yellow solid (from toluene), m.p. 152.5°C (dec.), trans to cis ratio 2.0 [6, 7] ^1H NMR (acetone-d_6): 3.05 (CH_3, cis isomer; J(^{207}Pb, H) = 26), 3.73 (CH_3, trans isomer; J(^{207}Pb, H) = 8), 5.17 (C_5H_5, trans isomer;

References on p. 16

Table 2 (continued)

No.	compound	method of preparation (yield) properties and remarks
		J(^{207}Pb, H) = 5), 5.40 (C_5H_5, cis isomer), 7.40 (m, C_6H_5) [6, 7] IR (cyclohexane): 1890, 1899, 1945, 1959 (ν(CO)), 2125 (ν(CN)) [7]
6	$C_5H_5Mo(CO)_2(C≡NCH_3)Cl$	II, for 8 h in THF (75%) [7] red solid (from toluene/hexane at −20°C), m.p. 105 to 108.5°C, trans to cis ratio 0.2 [7] ^{1}H NMR: 2.48 (CH_3, trans isomer), 4.93 (C_5H_5, trans isomer), 5.03 (C_5H_5, cis isomer) [7] IR (THF): 1900, 1920, 1975, 1995 (ν(CO)), 2190 (ν(CN)) [7] reduction with sodium amalgam in THF at 25°C affords $Na[C_5H_5Mo(CO)_2CNCH_3]$ [6, 7]
7	$C_5H_5Mo(CO)_2(C≡NCH_3)HgI$	II (61%) [7] orange solid (from C_6H_6), m.p. 107.5 to 109°C [7] ^{1}H NMR (C_6H_6): 2.16 (CH_3), 4.51 (C_5H_5) [7] IR (CH_2Cl_2): 1885, 1954 (ν(CO)), 2175 (ν(CN)) [7]
8	$C_5H_5Mo(CO)_2(C≡NCH_2C_6H_5)I$	II, see General Remarks m.p. 94 to 98°C (from CH_2Cl_2/hexane) [12, 13] ^{1}H NMR (C_6D_6): 3.94 (CH_2, trans isomer), 4.10 (CH_2, cis isomer), 4.75 (C_5H_5, trans isomer), 4.79 (C_5H_5, cis isomer), 7.02 (m, C_6H_5) [13] IR (C_6H_6): 1913, 1978 (ν(CO)), 2156 (ν(CN)) [12] mass spectrum (70 eV): $[M]^+$ [13]
*9	trans-$(CH_3)_5C_5Mo(CO)_2(C≡NC_4H_9$-t)H	II, in pentane for 1 h (99%) [19] m.p. 68°C (from pentane at −78°C) [19] ^{1}H NMR (acetone-d$_6$): −5.41 (MoH), 1.50 (t-C_4H_9), 1.99 ($C_5(CH_3)_5$) [19] ^{13}C NMR (acetone-d$_6$): 11.7 ($C_5(\mathbf{C}H_3)_5$), 31.3 ($C(\mathbf{C}H_3)_3$), 58.7 ($\mathbf{C}(CH_3)_3$), 103.0 ($\mathbf{C}_5(CH_3)_5$), 161.7 (CN), 235.8 (CO) [19] IR (pentane): 1887, 1943 (ν(CO)), 2097 (ν(CN)) [19] mass spectrum: $[M]^+$ [19] stirring in CCl_4 for 1 h affords quantitatively No. 13 [19]
10	$C_5H_5Mo(CO)_2(C≡NC_4H_9$-t)$Sn(C_6H_5)_3$	I, with $(C_6H_5)_3SnCl$ (52%) [18] m.p. 135 to 137°C (from light petroleum/CH_2Cl_2), cis to trans ratio 4:6 by ^{1}H NMR [18]

References on p. 16

Table 2 (continued)

No.	compound	method of preparation (yield) properties and remarks
10 (continued)		^{1}H NMR ($CDCl_3$): 0.97 (CH_3, cis isomer), 1.53 (CH_3, trans isomer), 5.07 (C_5H_5, trans isomer), 5.21 (C_5H_5, cis isomer), 7.30, 7.55, and 7.60 (m, C_6H_5) [18] ^{13}C NMR ($CDCl_3$, −50 °C): 29.7 (CH_3, cis isomer), 30.6 (CH_3, trans isomer), 57.6 (**C**$(CH_3)_3$, cis isomer), 58.1 (**C**$(CH_3)_3$, trans isomer), 88.1 and 88.3 (C_5H_5, not assigned), 125.7 (C_6H_5, C-4), 127.7 (C_6H_5, C-3), 136.6 (C_6H_5, C-2), 144.2 (C_6H_5, C-1), 229.4 (CO) [18] IR (CH_2Cl_2): 1865, 1928, 1936 (ν(CO)), 2113, 2125 (ν(CN)) [18]
11	$C_5H_5Mo(CO)_2(C{\equiv}NC_4H_9\text{-t})(\mu\text{-}SCH_3)W(CO)_5$	probably from $C_5H_5Mo(CO)_2(\mu\text{-}SCH_3)W(CO)_5$ and t-$C_4H_9N{\equiv}C$ [17] no details given [17] cyclic voltammetry (vs. $(C_5H_5)_2Fe/[(C_5H_5)_2Fe]^+$): $E^{red}_{2/p}$ = −1.58 (in propylene carbonate), −1.73 (in THF), −1.70 (in CH_2Cl_2) [17]
12	$C_5H_5Mo(CO)_2(C{\equiv}NC_4H_9\text{-t})Cl$	II, photolysis in C_6H_6 (64%) [10] red-orange solid (from CH_2Cl_2/hexane), m.p. 121 to 122 °C, cis to trans ratio 8:2 [10] ^{1}H NMR ($CDCl_3$): 1.49 (CH_3, trans isomer), 1.55 (CH_3, cis isomer), 5.27 (C_5H_5, trans isomer), 5.47 (C_5H_5, cis isomer) [10] ^{13}C NMR ($CDCl_3$): 30.5 (CH_3), 58.6 (**C**$(CH_3)_3$), 93.4 (C_5H_5, trans isomer), 94.5 (C_5H_5, cis isomer), 226.2 and 239.0 (CO, cis isomer), 230.4 (CO trans isomer) [10] IR (CH_2Cl_2): 1977, 1995 (ν(CO)), 2145 (ν(CN)) [10]
13	cis-$(CH_3)_5C_5Mo(CO)_2(C{\equiv}NC_4H_9\text{-t})Cl$	No. 9 was stirred in CCl_4 for 1 h (99%) [19] dark red solid, m.p. 68 °C (from CCl_4) [19] ^{1}H NMR ($CDCl_3$): 1.48 (t-C_4H_9), 1.83 (CH_3) [19] ^{13}C NMR ($CDCl_3$): 10.4 ((**C**$H_3)_5C_5$), 30.4 ((**C**$H_3)_3$C), 57.9 (**C**$(CH_3)_3$), 106.3 (**C**$_5(CH_3)_5$), 161.7 (CN), 244.8, 256.8 (CO) [19] IR (pentane): 1890, 1960 (ν(CO)), 2130 (ν(CN)) [19]

References on p. 16

Table 2 (continued)

No.	compound	method of preparation (yield) properties and remarks
14	$C_5H_5Mo(CO)_2(C{\equiv}NC_4H_9\text{-}t)Br$	II, photolysis in C_6H_6 [4, 10] red solid (from CH_2Cl_2/hexane), m.p. 133 °C, cis to trans ratio 2:8 [4, 10] ^1H NMR ($CDCl_3$): 1.50 (CH_3), 5.28 (C_5H_5, trans isomer), 5.44 (C_5H_5, cis isomer) [4] ^{13}C NMR ($CDCl_3$): 30.6 (CH_3), 58.8 (**C**$(CH_3)_3$, cis isomer), 59.9 (**C**$(CH_3)_3$, trans isomer), 93.7 (C_5H_5, trans isomer), 94.2 (C_5H_5, cis isomer), 229.0 and 238.0 (cis CO) [10] IR (CH_2Cl_2): 1923, 1998 (ν(CO)), 2143 (ν(CN)) [4] the isomer ratio depends on purification; a more soluble fraction with reverse cis/trans ratio was also observed [4]
15	$C_5H_5Mo(CO)_2(C{\equiv}NC_4H_9\text{-}t)I$	II, see General Remarks red solid (from CH_2Cl_2/hexane), m.p. 156 to 157 °C, cis to trans ratio 45:55 [4, 13] ^1H NMR ($CDCl_3$): 1.54 (t-C_4H_9), 5.29 (C_5H_5, trans isomer), 5.40 (C_5H_5, cis isomer) [4]; (C_6D_6): 0.85 (t-C_4H_9, cis isomer), 0.98 (t-C_4H_9, trans isomer), 4.79 (C_5H_5, cis isomer), 4.83 (C_5H_5, trans isomer) [13] IR (CH_2Cl_2): 1915, 1986 (ν(CO)), 2141 (ν(CN)) [4]; (benzene): 1902, 1915, 1973 (ν(CO)), 2138 (ν(CN)) [12] mass spectrum (70 eV): $[M]^+$ [13] the isomer ratio depends on the mode of purification: cis to trans ratios of 25:75 to 45:55 were obtained by crystallization and after sublimation, a 60:40 ratio was observed [4]
16	$C_5H_5Mo(CO)_2(C{\equiv}NC_6H_{11}\text{-cyclo})Cl$	formed as by-product of the reaction of $C_5H_5Mo(CO)_3Cl$ with cyclo-$C_6H_{11}N{\equiv}C$ in the presence of $LiN{=}C(C_6H_5)_2$ in ether; the main product was $C_5H_5Mo(CO)_2N{=}C(C_6H_5)_2$ [3] IR (hexane): 1918, 1987 (ν(CO)) [3]
*17	$C_5H_5Mo(CO)_2(C{\equiv}NC_6H_5)I$	II, with 8 equivalents of $C_6H_5N{\equiv}C$ in refluxing THF (19%) [1]; also formed by the reaction of $C_5H_5Mo(CO)_2(P(C_6H_5)_3)I$ with the isonitrile [2] red solid (from pentane), m.p. 75 °C [1]; pink solid, m.p. 78 °C (cis isomer), peach solid, m.p. 103 °C (trans isomer) [2]

References on p. 16

Table 2 (continued)

No.	compound	method of preparation (yield) properties and remarks
*17 (continued)		IR (CCl_4): 1942, 2004 (ν(CO)), 2070, 2141 (ν(CN)) [1]; 2098 (ν(CN)) [5] the isomers are difficultly to separate, because the complex equilibrates rapidly to a cis to trans ratio of 60:40 in C_6H_6 at 25°C [2] reduction with sodium amalgam (1%) in THF at 25°C affords $Na[C_5H_5Mo(CO)_2C{\equiv}NC_6H_5]$ [8, 9]
18	$C_5H_5Mo(CO)_2(C{\equiv}NC_6H_3(CH_3)_2\text{-}2,6)I$	II, see General Remarks m.p. 132 to 140°C (from CH_2Cl_2/hexane) [13] ^1H NMR (C_6D_6): 1.91 (CH_3, trans isomer), 2.21 (CH_3, cis isomer), 4.78 (C_5H_5, trans isomer), 4.80 (C_5H_5, cis isomer), 6.61 (C_6H_3) [13] IR (C_6H_6): 1912, 1924, 1977 (ν(CO)), 2115 (ν(NC)) [12] mass spectrum (70 eV): $[M]^+$ [13]
^{1}L = ylide		
19	trans-$C_5H_5Mo(CO)_2(C(C_6H_5)HP(OCH_3)_3)Sn(C_6H_5)_3$	from trans-$C_5H_5Mo(CO)_2({=}CHC_6H_5)Sn(C_6H_5)_3$ and $P(OCH_3)_3$ in THF solution at −78°C [20] ^1H NMR ($CDCl_3$): 3.34 (d, CH; J(P, H) = 14.5), 3.66 (d, OCH_3; J(P, H) = 10.5), 4.62 (s, C_5H_5), 6.97 to 7.06 (m, C_6H_5, H-4), 7.18 to 7.32 (m, H-2, 3 of CC_6H_5 and H-3, 4 of SnC_6H_5), 7.62 to 7.72 (m, H-2 of SnC_6H_5) [20] ^{13}C NMR ($CDCl_3$, −50°C): 1.7 (d, CH; J(P, C) = 113), 55.8 (d, OCH_3; J(P, C) = 8), 89.9 (C_5H_5), 123.5 (C-4 of C_6H_5), 126.5 to 128.2 (m, C-2, 3 of CC_6H_5 and C-3, 4 of C_6H_5Sn), 137.0 (C-2 of SnC_6H_5; J(Sn, C) = 33), 145.4 (C-1 of C_6H_5), 145.8 (C-1 of SnC_6H_5), 240.3, 241.7 (CO) [20] IR (CH_2Cl_2): 1800, 1880 (ν(CO)) [20] mass spectrum (field desorption): $[M + 1]^+$ [20]

*Further information:

trans-$(CH_3)_5C_5Mo(CO)_2(C{\equiv}NC_4H_9\text{-}t)H$ (Table **2**, No. **9**) crystallizes in the monoclinic space group $P2_1/c-C_{2h}^5$ (No. 14) with the unit cell parameters a = 9.422 (6), b = 17.609 (9), c = 11.804 (5) Å, β = 107.99 (8)°; Z = 4 molecules per unit cell. The molecular structure with the main bond distances and angles is shown in **Fig. 1** [19].

trans-$C_5H_5Mo(CO)_2(C{\equiv}NC_6H_5)I$ (Table **2**, No. **17**) crystallizes in the orthorhombic space group $Pbca-D_{2h}^{15}$ (No. 61) with the unit cell parameters a = 12.035 (4), b = 17.248 (5), c = 14.510 (5) Å; Z = 8 molecules per unit cell, D_{meas} = 1.95, and D_{calc} = 1.972 g/cm^3. The molecular structure with the main bond distances and angles is shown in **Fig. 2** [5, 11]. A crystal structure view along a is also shown in [11].

References on p. 16

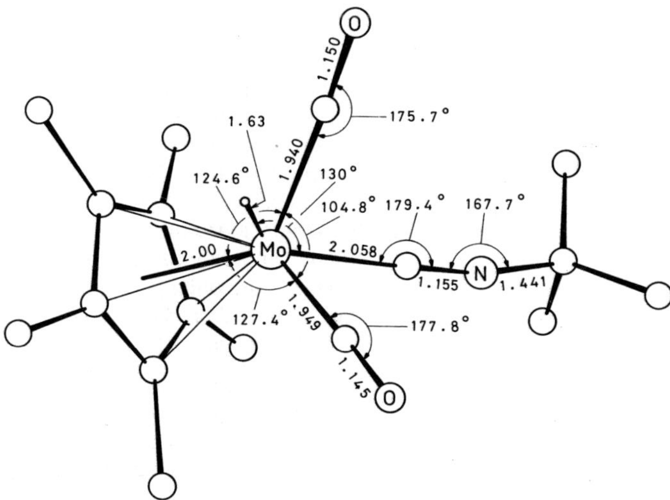

Fig. 1. Molecular structure of trans-$(CH_3)_5C_5Mo(CO)_2(C{\equiv}NC_4H_9\text{-}t)H$ [19].

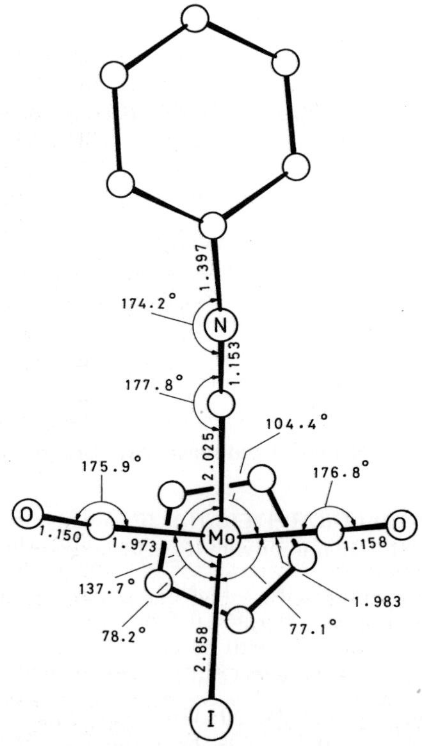

Fig. 2. Molecular structure of trans-$C_5H_5Mo(CO)_2(C{\equiv}NC_6H_5)I$ [11].

References on p. 16

References:

[1] Joshi, K. K.; Pauson, P. L.; Strubbs, W. H. (J. Organometal. Chem. **1** [1963] 51/7).
[2] Bolton, E. S.; Denker, M.; Knox, G. R.; Robertson, C. G. (Chem. Ind. **1969** 327/8).
[3] Keable, H. R.; Kilner, M.; Robertson, E. E. (J. Chem. Soc. Dalton Trans. **1974** 639/44).
[4] King, R. B.; Saran, M. S. (Inorg. Chem. **13** [1974] 364/7).
[5] Sim, G. A.; Sime, J. G.; Woodhouse, D. I.; Knox, G. R. (J. Organometal. Chem. **74** [1974] C 7/C 8).
[6] Adams, R. D. (J. Organometal. Chem. **88** [1975] C 38/C 40).
[7] Adams, R. D. (Inorg. Chem. **15** [1976] 169/74).
[8] Adams, R. D.; Chodosh, D. F. (J. Organometal. Chem. **122** [1976] C 11/C 14).
[9] Adams, R. D.; Chodosh, D. F. (J. Am. Chem. Soc. **99** [1977] 6544/50).
[10] King, R. B.; Saran, M. S.; McDonald, D. P.; Diefenbach, S. P. (J. Am. Chem. Soc. **101** [1979] 1138/42).
[11] Sim, G. A.; Sime, J. G.; Woodhouse, D. I.; Knox, G. R. (Acta Crystallogr. B **35** [1979] 2406/8).
[12] Coville, N. J. (J. Organometal. Chem. **190** [1980] C 84/C 86).
[13] Coville, N. J. (J. Organometal. Chem. **218** [1981] 337/49).
[14] Menzel, H.; Fehlhammer, W. P.; Beck, W. (Z. Naturforsch. **37b** [1982] 201/8).
[15] Coville, N. J.; Albers, M. O.; Singleton, E. (J. Chem. Soc. Dalton Trans. **1983** 947/53).
[16] Hooker, R. H.; Rest, A. J. (J. Chem. Soc. Dalton Trans. **1984** 761/70).
[17] Courtot-Coupez, J.; Guerchais, J. E.; Pétillon, F. Y.; Talarmin, J. (J. Chem. Soc. Dalton Trans. **1986** 1917/21).
[18] Adams, H.; Bailey, N. A.; Bannister, C.; Faers, M. A.; Fedorko, P.; Osborne, V. A.; Winter, M. J. (J. Chem. Soc. Dalton Trans. **1987** 341/8).
[19] Alt, H. G.; Engelhardt, H.; Frister, T.; Rogers, R. D. (J. Organometal. Chem. **366** [1989] 297/304).
[20] Winter, M. J.; Woodward, S. (J. Chem. Soc. Chem. Commun. **1989** 457/8).

1.5.1.3.2.1.4 Compounds of the Type $^5LMo(CO)_2(^1L)X$ (^1L = Carbene)

This section covers compounds with one carbene and various ligands X bonded to the $^5LMo(CO)_2$ fragment. ^5L ligands are mainly C_5H_5, $CH_3C_5H_4$ (No. 34), $CH_3C(O)C_5H_4$ (No. 35), $(CH_3)_3SiC_5H_4$ (No. 36), $(CH_3)_5C_5$ (Nos. 27, 28, 30, 32, and 38), or C_7H_9 (indenyl, No. 37). The carbene ligand and X can be arranged in trans or in cis position; X are H, halogens, pseudohalogens, various ER_3 groups (E = Si, Ge), and SR. For the cis isomers chirality at Mo is expected, see General Remarks. The compounds listed in Table 3 were obtained in most cases by the following methods:

Method I: $Na[C_5H_5Mo(CO)_2{=}CR_2]$ was treated with $(C_6H_5)_3SnCl$, CH_3COOH, or X_2 (X = Br, I) in THF [12, 16].

Method II: trans-$C_5H_5Mo(CO)_2({=}C(OR)R')E(C_6H_5)_3$ (E = Ge, Sn) was treated with $C_2H_5NH_2$ under pressure for 10 min or with $(C_2H_5)_2NH$ overnight [5].

Method III: $M[^5LMo(CO)_2(C(O)R)ER_3]$ (M = Li, $N(C_2H_5)_4$, or $N(CH_3)_4$; E = Ge, Sn) was treated with $[O(CH_3)_3]BF_4$ or $[O(C_2H_5)_3]PF_6$ in CH_2Cl_2 or water. After removal of the solvent the residue was extracted with ether [2, 5]. The compounds were purified by chromatography on Al_2O_3 with CH_3Cl/hexane [5].

Method IV: trans-$C_5H_5Mo(CO)_2({=}C(OC_2H_5)C_6H_5)Sn(C_6H_5)_3$ was treated with $Li[(C_2H_5)_3BH]$ at $-60\,°C$ or $LiCH_3$ at $-40\,°C$ in THF. Acetic acid was added to the reaction mixture, which was then allowed to warm up [19].

References on p. 32

Method V: a. $Na[C_5H_5Mo(CO)_2C{\equiv}NCH_3]$ was allowed to react with $I(CH_2)_nI$ (n = 3, 4) in THF [11, 14, 16, 17].
b. $Na[^5LMo(CO)_3]$ and $I(CH_2)_3I$ were stirred in THF [6, 8].

Method VI: $^5LMo(CO)_3(CH_2)_3Br$ and MX were stirred in solution to give $^5LMo(CO)_2({=}C(CH_2)_3O)X$ [1, 2, 6 to 8, 18]. Further information is given in the table.

General Remarks. The carbene compounds with two different ligands at the carbene carbon atom of which one is bonded with a heteroatom (O or N in most cases), can exist in two different conformers due to restrict rotation about the Mo=C bond. For the alkoxy carbene complex (Formulas I and II) a barrier of rotation of ca. 2830 J/mol was found. Mostly, both conformers are present in solution; however, in compounds with bulky ligands (C_4H_9-t in No. 13) or in amino carbene compounds, only one conformer is observed by IR spectroscopy. The type of conformer could not be determined [5], but see the different conformations in Fig. 3 to 7. A possible conformer III was excluded by IR studies on ^{13}C-enriched samples [5].

Nothing is reported about chirality of the cis isomers. However, the cis-configurated complex No. 30 (with the O atom directed to the ^5L ligand, see also Formula I) exhibits a temperature-dependent ^1H NMR spectrum involving $=CCH_2$ and $=COCH_2$ protons of the cyclic carbene ligand. Each of the two 317 K triplets reversibly splits into two sets of multiplets on lowering the temperature to 223 K. This averaging process was interpreted with a cis-cis isomerization, or racemization on elevated temperatures [18].

I II III ○ = Mo

Table 3
Compounds of the Type $^5LMo(CO)_2({=}CR_2)X$.
An asterisk indicates further information at the end of the table.
For explanations, abbreviations, and units see p. X.

No.	compound	method of preparation (yield) properties and remarks
compounds with a noncyclic carbene ligand		
1	$C_5H_5Mo(CO)_2({=}C(NHCH_3)CH_3)Sn(C_6H_5)_3$	I [12] no further information given
2	$C_5H_5Mo(CO)_2({=}C(NHCH_3)CH_3)I$	I [12] reduction with 2 equivalents of sodium naphthalide affords good yields of $Na[C_5H_5Mo(CO)_2{=}C(CH_3)NHCH_3]$ [12]

References on p. 32

Table 3 (continued)

No.	compound	method of preparation (yield) properties and remarks
*3	$C_5H_5Mo(CO)_2(=C(N(CH_3)C(C_6H_5)=NCH_3)C_6H_5)Cl$	see "Further information" dark brown prisms (isomer mixture), m.p. 152°C [9] conductivity (acetone, 22°C): $\Lambda = 1.95$ $cm^2 \cdot \Omega^{-1} \cdot mol^{-1}$ [9] ^1H NMR ($CDCl_3$, −10°C): cis isomer: 3.37 and 3.86 (s, CH_3), 5.20 (s, C_5H_5), 6.3 to 7.7 (m, C_6H_5); trans isomer: 3.50 and 3.92 (s, CH_3), 5.19 (s, C_5H_5), 6.3 to 7.7 (m, C_6H_5) [9] IR (KBr): 1497, 1658 (ν(CN)), 1851, 1954 (ν(CO)) [9] mass spectrum (field desorption, acetone): $[M]^+$ [9]
4	cis-$C_5H_5Mo(CO)_2(=C(NHCH_3)CH_3)I$	$Na[C_5H_5Mo(CO)_2C{\equiv}NCH_3]$ was treated with two equivalents of CH_3I in THF for 15 min; the reaction mixture was treated with an excess of LiI · 3 H_2O and CH_3COOH for 2 h (55%) [14]; similar conditions in [3] red [3], maroon solid [14], m.p. 121.5 to 123°C (from toluene/hexane) [3, 14] ^1H NMR (C_6D_6): 2.17 (CH_3), 2.75 (CH_3; J = 5), 5.00 (C_5H_5) [3]; ($CDCl_3$): 2.93 (qui, CCH_3; J(H, H) = 1.0), 3.25 (d of q, NCH_3; J(H, H) = 5.0, 1.0), 5.56 (s, C_5H_5), 8.12 (s br, NH) [14] ^{13}C NMR ($CDCl_3$, −50°C): 40.4 and 43.8 (CH_3), 95.9 (C_5H_5), 246.1, 251.9, and 263.7 (CO and =C) [14] IR (C_6H_6): 1550 (ν(CN)), 1865, 1955 (ν(CO)); (CH_3CN): 3255 (ν(NH)) [3]; (CH_2Cl_2): 1863, 1952 (ν(CO)) [14] mass spectrum (70 eV): $[M]^+$, $[M - CO]^+$, and other ions of which $[C_5H_5MoI]^+$ is the main peak [3]; (m/e) 403 ($[M]^+$) [14]
5	trans-$C_5H_5Mo(CO)_2(=C(NHC_2H_5)C_6H_5)Ge(C_6H_5)_3$	II (86%) [5] yellow solid, m.p. 192 to 193°C (from CH_2Cl_2/hexane) [5]

Table 3 (continued)

No.	compound	method of preparation (yield) properties and remarks
		^1H NMR: 0.83 to 1.33 (m, CH_3), 3.04 (CH_2), 4.68 (C_5H_5) [5] IR (KBr): 1500 (ν(CN)); (heptane): 1861, 1929 (ν(CO)) [5]
6	trans-$C_5H_5Mo(CO)_2(=C(NHC_2H_5)C_6H_5)Sn(C_6H_5)_3$	II (74%) [5] yellow solid, m.p. 165 to 166°C (from CH_2Cl_2/hexane) [5] ^1H NMR: 0.83 to 1.33 (CH_3), 3.06 (CH_2), 4.75 (C_5H_5) [5] IR (KBr): 1500 (ν(CN)); (heptane): 1855, 1919 (ν(CO)) [5]
7	trans-$C_5H_5Mo(CO)_2(=C(N(CH_3)_2)C_6H_5)Ge(C_6H_5)_3$	II (83%) [5] yellow solid, m.p. 214 to 215°C (from CH_2Cl_2/hexane) [5] ^1H NMR: 2.84 and 3.17 (CH_3), 4.63 (C_5H_5) [5] IR (heptane, 70% ^{13}CO-enriched): 1855.0, 1926.0 (ν($^{12}CO)_2$); 1829.5, 1910.5 (ν(^{12}CO, ^{13}CO)); 1814.0, 1881.5 (ν($^{13}CO)_2$); spectra are depicted [5] the interaction constants were calculated: k_2 = 14.438, k_{22} = 0.535 mdyn/Å [5]
8	trans-$C_5H_5Mo(CO)_2(=C(N(CH_3)_2)C_6H_5)Sn(C_6H_5)_3$	II (89%) [5] yellow solid, m.p. 189 to 191°C (from CH_2Cl_2/hexane) [5] ^1H NMR: 2.88 and 3.32 (CH_3), 4.74 (C_5H_5) [5] IR (KBr): 1500 (ν(CN)); (heptane): 1849, 1917 (ν(CO)) [5]
9	trans-$C_5H_5Mo(CO)_2(=C(OCH_3)CH_3)GeCl_3$	III (38%) [2] yellow-green crystals, m.p. 150 to 151°C (from CH_2Cl_2/heptane) [2] ^1H NMR ($CDCl_3$): 3.04 (CH_3), 4.45 (OCH_3), 5.60 (C_5H_5) [2] IR (heptane): 1914, 1987 (ν(CO)) [2]

References on p. 32

Table 3 (continued)

No.	compound	method of preparation (yield) properties and remarks
10	trans-$C_5H_5Mo(CO)_2(=C(OC_2H_5)CH_3)GeCl_3$	III (54%) [2] yellow-green crystals, m.p. 165 to 167 °C (from CH_2Cl_2/hexane) [2] ^{1}H NMR (CD_2Cl_2): 1.61 (t, **CH_3**CH_2), 3.07 (s, CH_3), 4.59 (m, CH_2), 5.57 (s, C_5H_5) [2] IR (heptane): 1912, 1986 (ν(CO)) [2]
11	trans-$C_5H_5Mo(CO)_2(=C(OCH_3)CH_3)Ge(C_6H_5)_3$	III (15%) [5] light green solid, m.p. 173 to 174 °C (from CH_2Cl_2/hexane) [5] ^{1}H NMR: 2.87 (CH_3), 4.05 (OCH_3), 5.27 (C_5H_5) [5] IR (heptane): 1877, 1944 (ν(CO)) [5]
12	trans-$C_5H_5Mo(CO)_2(=C(OC_2H_5)CH_3)Ge(C_6H_5)_3$	III (44%) [5] light green solid, m.p. 178 to 180 °C (from CH_2Cl_2/hexane) [5] ^{1}H NMR: 1.38 (**CH_3**CH_2), 2.88 (CH_3), 4.27 (CH_2), 5.26 (C_5H_5) [5] IR (heptane): 1875, 1943 (ν(CO)) [5]
13	trans-$C_5H_5Mo(CO)_2(=C(OC_2H_5)C_4H_9$-t$)Ge(C_6H_5)_3$	III (6%) [5] yellow solid, m.p. 176 to 178 °C (from CH_2Cl_2/heptane) [5] ^{1}H NMR: 1.00 (CH_3), 1.23 (t-C_4H_9), 4.09 (CH_2), 5.32 (C_5H_5) [5] IR (heptane): 1863, 1934 (ν(CO)) [5] only one conformer in solution due to the bulky t-butyl substituent [5]
14	trans-$C_5H_5Mo(CO)_2(=C(OCH_3)C_6H_5)Ge(C_6H_5)_3$	III (53%) [5] orange solid, m.p. 188 to 189 °C (from CH_2Cl_2/heptane) [5] ^{1}H NMR: 3.98 (CH_3), 5.03 (C_5H_5) [5] IR (heptane): 1880, 1898, 1945, 1967 (ν(CO)) [5]
15	trans-$C_5H_5Mo(CO)_2(=C(OCH_3)C_6H_5)Sn(C_6H_5)_3$	III (27%) [5] orange solid, m.p. 159 to 160 °C (from CH_2Cl_2/heptane) [5] ^{1}H NMR: 4.10 (CH_3), 5.08 (C_5H_5) [5] IR (heptane): 1872, 1891, 1934, 1949 (ν(CO)) [5]

References on p. 32

Table 3 (continued)

No.	compound	method of preparation (yield) properties and remarks
*16	trans-$C_5H_5Mo(CO)_2(=C(OC_2H_5)C_6H_5)Ge(C_6H_5)_3$	III (44%) [5] orange solid, m.p. 187 to 189°C (from CH_2Cl_2/heptane) [5] ^{1}H NMR: 1.30 (CH_3), 4.19 (CH_2), 5.00 (C_5H_5) [5] IR (heptane): 1896, 1897, 1945, 1960 (ν(CO)); observed and calculated ν(CO) frequencies in ^{13}CO-enriched compounds are also given in [5]; k_2 = 15.018, k_{22} = 0.501 (conformer I), k_2 = 14.758, k_{22} = 0.513 (conformer II) mdyn/Å [5]
17	trans-$C_5H_5Mo(CO)_2(=C(OC_2H_5)C_6H_5)Sn(C_6H_5)_3$	III (28%) [5] orange solid, m.p. 165 to 166°C (from CH_2Cl_2/heptane) [5] ^{1}H NMR: 1.30 (CH_3), 4.31 (CH_2), 5.07 (C_5H_5) [5] IR (heptane): 1870, 1880, 1934, 1950 (ν(CO)) [5] reaction with $Li[(C_2H_5)_3BH]$ or $LiCH_3$ (see Method IV) gives Nos. 18 and 19 [19]
*18	trans-$C_5H_5Mo(CO)_2(=CHC_6H_5)Sn(C_6H_5)_3$	IV ^{1}H NMR ($CDCl_3$): 5.73 (C_5H_5), 7.32 to 7.43 (m, H-3, 4 of C_6H_5), 7.61 to 7.68 (m, H-1 of C_6H_5), 15.21 (s, CH=; J(Sn, H) = 50.5) [19] ^{13}C NMR (CD_2Cl_2, −50°C): 96.6 (C_5H_5), 128.2 (s, C-3, 4 of SnC_6H_5; J(Sn, C) = 43), 128.7 (C-3 of C_6H_5), 132.1 (C-4 of C_6H_5), 132.3 (C-2 of C_6H_5), 136.4 (C-2 of C_6H_5Sn; J(Sn, C) = 35), 142.8 (C-1 of SnC_6H_5; J(Sn, C) = 352), 147.2 (C-1 of C_6H_5), 233.0 (CO), 312.1 (Mo=C) [19] IR (CH_2Cl_2): 1890, 1955 (ν(CO)) [19] mass spectrum (fast atom bombardment): $[M + 1]^+$ [19]
19	trans-$C_5H_5Mo(CO)_2(=C(CH_3)C_6H_5)Sn(C_6H_5)_3$	IV (39%) [19] rearranges over several hours in THF at ambient temperature into $C_5H_5Mo(CO)_2$-$(CH_2{=}CHC_6H_5)Sn(C_6H_5)_3$ [19]

References on p. 32

Table 3 (continued)

No.	compound	method of preparation (yield) properties and remarks
compounds with a cyclic carbene ligand		
$(=C(CH_2)_3NCH_3 =$ CH_3 N $)$		
20	$C_5H_5Mo(CO)_2(=C(CH_2)_3NCH_3)H$	I [16] pale yellow solid, cis to trans ratio 11:88 by ^1H NMR at $-50°C$ [16] ^1H NMR (toluene-d_8, $-50°C$): cis isomer: -5.09 (MoH), 2.72 (CH_3), 5.05 (C_5H_5), other signals not assigned due to overlap with the trans isomer; trans isomer: -5.31 (MoH), 0.73 (m, central CH_2), 2.28 (m, 2 CH_2), 3.07 (CH_3), 4.90 (C_5H_5); spectra are temperature-dependent and the coalescence temperature for the C_5H_5 signals are very close to ambient temperature, attributed to increasing cis/trans interconversion [16] ^{13}C NMR (toluene-d_8, $-50°C$): cis isomer: 21.2 (central CH_2), 40.9 (CH_3), 53.9 and 59.0 (CH_2), 92.1 (C_5H_5), 266.0 (Mo=C); trans isomer: 20.7 (central CH_2), 40.4 (CH_3), 53.8, 59.2 (CH_2), 90.2 (C_5H_5), 233.5 (CO), 260.5 (Mo=C) [16] IR (hexane): 1868, 1936 (ν(CO)) [16] mass spectrum (chemical ionization, NH_3): $[M + 1]^+$ [16] rearranges in THF at room temperature into $C_5H_5Mo(CO)_2CH(CH_2)_3NCH_3$ (see No. 20, p. 121); based on IR measurements, the isomerization follows first-order kinetics at 31°C with $t_{1/2} \approx 10$ h, indicating a unimolecular process [16]
21	trans-$C_5H_5Mo(CO)_2(=C(CH_2)_3NCH_3)Sn(C_6H_5)_3$	I (66%) [12] crystalline solid [12] ^1H NMR ($CDCl_3$): 1.91 (qui, central CH_2; J(H, H) = 8), 3.08 (s, CH_3), 3.40 and 3.68 (t, CH_2; J(H, H) = 8), 5.22 (s, C_5H_5), 7.28 (m, C_6H_5, 9H), 7.61 (m, C_6H_5, 6H) [12]

References on p. 32

Table 3 (continued)

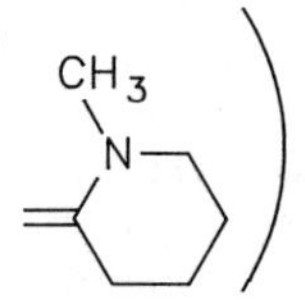

No.	compound	method of preparation (yield) properties and remarks
		^{13}C NMR ($CDCl_3$, −50°C): 21.0 (central CH_2), 40.5 (NCH_3), 55.1, 60.8 (CH_2), 90.4 (C_5H_5), 127.6, 136.8, and 144.8 (C_6H_5), 233.5 (CO), 257.7 (Mo=C) [12] IR (CH_2Cl_2): 1832, 1908 (ν(CO)) [12] mass spectrum (CI, NH_3): $[M + 1]^+$ [12]
22	cis-$C_5H_5Mo(CO)_2(=C(CH_2)_3NCH_3)Br$	I [12] no further information given
*23	cis-$C_5H_5Mo(CO)_2(=C(CH_2)_3NCH_3)I$	Va (62% by chromatography on Al_2O_3 with 1:3 CH_2Cl_2/light petroleum ether) [11, 14, 16, 17] red needles [10], maroon crystals [14], m.p. 129 to 131°C (from toluene) [10, 14] ^1H NMR ($CDCl_3$): 2.00 (m, central CH_2), 3.32 (t, NCH_3; J(H, H) = 1.4), 3.43 (d of t of sext, NCH_2, 1H; J(H, H) = 18.3, 7.0, 1.4), 3.53 (d of t of qui, NCH_2, 1H; J(H, H) = 18.3, 7.0, 1.4), 3.65 (d of t of t, Mo=CCH_2, 1H; J(H, H) = 11.7, 8.6, 1.4), 3.86 (d of t of d, Mo=CCH_2, 1H; J(H, H) = 11.7, 7.2, 1.4), 5.52 (s, C_5H_5) [10, 14] ^{13}C NMR (toluene-d_8, −60°C): 20.9 (central CH_2), 42.9 (CH_3), 53.6 (Mo=C**C**H_2), 60.3 (NCH_2), 95.4 (C_5H_5), 247.2, 253.2, and 254.3 (CO and Mo=C) [14]; similar data in [10] IR (CH_2Cl_2): 1860, 1950 (ν(CO)) [14]; similar data in THF [10] mass spectrum (m/e): 429 ($[M]^+$) [14]
	($=C(CH_2)_4NCH_3$ = [structure: N-methylpiperidin-2-ylidene])	
24	trans-$C_5H_5Mo(CO)_2(=C(CH_2)_4NCH_3)Sn(C_6H_5)_3$	I [12] no further information given
*25	cis-$C_5H_5Mo(CO)_2(=C(CH_2)_4NCH_3)I$	Va (21%) [11, 14, 17] maroon crystals, m.p. 117 to 119°C (from ether) [14] ^1H NMR ($CDCl_3$): 1.61 (m, $NCH_2C\mathbf{H_2}$), 1.85 (m, Mo=$CCH_2C\mathbf{H_2}$), 3.32 (d of t of q, NCH_2, 1H; J(H, H) = 19.0, 6.0, 1.0), 3.41 (d of t, Mo=CCH_2, 1H; J(H,H) = 14.0, 6.0), 3.44 (t, CH_3; J(H, H) =

References on p. 32

Table 3 (continued)

No.	compound	method of preparation (yield) properties and remarks
*25 (continued)		1.0), 3.54 (d of t of q, NCH_2, 1H; J(H, H) = 19.0, 6.5, 1.0), 3.62 (d of t, Mo=C$C\mathbf{H}_2$, 1H; J(H, H) = 14, 6.6), 5.52 (s, C_5H_5) [14] ^{13}C NMR (C_6D_6): 19.6, 21.5 (CH_2), 48.0 (CH_3), 52.6 (=$\mathbf{C}CH_2$), 54.1 (NCH_2), 96.1 (C_5H_5), 243.1 and 254.4 (CO or Mo=C) [14] IR (CH_2Cl_2): 1852, 1944 (ν(CO)) [14] mass spectrum: $[M]^+$ [14]
	$(=C(CH_2)_3O$ = O (ring))	
26	trans-$C_5H_5Mo(CO)_2(=C(CH_2)_3O)CN$	VI, with KCN in refluxing methanol for 1.5 h (62%) [1, 7] light yellow crystals (from ether/hexane) [1], yellow solid, m.p. 172 to 174°C (from CH_2Cl_2/light petroleum) [7] ^{1}H NMR ($CDCl_3$): 1.94 (qui, central CH_2; J(H, H) = 7), 3.69 (t, =CCH_2; J(H, H) = 7), 4.76 (t, OCH_2; J(H, H) = 7), 5.52 (s, C_5H_5) [7]; similar data in [1] ^{13}C NMR ($CDCl_3$): 22.0 (central CH_2), 59.5 (=$\mathbf{C}CH_2$), 83.9 (CH_2), 95.0 (C_5H_5), 225.6 (CO), 325.1 (Mo=C) [7] IR (KBr): 1192 (ν(C-O)), 1910, 2002 (ν(CO)), 2110 (ν(CN)) [1]; (CH_2Cl_2): 1919, 1998 (ν(CO)), 2115 (ν(CN)) [7] mass spectrum: $[M]^+$ [7]
27	trans-$(CH_3)_5C_5Mo(CO)_2(=C(CH_2)_3O)CN$	VI, with KCN in methanol for 3 h (31%) [18] orange crystals, m.p. 102 to 104°C (from CH_2Cl_2/light petroleum) [18] ^{1}H NMR ($CDCl_3$): 1.94 (qui, central CH_2), 2.03 (CH_3), 3.71 (t, CH_2; J(H, H) = 7), 4.87 (t, OCH_2; J(H, H) = 7) [18] ^{13}C NMR (CD_2Cl_2, 203 K): 10.7 ($C_5(\mathbf{C}H_3)_5$), 21.4 (central CH_2), 60.3 (CH_2), 84.9 (OCH_2), 106.8 ($\mathbf{C}_5(CH_3)_5$), 230.6 (CO), 322.3 (Mo=C) [18] IR (CH_2Cl_2): 1950, 1985 (ν(CO)), 2102 (ν(CN)) [18] mass spectrum: $[M + 1]^+$ [18]

References on p. 32

Table 3 (continued)

No.	compound	method of preparation (yield) properties and remarks
28	trans-$(CH_3)_5C_5Mo(CO)_2(=C(CH_2)_3O)GeCl_3$	VI, with $[N(C_2H_5)_4][GeCl_3]$ (26%) [2] bright yellow crystals, m.p. 202 to 204°C (from CH_2Cl_2/heptane) [2] ^{1}H NMR ($CDCl_3$): 2.01 (qui, central CH_2), 3.75 (t, CH_2; J(H, H) = 7.5), 4.88 (t, OCH_2; J(H, H) = 7.5), 5.56 (s, C_5H_5) [2] IR ($CHCl_3$): 1911, 1983 (ν(CO)) [2]
29	trans-$C_5H_5Mo(CO)_2(=C(CH_2)_3O)Sn(C_6H_5)_3$	I (> 50%) [12] no further information given
30	cis-$(CH_3)_5C_5Mo(CO)_2(=C(CH_2)_3O)NCS$	VI, with LiSCN in THF for 1 h (54%) [18] bright red crystals, m.p. 85°C (from CH_2Cl_2/hexane) [18] ^{1}H NMR ($CDCl_3$, 223 K): 1.89 (s, CH_3), 2.0 (m, central CH_2), 3.49 (d of t, 1H, CH_2; J(H, H) = 21, 10), 4.01 (d of d of d, CH_2, 1H; J(H, H) = 21, 8, 3), 4.68 (t of d, OCH_2, 1H; J(H, H) = 8, 10), 5.15 (t of d, OCH_2, 1H; J(H, H) = 9, 3); spectra are temperature-dependent and given as diagrams at 223, 297, and 317 K; $\Delta G^{\neq}$ = 57 kJ/mol for a cis-cis isomerization, see General Remarks [18] ^{13}C NMR ($CDCl_3$, 223 K): 10.4 ($C_5(\mathbf{C}H_3)_5$), 21.1 (central CH_2), 54.3 (CH_2), 85.0 (OCH_2), 106.4 ($\mathbf{C}_5(CH_3)_5$), 146.6 (SCN), 236.1, 247.4 (CO), 328.5 (Mo=C) [18] IR (CH_2Cl_2): 1905, 1980 (ν(CO)), 2092 (ν(SCN)) [18] mass spectrum: $[M]^+$ [18]
31	trans-$C_5H_5Mo(CO)_2(=C(CH_2)_3O)SC_6H_5$	VI, with KSC_6H_5 in methanol for 3.5 h and chromatography on Al_2O_3 with CH_2Cl_2 (24%) [8] yellow-brown solid (from CH_2Cl_2/light petroleum), m.p. 93 to 96°C (dec.) [8] ^{1}H NMR ($CDCl_3$): 1.87 (qui, central CH_2; J(H, H) = 7), 3.70 (t, CH_2; J(H, H) = 7), 4.60 (t, OCH_2; J(H, H) = 7), 5.47 (s, C_5H_5), 7.00 (d, C_6H_5, H-4; J(H, H) = 7), 7.11 (t, C_6H_5, H-3; J(H, H) = 7), 7.43 (d, C_6H_5, H-2; J(H, H) = 7) [8] IR (CH_2Cl_2): 1900, 1978 (ν(CO)) [8] mass spectrum: $[M]^+$ [8]

References on p. 32

Table 3 (continued)

No.	compound	method of preparation (yield) properties and remarks
32	$(CH_3)_5C_5Mo(CO)_2(=C(CH_2)_3O)Br$	VI, with 4.5 equivalents of LiBr in THF for 3 h [18] lustrous dark red crystals (from CH_2Cl_2/light petroleum), m.p. 102°C (dec.), cis to trans ratio 88:12 at 250 K by ^{1}H NMR [18] ^{1}H NMR ($CDCl_3$): cis isomer: 1.87 (s and m, CH_3 and central CH_2), 3.81 (d of d of d, CH_2, 1H; J(H, H) = 21, 9, 5), 3.99 (d of t, CH_2, 1H; J(H, H) = 21, 9), 4.67 (d of t, OCH_2, 1H; J(H, H) = 9, 7), 5.06 (d of t, OCH_2, 1H; J(H, H) = 9, 5); trans isomer: 1.90 (s and m, CH_3 and central CH_2), 3.79 (t, CH_2; J(H, H) = 7), 4.67 (t, OCH_2; J(H, H) = 7) [18] ^{13}C NMR ($CDCl_3$, 233 K): cis isomer: 10.6 ($C_5(\mathbf{C}H_3)_5$), 21.8 (central CH_2), 58.2 (CH_2), 85.5 (OCH_2), 107.0 ($\mathbf{C}_5(CH_3)_5$), 236.4 and 249.3 (CO), 273.5 (Mo=C) [18] IR (CH_2Cl_2): 1871, 1892, 1967 (ν(CO)) [18] mass spectrum: $[M]^+$ [18]
*33	trans-$C_5H_5Mo(CO)_2(=C(CH_2)_3O)I$	Vb (55 to 90%); VI, with LiI · 3 H_2O, LiI, or $[N(C_4H_9\text{-}n)_4]I$ in refluxing THF for 1 h (90%) [6, 8] orange-brown crystals, m.p. 98 to 100°C (from CH_2Cl_2/light petroleum) [6, 8] ^{1}H NMR ($CDCl_3$): 1.92 (qui, central CH_2; J(H, H) = 7), 3.60 (t, CH_2; J(H, H) = 7), 4.60 (t, OCH_2; J(H, H) = 7), 5.57 (s, C_5H_5) [6, 8] ^{13}C NMR ($CDCl_3$): 22.5 (central CH_2), 58.8 (CH_2), 82.0 (OCH_2), 96.9 (C_5H_5), 223.4 (CO), 316.6 (Mo=C) [6, 8] IR (CH_2Cl_2): 1909, 1985 (ν(CO)) [6, 8] mass spectrum: $[M]^+$ [8]
34	trans-$CH_3C_5H_4Mo(CO)_2(=C(CH_2)_3O)I$	Vb, reflux for 45 min (59%) [8] orange crystals, m.p. 94 to 96°C (from CH_2Cl_2/petroleum ether) [8] ^{1}H NMR ($CDCl_3$): 1.91 (qui, central CH_2; J(H, H) = 7), 2.18 (s, CH_3), 3.56 (t, CH_2; J(H, H) = 7), 4.62 (t, OCH_2; J(H, H) = 7), 5.35 and 5.44 (m, C_5H_4, 2 H each) [8] ^{13}C NMR ($CDCl_3$): 14.5 (CH_3), 22.6 (central CH_2), 59.0 (CH_2), 81.7 (OCH_2), 94.0, 100.6 (C_5H_4), 111.7 (C on C_5H_4), 224.1 (CO), 316.9 (Mo=C) [8] IR (CH_2Cl_2): 1909, 1988 (ν(CO)) [8] mass spectrum: $[M]^+$ [8]

References on p. 32

Table 3 (continued)

No.	compound	method of preparation (yield) properties and remarks
		reduction with $Li[(C_2H_5)_3BH]$ in THF at −78°C affords ca. 30% $CH_3C_5H_4Mo(CO)_2CH(CH_2)_3O$ (C, O-bonded) besides ca. 10% $CH_3C_5H_4Mo(CO)_2CH_2CHCHCH_3$-$\eta^3$ [7, 13]
35	trans-$CH_3C(O)C_5H_4Mo(CO)_2({=}C(CH_2)_3O)I$	Vb, reflux for 1 h (74%) [8] orange crystals, m.p. 119 to 120°C (from CH_2Cl_2/petroleum ether) [8] ^1H NMR ($CDCl_3$): 1.94 (qui, central CH_2; J(H, H) = 7), 3.36 (s, CH_3), 3.57 (t, CH_2; J(H, H) = 7), 4.68 (t, OCH_2; J(H, H) = 7), 5.59 and 5.89 (m, C_5H_4, 2 H each) [8] ^{13}C NMR ($CDCl_3$): 22.3 (central CH_2), 27.6 (CH_3), 59.3 (CH_2), 83.0 (OCH_2), 96.1, 102.3 (C_5H_4), 104.6 (C on C_5H_4), 193.6 (C=O), 221.8 (CO), 316.3 (Mo=C) [8] IR (CH_2Cl_2): 1678 (ν(C=O)), 1924, 1998 (ν(CO)) [8] mass spectrum: $[M]^+$ [8]
36	trans-$(CH_3)_3SiC_5H_4Mo(CO)_2({=}C(CH_2)_3O)I$	Vb, for 19 h (15%); contaminated with some No. 33 [8] red crystals, m.p. 75 to 77°C (from CH_2Cl_2/petroleum ether) [8] ^1H NMR (C_6D_6): 0.37 (s, $SiCH_3$), 0.94 (qui, central CH_2; J(H, H) = 7), 3.04 (t, CH_2; J(H, H) = 7), 3.54 (t, OCH_2; J(H, H) = 7), 5.31 and 5.47 (m, C_5H_4, 2H) [8] ^{13}C NMR (C_6D_6): 0.3 ($SiCH_3$), 24.3 (central CH_2), 58.6 (CH_2), 81.3 (OCH_2), 96.6 (CH on C_5H_4), 100.6 (C on C_5H_4), 107.8, (CH on C_5H_4), 224.1 (CO), 315.6 (Mo=C) [8] IR (CH_2Cl_2): 1909, 1984 (ν(CO)) [8] mass spectrum: $[M]^+$ [8]
37	trans-$C_9H_7Mo(CO)_2({=}C(CH_2)_3O)I$ (C_9H_7 = indenyl)	Vb, for 19 h followed by rapid chromatography on Al_2O_3 with CH_2Cl_2/light petroleum ether (1:1) [8] brown crystals, m.p. 84 to 86°C (from CH_2Cl_2/light petroleum ether) [8] ^1H NMR ($CDCl_3$): 1.91 (qui, central CH_2; J(H, H) = 7), 3.45 (t, CH_2; J(H, H) = 7), 4.75 (t, OCH_2; J(H, H) = 7), 5.77 (t, CH, 1H; J(H, H) = 3), 6.05

References on p. 32

Table 3 (continued)

No.	compound	method of preparation (yield) properties and remarks
37 (continued)		(d, 5-membered ring, 2H; J(H, H) = 3), 7.16 and 7.50 (d of d, 6-membered ring, 2H; J(H, H) = 7) [8] ^{13}C NMR ($CDCl_3$): 22.7 (central CH_2), 58.9 (CH_2), 82.2 (OCH_2), 92.0 (2 C on 5-membered ring), 95.0 (CH on 5-membered ring), 112.4 (C on C_9H_7), 125.9 (CH on 6-membered ring), 223.9 (CO), 315.7 (Mo=C) [8] IR (CH_2Cl_2): 1912, 1987 (ν(CO)) [8] mass spectrum: $[M]^+$ [8]
*38	$(CH_3)_5C_5Mo(CO)_2(=C(CH_2)_3O)I$	VI, with two equivalents of LiI in THF for 2 h (67%); ca. 8% of cis-$(C_5(CH_3)_5MoO)_2O$-μ was obtained as side product [18] small red crystals (pure trans isomer), large red-black crystals (predominantly cis isomer), m.p. of mixture 116 to 118°C (from CH_2Cl_2/light petroleum ether); crystals can be separated manually [18] ^{1}H NMR ($CDCl_3$): cis isomer: 1.86 (qui, central CH_2; J(H, H) = 7.5), 1.99 (s, CH_3), 3.68, 4.40 (both d of t, CH_2, 1H; J(H, H) = 19.5, 7.5), 4.83, 4.93 (both d of t, OCH_2, 1H; J(H, H) = 9, 7.5); trans isomer: 1.88 (qui, central CH_2), 2.03 (s, CH_3), 3.67 (t, CH_2; J(H, H) = 7), 4.73 (t, OCH_2; J(H, H) = 7) [18] ^{13}C NMR ($CDCl_3$, 223 K): cis isomer: 11.2 ($C_5(\mathbf{CH_3})_5$), 22.5 (central CH_2), 63.9 (CH_2), 86.1 (OCH_2), 107.3 ($\mathbf{C_5}(CH_3)_5$), 236.4, 248.7 (CO), 331.4 (Mo=C); trans isomer: 11.9 ($C_5(\mathbf{CH_3})_5$), 22.5 (central CH_2), 60.4 (CH_2), 82.7 (OCH_2), 108.2 ($\mathbf{C_5}(CH_3)_5$), 228.2 (CO), 316.3 (Mo=C) [18] IR (CH_2Cl_2): cis isomer: 1885, 1973 (ν(CO)); trans isomer: 1867, 1957 (ν(CO)) [18] mass spectrum: $[M]^+$ [18]

*Further information:

$C_5H_5Mo(CO)_2(=C(C_6H_5)N(CH_3)C(C_6H_5)=NCH_3)Cl$ (Table **3**, No. **3**) was obtained by dropwise addition of two equivalents of $C_6H_5C(Cl)=NCH_3$ to $Na[C_5H_5Mo(CO)_3]$ in THF solution followed by heating of the mixture for 0.5 h at 50°C; recrystallization from CH_2Cl_2/ether at −35°C gives 27% of the cis/trans isomer mixture. Fractional crystallization at −35°C enriches one isomer (probably cis) to 58%, but above −10°C an equilibrium between 43% cis and 57% trans is established. The mechanism of formation via a salt like imide chloride dimer is discussed [9].

It rearranges in refluxing $CHCl_3$ to form unstable $[C_5H_5Mo(CO)_2C(C_6H_5)N(CH_3)C(C_6H_5)=N\text{-}CH_3]Cl$ (Formula IV, Mo is $C_5H_5Mo(CO)_2^+$). The corresponding PF_6 salt is obtained by treatment of No. 3 with NH_4PF_6 in aqueous ethanol. Refluxing in THF affords $C_5H_5Mo(CO)(C(C_6H_5)N(CH_3)C(C_6H_5)=NCH_3)Cl$ (Formula IV, Mo is $C_5H_5Mo(CO)Cl$) [9].

IV V

trans-$C_5H_5Mo(CO)_2(=C(OC_2H_5)C_6H_5)Ge(C_6H_5)_3$ (Table **3**, No. **16**) crystallizes in the monoclinic space group $P2_1/n-C^5_{2h}$ (No. 14) with the unit cell parameters a = 10.611 (2), b = 21.247 (3), c = 13.040 (2) Å, β = 96.32 (1)°; Z = 4 molecules per unit cell, $D_{calc} = D_{meas} = 1.47$ g/cm³. The molecular structure with the main bond distances and angles is given in **Fig. 3** [10, 14].

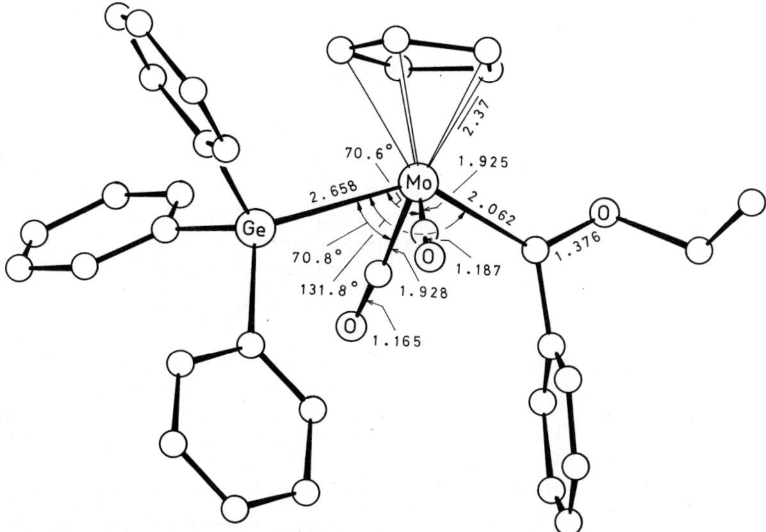

Fig. 3. Molecular structure of trans-$C_5H_5Mo(CO)_2(=C(OC_2H_5)C_6H_5)Ge(C_6H_5)_3$ [4].

$C_5H_5Mo(CO)_2(=CHC_6H_5)Sn(C_6H_5)_3$ (Table **3**, No. **18**) rearranges reversibly over a period of 1 h at ambient temperature in THF or $CDCl_3$ into an equilibrium mixture with the η^3-bonded benzyl complex V by migration of the $Sn(C_6H_5)_3$ group to the carbene carbon atom. The equilibrium is achieved at a 3:1 ratio (in $CDCl_3$) in favor of the allyl complex. Crystallization of the equilibrium mixture gives pure title complex and reaction of the mixture with $P(OCH_3)_3$ at −78°C in THF produces the ylide complex $C_5H_5Mo(CO)_2(C(C_6H_5)HP(OCH_3)_3)Sn(C_6H_5)_3$ (Table 2, No. 19) [19].

cis-$C_5H_5Mo(CO)_2(=C(CH_2)_nNCH_3)I$ (Table **3**, Nos. **23**, **25**; n = 3, 4). Compound No. 23 (n = 3) crystallizes in the triclinic space group $P\bar{1}-C^1_i$ (No. 2) with the unit cell parameters a = 7.824 (3), b = 13.363 (6), c = 14.559 (6) Å, α = 78.62 (4)°, β = 77.37 (3)°, γ = 80.99 (4)°; Z =

References on p. 32

4 molecules per unit cell, and D_{calc} = 1.962 g/cm^3. The molecular structure with the main bond distances and angles for one of the two observed independent molecules is given in **Fig. 4** [10, 14].

No isomerization to the trans isomer was observed in agreement with the lack of steric requirements for this isomerization [10, 11, 14, 17]. Protonation with H_2SO_4 affords N-methylpyrrolidine (n = 3) or N-methylpiperidine (n = 4) and an unidentified metal complex [17]. Reduction with 2 equivalents of sodium naphthalide affords $Na[C_5H_5Mo(CO)_2{=}C(CH_2)_nNCH_3]$ in good yields [12] while with $Li[(C_2H_5)_3BH]$, $C_5H_5Mo(CO)_2CH(CH_2)_nNCH_3$ (C, N-bonded) is obtained; see on p. 121 [16, 17].

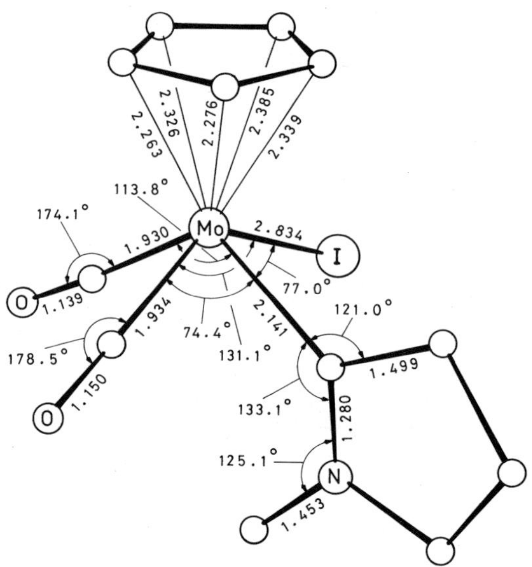

Fig. 4. Molecular structure of cis-$C_5H_5Mo(CO)_2({=}C(CH_2)_3NCH_3)I$ [14].

trans-$C_5H_5Mo(CO)_2({=}C(CH_2)_3O)I$ (Table **3**, No. **33**) crystallizes in the monoclinic space group $P2_1/n-C^5_{2h}$ (No. 14) with the unit cell parameters a = 6.372 (2), b = 14.168 (5), c = 14.390 (3) Å, β = 98.12 (2)°; Z = 4 molecules per unit cell, D_{meas} = 2.15, and D_{calc} = 2.138 g/cm^3. The molecular structure with the main bond distances and angles is shown in **Fig. 5** [6, 8].

Reduction with two equivalents of sodium naphthalide in THF at −78°C affords $Na[C_5H_5Mo(CO)_2{=}C(CH_2)_3O]$ [12] while with $Li[(C_2H_5)_3BH]$, $C_5H_5Mo(CO)_2CH(CH_2)_3O$ (C, O-bonded) is obtained [7, 15]. At −78°C in THF solution, a 31% yield of the former is obtained besides 11% of the allyl complex $C_5H_5Mo(CO)_2CH_2CHCHCH_3$-$\eta^3$ [13]. No. 33 decomposes on reaction with $[C_5H_5M(CO)_3]^-$ (M = Mo, W), $[C_5H_5Fe(CO)_2]^-$, or $[Mn(CO)_5]^-$ [8].

$(CH_3)_5C_5Mo(CO)_2({=}C(CH_2)_3O)I$ (Table **3**, No. **38**). The cis isomer crystallizes from CH_2Cl_2/light petroleum as very dark polyhedral fragments, which contains some trans isomer, in the monoclinic space group $P2_1/c-C^5_{2h}$ (No. 14) with the unit cell parameters a = 12.73 (4), b = 9.26 (3), c = 16.09 (4) Å, β = 108.68 (9)°; Z = 4 molecules per unit cell, D_{calc} = 1.791 g/cm^3. The molecular structure with the main bond distances and angles is given in **Fig. 6**. The structure is extensively disordered with two different cis components (refined population

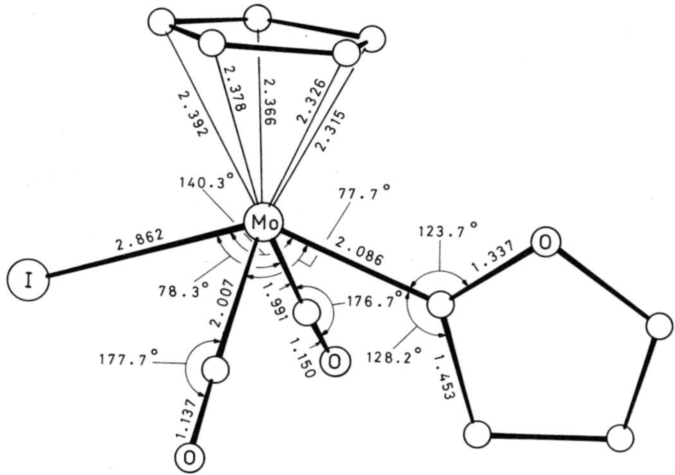

Fig. 5. Molecular structure of trans-$C_5H_5Mo(CO)_2(=C(CH_2)_3O)I$ [6, 8].

68.5% and 7.1%) and 24.4% trans isomer. The $C_5(CH_3)_5$ ligand is also rotationally disordered in about 2:1 ratio [18]. The trans isomer crystallizes from CH_2Cl_2/light petroleum as red hexagonal plates in the orthorhombic space group Pbca$-D_{2h}^{15}$ (No. 61) with the unit cell parameters a = 15.38 (6), b = 15.65 (8), c = 14.78 (6) Å; Z = 8 molecules per unit cell, D_{calc} = 1.808 g/cm^3. The molecular structure with the main bond distances and angles is given in **Fig. 7**, p. 32 [18].

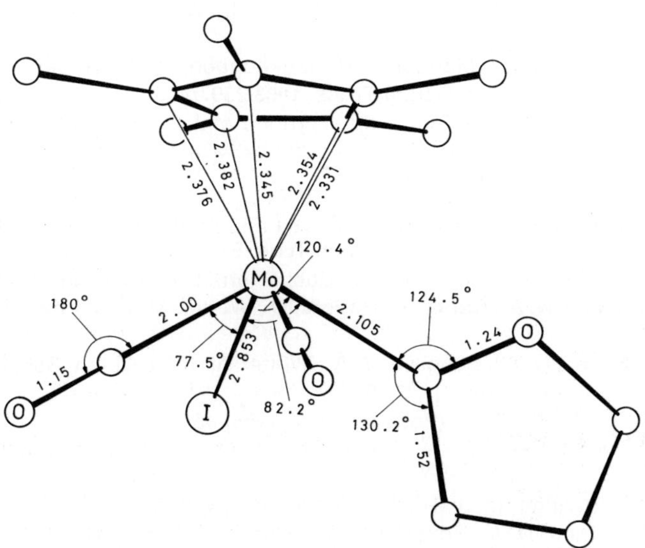

Fig. 6. Molecular structure of cis-$(CH_3)_5C_5Mo(CO)_2(=C(CH_2)_3O)I$ [18].

References on p. 32

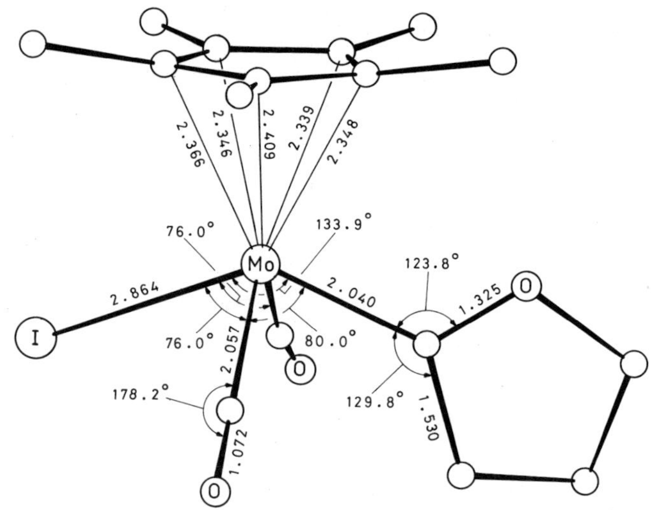

Fig. 7. Molecular structure of trans-$(CH_3)_5C_5Mo(CO)_2(=C(CH_2)_3O)I$ [18].

References:

[1] Kruck, T.; Liebig, L. (Chem. Ber. **106** [1973] 1055/61).

[2] Dean, W. K.; Graham, W. A. G. (J. Organometal. Chem. **120** [1976] 73/86).

[3] Adams, R. D.; Chodosh, D. F. (J. Am. Chem. Soc. **99** [1977] 6544/50).

[4] Chan, L. Y. Y.; Dean, W. K.; Graham, W. A. G. (Inorg. Chem. **16** [1977] 1067/71).

[5] Dean, W. K.; Graham, W. A. G. (Inorg. Chem. **16** [1977] 1061/7).

[6] Bailey, N. A.; Chell, P. L.; Mukhopadhyay, A.; Tabbron, H. E.; Winter, M. J. (J. Chem. Soc. Chem. Commun. **1982** 215/7).

[7] Adams, H.; Bailey, N. A.; Cahill, P.; Rogers, D.; Winter, M. J. (J. Chem. Soc. Chem. Commun. **1983** 831/3).

[8] Bailey, N. A.; Chell, P. L.; Manuel, C. P.; Mukhopadhyay, A.; Rogers, D.; Tabbron, H. E.; Winter, M. J. (J. Chem. Soc. Dalton Trans. **1983** 2397/403).

[9] Brunner, H.; Meyer, W.; Wachter, J. (J. Organometal. Chem. **243** [1983] 437/41).

[10] Adams, H.; Bailey, N. A.; Osborne, V. A.; Winter, M. J. (J. Organometal. Chem. **284** [1984] C 1/C 4).

[11] Adams, H.; Bailey, N. A.; Osborne, V. A.; Winter, M. J. (Chem. Uses Molybdenum Proc. 5th Intern. Climax Conf., Newcastle Upon Tyne, Engl., 1985, pp. 129/30, poster 1.17).

[12] Osborne, V. A.; Winter, M. J. (J. Chem. Soc. Chem. Commun. **1985** 1744/5).

[13] Adams, H.; Bailey, N. A.; Cahill, P.; Rogers, D.; Winter, M. J. (J. Chem. Soc. Dalton Trans. **1986** 2119/26).

[14] Adams, H.; Bailey, N. A.; Osborne, V. A.; Winter, M. J. (J. Chem. Soc. Dalton Trans. **1986** 2127/35).

[15] Gauntlett, J. T.; Winter, M. J. (Polyhedron **5** [1986] 451/59).

[16] Osborne, V. A.; Parker, C. A.; Winter, M. J. (J. Chem. Soc. Chem. Commun. **1986** 1185/6).

[17] Osborne, V. A.; Winter, M. J. (Polyhedron **5** [1986] 435/7).

[18] Bailey, N. A.; Dunn, D. A.; Foxcroft, C. N.; Harrison, G. R.; Winter, M. J.; Woodward, S. (J. Chem. Soc. Dalton Trans. **1988** 1449/56).

[19] Winter, M. J.; Woodward, S. (J. Chem. Soc. Chem. Commun. **1989** 457/8).

1.5.1.3.2.1.5 Compounds of the Type $^5LMo(CO)_2{\equiv}CR$

This section covers compounds with a carbyne ligand bonded to the $^5LMo(CO)_2$ fragment. The compounds listed in Table 4 were prepared in most cases by the following method.

Method I: Solid $NaC_5H_5 \cdot CH_3OC_2H_4OCH_3$ was added to a solution of $(CO)_4Mo({\equiv}CR)O_2CCF_3$ in ether (obtained in situ from $Mo(CO)_6$ and LiR in ether, followed by addition of $O(C(O)CF_3)_2$ at −78°C) cooled to ca. −20°C. The reaction mixture was allowed to warm up. After the reaction had ceased (monitored by IR), the volume was reduced and the residue was extracted with CH_2Cl_2/light petroleum. The extracts were chromatographed on an alumina column [9].

Table 4
Compounds of the Type $^5LMo(CO)_2{\equiv}CR$.
An asterisk indicates further information at the end of the table.
For explanations, abbreviations, and units see p. X.

No.	compound	method of preparation (yield) properties and remarks
*1	$C_5H_5Mo(CO)_2{\equiv}CCH_3$	I [1, 3]
2	$C_5H_5Mo(CO)_2{\equiv}CCH_2C_4H_9$-t	from sequential displacement of the phosphite ligands of $C_5H_5Mo(P(OCH_3)_3)_2{\equiv}CCH_2C_4H_9$-t in solution by CO [2] ^1H NMR (C_6D_6): 0.94 (CH_3), 2.15 (CH_2), 5.04 (C_5H_5) [2] ^{13}C NMR (C_6D_6): 29.5 (CH_3), 33.2 ($\mathbf{C}(CH_3)_3$), 64.8 (CH_2), 92.4 (C_5H_5), 229.8 (CO), 332.8 (Mo≡C) [2] IR (hexane): 1928, 1998 (ν(CO)) [2] oxidation with E (E = S, Se) in THF at 25°C for 12 h affords $C_5H_5Mo(CO)_2E_2CCH_2C_4H_9$-t [4]
*3	$C_5H_5Mo(CO)_2{\equiv}CC_6H_4CH_3$-2	I (75%) [9] orange solid (from CH_2Cl_2/light petroleum) [9] ^1H NMR (CD_2Cl_2): 2.50 (s, CH_3), 5.53 (s, C_5H_5), 6.90 to 7.20 (m, C_6H_4) [9] ^{13}C NMR ($CDCl_3$): 20.1 (CH_3), 92.1 (C_5H_5), 124.6, 128.5, 128.9, 129.9, and 138.0 (C_6H_4), 144.6 (C_6H_4, C-1), 227.5 (CO), 309.5 (Mo≡C) [9] IR (CH_2Cl_2): 1920, 1995 (ν(CO)) [9]
*4	$C_5H_5Mo(CO)_2{\equiv}CC_6H_4CH_3$-4	I (high) [7] orange-red crystals (from CH_2Cl_2/pentane) [13] mass spectrum: $[M]^+$ [13]
*5	$C_5H_5Mo(CO)_2{\equiv}CC_6H_4N(CH_3)_2$-4	I (60%) [9] orange solid (from CH_2Cl_2/light petroleum ether) [9] ^1H NMR (CD_2Cl_2): 2.97 (s, CH_3), 5.56 (s, C_5H_5), 6.44, 7.38 ($(AB)_2$ system, C_6H_4; J(AB) = 9) [9]

References on pp. 38/9

Table 4 (continued)

No.	compound	method of preparation (yield) properties and remarks
*5 (continued)		^{13}C NMR ($CDCl_3$): 39.8 (CH_3), 91.9 (C_5H_5), 109.7, 131.3, and 135.5 (C_6H_4), 149.8 (C_6H_4, C-1), 228.6 (CO), 312.7 (Mo$\equiv$C) [9] IR (CH_2Cl_2): 1924, 1991 (ν(CO)) [9]
*6	$C_5H_5Mo(CO)_2{\equiv}CC_6H_4OCH_3$-2	I (80%) [9] orange solid (from CH_2Cl_2/light petroleum) [9] ^{1}H NMR (CD_2Cl_2): 3.83 (s, CH_3), 5.60 (s, C_5H_5), 6.71 to 7.37 (m, C_6H_4) [9] ^{13}C NMR ($CDCl_3$): 55.5 (CH_3), 92.6 (C_5H_5), 110.8, 119.8, 130.6, 131.6, and 135.9 (C_6H_4), 160.0 (C_6H_4, C-1), 229.2 (CO), 305.9 (Mo$\equiv$C) [9] IR (CH_2Cl_2): 1919, 1995 (ν(CO)) [9]
*7	$C_5H_5Mo(CO)_2{\equiv}C$–$C_6H_4(OCH_3)$–$Cr(CO)_3$ (structure: OCH3; Mo$\equiv$C; OC; OC; Cr(CO)3)	similar to I with $(CO)_2Mo(C_{10}H_8N_2)(O_2CCF_3){\equiv}C$-$(C_6H_4OCH_3$-2)$Cr(CO)_3$ ($C_{10}H_8N_2$ = bipyridine) (46%) [10] red solid (from CH_2Cl_2/light petroleum 1:1 at −78°C) [10] ^{1}H NMR ($CDCl_3$): 3.77 (s, CH_3), 4.71 (t, C_6H_4, 1H; J(H, H) = 6), 4.89 (d, C_6H_4, 1H; J(H, H) = 6), 5.51 (t, C_6H_4, 1H; J(H, H) = 6), 5.62 (s, C_5H_5), 5.83 (d, C_6H_4, 1H; J(H, H) = 6) [10] ^{13}C NMR (CD_2Cl_2/CH_2Cl_2): 55.9 (CH_3), 72.2, 83.3 (C_6H_4), 93.0 (C_5H_5), 93.3, 97.8, and 99.8 (C_6H_4), 143.7 (C_6H_4, C-1), 228.3 and 228.7 (MoCO), 232.0 (CrCO), 295.6 (Mo$\equiv$C) [10] IR (ether): 1903, 1932, 1967, 2009 (ν(CO)) [10]
*8	$C_5H_5Mo(CO)_2{\equiv}CC_6H_3(CH_3)_2$-2,6	I (75%) [9] orange solid (from CH_2Cl_2/light petroleum) [9] ^{1}H NMR (CD_2Cl_2): 2.53 (s, CH_3), 5.61 (s, C_5H_5), 6.90 to 7.13 (m, C_6H_3) [9] ^{13}C NMR ($CDCl_3$): 20.9 (CH_3), 92.1 (C_5H_5), 126.7, 128.1, and 139.3 (C_6H_3), 143.6 (C_6H_4, C-1), 228.7 (CO), 310.5 (Mo$\equiv$C) [9] IR (CH_2Cl_2): 1919, 1992 (ν(CO)) [9]

*Further information:

$C_5H_5Mo(CO)_2{\equiv}CCH_3$ (Table **4**, No. **1**). Reaction with two equivalents of $P(CH_3)_3$ affords $C_5H_5Mo(CO)(P(CH_3)_3)_2C(CH_3){=}C{=}O$, which loses one $P(CH_3)_3$ ligand reversibly at ambient temperature to yield $C_5H_5Mo(CO)(P(CH_3)_3)O{\cdots}C{\cdots}CCH_3$-$\eta^3$ [1]. With one equivalent of $[(C_2H_5)_3NH][Fe_3(\mu$-$H)(CO)_{11}]$ in THF at 80°C for 4 d, $C_5H_5Mo(CO)_2(\mu$-$C{\equiv}CCH_3)(Fe(CO)_3)_2$ (Formula I) is obtained [3]. Treatment with one equivalent of $C_5H_5W(CO)(P(CH_3)_3)(\mu$-$CO)(\mu$-$CC_6H_4CH_3$-4)$PtC_8H_{12}$ (C_8H_{12} = cycloocta-1,5-diene) in THF at 25°C affords the trinuclear compound shown in Formula II (M = Pt, M′ = W, M″ = Mo, R = $C_6H_4CH_3$-4, R′ = CH_3) [14].

References on pp. 38/9

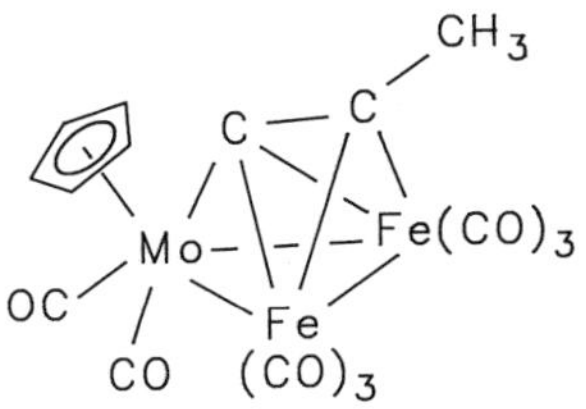

$C_5H_5Mo(CO)_2{\equiv}CC_6H_4CH_3$-2 (Table **4**, No. **3**) was treated with one equivalent of $Fe_2(CO)_9$ in ether for 4 h at 25°C to give $C_5H_5Mo(CO)_2(\mu\text{-}CC_6H_4CH_3\text{-}2)Fe(CO)_4$ (Formula III; R = $C_6H_4CH_3$-2) [9].

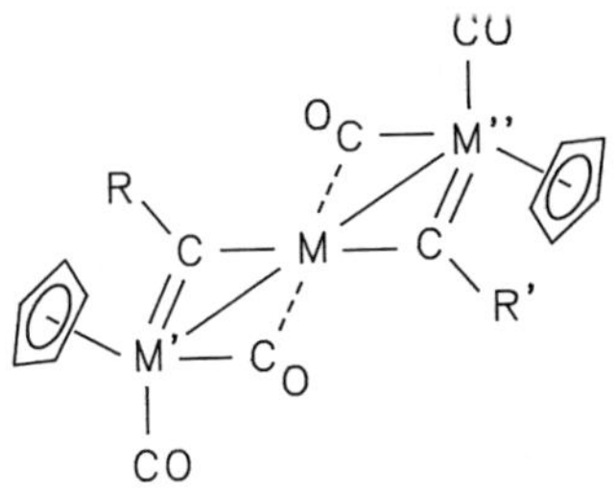

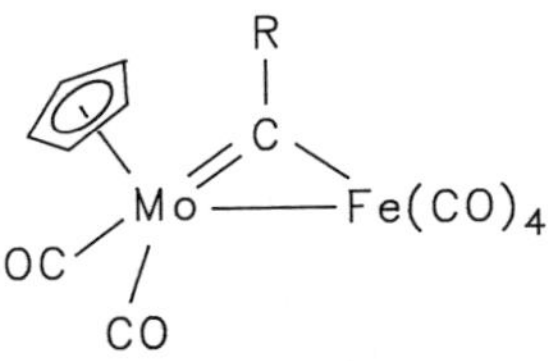

$C_5H_5Mo(CO)_2{\equiv}CC_6H_4CH_3$-4 (Table **4**, No. **4**) was also prepared by treating $(CO)_4Mo({\equiv}C\text{-}C_6H_4CH_3\text{-}2)Cl$ with one equivalent NaC_5H_5 in ether at −78°C [13].

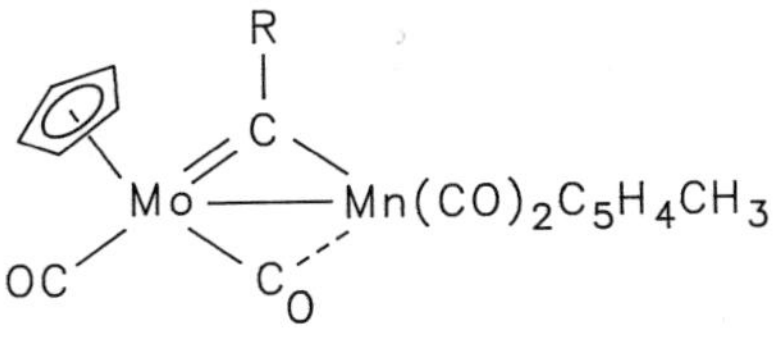

The results of the reaction with $P(CH_3)_3$ depend on the conditions. With two equivalents in solution, $C_5H_5Mo(CO)(P(CH_3)_3)_2C(C_6H_4CH_3\text{-}4){=}C{=}O$ is obtained, which loses one $P(CH_3)_3$ ligand reversibly at ambient temperature to yield $C_5H_5Mo(CO)(P(CH_3)_3)O{\cdots}C{\cdots}CC_6H_4CH_3\text{-}4\text{-}\eta^3$ [1]. With one equivalent of $P(CH_3)_3$, $C_5H_5Mo(CO)(P(CH_3)_3){\equiv}CC_6H_4CH_3$-4 is formed [13]. Treating No. 4 with HX (X = Cl, CF_3COO) leads to the formation of $C_5H_5Mo(CO)(C(O)CH_2C_6H_4CH_3\text{-}4)X_2$ [8]. No. 4 is used as parent compound by the preparation of various heteronuclear compounds. These reactions with other organometallic compounds are compiled in the following table.

reaction with	products (R = $C_6H_4CH_3$-4) conditions and remarks
$CH_3C_5H_4Mn(CO)_2 \cdot THF$	$C_5H_5Mo(CO)(CO\text{-}\mu)(\mu\text{-}CR)Mn(CO)_2C_5H_4CH_3$ (Formula IV) in THF for 12 h [15]
$Fe_2(CO)_9$	$(C_5H_5MoCO)_2(CO\text{-}\mu)(RC{\equiv}CR\text{-}\mu)Fe(CO)_3$ (Formula V) (5%) and the complex shown in Formula III (80%) 1:1 mole ratio in ether for 1 h [7]

References on pp. 38/9

reaction with	products ($R = C_6H_4CH_3$-4) conditions and remarks
$Fe_2(CO)_9$	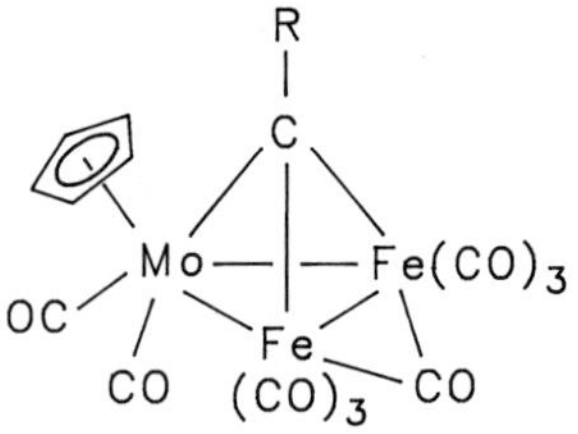 [structure: C_5H_5(CO)$_2$Mo–Fe(CO)$_3$–Fe(CO)$_3$ cluster with μ_3-CR and bridging CO] 1:3 mole ratio in ether for 12 h [7]
$Fe_2(CO)_9$	$(C_5H_5MoCO)_2(CO\text{-}\mu)(RC{\equiv}CR\text{-}\mu)Fe(CO)_3$ (Formula V) (75%) 2:1 mole ratio in ether for 6 h [7]
$(CO)_3Fe(C_8H_{14}\text{-cyclo})_2$	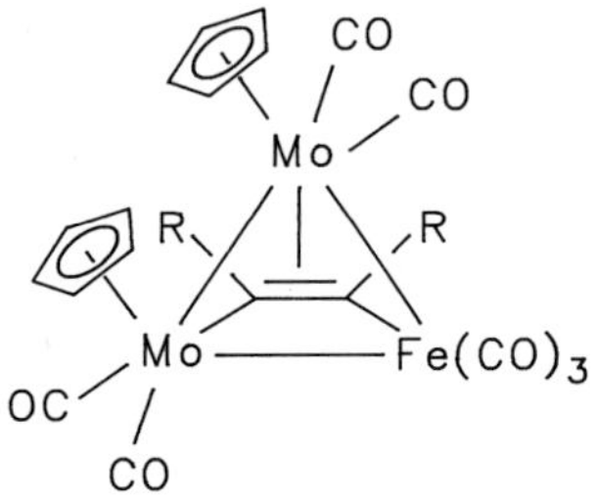 [structure: C_5H_5Mo(CO)$_2$, C_5H_5Mo(CO)$_2$ and Fe(CO)$_3$ bridged by RC=CR] with 0.5 equivalents in light petroleum at −40°C for 4 h [7]
$(CO)_8Fe_2(\mu\text{-}CH_2)$	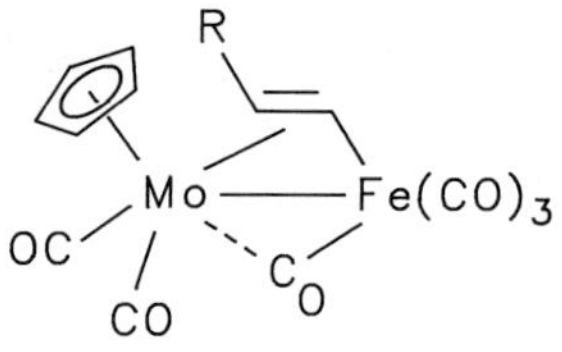 [structure: C_5H_5(CO)$_2$Mo–Fe(CO)$_3$ with bridging RCH=CH and semibridging CO] in THF for 12 h [12]
$C_9H_7M(CO)_{2-n}(P(CH_3)_3)_n$ (C_9H_7 = indenyl; M = Rh, n = 0, 1; M = Ir, n = 0)	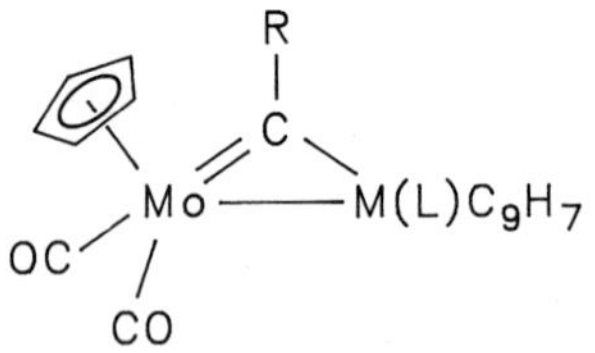 [structure: C_5H_5(CO)$_2$Mo(=CR)–M(L)C_9H_7] (M = Rh, L = CO, $P(CH_3)_3$; M = Ir, L = CO) in light petroleum for 8 h [6]
$Ni(C_8H_{12})_2$ (C_8H_{12} = cycloocta-1,5-diene)	compound II; M = Ni, M′ = M″ = Mo in THF for 1 h [14]

References on pp. 38/9

reaction with	products ($R = C_6H_4CH_3$-4) conditions and remarks
$Pt(C_8H_{12})_2$ or $Pt(CH_2{=}CH_2)_2$ (C_8H_{12} = cycloocta-1,5-diene)	compound II; M = Pt, M′ = M″ = Mo in light petroleum at 0 to 25°C for 1h [14]
$C_5H_5W(CO)(P(CH_3)_3)(CO\text{-}\mu)(CR'\text{-}\mu)PtC_8H_{12}$ ($R' = C_6H_5, C_6H_4CH_3$-4) (C_8H_{12} = cycloocta-1,5-diene)	compound II (M = Pt, M′ = W, M″ = Mo, $R' = C_6H_5$, $C_6H_4CH_3$-4) in THF [14]

$C_5H_5Mo(CO)_2{\equiv}CC_6H_4N(CH_3)_2$-4 (Table **4**, No. **5**) reacts with two equivalents of $C_5H_4CH_3Mn(CO)_2 \cdot THF$ in THF at 25°C for 12 h to afford compound IV ($R = C_6H_4N(CH_3)_2$-4) [15]. With one equivalent of $Fe_2(CO)_9$ in ether at 25°C for 4 h a mixture of 85% compound III ($R = C_6H_4N(CH_3)_2$-4) and 11% compound V ($R = C_6H_4(CH_3)_2$-4) is obtained [9].

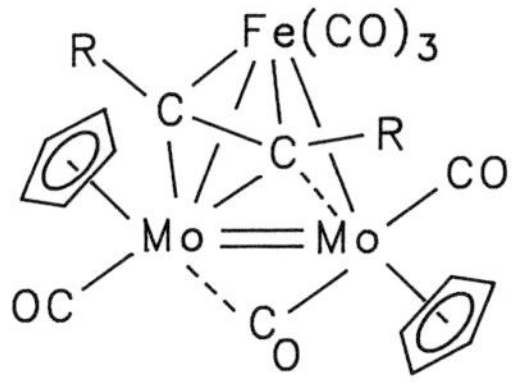

V 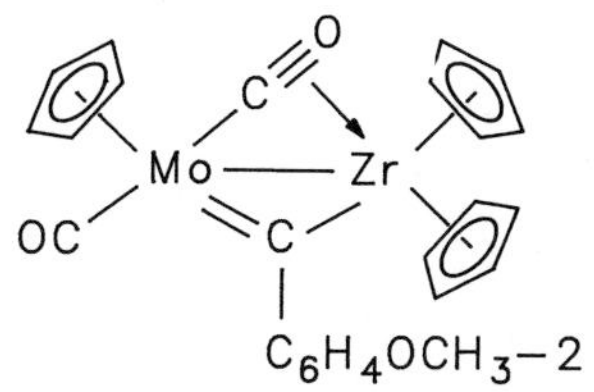 VI

$C_5H_5Mo(CO)_2{\equiv}CC_6H_4OCH_3$-2 (Table **4**, No. **6**) reacts with $(C_5H_5)_2Zr(C_4H_9\text{-}n)_2$ in ether at −10°C forming $C_5H_5Mo(CO)(\mu\text{-}CO)(\mu\text{-}CC_6H_4OCH_3\text{-}2)Zr(C_5H_5)_2$ (Formula VI) [11]. Reaction with 2 equivalents of $C_5H_4CH_3Mn(CO)_2 \cdot THF$ in THF at 25°C for 12 h affords compound IV ($R = C_6H_4OCH_3$-2) [15]. With one equivalent of $Fe_2(CO)_9$ in ether at 25°C for 4 h, compound III ($R = C_6H_4OCH_3$-2) is obtained in 76% yield [9].

$C_5H_5Mo(CO)_2({\equiv}CC_6H_4OCH_3\text{-}2)Cr(CO)_3$ (Table **4**, No. **7**) reacts with one equivalent of $Co_2(CO)_8$ in ether at 25°C to afford compound VII ($R = (\eta^6\text{-}C_6H_4OCH_3\text{-}2)Cr(CO)_3$) while one equivalent of $C_9H_7Rh(CO)_2$ in petroleum ether for 0.5 h affords compound VIII ($R = (\eta^6\text{-}C_6H_4OCH_3\text{-}2)Cr(CO)_3$) in 67 and 75% yield, respectively [10].

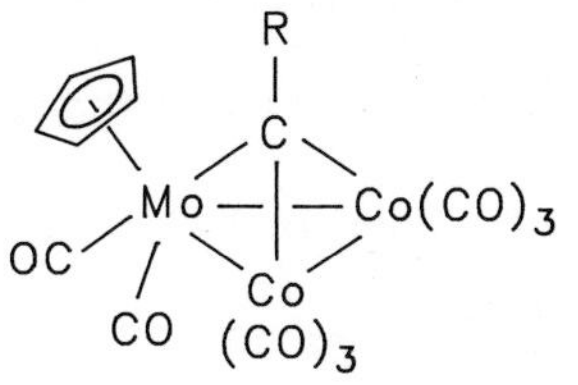

VII 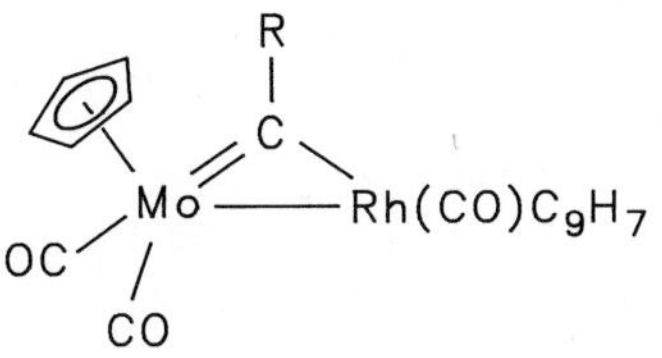 VIII

References on pp. 38/9

$C_5H_5Mo(CO)_2{\equiv}CC_6H_3(CH_3)_2$-2,6 (Table **4**, No. **8**). The reaction with $C_5H_4CH_3Mn(CO)_2 \cdot$ THF in THF failed, possibly for steric reasons. With $C_9H_7Rh(CO)_2$, a mixture containing 15% of unstable IX (see also Formula VIII, R = $C_6H_3(CH_3)_2$-2,6; formulation without a semi-bridging CO group) and 45% of X is obtained [15]. Reaction with one equivalent of $P(C_6H_5)_3$ under UV irradiation in petroleum ether affords $C_5H_5Mo(CO)(P(C_6H_5)_3)O{\cdots}C{\cdots}CC_6H_3(CH_3)_2$-2,6-$\eta^3$ with a metallacyclopropene ring, while refluxing in THF affords $C_5H_5Mo(CO)(P(C_6H_5)_3){\equiv}CC_6H_3(CH_3)_2$-2,6. With 1.3 equivalents of $Fe_2(CO)_9$ in THF for 5 h at 25°C in the dark, C_5H_5-$Mo(CO)_2(CC_6H_3(CH_3)_2$-2,6-$\mu)Fe(CO)_4$ (Formula III) is formed. Stirring with one equivalent of $Co_2(CO)_8$ in petroleum ether for 2 h at 25 °C affords $C_5H_5Mo(CO)_2(\mu_3$-$CC_6H_3(CH_3)_2$-2,6$)(Co(CO)_3)_2$ (Formula VII, R = $C_6H_3(CH_3)_2$-2,6) [9].

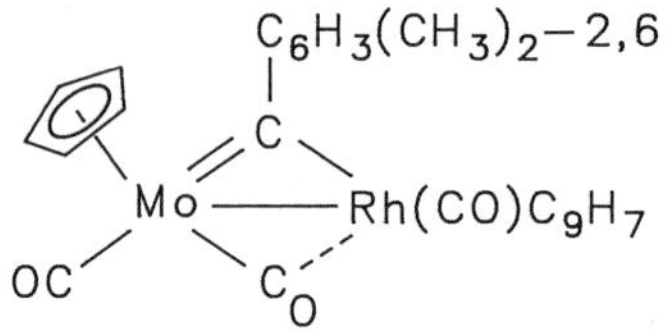

IX

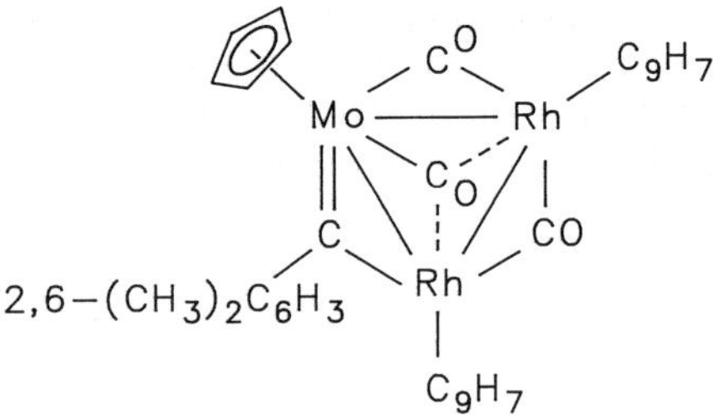

X

References:

[1] Uedelhoven, W.; Eberl, K.; Kreißl, F. R. (Chem. Ber. **112** [1979] 3376/89).

[2] Gill, D. S.; Baker, P. K.; Green, M.; Paddick, K. E.; Murray, M. (J. Chem. Soc. Chem. Commun. **1981** 986/8).

[3] Green, M.; Marsden, K.; Salter, I. D.; Stone, F. G. A.; Woodward, P. (J. Chem. Soc. Chem. Commun. **1983** 446/7).

[4] Gill, D. S.; Green, M.; Marsden, K.; Moore, I.; Orpen, A. G.; Stone, F. G.; Williams, I. D.; Woodward, P. (J. Chem. Soc. Dalton Trans. **1984** 1343/7).

[5] Garcia, M. E.; Jeffery, J. C.; Sherwood, P.; Stone, F. G. A. (J. Chem. Soc. Chem. Commun. **1986** 802/4).

[6] Abad, J. A.; Delgado, E.; Garcia, M. E.; Grosse-Ophoff, M. J.; Hart, I. J.; Jeffery, J. C.; Simmons, M. S.; Stone, F. G. A. (J. Chem. Soc. Dalton Trans. **1987** 41/50).

[7] Garcia, M. E.; Jeffery, J. C.; Sherwood, P.; Stone, F. G. A. (J. Chem. Soc. Dalton Trans. **1987** 1209/14).

[8] Kreißl, F. R.; Sieber, W. J.; Keller, H.; Riede, J.; Wolfgruber, M. (J. Organometal. Chem. **320** [1987] 83/90).

[9] Dossett, S. J.; Hill, A. F.; Jeffery, J. C.; Marken, F.; Sherwood, P.; Stone, F. G. A. (J. Chem. Soc. Dalton Trans. **1988** 2453/65).

[10] Fernández, J. R.; Stone, F. G. A. (J. Chem. Soc. Dalton Trans. **1988** 3035/40).

[11] Hill, A. F.; Hönig, H. D.; Stone, F. G. A. (J. Chem. Soc. Dalton Trans. **1988** 3031/4).

[12] Jeffery, J. C.; Parrott, M. J.; Stone, F. G. A. (J. Chem. Soc. Dalton Trans. **1988** 3017/30).

[13] Kreißl, F. R.; Uedelhoven, W.; Neugebauer, D. (J. Organometal. Chem. **344** [1988] C 27/C 30).
[14] Davis, S. J.; Stone, F. G. A. (J. Chem. Soc. Dalton Trans. **1989** 785/95).
[15] Hill, A. F.; Marken, F.; Nasir, B. A.; Stone, F. G. A. (J. Organometal. Chem. **363** [1989] 311/23).

1.5.1.3.2.1.6 Compounds of the Type $^5LMo(CO)_2(\mu\text{-}CO)M$

This section deals with compounds with an isocarbonyl linkage between Mo and M (Mo-(μ-CO)M; μ-CO means here end-to-end bonded) without a metal-metal bond. The compounds can be considered as carbynes, $C_5H_5Mo(CO)_2{\equiv}COR$, in which R is replaced by a metal fragment M, ($C_5H_5Mo(CO)_2{\equiv}COM$), or as adducts of the Lewis base $[C_5H_5Mo(CO)_3]^-$ at a Lewis acid, M^+, via the oxygen atom of one carbonyl group of the base ($C_5H_5(CO)_2MoCO$-M); alternatively, adduct formation can occur via a metal-metal bond. However, isocarbonyl linkages are formed with oxophilic metalcenters ("hard" Lewis acids) such as highly oxidized early transition metals or actinides; in one case, with the neutral $Al(CH_3)_3$, a group 13 metal plays the role of the hard Lewis acid. Only neutral compounds of this type are known; the ^5L ligand at Mo is C_5H_5, $(CH_3)_5C_5$, or $[(C_6H_5)_3PC_5H_4]^+$. The first complex in this series is paramagnetic with one unpaired electron.

Mentioned without details or data were the complexes **$CH_3C_5H_4Mo(CO)_2(\mu\text{-}CO)Zr(C_5H_4\text{-}CH_3\text{-}4)_2CH_3$**, **$CH_3C_5H_4Mo(CO)_2(\mu\text{-}CO)(\mu\text{-}C(CH_3){=}O)Zr(C_5H_4CH_3)_2$**, and **$CH_3C_5H_4Mo(CO)_2(\mu\text{-}CO)Zr(C_5H_4CH_3)_2(\mu\text{-}(CH_3)_2C{=}O)Zr(C_5H_4CH_3)_2CH_3$**, of which the last member was described to be well soluble in hydrocarbons [9].

General Remarks. In solution at room temperature a rapid exchange between the two terminal CO groups and the μ-CO group takes place as shown by temperature-dependent ^{13}C NMR studies [3]. The stretching frequencies of end to end bridging carbonyl groups are between 1550 to 1650 cm^{-1}. The CO stretching frequencies of THF solutions of the complexes in which M is bonded by uranium or thorium suggest a dissociation with formation of $[^5LMo(CO)_3]^-$ and solvated cations [6, 7].

$C_5H_5Mo(CO)_2(\mu\text{-}CO)Ti(C_5H_5)_2 \cdot THF$. $(C_5H_5)_2Ti(CO)_2$ was added to a solution of $C_5H_5Mo(CO)_2{\equiv}Mo(CO)_2C_5H_5$ in THF and the mixture was stirred for 18 h. After removal of the solvent, the residue was extracted with toluene to remove $(C_5H_5Mo(CO)_3)_2$ and a high yield of a bright green solid was obtained. A similar reaction in toluene gave only $(C_5H_5Mo(CO)_3Ti(C_5H_5)_2)_n$ and no reaction took place with $(C_5H_5Mo(CO)_3)_2$ in THF [5].

The solvate complex is paramagnetic with $\mu_{eff} = 1.65$ at 34 °C, obtained by the Evans NMR method. IR spectrum (THF): 1650, 1830, 1920 (ν(CO)) cm^{-1} [5].

The complex crystallizes in the monoclinic space group $P2_1/c - C^5_{2h}$ (No. 14) with the unit cell parameters a = 14.333 (2), b = 10.783 (3), c = 14.051 (3) Å, β = 109.19 (2)°; Z = 4 molecules per unit cell, and D_{calc} = 1.601 g/cm. The molecular structure with the main bond distances and angles is shown in **Fig. 8**. The bridging carbonyl has a shorter MoC distance and a longer CO bond than the two terminal CO groups [5].

The compound is very soluble in THF, slightly soluble in toluene, and insoluble in aliphatic hydrocarbons. The complex is highly air- and moisture-sensitive, yielding $(C_5H_5Mo(CO)_3)_2$ as the only carbonyl species on exposure to air. Protonation with acetic acid affords $C_5H_5Mo(CO)_3H$ and $((C_5H_5)_2TiO_2CCH_3)_2$. Reaction with CH_3I gives $C_5H_5Mo(CO)_3CH_3$ and $((C_5H_5)_2TiI)_2$. There appears to be no reaction with H_2 or CO at ambient temperature [5].

References on pp. 44/5

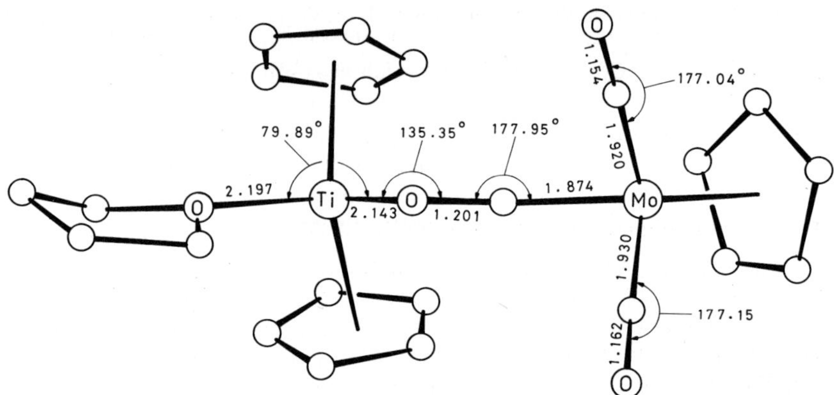

Fig. 8. The molecular structure of $C_5H_5Mo(CO)_2(\mu\text{-}CO)Ti(C_5H_5)_2 \cdot THF$ [5].

$C_5H_5Mo(CO)_2(\mu\text{-}CO)Ti(C_5(CH_3)_5)_2CH_3 \cdot 0.5\ C_6H_6$. Addition of $C_5H_5Mo(CO)_3H$ to a toluene solution of $(CH_3)_5C_5Ti(C_5(CH_3)_4{=}CH_2)CH_3$ resulted in a color change from turquoise to dark red-brown. Filtration and removal of the solvent gave an 80% yield of a dark red-brown powder [2].

IR spectrum (Nujol): 1623 ($\nu(\mu\text{-CO})$), 1830, 1849, 1918, 1927 (ν(CO)) cm^{-1}. Mass spectrum (field desorption): $[M]^+$ [2].

The complex crystallizes from a saturated benzene solution in the monoclinic space group $P2_1/n-C_{2h}^5$ (No. 14) with the unit cell parameters a = 10.109 (3), b = 18.106 (5), c = 16.648 (4) Å, β = 104.21 (2)°, and Z = 4 molecules per unit cell. The molecular structure with the main

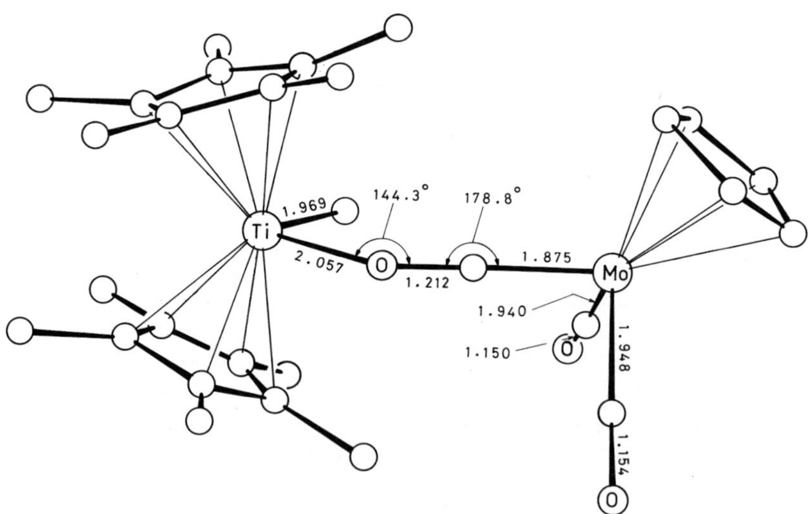

Fig. 9. The molecular structure of $C_5H_5Mo(CO)_2(\mu\text{-}CO)Ti(C_5(CH_3)_5)_2CH_3 \cdot 0.5\ C_6H_6$ [2].

bond distances and angles is shown in **Fig. 9** (solvent molecule not shown). The bonding of the μ-CO fragment is explained as a Lewis acid interaction of the Ti^{IV} center with the carbonyl oxygen atom and is further supported by results presented and discussed of a Fenske-Hall MO calculation [2].

$C_5H_5Mo(CO)_2(\mu$-$CO)Zr(C_5H_5)_2CH_3$. $C_5H_5Mo(CO)_3H$ and $(C_5H_5)_2Zr(CH_3)_2$ were stirred in THF for 1 h [1, 4]. Removal of the solvent and crystallization of the oil obtained from toluene/hexane at −20°C gave a yellow-orange precipitate in 63% yield [4]. The compound was first believed to contain an Mo-Zr bond [1], which was later corrected [4, 9].

^1H NMR spectrum (C_6D_6): δ = 0.45 (s, CH_3), 5.2 (s, C_5H_5Mo), 5.72 (s, C_5H_5Zr) ppm [4]; related data in [1]. ^{13}C NMR spectrum (toluene-d_8): δ = 35.1 (CH_3), 89.9 (C_5H_5Mo), 113.5 (C_5H_5Zr), 236.1 (CO) ppm. At −80°C in toluene-d_8, two CO resonances were observed at δ = 232.5 and 245.4 ppm in the relative ratio 2:1 for the terminal and the bridging CO [4]. The barrier to migration of the bridging CO to the equivalent terminal CO sites is $\Delta G^{\neq}$ = 9.6 kcal/mol [3]. IR spectrum (toluene): 1545 ($\nu(\mu$-CO)), 1863, 1948 (ν(CO)) cm^{-1} [4].

The complex is well soluble in aromatic solvents and THF and is extremely air-sensitive. Reaction with 1 atm CO pressure affords $C_5H_5Mo(CO)_2(\mu$-$CO)Zr(C_5H_5)_2(C(O)CH_3)$ in high yield (Formula III; see below). This reaction is partially reversible; 20 freeze/pump/thaw cycles afford 4% of the parent compound [4, 9]. Reaction with HCl affords $C_5H_5Mo(CO)_3H$ and $(C_5H_5)_2ZrCl_2$; with CCl_4, $C_5H_5Mo(CO)_3Cl$ and $(C_5H_5)_2ZrCl_2$ are obtained. Treatment with $BrCH_2CH_2Br$ yields $(C_5H_5Mo(CO)_3)_2$ and $(C_5H_5)_2Zr(Br)CH_3$ [1].

$C_5H_5Mo(CO)_2(\mu$-$CO)Zr(C_5H_5)_2C(O)CH_3$ (Formula III). $C_5H_5Mo(CO)_2(\mu$-$CO)Zr(C_5H_5)_2CH_3$ was stirred in toluene under a CO atmosphere for 30 min. Yellow crystals separated in 67% yield. The complex was also obtained in 50% yield by the reaction of $C_5H_5Mo(CO)_3H$ with $(C_5H_5)_2Zr(C(O)CH_3)CH_3$ in THF at room temperature. The product prepared by this method contains ca. 10% of an unstated impurity [4].

Conductivity (CH_3CN): Λ = 115 (1.9×10^{-3} M), 96 (6.9×10^{-3} M) $cm^2\cdot\Omega^{-1}\cdot mol^{-1}$ [4, 8]. ^1H NMR spectrum (C_6D_6): δ = 2.20 (CH_3), 5.37 (C_5H_5Mo), 5.55 (C_5H_5Zr); (CD_3CN): δ = 3.08 (CH_3), 5.07 (C_5H_5Mo), 6.02 (C_5H_5Zr) ppm. The shift of the acyl-methyl group varies from δ = 2.05 (0.01 M solution) to 2.35 (saturated solution) ppm. ^{13}C NMR spectrum (toluene-d_8): δ = 33.0 (CH_3), 89.1 (C_5H_5Mo), 110.3 (C_5H_5Zr), 232.6 (2 CO), 248.4 (CO), 316.5 (C=O) ppm; the spectrum is temperature-independent to −45°C. IR spectrum (toluene): 1600 (ν(C=O)), 1831, 1935 (ν(CO)) cm^{-1} [4]. Attempts to obtain a mass spectrum gave only the spectra of decarbonylation and rearrangement products [3, 4, 9].

The complex crystallizes in the triclinic space group $P\bar{1}-C_i^1$ (No. 2) with the unit cell parameters a = 14.548 (3), b = 12.800 (3), c = 10.432 (2) Å, α = 85.24 (1)°, β = 90.06 (1)°, γ = 73.87 (1)°; Z = 4 molecules per unit cell, and D_{calc} = 1.81 g/cm^3. The molecular structure with the main bond distances and angles for one of the two observed conformers is shown in **Fig. 10** [4].

Hydrolysis in benzene solution affords $((C_5H_5)_2ZrO)_3$, $CH_3C(O)H$, and $C_5H_5Mo(CO)_3H$. Protonation with two equivalents of CF_3COOH in C_6D_6 yields $C_5H_5Mo(CO)_3H$, $CH_3C(O)H$, and $(C_5H_5)_2Zr(O_2CCF_3)_2$. Dissolving in CH_3CN leads to the formation of the ions $[C_5H_5Mo(CO)_3]^-$ and $[(C_5H_5)_2Zr(OCCH_3)NCCH_3]^+$ [4]. Reaction with either CH_3Li (at −80°C) or $LiN(C_3H_7$-$i)_2$ (at −42°C) in ether affords $Li[C_5H_5Mo(CO)_3]$ and $(C_5H_5)_2ZrOC{=}CH_2$. Reaction with one equivalent of $(C_5H_5)_2Zr(CH_3)_2$ in toluene at 25°C affords $(C_5H_5)_2Zr(CH_3)(\mu$-$O{=}C(CH_3)_2)Zr(C_5H_5)_2OC$-$Mo(CO)_2C_5H_5$ (Formula I). Slow decarbonylation in toluene or THF leads to the formation of $C_5H_5Mo(CO)(\mu$-$CO)({=}C(CH_3)O)Zr(C_5H_5)_2$ (Formula II) [4, 9]; kinetic data and mechanism are given [9].

References on pp. 44/5

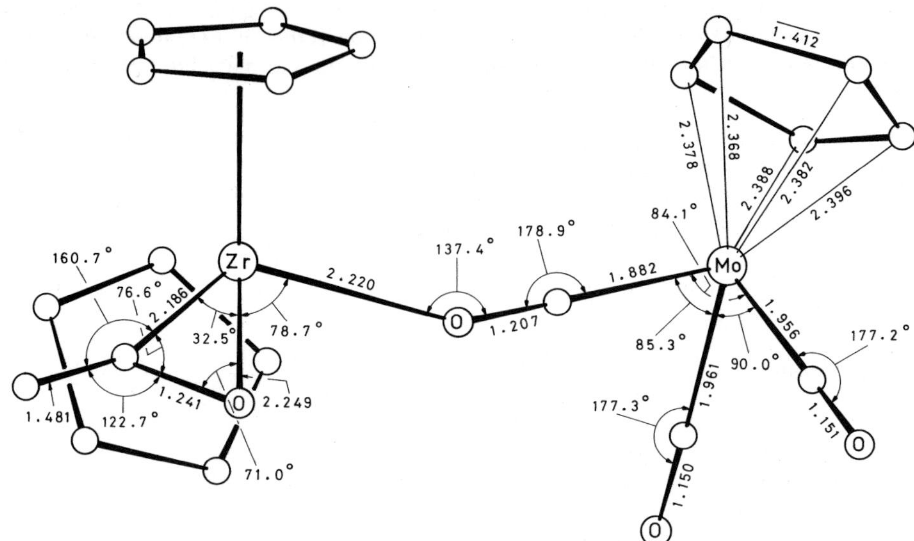

Fig. 10. The molecular structure of $C_5H_5Mo(CO)_2(\mu\text{-}CO)Zr(C_5H_5)_2(C(O)CH_3)$ [4].

$(CH_3)_2C$—$Zr(C_5H_5)_2$—OC—$Mo(CO)_2C_5H_5$ (C and Zr bridged by O; O bonded to $(C_5H_5)_2ZrCH_3$)

I

$(C_5H_5)_2Zr$—$Mo(CO)C_5H_5$ (bridging CO: O≡C; Zr—O—C(CH₃)=Mo ring)

II

$C_5H_5Mo(CO)_2(\mu\text{-}CO)Zr(C_5H_4CH_3)_2C(O)CH_3$ (Formula III; C_5H_5 at Zr replaced by $C_5H_4CH_3$). $C_5H_5Mo(CO)_3H$ was treated with $(C_5H_4CH_3)_2Zr(CH_3)_2$ in THF. The complex obtained, **$C_5H_5Mo(CO)_2(\mu\text{-}CO)Zr(C_5H_4CH_3)_2CH_3$**, was stirred in toluene under CO to give after isolation a tan-colored solid [9].

$(C_5H_5)_2Zr$—OC—$Mo(CO)_2C_5H_5$ (Zr bonded to C and O of H_3C—C=O)

III

^1H NMR spectrum (C_6D_6): $\delta = 1.71$ (s, $CH_3C_5H_4$), 2.18 (s, $CH_3C{=}O$), 5 to 6 (m, C_5H_4), 5.38 (s, C_5H_5) ppm [9].

Addition of one equivalent of $(C_5H_4CH_3)_2Zr(CH_3)_2$ in toluene affords in good yield $C_5H_5Mo(CO)_2(\mu\text{-}CO)Zr(C_5H_4CH_3)_2(\mu\text{-}O{=}C(CH_3)_2)Zr(C_5H_4CH_3)_2CH_3$ (compare Formula I) [9].

$C_5H_5Mo(CO)_2(\mu\text{-}CO)Zr(C_5H_5)_2(\mu\text{-}O{=}C(CH_3)_2)Zr(C_5H_5)_2CH_3$ (Formula I) was obtained as bright yellow crystals in 81% yield by the reaction of $C_5H_5Mo(CO)_2(CO\text{-}\mu)Zr(C_5H_5)_2C(O)CH_3$ (Formula

III) with one equivalent of $(C_5H_5)_2Zr(CH_3)_2$ in toluene for 1 h. The 78% ^{13}C-enriched species $C_5H_5Mo(CO)_2(\mu\text{-}CO)Zr(C_5H_5)_2(\mu\text{-}O{=}^{13}C(CH_3)_2)Zr(C_5H_5)_2CH_3$ was similarly obtained as a tan-colored solid in 61% yield from $C_5H_5Mo(CO)_2(\mu\text{-}CO)Zr(C_5H_5)_2{}^{13}C(O)CH_3$ (prepared in situ from ^{13}CO insertion into the $ZrCH_3$ bond of $C_5H_5Mo(CO)_2(\mu\text{-}CO)Zr(C_5H_5)_2CH_3$) [9].

^1H NMR spectrum (C_6D_6): δ = 0.36 (s, $ZrCH_3$), 1.36 (s, $(CH_3)_2C{=}O$; d, $^2J(^{13}C, H)$ = 4.4 Hz from the ^{13}C-enriched species), 5.36 (s, C_5H_5Mo), 5.94 (C_5H_5Zr) ppm. IR spectrum (toluene): 1163.5 (ν(C-O), $(CH_3)_2CO$), 1587, 1849, 1935 (ν(CO)) cm^{-1}. The IR spectrum of the ^{13}C-enriched complex is identical with exception of the labeled ketone stretch vibration at 1147 cm^{-1} [9].

Treatment with gaseous HCl in ether suspension affords i-C_3H_7OH as the major volatile organic fragment [9].

$C_5H_5Mo(CO)_2(\mu\text{-}CO)Zr(C_5H_4CH_3)_2(\mu\text{-}O{=}C(CH_3)_2)Zr(C_5H_4CH_3)_2CH_3$ (compare Formula I) was obtained as orange crystals by the reaction of $C_5H_5Mo(CO)_2(\mu\text{-}CO)Zr(C_5H_4CH_3)_2CH_3$ with one equivalent of $(C_5H_4CH_3)_2Zr(CH_3)_2$ in toluene for 1 h [9].

^1H NMR spectrum (C_6D_6): δ = 0.29 (s, $ZrCH_3$), 1.37 (s, $(CH_3)_2C{=}O$), 1.83 and 2.21 (s, $CH_3C_5H_4$), 5.2 to 6.4 (m, C_5H_4), 5.38 (C_5H_5Mo) ppm [9].

The complex crystallizes in the monoclinic space group $P2_1/c-C^5_{2h}$ (No. 14) with the unit cell parameters (at −40°C) a = 8.683 (2), b = 17.814 (3), c = 21.852 (6) Å, β = 93.53 (2)°; Z = 4 molecules per unit cell, and D_{calc} = 1.61 g/cm^3. The molecular structure with the main bond distances and angles is shown in **Fig. 11** [9].

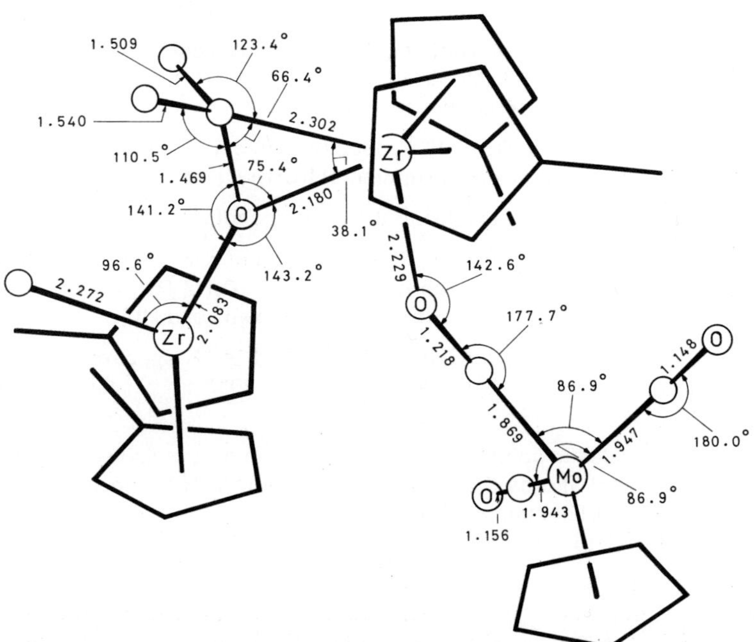

Fig. 11. The molecular structure of $C_5H_5Mo(CO)_2(\mu\text{-}CO)Zr(C_5H_4CH_3)_2(\mu\text{-}O{=}C(CH_3)_2)$-$Zr(C_5H_4CH_3)_2CH_3$ [9].

References on pp. 44/5

$C_5H_5Mo(CO)_2(\mu\text{-}CO)Th(N(Si(CH_3)_3)_2)_3$ and **$C_5H_5Mo(CO)_2(\mu\text{-}CO)U(N(Si(CH_3)_3)_2)_3$** were obtained as white microcrystalline solids by the reaction of $C_5H_5Mo(CO)_3H$ with one equivalent of the metallacyclic compound $(((CH_3)_3Si)_2N)_2MCH_2Si(CH_3)_2NSi(CH_3)_3$ (M = Th or U) in THF at −30°C [6].

IR spectra in the ν(CO) region (Nujol): 1586, 1820, 1929 (M = Th) and 1579, 1820, 1929 (M = U) cm^{-1}. The IR spectra of THF solutions exhibit strong CO stretching peaks at 1745, 1794, and 1900 cm^{-1} which are identical to those of free carbonyl anions, suggesting a dissociation to an ionic structure [6].

The compounds are only sparingly soluble in noncoordinating solvents such as toluene or heptane but highly soluble in THF [6].

$C_5H_5Mo(CO)_2(\mu\text{-}CO)U(C_5H_5)_3$ was obtained as a reddish brown powder or brown crystals by the following reactions. $Na[C_5H_5Mo(CO)_3]$ reacted in THF at −50°C with $(C_5H_5)_3UCl$ to give a 96% yield. To a toluene solution of $(C_5H_5)_3UCH_3$ cooled to −50°C was added $C_5H_5Mo(CO)_3H$ in heptane. The reaction mixture was stirred for 1 h at −25°C to give brown crystals in 97% yield. $C_5H_5Mo(CO)_3H$ also reacted either with $(C_5H_5)_3UN(C_2H_5)_2$ or 2/3 equivalents of $(C_5H_5)_2U(N(C_2H_5)_2)_2$ in THF or toluene solution to give in both cases an 81% yield [7].

^1H NMR spectrum (THF): δ = 6.80 (UC_5H_5) ppm. IR spectrum (Nujol): 1634 (ν(μ-CO)), 1740, 1900 (ν(CO)) cm^{-1}. The IR spectrum suggests a dissociation in THF with formation of $[C_5H_5Mo(CO)_3]^-$ [7].

Poorly soluble in noncoordinating solvents such as heptane or toluene; all attempts to recrystallize the complex affords only $C_5H_5Mo(CO)_3H$ and unidentified uranium compounds. Protonation with HCl in toluene at −70°C affords $C_5H_5Mo(CO)_3H$, $C_5H_5Mo(CO)_3Cl$, and $(C_5H_5)_3UCl$. With CCl_4 under the same conditions, $C_5H_5Mo(CO)_3Cl$ and $(C_5H_5)_3UCl$ are obtained [7].

$(CH_3)_5C_5Mo(CO)_2(\mu\text{-}CO)U(C_5H_5)_3$ was obtained either by the reaction of $Na[(CH_3)_5C_5Mo(CO)_3]$ with $(C_5H_5)_3UCl$ in THF or of $(CH_3)_5C_5Mo(CO)_3H$ with $(C_5H_5)_3UCH_3$ in toluene at −50°C for 1 h. In both cases, the complex was isolated in ca. 95% yield [7].

IR spectrum (Nujol): 1645 (ν(μ-CO)), 1737, 1902 (ν(CO)) cm^{-1}. The CO stretching frequencies in THF suggest a dissociation with formation of $[(CH_3)_5C_5Mo(CO)_3]^-$ [7].

The complex shows a behavior similar to the corresponding complex with a C_5H_5Mo unit [7].

$(C_6H_5)_3PC_5H_4Mo(CO)_2(\mu\text{-}CO)Al(CH_3)_3$ was obtained from $(C_6H_5)_3PC_5H_4Mo(CO)_3$ and excess $Al_2(CH_3)_6$ in toluene; addition of hexane precipitated a brown solid [10].

A structure with an isocarbonyl linkage was deduced by IR spectroscopy. In Nujol, the vibrations of the terminal CO groups appear at 1845 and 1932 cm^{-1}; the μ-CO group has a band at 1665 cm^{-1} [10].

The complex dissociates in solution and decomposition occurs on repeated recrystallization. Reaction with ethanol affords CH_4 and with excess of $N(CH_3)_3$, the starting Mo complex along with $(CH_3)_3AlN(CH_3)_3$ is obtained [10].

References:

[1] Renaut, P.; Tainturier, G.; Gautheron, B. (J. Organometal. Chem. **150** [1978] C 9/C 10).
[2] Hamilton, D. M., Jr.; Willis, W. S.; Stucky, G. D. (J. Am. Chem. Soc. **103** [1981] 4255/6).
[3] Longato, B.; Norton, J. R.; Huffman, J. C.; Marsella, J. A.; Caulton, K. G. (J. Am. Chem. Soc. **103** [1981] 209/10).

[4] Marsella, J. A.; Huffman, J. C.; Caulton, K. G.; Longato, B.; Norton, J. R. (J. Am. Chem. Soc. **104** [1982] 6360/8).
[5] Merola, J. S.; Gentile, R. A.; Ansell, G. B.; Modrick, M. A.; Zentz, S. (Organometallics **1** [1982] 1731/3).
[6] Dormond, A.; El Bouadili, A. A.; Moise, C. (J. Chem. Soc. Chem. Commun. **1985** 914/6).
[7] Dormond, A.; Moise, C. (Polyhedron **4** [1985] 595/8).
[8] Longato, B.; Martin, B. D.; Norton, J. R.; Anderson, O. P. (Inorg. Chem. **24** [1985] 1389/94).
[9] Martin, B. D.; Matchett, S. A.; Norton, J. R.; Anderson, O. P. (J. Am. Chem. Soc. **107** [1985] 7952/9).
[10] Kotz, J. C.; Tunipseed, C. D. (J. Chem. Soc. Chem. Commun. **1970** 41/2).

1.5.1.3.2.1.7 Compounds of the Type $^5LMo(CO)_2(^2D)C(X)R$ (X = O, NR′)

This section covers compounds of the type $^5LMo(CO)_2(^2D)C(O)R$ and $^5LMo(CO)_2(^2D)$-$C(=NR')R$. $C_5H_5Mo(CO)_2(^2D)C(O)CH_3$ in which ^2D is probably a solvent molecule, was mentioned to be formed by the hydrolysis of $C_5H_5Mo(CO)_2C(CH_3)OAl(Br)_2Br$-μ under mild conditions. It decomposes to $C_5H_5Mo(CO)_3CH_3$ and other unidentified products [50]. Further metallacarboxyclic acids $C_5H_5Mo(CO)_2(^2D)C(O)OH$ (^2D is a phosphane or phosphite) were mentioned in [62] without specification. The compounds listed in Table 5 were obtained in most cases by one of the following methods. The formation of $C_5H_5Mo(CO)_2(P(C_6H_5)_3)C(O)R$ in reactions between $C_5H_5Mo(CO)_3R$ (R = mono- or disubstituted benzyl) and $P(C_6H_5)_3$ was studied in CH_3CN at 30 °C. The results were interpreted in terms of the steric demands of the reaction [20].

Cis-trans isomerism arises from the arrangement of the four ligands, 2CO, PR_3, and C(X)R, in the basal plane of the molecule. The metal of the cis isomer is a center of chirality; however, nothing is reported about optical isomers.

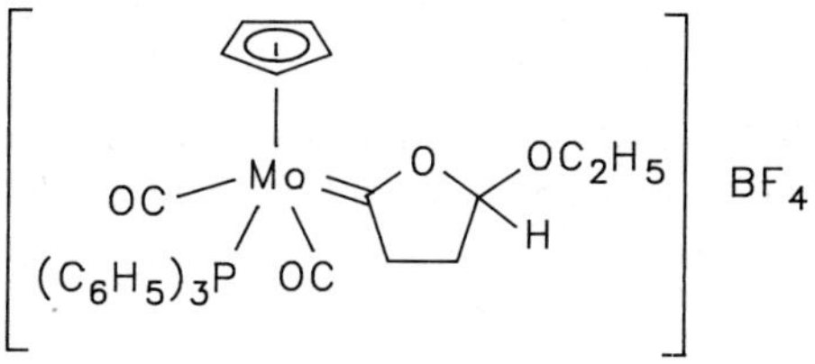

I

Method I: $[^5LMo(CO)_3{}^2D]X$ (X = BF_4 [59, 67] or PF_6 [49, 56, 64]) was reduced with one equivalent of $Li[(C_2H_5)_3BH]$ in CH_2Cl_2 at −41 °C [49, 56]. A better method was the reduction with one equivalent of $NaBH_4$ [54, 59, 64, 67] or $[N(C_2H_5)_4][BH_4]$ [59] in methanol at −60 to −30 °C. Nos. 3 and 4 precipitated directly from the reaction mixture. No. 5 was isolated by slow warmup to 0 °C, removal of the solvent, and extraction of the residue with ether/hexane (1:4).

Method II: $C_5H_5Mo(CO)_3R$ was allowed to react with the ^2D ligand as solvent or in solution. The most used solvents were CH_3CN and THF at ambient temperature to reflux. Further information is given in Table 5.

Method III: Reactions of $[C_5H_5Mo(CO)_2(P(C_6H_5)_3)COCH_2CH_2C(OC_2H_5)H]BF_4$ (Formula I).

a. The complex was dissolved in dimethyl sulfoxide and the solution was added to a mixture of H_2O/ether. After 5 min the ether layer was separated and dried over $MgSO_4$ [42].

References on pp. 75/7

b. $NaOCH_3$ was added to solution in CH_2Cl_2 and the mixture was stirred for 16 h. The solvent was removed under reduced pressure and the residue was extracted with ether [42].

Method IV: $Na[^5LMo(CO)_2{}^2D]$ was allowed to react with RC(O)Cl (R = C_6H_5, C_6H_4F-4) [21] in THF [31].

Method V: $^5LMo(CO)_2C(R')$=NR (Formula II) was allowed to react with the ^2D ligand in THF or hexane [44].

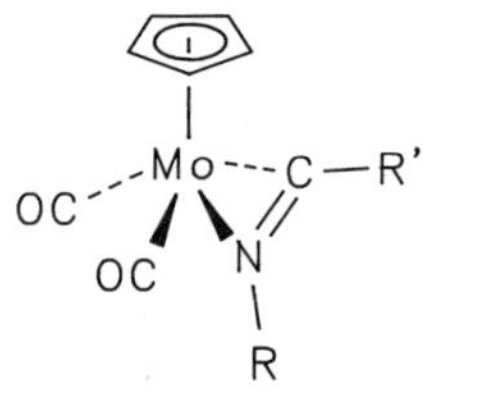

III 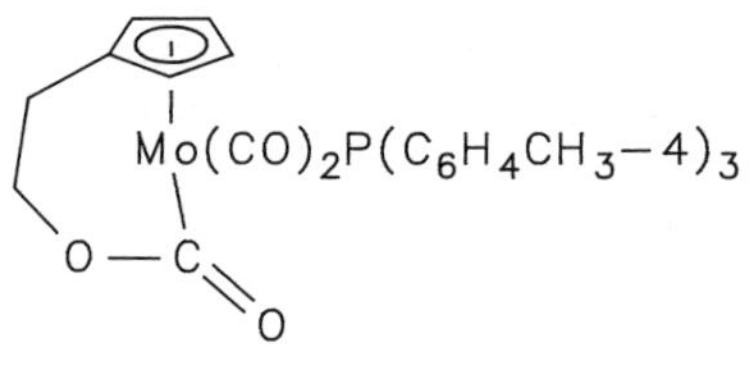 IV

Method VI: $[C_5H_5Mo(CO)_3{}^2D]X$ (^2D = NH_3, X = Cl; ^2D = $P(C_6H_5)_3$, X = PF_6) was allowed to react with liquid NH_3 [17, 46].

Method VII: $[C_5H_5Mo(CO)_3P(C_6H_5)_3]BF_4$ was allowed to react with $[N(C_2H_5)_4]OH$ in methanol or with $NaOCH_3$ in acetone [58, 59].

The formyl complexes, $(CH_3)_5C_5Mo(CO)_2(^2D)C(O)H$ (Nos. 1, 2, 4, and 6), can be stored at −20 °C for several months, but decomposition occurs in 1 to 5 h at room temperature [54]. The ^1H NMR spectra show the quantitive conversion to the hydride analogues, $(CH_3)_5C_5Mo(CO)_2(^2D)H$ which is complete in toluene solution after 72 h at room temperature. A solution in toluene-d_8 cooled to −80 °C was allowed to warm up monitored by ^1H NMR. Upon warming the trans isomer decomposed first in $(CH_3)_5C_5Mo(CO)_3H$ and free phosphane; whereas the cis isomers converted, at higher temperatures, in cis-$(CH_3)_5C_5Mo(CO)_2(^2D)H$ which additionally is formed from the first decomposition product, $(CH_3)_5C_5Mo(CO)_3H$, and the free phosphane. The similar decomposition of $(CH_3)_5C_5Mo(CO)_2(P(OC_6H_5)_3)C(O)H$ (No. 6) into the tricarbonyl hydride starts at 20 °C for both isomers with equal ratios, indicative for a conversion of the cis formed isomer into the trans one; the consecutive reaction of the tricarbonylhydride with the free phosphite to cis-$(CH_3)_5C_5Mo(CO)_2(P(OC_6H_5)_3)H$ occurs at 25 °C [54].

$C_5H_5Mo(CO)_2(P(C_6H_5)_3)C(O)H$ (No. 3) is labile in solution at room temperature, converting cleanly to $C_5H_5Mo(CO)_3H$ within a few minutes and NMR and IR spectra of C_5H_5-$Mo(CO)_2(P(C_6H_5)_3)C(O)H$ in solution at ambient temperature could only be obtained upon adding either $P(C_6H_5)_3$ or $[N(C_2H_5)_4]BH_4$ to the solvent prior to the addition of the formed complex [59]. trans-$(CH_3)_5C_5Mo(CO)_2(P(C_6H_5)_3)C(O)H$ (No. 4) decomposes in CH_3OH at ca. −40 °C and the cis isomer at 0 °C [67]. The cis isomer of No. 4 is stable up to 50 °C and its decomposition occurs at 60 to 80 °C during 1 h [64].

cis-$(CH_3)_5C_5Mo(CO)_2(P(OCH_3)_3)C(O)H$ (No. 5) is stable in polar solvents, e.g., methanol or ethanol, for several days. A 0.3 M solution in C_6H_6 decomposes to the corresponding hydrido complex with a half-life of 4 h [67].

The acetyl complexes, $C_5H_5Mo(CO)_2(^2D)C(O)CH_3$, undergo decarbonylation in solution with formation of $C_5H_5Mo(CO)_2(^2D)CH_3$. The observed first-order rate constants in CH_3CN (0.01 M solutions) in the presence of a 5- to 10-fold excess of ^2D ligand depend markedly on the organic groups attached to phosphorus, decreasing in the order $P(C_6H_{11}\text{-cyclo})_3$ > $P(C_6H_5)_2C_3H_7\text{-i}$ > $P(C_6H_5)_3$ > $P(C_6H_4CH_3\text{-4})_3$ > $P(C_6H_4OCH_3\text{-4})_3$ > $P(C_6H_5)_2CH_3$ > $P(C_4H_9\text{-i})_3$ > $P(CH_3)_2C_6H_5$ > $P(C_4H_9\text{-n})_3$. A good correlation between decarbonylation rate constant and ligand cone angle was observed. A proposed mechanism was given [35]. In refluxing CH_3CN a dissociation into $C_5H_5Mo(CO)_3CH_3$ and free phosphane is observed, the rate of dissociation depending on the nature of the phosphane [35].

Table 5
Compounds of the Type $^5LMo(CO)_2(^2D)C(X)R$ (X = O, NR′).
An asterisk indicates further information at the end of the table.
For explanations, abbreviations, and units see p. X.

No.	compound	method of preparation (yield) properties and remarks
compounds of the type $^5LMo(CO)_2(^2D)C(O)H$		
*1	$(CH_3)_5C_5Mo(CO)_2(P(CH_3)_3)C(O)H$	I (90 to 95%) [54] yellow microcrystals with a cis to trans ratio of 95:5 (from ether at low temperatures) [54] ^1H NMR (toluene-d_8, 0°C): cis isomer: 1.34 (d, PCH_3; J(P, H) = 9.0), 1.91 ($C_5(CH_3)_5$), 14.68 (d, C(O)H; J(P, H) = 1.2); trans isomer: 1.34 (d, PCH_3; J(P, H) = 8.9), 1.91 ($C_5(CH_3)_5$), 14.42 (d, C(O)H; J(P, H) = 5.2) [54] ^{13}C NMR (CD_2Cl_2, −30°C): cis isomer: 11.2 (q, $C_5(\mathbf{C}H_3)_5$; J(H, C) = 127), 17.1 (d of q, PCH_3; J(P, C) = 30; J(H, C) = 127), 105.8 ($\mathbf{C}_5(CH_3)_5$), 242.1 (CO), 249.5 (d, CO; J(P, C) = 29), 287.1 (d of d, C(O)H; J(P, C) = 27.1, J(H, C) = 137.3) [54] ^{31}P NMR (CD_2Cl_2): cis isomer: 18.8 (at −30°C); trans isomer: 22.4 (at 0°C) [54] IR (Nujol): 1600 (ν(C=O)), 1890, 1935 (ν(CO)) [54] reaction with CH_3SO_3F in CH_2Cl_2 at −70°C followed by anion exchange affords $[(CH_3)_5C_5Mo(CO)_3P(CH_3)_3]PF_6$ [54]
2	$(CH_3)_5C_5Mo(CO)_2(P(C_6H_5)_2CH_3)C(O)H$	I (90 to 95%) [54] yellow microcrystals with a cis to trans ratio of 30:70 (from ether at low temperatures) [54] ^1H NMR (toluene-d_8, −30°C): cis isomer: 1.56 ($C_5(CH_3)_5$), 2.09 (d, PCH_3; J(P, H) = 8.7), 7.50 (m, PC_6H_5), 14.68 (d, C(O)H; J(P, H) = 4.6); trans isomer: 1.46 ($C_5(CH_3)_5$), 2.07 (d, PCH_3; J(P, H) = 8.5), 7.50 (m, PC_6H_5), 14.91 (d, C(O)H; J(P, H) = 8.4) [54] ^{13}C NMR (CD_2Cl_2, −30°C): cis isomer: 10.4 (q, $C_5(\mathbf{C}H_3)_5$; J(H, C) = 127.4), 18.2 (d of q, PCH_3;

References on pp. 75/7

Table 5 (continued)

No.	compound	method of preparation (yield) properties and remarks
2 (continued)		J(P, C) = 33.3, J(H, C) = 132), 105.2 ($\mathbf{C}_5(CH_3)_5$), 132.5 (m, PC_6H_5), 242.0 (CO), 249.7 (d, CO; J(P, C) = 26), 283.6 (d of d, C(O)H; J(P, C) = 25, J(H, C) = 135.0); trans isomer: 10.1 (q, C_5-$(\mathbf{C}H_3)_5$; J(H, C) = 127.5), 16.9 (d of q, PCH_3; J(P, C) = 29.3, J(H, C) = 131.1), 105.1 ($\mathbf{C}_5(CH_3)_5$), 132.5 (m, PC_6H_5), 237.5 (d, CO; J(P, C) = 26), 273.1 (d of d, C(O)H; J(P, C) = 9.6, J(H, C) = 149.0) [54] ^{31}P NMR (CD_2Cl_2, −30°C): trans isomer: 43.4; cis isomer: 49.5 [54] IR (Nujol): 1600 (ν(C=O)), 1862, 1940 (ν(CO)) [54] alkylation with 1.2 equivalents of $CF_3SO_3Si(CH_3)_3$ in CH_2Cl_2 at 0°C affords thermally unstable $[(CH_3)_5C_5Mo(CO)_2(P(C_6H_5)_2CH_3)$=CH(O-$Si(CH_3)_3)]O_3SCF_3$ [54]
3	trans-$C_5H_5Mo(CO)_2(P(C_6H_5)_3)C(O)H$	I (41% by ^{1}H NMR, not isolated) [56], (89%) [59] yellow powder, m.p. 155 to 160°C (dec.) [59] ^{1}H NMR (CD_2Cl_2 in the presence of $[N(C_2H_5)_4]BH_4$): 5.16 (d, C_5H_5; J(P, H) = 1.0), 7.31 (s, C_6H_5), 14.89 (d, C(O)H; J(P, H) = 4.0) [59]; (CH_2Cl_2, −41°C): 14.70 (d, C(O)H; J(P, H) = 4) [49, 56] IR (Nujol): 1596 (ν(C=O)), 1854, 1945 (ν(CO)) [59] reaction with an excess of CCl_4 affords $C_5H_5Mo(CO)_3Cl$ [56]
*4	$(CH_3)_5C_5Mo(CO)_2(P(C_6H_5)_3)C(O)H$	I (90% with a cis to trans ratio of 9:1) [54, 64] (55% only cis isomer) [67] yellow air-sensitive powder [67], m.p. 85°C [64] ^{1}H NMR (CD_2Cl_2, −80°C): 1.70 (s, CH_3), 14.28 (d, trans-C(O)H; J(P, H) = 7.3), 14.60 (d, cis-C(O)H; J(P, H) = 5.1) [64]; (CD_3OD, −40°C): 1.82 (CH_3), 7.30 to 7.60 (m, C_6H_5), 14.40 (d, cis-C(O)H; J(P, H) = 14), 14.54 (d, trans-C(O)H; J(P, H) = 3) [67]; similar data in toluene-d_8 at −30°C [54] ^{13}C NMR (toluene-d_8, −50°C): 10.7 ($(\mathbf{C}H_3)_5C_5$; J(H, C) = 61?), 105.0 ($\mathbf{C}_5(CH_3)_5$), 250.4 (d, CO; J(P, C) = 26), 277.8 (d of d, C(O)H; ^{2}J(P, C) = 19.7, J(H, C) = 144.5) [64]; (CD_2Cl_2, −30°C): trans isomer: 10.7 (q, $C_5(\mathbf{C}H_3)_5$; J(H, C) = 128.0), 105.0 ($\mathbf{C}_5(CH_3)_5$), 129.0 (m, C_6H_5), 250.4 (d, CO; J(P, C) = 26), 277.8 (d of d, C(O)H; J(P, C) = 9.6, J(H, C) = 144.5) [54]

Table 5 (continued)

No.	compound	method of preparation (yield) properties and remarks
		^{31}P NMR (toluene-d_8, −50°C): trans isomer: 65.4; cis isomer: 71.7 [64]; same resonances in CD_2Cl_2 at −30°C [54] IR (Nujol): 1610 (ν(C=O)), 1860, 1930 (ν(CO)), 2510, 2640 [67]; similar data in [54, 64] mass spectrum (170°C, 70 eV): $[M]^+$ (metastable), $[M - CO]^+$ [64]
5	$(CH_3)_5C_5Mo(CO)_2(P(OCH_3)_3)C(O)H$	I (65% cis isomer; the trans isomer was only observed when preparation was carried out at 0°C) [67] yellow crystals (from ether/hexane at −78°C) [67] ^{1}H NMR (CD_3OD): cis isomer: 2.00 (CH_3), 3.70 (d, OCH_3; J(P, H) = 12), 14.67 (d, C(O)H; J(P, H) = 6); trans isomer at 0°C: 1.98 (CH_3), 3.62 (d, OCH_3; J(P, H) = 12), 14.62 (d, C(O)H; J(P, H) = 14) [67] IR (Nujol): cis isomer: 1600 (ν(C=O)), 1870, 1950 (ν(CO)), 2510, 2640 [67]
6	$(CH_3)_5C_5Mo(CO)_2(P(OC_6H_5)_3)C(O)H$	I (90 to 95%) [54] yellow microcrystals with a cis to trans ratio of 1:1 [54] ^{1}H NMR (toluene-d_8, −30°C): cis isomer: 1.69 (CH_3), 7.50 (m, C_6H_5), 14.73 (d, C(O)H; J(P, H) = 5.7); trans isomer: 1.67 (CH_3), 7.50 (m, C_6H_5), 14.64 (d, C(O)H; J(P, H) = 5.3) [54] ^{13}C NMR (CD_2Cl_2, −30°C): cis isomer: 10.7 (q, $C_5(\mathbf{C}H_3)_5$), 106.4 ($\mathbf{C}_5(CH_3)_5$), 137.5 (m, C_6H_5), 234.7 (d, CO; J(P, C) = 38.0), 235.9 (CO), 276.4 (d of d, C(O)H; J(P, C) = 36.2, J(H, C) = 141.5); trans isomer: 10.5 (q, $C_5(\mathbf{C}H_3)_5$; J(H, C) = 127.3), 106.1 ($\mathbf{C}_5(CH_3)_5$), 137.5 (m, C_6H_5), 243.6 (d, CO; J(P, C) = 35.5), 268.6 (d of d, C(O)H; J(P, C) = 9.6, J(H, C) = 151.0) [54] ^{31}P NMR (CD_2Cl_2, −30°C): cis isomer: 166.0; trans isomer: 184.5 [54] IR (Nujol): 1605 (ν(C=O)), 1890, 1960 (ν(CO)) [54]
compounds of the type ^{5}LMo(CO)$_2$(^{2}D)C(O)CX$_3$ (X = H, F)		
7	trans-$C_5H_5Mo(CO)_2(P(CH_3)_3)C(O)CH_3$	II, in CH_3CN for 18 h (92%) [53] yellow crystals, m.p. 118°C (from benzene/pentane 2:3) [53] ^{1}H NMR (C_6D_6): 1.08 (d, PCH_3; J(P, H) = 9.2), 2.90 (s, CH_3), 4.93 (d, C_5H_5; J(P, H) = 1.6) [53]

Table 5 (continued)

No.	compound	method of preparation (yield) properties and remarks
7 (continued)		^{31}P NMR (C_6D_6): 22.1 (s) [53] IR (C_6H_6): 1625 (ν(C=O)), 1847, 1929 (ν(CO)) [53] reaction with two equivalents of $(CH_3)_3P{=}CH_2$ in C_6H_6 yields $[P(CH_3)_4][C_5H_5Mo(CO)_2P(CH_3)_3]$ and $(CH_3)_3P{=}CHC(O)CH_3$ [53]
8	$C_5H_5Mo(CO)_2(P(C_2H_5)_3)C(O)CH_3$	II, at 20 to 60°C (quantitative) [1] yellow crystals, m.p. 98°C [1] ^{1}H NMR (CCl_4): 2.45 (CH_3), 5.08 (C_5H_5) [1] IR (hexadecane): 1636 (ν(C=O)), 1854, 1934 (ν(CO)) [1]
*9	trans-$C_5H_5Mo(CO)_2(P(C_4H_9\text{-}n)_3)C(O)CH_3$	II [15], in THF or toluene [4], in refluxing THF or hexane [12], in CH_3CN with two equivalents of $P(C_4H_9\text{-}n)_3$ (60%) [11], or in CH_3CN for 2 to 3 h (75 to 85%) [35] yellow oil [4], yellow solid, m.p. 89°C (from CH_2Cl_2/hexane) [11] ^{1}H ($CDCl_3$): 2.49 (CH_3), 5.00 (d, C_5H_5; J(P, H) = 1.0) [11]; similar in [35]; (CD_3CN): 0.9, 1.5 (m, n-C_4H_9), 2.47 (s, CH_3), 5.12 (d, C_5H_5; J(P, H) = 1.0) [12] IR (THF): 1600 (ν(C=O)), 1860, 1940 (ν(CO)) [4]; similar data in [12, 15]; (cyclohexane): 1602 (ν(C=O)), 1847, 1936 (ν(CO)) [11]
10	trans-$C_9H_7Mo(CO)_2(P(C_4H_9\text{-}n)_3)C(O)CH_3$ (C_9H_7 = indenyl)	II, in THF [15] IR (THF): 1637 (ν(C=O)), 1854, 1940 (ν(CO)) [15] first-order rate constants for the formation in THF at 33°C, activation parameters, and a proposed mechanism are given [15]
11	trans-$C_5H_5Mo(CO)_2(P(C_4H_9\text{-}i)_3)C(O)CH_3$	II [25, 35] ^{1}H NMR ($CDCl_3$): 2.58 (s, CH_3), 5.11 (d, C_5H_5; J(P, H) = 1.5 to 2) [35] rate constant for decarbonylation in CH_3CN at 60°C: $k = 5.6 \times 10^{-6}\ s^{-1}$ [25, 35]
12	trans-$C_5H_5Mo(CO)_2(P(C_6H_{11}\text{-cyclo})_3)C(O)CH_3$	II [25, 35] ^{1}H NMR ($CDCl_3$): 2.55 (s, CH_3), 5.20 (d, C_5H_5; J(P, H) = 1.5 to 2) [35] rate constant for decarbonylation in CH_3CN at 60°C: $k = 46.4 \times 10^{-6}\ s^{-1}$ [35]

References on pp. 75/7

Table 5 (continued)

No.	compound	method of preparation (yield) properties and remarks
		equilibrium constant for dissociation (compare further information on No. 9): $K_{equ} = 50 \times 10^{-4}$ with 50% dissociation [35] alkylation with $[O(C_2H_5)_3]BF_4$ in CH_2Cl_2 affords $[C_5H_5Mo(CO)_2(P(C_6H_{11}\text{-cyclo})_3)=C(OC_2H_5)\text{-}CH_3]BF_4$ [23]
*13	trans-$C_5H_5Mo(CO)_2(P(C_6H_5)_3)C(O)CH_3$	II, see "Further information" yellow solid [3 to 5, 11, 13], m.p. 147°C (from $CHCl_3$/hexane) [3], 153 to 154°C (from $CHCl_3$/petroleum ether) [4], 155 to 156°C [13], 158°C (from CH_2Cl_2/hexane) [11] ^{1}H NMR ($CDCl_3$): 2.62 (s, CH_3), 5.03 (d, C_5H_5; J(P, H) = 1.5 to 2), 7.43 (m, C_6H_5) [3, 11, 35, 41], similar data in C_6D_6 [43]; (CD_3CN): 2.81 (s, CH_3), 4.79 (d, C_5H_5; J(P, H) = 1), 7.3 (m, C_6H_5) [12] ^{13}C NMR ($CDCl_3$ or CD_2Cl_2): 50.51 (CH_3), 96.19 (C_5H_5), 238.1 (d, CO; J(P, C) = 24), 263.2 (d, C=O; J(P, C) = 10.6) [48] IR: 237 (ν(MoP)), 351, 360 (MoC_5H_5), 440, 468 (ν(MoC)), 540, 546, 675, 584 (δ(MoCO)), 1600 (ν(C=O)), 1850, 1930 (ν(CO)); other data between 90 and 512 also given in [18]; ($CHCl_3$): 1603 (ν(C=O)), 1858, 1943 (ν(CO)) [3, 4, 12, 41]; similar ν(CO) and ν(CO) bands in cyclohexane [11] and THF [4, 15] Raman spectrum (solid): 237 (ν(MoP)), 353, 364 (MoC_5H_5), 440, 470 (ν(MoC)), 548, 588 (δ(MoCO)), 1603 (ν(C=O)), 1847, 1935 (ν(CO)); other data between 200 and 517 also given in [18]
14	$C_9H_7Mo(CO)_2(P(C_6H_5)_3)C(O)CH_3$ (C_9H_7 = indenyl)	II, with excess phosphane in CH_2Cl_2/petroleum ether (1:2) for 24 h; the cis isomer separates and is isolated from the mother liquid after addition of petroleum ether [15] first-order rate constants of the formation in THF at 33°C are given [15] fine yellow crystals (trans isomer), yellow-brown prisms (cis isomer) [15] IR (THF): trans isomer: 1640 (ν(C=O)), 1868, 1944 (ν(CO)); cis isomer: 1670 (ν(C=O)), 1910 (ν(CO)) [15]

References on pp. 75/7

Table 5 (continued)

No.	compound	method of preparation (yield) properties and remarks
15	trans-$C_5H_5Mo(CO)_2(P(C_6H_4CH_3\text{-}2)_3)C(O)CH_3$	II, in CH_3CN for 2 to 3 h [25, 35] refluxing of a 0.01 M CH_3CN leads to a quantitative dissociation into $C_5H_5Mo(CO)_3CH_3$ and the phosphane within 1 to 2 h [25, 35]
16	trans-$C_5H_5Mo(CO)_2(P(C_6H_4CH_3\text{-}4)_3)C(O)CH_3$	II, in CH_3CN for 2 to 3 h (75 to 85%) [25, 35]; recrystallization from $CHCl_3$/hexane [35] rate constant for decarbonylation in CH_3CN: $k = 10.9 \times 10^{-6}\ s^{-1}$ [35] equilibrium constant for dissociation in refluxing CH_3CN after 1 to 2 h: $K_{equ} = 6.1 \times 10^{-4}$ with 20% dissociation [35]
17	trans-$C_5H_5Mo(CO)_2(P(C_6H_4OCH_3\text{-}4)_3)C(O)CH_3$	II, in CH_3CN for 2 to 3 h (75 to 85%) [25, 35]; recrystallization from $CHCl_3$/hexane [35] rate constant for decarbonylation in CH_3CN: $k = 9.6 \times 10^{6}\ s^{-1}$ [35] equilibrium constant for dissociation in refluxing CH_3CN after 1 to 2 h: $K_{equ} = 5.3 \times 10^{-4}$ with 20% dissociation [35]
18	$C_5H_5Mo(CO)_2(P(N(CH_3)_2)_3)C(O)CH_3$	II, without solvent at 20 to 60°C (quantitative) [1] yellow crystals, m.p. 120°C [1] ^1H NMR (CCl_4): 2.43 (CH_3), 5.10 (C_5H_5) [1] IR (hexadecane): 1636 (ν(C=O)), 1856, 1934 (ν(CO)) [1]
*19	trans-$C_5H_5Mo(CO)_2(P(OCH_3)_3)C(O)CH_3$	II, without solvents between 20 and 60°C (quantitative) [1], in CH_3CN with 2 equivalents of $P(OCH_3)_3$ (20%) [11], in CD_3CN at 60°C (not isolated) [12], or in THF with 10 equivalents of $P(OCH_3)_3$ [15] yellow crystals, m.p. 61°C [1], low melting solid (from CH_2Cl_2/hexane) [11] ^1H NMR (CCl_4): 2.50 (CH_3), 5.22 (C_5H_5) [1]; ($CDCl_3$): 2.53 (s, CH_3), 5.14 (d, C_5H_5; J(P, H) = 1) [11]; (CD_3CN): 2.47 (s, CH_3), 3.62 (d, OCH_3; J(P, H) ≈ 10), 5.28 (d, C_5H_5; J(P, H) = 1.5) [12] IR (hexadecane): 1636 (ν(C=O)), 1873, 1927 (ν(CO)); (cyclohexane): 1605 (ν(C=O)), 1868, 1948 (ν(CO)) [11]; similar data in [12]; (THF): 1640 (ν(C=O)), 1874, 1950 (ν(CO)) [15]

Table 5 (continued)

No.	compound	method of preparation (yield) properties and remarks
20	$C_9H_7Mo(CO)_2(P(OCH_3)_3)C(O)CH_3$ (C_9H_7 = indenyl)	II, in THF or hexane with 10 equivalents of $P(OCH_3)_3$ (not isolated); kinetics of the formation in [15] IR (THF): trans isomer: 1640 (ν(C=O)), 1875, 1950 (ν(CO)); cis isomer: 1686 (ν(C=O)), 1849, 1914 (ν(CO)) [15]
21	5,6-$(CH_3O)_2C_9H_5Mo(CO)_2(P(OCH_3)_3)C(O)CH_3$ ([5]L is 5,6-dimethoxyindenyl)	II, in THF or hexane with 10 equivalents of $P(OCH_3)_3$ (not isolated); kinetics of formation in [15]
22	trans-$C_5H_5Mo(CO)_2(P(OC_2H_5)_3)C(O)CH_3$	II, in CH_3CN with 2 equivalents of $P(OC_2H_5)_3$ (35%) [11] m.p. 35°C (from CH_2Cl_2/hexane) [11] ^1H NMR ($CDCl_3$): 2.48 (s, CH_3), 5.09 (d, C_5H_5; J(P, H) = 1) [11] IR (cyclohexane): 1598 (ν(C=O)), 1863, 1932 (ν(CO)) [11]
23	trans-$C_5H_5Mo(CO)_2(P(OC_4H_9\text{-}n)_3)C(O)CH_3$	II, in THF with excess $P(OC_4H_9)_3$; kinetics of the formation in [4] yellow oil [4] IR (THF): 1632 (ν(C=O)), 1860, 1941 (ν(CO)) [4]
*24	trans-$C_5H_5Mo(CO)_2(P(OC_6H_5)_3)C(O)CH_3$	II, in $P(OC_6H_5)_3$ [4], in CH_3CN with 2 equivalents of $P(OC_6H_5)_3$ for 2 h (80%) [11], in CD_3CN at 60°C (not isolated) [12], and in THF or n-hexane with 10 equivalents of $P(OC_6H_5)_3$ [15] yellow oil [4], solid, m.p. 85°C (from CH_2Cl_2/hexane) [11] ^1H NMR ($CDCl_3$): 2.13 (s, CH_3), 4.74 (d, C_5H_5; J(P, H) = 1) [11]; (CD_3CN): 2.20 (s, CH_3), 4.86 (d, C_5H_5; J(P, H) = 1), 7.3 (m, C_6H_5) [12] IR (THF): 1650 (ν(C=O)), 1857 [4] or 1876 [15], 1950 (ν(CO)); (cyclohexane): 1588 (ν(C=O)), 1878, 1956 (ν(CO)) [11]; similar data in [12]
25	$C_9H_7Mo(CO)_2(P(OC_6H_5)_3)C(O)CH_3$ (C_9H_7 = indenyl)	II, in THF or n-hexane with 10 equivalents of $P(OC_6H_5)_3$ (not isolated); kinetics of the formation in [15] IR (THF): trans isomer: 1650 (ν(C=O)), 1887, 1945 (ν(CO)); cis isomer: 1670 (ν(C=O)), 1899, 1958 (ν(CO)) [15]

References on pp. 75/7

Table 5 (continued)

No.	compound	method of preparation (yield) properties and remarks
26	trans-$C_5H_5Mo(CO)_2(P(OCH_2)_3CCH_3)C(O)CH_3$	II, in CH_3CN with 2 equivalents of $P(OCH_2)_3CCH_3$ for 2 h (70%); kinetics of the formation in [11] m.p. 170.5°C (from CH_2Cl_2/hexane) [11] ^1H NMR ($CDCl_3$): 2.52 ($CH_3C(O)$), 5.21 (d, C_5H_5; J(P, H) = 1.5) [11] IR: 351, 366 (MoC_5H_5), 442, 458 (ν(MoC)), 535, 541, 581 (δ(MoCO)), 1616 (ν(C=O)), 1860, 1956 (ν(CO)); other bands between 271 and 512 also given [18]; (cyclohexane): 1618 (ν(C=O)), 1878, 1983 (ν(CO)) [11] Raman spectrum (solid): 210 (ν(MoP)), 349, 366 (MoC_5H_5), 442, 462 (ν(MoC)), 544, 582 (δ(MoCO)), 1614 (ν(C=O)), 1861, 1961 (ν(CO)); other bands between 153 and 510 also given [18]
27	trans-$C_5H_5Mo(CO)_2(P(CH_3)_2C_6H_5)C(O)CH_3$	II, in CH_3CN for several hours [35, 36] (85%); recrystallized from CH_2Cl_2/hexane [26] ^1H NMR ($CDCl_3$): 1.86 (d, PCH_3; J(P, H) = 9.1), 2.59 (CH_3C), 4.99 (d, C_5H_5; J(P, H) = 1.2); spectrum is temperature-invariant down to −40°C [36], similar data in [26, 34, 35] ^{13}C NMR ($CDCl_3$ or CD_2Cl_2): 51.52 (**C**H_3C), 95.40 (C_5H_5), 236.9 (d, CO; J(P, C) = 23.6), 266.8 (d, C=O; J(P, C) = 12.5) [48] IR (CS_2): 1828, 1919 (ν(CO)) [26]; (cyclohexane): 1638 (ν(C=O)), 1852, 1931 (ν(CO)); spectrum given as a diagram in [36] [34, 36] Raman spectrum (solid): 1841, 1924 (ν(CO)) [34, 36] rate constant of the decarbonylation in CH_3CN at 60°C: $k = 3.6 \times 10^{-6}\ s^{-1}$ [35] equilibrium constant for dissociation in CH_3CN after 1 to 2 h at 60°C: $K_{equ} = 0.1 \times 10^{-4}$ with 3% dissociation [35]
28	trans-$C_5H_5Mo(CO)_2(P(CH_3)_2C_6H_5)^{14}C(O)CH_3$	prepared by treating $Na[C_5H_5Mo(CO)_2P(CH_3)_2C_6H_5]$ with $CH_3{}^{14}C(O)Cl$ [35] undergoes rapid scrambling of the ^{14}C label among the terminal and acetyl CO groups at 60°C [35]

Table 5 (continued)

No.	compound	method of preparation (yield) properties and remarks
29	trans-$C_5H_5Mo(CO)_2(P(C_2H_5)_2C_6H_5)C(O)CH_3$	II, in CH_3CN for several hours (high) [34, 36] ^{1}H NMR ($CDCl_3$): 1.16 (m, **C**H_3CH_2; J(H, H) = 7.5, J(P, H) = 6.5), 2.25 (m, PCH_2; J(H, H) = 7.5), 2.61 ($CH_3C(O)$), 4.97 (d, C_5H_5; J(P, H) = 1.2) [34, 36] IR (cyclohexane): 1642 (ν(C=O)), 1855, 1934 (ν(CO)) [34, 36] Raman spectrum (powder): 1854, 1927 (ν(CO)) [34, 36] decarbonylates slowly in solution with formation of $C_5H_5Mo(CO)_3CH_3$ [34, 36]
30	trans-$C_5H_5Mo(CO)_2(P(C_6H_5)_2CH_3)C(O)CH_3$	II, in CH_3CN for several hours (high) [34 to 36] ^{1}H NMR ($CDCl_3$): 2.17 (d, PCH_3; J(P, H) = 9), 2.57 (CH_3C), 5.00 (d, C_5H_5; J(P, H) = 1.2) [34 to 36] IR (cyclohexane): 1639 (ν(C=O)), 1857, 1935 (ν(CO)) [34, 36] Raman spectrum (solid): 1839, 1922 [34, 36] rate constant for decarbonylation in CH_3CN at 60°C: $k = 6.8 \times 10^{-6}\ s^{-1}$ [35] equilibrium constant for the dissociation in CH_3CN after 1 to 2 h at 60°C: $K_{equ} = 1.3 \times 10^{-4}$ with 10% dissociation [35]
31	trans-$C_5H_5Mo(CO)_2(P(C_6H_5)_2C_2H_5)C(O)CH_3$	II, in CH_3CN for several hours (high) [34, 36] ^{1}H NMR ($CDCl_3$): 1.16 (2 overlapping t, **C**H_3CH_2; J(H, H) = 7.8, J(P, H) = 18.0), 2.62 ($CH_3C(O)$), 2.66 (m, PCH_2; J(H, H) = 7.8), 4.92 (d, C_5H_5; J(P, H) = 1.2) [34, 36] IR (cyclohexane): 1640 (ν(CO)), 1858, 1938 (ν(CO)) [34, 36] Raman spectrum (solid): 1854, 1926 (ν(CO)) [34, 36] partially decarbonylates in solution on standing to $C_5H_5Mo(CO)_3CH_3$ [34, 36]
32	trans-$C_5H_5Mo(CO)_2(P(C_6H_5)_2C_3H_7\text{-}i)C(O)CH_3$	II, in CH_3CN for 2 to 3 h (75 to 85%) [35] ^{1}H NMR ($CDCl_3$): 2.63 ($CH_3C(O)$), 4.83 (d, C_5H_5; J(P, H) = 1.5 to 2) [35] rate constant for decarbonylation in CH_3CN at 60°C: $k = 23.9 \times 10^{-6}\ s^{-1}$ [35] equilibrium constant for the dissociation in CH_3CN at 60°C after 1 to 2 h: $K_{equ} = 90 \times 10^{-4}$ with 60% dissociation [25, 35]

References on pp. 75/7

Table 5 (continued)

No.	compound	method of preparation (yield) properties and remarks
33	trans-$C_5H_5Mo(CO)_2(P(OC(CH_3)_2CH_2)_2N)C(O)CH_3$ (^2D= 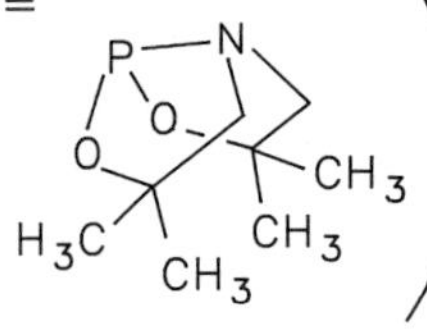	II, in refluxing hexane for 4 h (82%) [61] yellow parallelepipedic crystals (from toluene/pentane 7:3), m.p. 135 to 137°C (dec.) [61] ^1H NMR (C_6D_6): 2.99 ($CH_3C(O)$), 5.20 (d, C_5H_5; J(P, H) = 0.9) [61] ^{31}P NMR (toluene): 223.7 [61] IR: 1634 (ν(C=O)), 1878, 1954 (ν(CO)) [61] mass spectrum (chemical ionization): $[M + H]^+$, $[M - n\ CO]^+$ (n = 0 to 2), other fragments also given in [61] soluble in the usual solvents except hydrocarbons [61]
34	$C_5H_5Mo(CO)_2(PF_2N(CH_3)_2)C(O)CH_3$	II, in CH_3CN with 5 equivalents of $PF_2N(CH_3)_2$ for 2 h; purified by chromatography on Florisil with petroleum ether (40%) [28] yellow solid, m.p. 61°C (from pentane at −78°C) [28] ^1H NMR ($CDCl_3$): 2.50 (CH_3C), 2.81 (d of t, NCH_3; J = 10, 3), 5.18 (d, C_5H_5; J(P, H) = 2) [28] ^{19}F NMR (CH_2Cl_2): 26.5 (d; J(P, F) = 1131) [28] IR (KBr): 720, 778 (ν(PF)); (cyclohexane): 1652 (ν(C=O)), 1892, 1964 (ν(CO)) [28]
35	$C_5H_5Mo(CO)_2(P(C_6H_5)_2C{\equiv}CC_6H_5)C(O)CH_3$	II, in CH_3CN for 72 h (14% after chromatography on Al_2O_3) [19] yellow solid, m.p. 114 to 115°C (from CH_2Cl_2/hexane) [19] ^1H NMR ($CDCl_3$): 2.59 (s, CH_3), 5.19 (d, C_5H_5; J(P, H) = 2) [19] IR (CH_2Cl_2): 1606 (ν(C=O)), 1856, 1943 (ν(CO)), 2174.5 (ν(C≡C)) [19]
36	$C_5H_5Mo(CO)_2(P(C_6H_5)_2C_2H_4As(C_6H_5)_2)C(O)CH_3$	II, in CH_3CN for 40 h (55% by chromatography on Al_2O_3) [27] the ^2D ligand is only coordinated through P [27] yellow crystals, m.p. 132 to 133°C (from CH_2Cl_2/hexane) [27] ^1H NMR ($CDCl_3$): 2.1 (br, CH_2), 2.61 (s, CH_3), 4.85 (s, C_5H_5), 7.28, 7.35 (m, C_6H_5) [27] IR (CH_2Cl_2): 1607, 1615 (ν(C=O)), 1846, 1935 (ν(CO)) [27]

References on pp. 75/7

Table 5 (continued)

No.	compound	method of preparation (yield) properties and remarks
37	$C_5H_5Mo(CO)_2(P(C_6H_5)(C_2H_4As(C_6H_5)_2)_2)C(O)CH_3$	II, in CH_3CN for 100 h (53% by chromatography on Al_2O_3 with CH_2Cl_2/acetone/hexane) [27] yellow solid, m.p. 93 to 95°C [27] ^1H NMR ($CDCl_3$): 2.1 (br, CH_2), 2.57 (CH_3), 4.84 (C_5H_5), 7.32 (m, C_6H_5) [27] IR (CH_2Cl_2): 1604, 1620 (ν(C=O)), 1848, 1935 (ν(CO)) [27]
38	$C_5H_5Mo(CO)_2(As(C_6H_5)_3)C(O)CH_3$	II, in CH_3CN at 25°C for 12 to 24 h or at 35°C for 3 h with 2 equivalents of $As(C_6H_5)_3$ (26% to 39%) [37, 41] bright yellow crystals, m.p. 118°C (from CH_2Cl_2/hexane or $CHCl_3$/petroleum ether) [37, 41] ^1H NMR ($CDCl_3$): 2.64 (CH_3), 5.04 (C_5H_5), 7.3 to 7.6 (m, C_6H_5) [37, 41] IR ($CHCl_3$): 1605 (ν(C=O)), 1860, 1946 (ν(CO)) [41]; similar data in [37] heating in CH_3CN at 40°C for 1 h affords $C_5H_5Mo(CO)_3CH_3$; UV irradiation in THF for 1 h at 25°C yields small amounts of $C_5H_5Mo(CO)_2(As(C_6H_5)_3)CH_3$ [41]
39	$C_5H_5Mo(CO)_2(Sb(C_6H_5)_3)C(O)CH_3$	II, in CH_3CN for 27 h (37% after chromatography on Al_2O_3 with $CHCl_3$/petroleum ether 1:1) [41] bright yellow crystals, m.p. 114°C (from $CHCl_3$/petroleum ether) [41] ^1H NMR ($CDCl_3$): 2.66 (CH_3), 5.17 (C_5H_5), 7.2 to 7.6 (m, C_6H_5) [41] IR ($CHCl_3$): 1602 (ν(C=O)), 1860, 1940 (ν(CO)) [41]
40	$C_5H_5Mo(CO)_2(P(C_6H_5)_3)C(O)CF_3$	$C_5H_5Mo(CO)_3C(O)CF_3$ was treated with 1.2 equivalents of $Rh(P(C_6H_5)_3)_3Cl$ in CH_2Cl_2 for 1 h (70% by chromatography on Al_2O_3 with $CHCl_3$) [32] orange solid (from $CHCl_3$/hexane), m.p. 149 to 151°C (dec.) [32] IR (CH_2Cl_2): 1609 (ν(C=O)), 1893, 1965 (ν(CO)) [41] the mechanism of CO/D exchange was discussed [32]
	compound of the type $C_5H_5Mo(CO)_2(^2D)C(O)CH{=}PR_3$	
41	$C_5H_5Mo(CO)_2(P(CH_3)_3)C(O)CH{=}P(CH_3)_3$	$[C_5H_5Mo(CO)_3P(CH_3)_3]X$ was allowed to react with 2 equivalents of $(CH_3)_3P{=}CH_2$ [55]

References on pp. 75/7

Table 5 (continued)

No.	compound	method of preparation (yield) properties and remarks
compounds of the type $^5LMo(CO)_2(^2D)C(O)C_2H_5$		
42	trans-$C_5H_5Mo(CO)_2(P(CH_3)_3)C(O)C_2H_5$	II, in refluxing heptane for 100 h with excess $P(CH_3)_3$ (47% by chromatography on silica with $CHCl_3$) [51] m.p. 156°C (from hexane) [51] ^{1}H NMR (acetone-d_6): 0.81 (t, **CH_3**CH_2; J(H, H) = 7.4), 1.65 (d, PCH_3; J(P, H) = 8.8), 2.83 (q, CH_2; J(H, H) = 7.4), 5.18 (d, C_5H_5; J(P, H) = 1.6) [51] ^{13}C NMR (acetone-d_6): 10.1 (**C**H_3CH_2), 21.1 (d, PCH_3; J(P, C) = 31.6), 58.2 (CH_2), 95.8 (C_5H_5), 209.7 (C=O), 239.4 (d, CO; J(P, C) = 24.8) [51] ^{31}P NMR (acetone-d_6): 22.3 (s) [51] IR (THF): 1615 (ν(C=O)), 1838, 1921 (ν(CO)) [51]
43	trans-$C_5H_5Mo(CO)_2(P(C_4H_9\text{-}n)_3)C(O)C_2H_5$	II, in CH_3CN for 2 h with 2 equivalents of $P(C_4H_9\text{-}n)_3$ (60%) [11] m.p. 58°C (from CH_2Cl_2/hexane) [11] ^{1}H NMR ($CDCl_3$): 3.01 (q, **CH_2**CH_3; J(H, H) = 7.2), 5.02 (d, C_5H_5; J(P, H) = 1.0) [11] IR (cyclohexane): 1600 (ν(C=O)), 1842, 1928 (ν(CO)) [11]
44	trans-$C_5H_5Mo(CO)_2(P(C_6H_5)_3)C(O)C_2H_5$	II, in CH_3CN for 2 h with 2 equivalents of $P(C_6H_5)_3$ (85%); kinetics of the formation in $CHCl_3$ in [11] m.p. 157.5°C (from CH_2Cl_2/hexane) [11] ^{1}H NMR ($CDCl_3$): 0.91 (t, CH_3; J(H, H) = 7.0), 3.02 (q, CH_2), 4.95 (d, C_5H_5; J(H, H) = 1.5) [11] IR: 235 (ν(MoP)), 319, 346, 370 (MoC_5H_5), 442, 466 (ν(MoC)), 525, 535, 560, 681 (δ(MoCO)), 1611 (ν(C=O)), 1836, 1922 (ν(CO)), other bands between 108 and 514 also given [18]; (cyclohexane): 1613 (ν(C=O)), 1853, 1936 (ν(CO)) [11] Raman spectrum (solid): 236 (ν(MoP)), 322, 347, 370 (MoC_5H_5), 449, 468 (ν(MoC)), 523, 540, 561, 580 (δ(MoCO)),1843, 1925 (ν(CO)), other bands between 206 and 498 also given [18]
45	trans-$C_5H_5Mo(CO)_2(P(OCH_3)_3)C(O)C_2H_5$	II, in CH_3CN for 2 h with 2 equivalents of $P(OCH_3)_3$ (30%); kinetics of the formation in $CHCl_3$ in [11] low-melting solid (from CH_2Cl_2/hexane) [11]

Table 5 (continued)

No.	compound	method of preparation (yield) properties and remarks
		^{1}H NMR ($CDCl_3$): 0.60 (t, **CH_3**CH_2; J(H, H) = 7.2), 2.94 (q, CH_2), 5.13 (d, C_5H_5; J(P, H) = 1.1) [11] IR (cyclohexane): 1611 (ν(C=O)), 1860, 1946 (ν(CO)) [11]
46	trans-$C_5H_5Mo(CO)_2(P(OC_2H_5)_3)C(O)C_2H_5$	II, in CH_3CN for 2 h with 2 equivalents of $P(OC_2H_5)_3$ (27%) [11] m.p. 69 °C (from CH_2Cl_2/hexane) [11] ^{1}H NMR ($CDCl_3$): 0.94 (t, **CH_3**CH_2C; J(H, H) = 7.4), 3.00 (q, CH_2C), 5.19 (d, C_5H_5; J(P, H) = 1.1) [11] IR (cyclohexane): 1612 (ν(C=O)), 1862, 1944 (ν(CO)) [11]
47	trans-$C_5H_5Mo(CO)_2(P(OC_6H_5)_3)C(O)C_2H_5$	II, in CH_3CN for 2 h with 2 equivalents of $P(OC_6H_5)_3$ (80%) [11] m.p. 93 °C (from CH_2Cl_2/hexane) [11] ^{1}H NMR ($CDCl_3$): 0.72 (t, CH_3; J(H, H) = 7.0), 2.50 (q, CH_2), 4.72 (d, C_5H_5; J(P, H) = 1.0) [11] IR (cyclohexane): 1592 (ν(C=O)), 1877, 1955 (ν(CO)) [11]
48	trans-$C_5H_5Mo(CO)_2(P(OCH_2)_3CCH_3)C(O)C_2H_5$	II, in CH_3CN for 2 h with 2 equivalents of $P(OCH_2)_3CCH_3$ (62%); kinetics of the formation in $CHCl_3$ in [11] m.p. 149 °C (from CH_2Cl_2/hexane) [11] ^{1}H NMR ($CDCl_3$): 0.85 (t, **CH_3**CH_2; J(H, H) = 7.3), 2.91 (q, $CH_2C(O)$), 5.15 (d, C_5H_5; J(P, H) = 1.0) [11] IR: 202 (ν(MoP)), 306, 344, 368 (MoC_5H_5), 446, 459 (ν(MoC)), 535, 558, 580 (δ(MoCO)), 1620 (ν(C=O)), 1861, 1948 (ν(CO)), other bands between 104 and 512 also given [18]; (cyclohexane): 1622 (ν(C=O)), 1881, 1957 (ν(CO)) [11] Raman spectrum (solid): 203 (ν(MoP)), 307, 351, 372 (MoC_5H_5), 448, 462 (ν(MoC)), 521, 539, 559, 583 (δ(MoCO)), 1625 (ν(C=O)), 1869, 1949 (ν(CO)); other bands between 177 and 412 also given [18]
49	trans-$C_5H_5Mo(CO)_2(P(CH_3)_2C_6H_5)C(O)C_2H_5$	II, in CH_3CN for several hours (40%) [26, 34] ^{1}H NMR ($CDCl_3$): 0.87 (t, **CH_3**CH_2; J(H, H) = 7.5), 1.88 (d, PCH_3; J(P, H) = 9.0), 3.00 (q, CH_2), 4.87 (d, C_5H_5; J(P, H) = 1.5); the spectrum is temperature-independent up to −40 °C [26, 34, 36]

References on pp. 75/7

Table 5 (continued)

No.	compound	method of preparation (yield) properties and remarks
49 (continued)		IR (cyclohexane): 1635 (ν(C=O)), 1854, 1940 (ν(CO)) [34, 36]; (CS_2): 1827, 1920 (ν(CO)) [26]
50	trans-$C_5H_5Mo(CO)_2(P(C_2H_5)_2C_6H_5)C(O)C_2H_5$	II, in CH_3CN for several hours [34, 36] ^{1}H NMR ($CDCl_3$): 0.90 (t, **CH_3**$CH_2C(O)$; J(H, H) = 6.8), 1.18 (t of t, **CH_3**CH_2P; J(H, H) = 7.5, J(P, H) = 17.3), 2.18 (m, CH_2P), 3.05 (q, $CH_2C(O)$), 4.96 (d, C_5H_5; J(P, H) = 1.2) [34, 36] IR (cyclohexane): 1636 (ν(C=O)), 1854, 1933 (ν(CO)) [34, 36] Raman spectrum (solid): 1845, 1919 (ν(CO)) [34, 36]
51	trans-$C_5H_5Mo(CO)_2(P(C_6H_5)_2CH_3)C(O)C_2H_5$	II, in CH_3CN for several hours [34, 36] ^{1}H NMR ($CDCl_3$): 0.92 (t, **CH_3**CH_2; J(H, H) = 7.5), 2.18 (d, PCH_3; J(P, H) = 8), 2.99 (q, CH_2), 4.98 (d, C_5H_5; J(P, H) = 1.2) [34, 36] IR (cyclohexane): 1637 (ν(C=O)), 1858, 1939 (ν(CO)) [34, 36]
52	trans-$C_5H_5Mo(CO)_2(P(C_6H_5)_2C_2H_5)C(O)C_2H_5$	II, in CH_3CN for several hours [34, 36] ^{1}H NMR ($CDCl_3$): 0.91 (t, **CH_3**CH_2C; J(H, H) = 7.5), 1.22 (m, **CH_3**CH_2P; J(H, H) = 7.5, J(P, H) = 19.5), 2.70 (m, PCH_2), 3.06 (q, CH_2C), 4.91 (d, C_5H_5; J(P, H) = 1.2) [34, 36] IR (cyclohexane): 1642 (ν(C=O)), 1863, 1946 (ν(CO)) [34, 36]

compounds of the type $^5LMo(CO)_2(^2D)C(O)CH_2CH{=}CH_2$ (^1L = $C(O)CH^3_2$ with ^1H on the same carbon, =C with H^2 and $H^{2'}$)

No.	compound	method of preparation (yield) properties and remarks
53	trans-$C_5H_5Mo(CO)_2(P(C_6H_5)_3)C(O)C_3H_5$	II, in CH_3CN for 22 h (38%); kinetics of the formation in [14] yellow solid, m.p. 117 °C (from CH_2Cl_2/hexane) [14] ^{1}H NMR ($CDCl_3$): 3.69 (d, H-3; J(H-1,3) = 6.9), 4.75 (m, H-2; J(H-2,2′) = 2.4, J(H-1,2) = 10.8), 4.86 (m, H-2′; J(H-2,2′) = 2.4, J(H-1,2′) = 17.0),

Table 5 (continued)

No.	compound	method of preparation (yield) properties and remarks
		4.92 (d, C_5H_5; J(P, H) = 1.0), 5.80 (m, H-1; J(H-1,2) = 10.8, 17.0, J(H-1,3) = 6.9), 7.36 (m, C_6H_5) [14] IR: 235 (ν(MoP)), 312, 349, 368 (MoC_5H_5), 453, 462 (ν(MoC)), 534, 540, 559, 579 (δ(MoCO)), 1613 (ν(C=O)), 1830, 1924 (ν(CO)), other bands between 106 and 513 also given in [18]; (cyclohexane): 1646 (ν(C=O)), 1866, 1943 (ν(CO)) [14] Raman spectrum (solid): 315, 348 (MoC_5H_5), 449, 460 (ν(MoC)), 1835, 1929 (ν(CO)) [18]
54	trans-$C_5H_5Mo(CO)_2(P(OCH_2)_3CC_2H_5)C(O)C_3H_5$	II, in CH_3CN at 70°C for 4 h (20%) [14] yellow solid, m.p. 92°C (three times from CH_2Cl_2/hexane at −78°C) [14] ^1H NMR ($CDCl_3$): 1.07 (m, C_2H_5, 5H), 3.84 (d, H-3; J(H-1,3) = 7.5), 4.51 (d, $POCH_2$; J(P, H) = 5.2), 4.71 (H-2, 1 H; J(H-2,2′) = 2.1, J(H-2,1) = 10.0), 4.90 (H-2, 1 H; J(H-2,2′) = 2.1, J(H-1,2) = 17), 5.52 (d, C_5H_5; J(P, H) = 1.2), 5.86 (m, H-1; J(H-2,1) = 10, 17, J(H-3,1) = 7.5) [14] IR (cyclohexane): 1652 (ν(C=O)), 1898, 1965 (ν(CO)) [14]
55	trans-$C_5H_5Mo(CO)_2(P(OC_3H_5)_2C_6H_5)C(O)C_3H_5$	formed in very small yields as a by-product of the reaction of $(C_5H_5Mo(CO)_3)_2$ with $P(OC_3H_5)_2C_6H_5$ in C_6H_6 (not isolated) [24] IR (CH_2Cl_2): 1606 (ν(C=O)), 1862, 1940 (ν(CO)) [24]
	compounds of the type $C_5H_5Mo(CO)_2(P(C_6H_5)_3)C(O)(CH_2)_nCH{=}CHR$	
56	trans-$C_5H_5Mo(CO)_2(P(C_6H_5)_3)C(O)CH^1_2CH^2{=}CH^3CH^4_3$ (cis)	II, in CH_3CN at 25°C for 21 h with 1.2 equivalents of $P(C_6H_5)_3$ (53%) [14] yellow solid, m.p. 108°C (from pentane at −78°C) [14] ^1H NMR ($CDCl_3$): 1.61 (H-4; J(3,4) = 6.0), 3.68 (H-1; J(2,1) = 5.0), 4.95 (d, C_5H_5; J(P, H) = 1.0), 5.19 (H-3; J(2,3) = J(2,4) = 6.0), 5.39 (H-2; J(2,3) = 6.0, J(2,1) = 5.0), 7.40 (m, C_6H_5) [14] IR (cyclohexane): 1631 (ν(C=O)), 1865, 1941 (ν(CO)) [14]

References on pp. 75/7

Table 5 (continued)

No.	compound	method of preparation (yield) properties and remarks
57	$C_5H_5Mo(CO)_2(P(C_6H_5)_3)C(O)CH^1_2CH^2_2CH^3{=}CH^4_2$	II, in THF at 60°C for 4 h with 0.7 equivalents of $P(C_6H_5)_3$ (10% by chromatography on Al_2O_3 with CH_2Cl_2) [29] yellow-orange crystals, m.p. 132°C (from CH_2Cl_2/hexane) [29] ^1H NMR (cyclohexane): 2.20 (m, H-2; J(1,2) = 7.0), 3.15 (m, H-1), 4.90 (m, H-4), 4.98 (d, C_5H_5; J(P, H) = 1.2), 5.60 (m, H-3) [29] IR (KBr): 905, 995; 1615 (ν(C=O)), 1850, 1932 (ν(CO)) [29]
58	$C_5H_5Mo(CO)_2(P(C_6H_5)_3)C(O)CH^1_2CH^2_2CH^3{=}CH^4CH_3$ (trans olefin)	II, in THF at 60°C for 4 h with 0.7 equivalents of $P(C_6H_5)_3$ (34% by chromatography on Al_2O_3 with CH_2Cl_2) [29] yellow-orange crystals, m.p. 161°C (from CH_2Cl_2/hexane) [29] ^1H NMR (cyclohexane): 1.60 (m, CH_3), 2.10 (m, H-2; J(1,2) = 7.0), 3.10 (m, H-1), 4.98 (d, C_5H_5; J(P, H) = 1.2), 5.35 (m, H-3,4) [29] IR (KBr): 965; 1615 (ν(C=O)), 1850, 1930 (ν(CO)) [29]
	compounds of the type $^5LMo(CO)_2(^2D)C(O)(CH_2)_nR$ (R is a functional group)	
59	trans-$C_5H_5Mo(CO)_2(P(C_6H_5)_3)C(O)CH_2CH_2C(O)H$	IIIa (25% by chromatography on silica with C_6H_6/ether 1:1) [42] bright yellow solid, dec. 121 to 123°C [42] ^1H NMR ($CDCl_3$): 2.38 (d of t, $\mathbf{CH_2}C(O)H$; J = 7.5, 2.1), 3.35 (t, $C(O)CH_2$), 4.96 (d, C_5H_5; J(P, H) ≈ 1.3), 7.39 (m, C_6H_5), 9.61 (t, CH; J(H, H) = 2.1) [42] IR (CH_2Cl_2): 1610 (ν(C=O)), 1720 (ν(C(O)H)), 1855, 1940 (ν(CO)) [42]
60	trans-$C_5H_5Mo(CO)_2(P(C_6H_5)_3)C(O)CH_2CH_2CH(OC_2H_5)OCH_3$	IIIb (40%) [42] yellow needles, dec. 124 to 125°C (from CH_2Cl_2/hexane 2:5 at −20°C) [42] ^1H NMR ($CDCl_3$): 1.12 (t, $\mathbf{CH_3}CH_2$), 1.71 (q, CH_2), 3.03 (t, CCH_2), 3.18 (s, OCH_3), 3.46 (qui, OCH_2), 4.27 (t, CH), 4.92 (d, C_5H_5; J(P, H) ≈ 1.3), 7.34 (m, C_6H_5) [42] IR (CH_2Cl_2): 1610 (ν(C=O)), 1850, 1937 (ν(CO)) [42]

References on pp. 75/7

Table 5 (continued)

No.	compound	method of preparation (yield) properties and remarks
61	trans-$C_5H_5Mo(CO)_2(P(C_6H_5)_3)C(O)CH_2CH_2CH(OC_2H_5)_2$	II, in CH_3CN for 2 h (31%) [42] yellow-orange solid, m.p. 123.4 to 125°C (from ether at −20°C) [42] ^1H NMR ($CDCl_3$): 1.12 (t, CH_3), 1.68 (t, $CH_2C(O)$), 3.05 (q, CH_2), 3.4 (qui, OCH_2), 4.33 (t, CH), 4.92 (d, C_5H_5; J(P, H) ≈ 1.3), 7.35 (m, C_6H_5) [42] IR (CH_2Cl_2): 1618 (ν(C=O)), 1850, 1940 (ν(CO)) [42]
62	trans-$C_5H_5Mo(CO)_2(P(C_6H_5)_3)C(O)(CH_2)_3Cl$	II, in CH_3CN with 3 equivalents of $P(C_6H_5)_3$ for 15 min (87%) [33] light yellow powder, m.p. 131 to 133°C (from CH_2Cl_2/pentane) [33] ^1H NMR ($CDCl_3$): 1.93 (br m, CH_2), 3.10 (t, CH_2; J(H, H) ≈ 7), 3.40 (t, CH_2; J(H, H) ≈ 7), 6.60 (d, C_5H_5; J(P, H) ≈ 1.5), 7.36 (br m, C_6H_5) [33] IR (CH_2Cl_2): 1605 (ν(C=O)), 1850, 1930 (ν(CO)) [33]
*63	trans-$C_5H_5Mo(CO)_2(P(C_6H_5)_3)C(O)(CH_2)_4Br$	II, in CH_3CN with 3 equivalents of $P(C_6H_5)_3$ for 20 min (55%, after recrystallization 34%); kinetics of the formation in [33] pale lime plates (from CH_3CN at −15°C), m.p. 133 to 135°C (dec.) [33] ^1H NMR (CS_2): 1.53 (br m, 2 CH_2), 2.80 (t, CH_2; J(H, H) ≈ 7), 3.13 (t, CH_2; J(H, H) ≈ 7), 4.93 (d, C_5H_5; J(P, H) ≈ 1.3), 7.33 (br m, C_6H_5) [33] IR (CH_2Cl_2): 1605 (ν(C=O)), 1850, 1930 (ν(CO)) [33]
*64	trans-$C_5H_5Mo(CO)_2(P(C_6H_5)_3)C(O)(CH_2)_4I$	II, in CH_3CN with 2 equivalents of $P(C_6H_5)_3$ for 8 min (49%) [33] lemon-yellow powder, m.p. 123.5 to 125.5°C ^1H NMR ($CDCl_3$): 1.56 (br m, 2 CH_2), 3.00 (br m, 2 CH_2), 4.83 (d, C_5H_5; J(P, H) ≈ 1.5), 7.33 (m, C_6H_5) [33] IR (CH_2Cl_2): 1605 (ν(C=O)), 1850, 1930 (ν(CO)) [33]
65	trans-$C_5H_5Mo(CO)_2(P(C_6H_5)_3)C(O)(CH_2)_5Br$	II, in CH_3CN with 2 equivalents of $P(C_6H_5)_3$ for 4.5 h (91%) [33] lemon yellow powder, m.p. 115 to 116°C (from CS_2 at −15°C) [33]

References on pp. 75/7

Table 5 (continued)

No.	compound	method of preparation (yield) properties and remarks
65 (continued)		^{1}H NMR (CS_2): 1.30 (br m, 3 CH_2), 1.80 (br t, CH_2; J(H, H) ≈ 6), 3.20 (br t, CH_2; J(H, H) ≈ 7), 4.80 (d, C_5H_5; J(P, H) ≈ 1.5), 7.40 (m, C_6H_5) [33] IR (CH_2Cl_2): 1610 (ν(C=O)), 1845, 1930 (ν(CO)) [33]
	compounds of the type ^{5}LMo(CO)$_2$(^{2}D)C(O)CH$_2$R (R is a cyclic group)	
66	$C_5H_5Mo(CO)_2(P(C_6H_5)_3)C(O)CH_2C_3H_5$-cyclo	II, in refluxing THF with 1.5 equivalents of $P(C_6H_5)_3$ for 40 min (60%); heating in hexane for 14 h gave the by-products $C_5H_5Mo(CO)_2(P(C_6H_5)_3)C_4H_7$ (C_4H_7 = crotyl) and small amounts of $C_5H_5Mo(CO)_2(P(C_6H_5)_3)CH_2C_3H_5$-cyclo; kinetic studies of the formation in THF in [65] ^{1}H NMR ($CDCl_3$): 0.0 to 1.4 (m, C_3H_5-cyclo), 2.8 (d, CH_2; J = 7.5), 4.9 (C_5H_5), 7.3 (m, C_6H_5) [65] IR (CH_2Cl_2): 1622 (ν(C=O)), 1851, 1934 (ν(CO)) [65]
67	trans-$C_5H_5Mo(CO)_2(P(C_4H_9\text{-}n)_3)C(O)CH_2C_6H_5$	II, in CH_3CN with 1.2 equivalents of the phosphane for 21 h (14%) [14] lemon-yellow solid, m.p. 75°C (from CH_2Cl_2/hexane) [14] ^{1}H NMR ($CDCl_3$): 0.68 to 2.19 (m, C_4H_9-n), 4.19 ($CH_2C(O)$), 5.05 (d, C_5H_5; J(P, H) = 1.1), 7.05 (m, C_6H_5) [14] IR (cyclohexane): 1641 (ν(C=O)), 1849, 1933 (ν(CO)) [14]
*68	trans-$C_5H_5Mo(CO)_2(P(C_6H_5)_3)C(O)CH_2C_6H_5$	II, in CH_3CN with 1.2 equivalents of the phosphane for 21 h (85%) [14]; also in dimethyl sulfoxide [52] yellow solid, m.p. 124 to 126°C (from CH_2Cl_2/hexane) [14] ^{1}H NMR ($CDCl_3$): 4.30 (s, CH_2), 4.98 (d, C_5H_5; J(P, H) = 1.2), 7.42 (m, C_6H_5) [14]; (CD_3CN): ca. 4.3 (CH_2), 5.05 (d, C_5H_5; J(P, H) = 1) [52] IR: 231 (ν(MoP)), 316, 358 (MoC_5H_5), 440, 471 (ν(MoC)), 538, 564, 580 (δ(MoCO)), 1609 (ν(C=O)), 1841, 1927 (ν(CO)); also other bands between 11 and 511 [18]; (cyclohexane): 1643 (ν(C=O)), 1866, 1943 (ν(CO)) [14]; (CH_3CN): 1620 (ν(C=O)), 1850, 1936 (ν(CO)) [52]

References on pp. 75/7

Table 5 (continued)

No.	compound	method of preparation (yield) properties and remarks
		Raman spectrum (solid): 229 (ν(MoP)), 318, 347, 366 (MoC_5H_5), 442, 471 (ν(MoC)), 540, 561, 584 (δ(MoCO)), 1851, 1931 (ν(CO)), also other bands between 174 and 510 [18]
69	trans-$C_5H_5Mo(CO)_2(P(OC_6H_5)_3)C(O)CH_2C_6H_5$	II, in CH_3CN with 1.2 equivalents of the phosphite for 5 h (60%); kinetic studies of the formation in [14] light yellow solid, m.p. 42°C (five times from CH_2Cl_2/hexane) [14] ^1H NMR ($CDCl_3$): 3.79 (CH_2), 4.92 (d, C_5H_5; J(P, H) = 1.2), 7.36 (m, C_6H_5) [14] IR (cyclohexane): 1597 (ν(C=O)), 1882, 1956 (ν(CO)) [14]
70	trans-$C_5H_5Mo(CO)_2(P(OCH_2)_3CC_2H_5)C(O)CH_2C_6H_5$	II, in CH_3CN with 1.2 equivalents of the phosphite for 4.5 h (92%) [14] yellow solid, m.p. 136 to 138°C (from CH_2Cl_2/hexane at −78°C) [14] ^1H NMR ($CDCl_3$): 0.81 (m, $\mathbf{CH_3}CH_2$), 1.84 (m, $\mathbf{CH_2}CH_3$), 4.15 ($CH_2C(O)$), 4.33 (d, OCH_2; J(P, H) = 5.0), 5.25 (d, C_5H_5; J(P, H) ≈ 1.5), 7.29 (m, C_6H_5) [14] IR (cyclohexane): 1642 (ν(C=O)), 1892, 1963 (ν(CO)) [14]
71	trans-$C_5H_5Mo(CO)_2(P(CH_3)_2C_6H_5)C(O)CH_2C_6H_5$	II; the kinetics of the formation in the presence of a 10-fold excess of phosphane in CH_3CN at 30°C were studied by either ^{31}P NMR or IR spectroscopy [60]
72	trans-$CH_3C_5H_4Mo(CO)_2(P(CH_3)_2C_6H_5)C(O)CH_2C_6H_5$	II; the kinetics of the formation in the presence of a 10-fold excess of phosphane in CH_3CN at 30°C were studied by either ^{31}P NMR or IR spectroscopy [60]
*73	$(CH_3)_5C_5Mo(CO)_2(P(CH_3)_2C_6H_5)C(O)CH_2C_6H_5$	II; the kinetics of the formation in the presence of a 10-fold excess of phosphane in CH_3CN at 30°C were studied by either ^{31}P NMR or IR spectroscopy [60]

References on pp. 75/7

Table 5 (continued)

No.	compound	method of preparation (yield) properties and remarks
*73 (continued)		^{31}P NMR (CH_3CN, rel. to the free phosphane): cis isomer: 65.31; trans isomer: 78.96 [60] IR (cyclohexane): cis isomer: 1838, 1923 (ν(CO)); trans isomer: 1858, 1936 (ν(CO)) [60]
74	$C_5H_5Mo(CO)_2(P(C_6H_5)_2C{\equiv}CC_6H_5)C(O)CH_2C_6H_5$	II, in CH_3CN for 80 h (15% by chromatography on Al_2O_3 with CH_3CN) [19] yellow solid, m.p. 123 to 124°C (from CH_2Cl_2/hexane) [19] ^1H NMR ($CDCl_3$): 4.28 (s, CH_2), 5.09 (d, C_5H_5; J(P, H) = 2), 7.2 (m, three apparent peaks) [19] IR (CH_2Cl_2): 1614 (ν(C=O)), 1859, 1944 (ν(CO)), 2174.5 (ν(C≡C)) [19]
75	$C_5H_5Mo(CO)_2(P(C_6H_5)_3)C(O)CH_2C_6H_4OCH_3$-4	II, in CH_3CN or dimethyl sulfoxide with excess $P(C_6H_5)_3$; the kinetics of the formation were studied [52] ^1H NMR (CD_3CN): ca. 4.3 ($CH_2C(O)$), ca. 5.05 (d, C_5H_5; J(P, H) ≈ 1) [52] IR (CH_3CN): ca. 1620 (ν(C=O)), 1850, 1936 (ν(CO)) [52] rates of the decarbonylation to $C_5H_5Mo(CO)_2(P(C_6H_5)_3)CH_2C_6H_4R$ in CH_3CN at 60°C are: $k = 10.4 \times 10^{-6}$ (R = CF_3-4), 4.0×10^{-6} (R = OCH_3-4) s^{-1} [52]
76	$C_5H_5Mo(CO)_2(P(C_6H_5)_3)C(O)CH_2C_6H_4CF_3$-3	like No.75
77	$C_5H_5Mo(CO)_2(P(C_6H_5)_3)C(O)CH_2C_6H_4OCH_3$-3	like No. 75
78	$C_5H_5Mo(CO)_2(P(C_6H_5)_3)C(O)CH_2C_6H_4F$-3	like No. 75
79	$C_5H_5Mo(CO)_2(P(C_6H_5)_3)C(O)CH_2C_6H_4F$-4	like No. 75
compounds of the type $^5LMo(CO)_2(^2D)C(O)CH_2M$ (M = organometallic group)		
80	$C_5H_5Mo(CO)_2(P(C_6H_5)_3)C(O)CH_2Fe(C_5H_5)(CO)_2 \cdot CH_2Cl_2$	$Na[C_5H_5Mo(CO)_2P(C_6H_5)_3]$ and 1 equivalent of $C_5H_5Fe(CO)_2C(O)CH_2Cl$ were stirred in THF [63] at −78°C for 1 h followed by warming up to 25°C (35% by chromatography with CH_2Cl_2/hexane 1:4 to 1:3); $(C_5H_5Fe(CO)_2)_2$ was obtained in the first fraction [66]

Table 5 (continued)

No.	compound	method of preparation (yield) properties and remarks
		yellow [66] to orange [63] crystals ^{1}H NMR (C_6D_6, 27 °C): 3.25 (s, CH_2Fe), 4.19 (s, FeC_5H_5), 4.97 (d, MoC_5H_5; J(P, H) = 1.2), 6.97 to 7.15, 7.37 to 7.72 (m, C_6H_5) [66]; partial data in [63] ^{13}C NMR (C_6D_6, 27 °C): 32.23 (t, CH_2Fe; J(H, C) = 135), 85.07 (d, FeC_5H_5; J(H, C) = 178), 96.64 (d, MoC_5H_5; J(H, C) = 177), 130.13 to 136.97 (m's, C_6H_5), 217.81 (FeCO), 239.88 (d, MoCO; J(P, C) = 23), 262.82 (d, C=O; J(P, C) = 11) [66]; partial data in [63] IR (CH_2Cl_2): 1585 (ν(C=O)) [63], 1843, 1930 (ν(MoCO)), 1957, 2006 (ν(FeCO)) [66]
81	$C_5H_5Mo(CO)_2(P(C_6H_5)_3)C(O)CH_2CH_2CH_2W(C_5H_5)(CO)_3$	$[C_5H_5Mo(CO)_2(P(C_6H_5)_3)=C(CH_2)_3O\text{-cyclo}]Br$ was allowed to react with $Na[C_5H_5W(CO)_3]$ in THF for 19 h (62%) [57] yellow solid (from CH_2Cl_2), darkening > 127 °C and gas evolution > 135 °C [57] ^{1}H NMR (C_6D_6): 1.77 and 2.05 (m, $2CH_2$), 3.57 (t, $CH_2C(O)$; J = 7), 4.64 (WC_5H_5), 4.89 (d, MoC_5H_5; J(P, H) = 1), 7.01 (m, C_6H_5) [57] ^{13}C NMR (C_6D_6): −9.4 (CH_2W), 33.9 (central CH_2), 71.5 (**CH_2**C(O)), 91.5, 96.6 (C_5H_5), 128.6 to 136.6 (m's, C_6H_5), 217.5 (WCO), 230.1 (WCO trans), 239.2 (d, MoCO; J(P, C) = 24), 261.2 (C=O) [57] IR (CH_2Cl_2): 1610 (ν(C=O)), 1851, 1912, 1933, 2012 (ν(CO)) [57] mass spectrum: the molecular ion was not observed [57]

compounds of the type $^5LMo(CO)_2(^2D)C(O)R$ (R = aryl)

No.	compound	method of preparation (yield) properties and remarks
82	$C_5H_5Mo(CO)_2(P(C_6H_5)_3)C(O)C_6H_5$	II, in refluxing C_6H_6 with 3 equivalents of $P(C_6H_5)_3$ for 1 h (36% by chromatography on Al_2O_3 with petroleum ether/C_6H_6 8:2 as last fraction) [30]; IV [21] yellow crystalline solid (from $CHCl_3/CH_3OH$ 1:1) [21, 30] ^{1}H NMR ($CDCl_3$): 5.09 (d, C_5H_5; J(P, H) = 1.3), 7.16 to 7.53 (m, C_6H_5) [21] IR ($CHCl_3$): 1603 (ν(C=O)), 1860, 1943 (ν(CO)) [21] air-stable in the solid state but air-sensitive in solution [21]

References on pp. 75/7

Table 5 (continued)

No.	compound	method of preparation (yield) properties and remarks
82 (continued)		treatment with $Rh(P(C_6H_5)_3)_3Cl$ in CH_2Cl_2 for 8 h affords a 10% yield of $C_5H_5Mo(CO)_2(P(C_6H_5)_3)C_6H_5$ besides $Rh(P(C_6H_5)_3)_2(CO)Cl$ [21]
83	$C_5H_5Mo(CO)_2(P(OC_6H_5)_3)C(O)C_6H_5$	II, in refluxing C_6H_6 for 2 h (33% by chromatography on Al_2O_3 with $CHCl_3/CH_3OH$ 1:1 as last fraction) [30] ^1H NMR ($CDCl_3$): 4.87 (d, C_5H_5; J(P, H) = 0.9), 7.18 to 7.62 (m, C_6H_5) [30] IR ($CHCl_3$): 1590 (ν(C=O)), 1895, 1971 (ν(CO)) [30]
84	trans-$C_5H_5Mo(CO)_2(P(C_6H_5)_3)C(O)C_6H_4F$-4	IV (33%) [31] m.p. 142 to 143°C [31] ^1H NMR ($CDCl_3$): 5.12 (d, C_5H_5; J(P, H) = 1.4), 7.34 to 7.54 (m, C_6H_4 and C_6H_5) [31] air-stable in the solid state but air-sensitive in solution treatment with $Rh(P(C_6H_5)_3)_3Cl$ in C_6H_6 at 70°C for 4 h affords a 75% of $C_5H_5Mo(CO)_2(P(C_6H_5)_3)C_6H_4F$-p [31]
85	$C_5H_5Mo(CO)_2(P(C_6H_5)_3)C(O)C_6F_5$	$[C_5H_5Mo(CO)_3P(C_6H_5)_3]PF_6$ was allowed to react with 1.5 equivalents of LiC_6F_5 in THF for 1 h (14.5% by chromatography on Al_2O_3 with C_6H_6) [8] yellow crystalline solid, m.p. 115 to 120°C (dec.) [8] IR (Nujol): 1600 (ν(C=O)), 1865, 1890, 1953 (ν(CO)), other bands between 495 and 1295 also given [8] decomposes slowly when exposed to air and more quickly in solution [8]
compounds of the type $^5LMo(CO)_2(^2D)C(=NR)R'$		
86	$C_5H_5Mo(CO)_2(P(C_6H_5)_3)C(=NC_6H_5)CH_3$	V, in hexane (75%) [44] yellow solid, m.p. 128 to 130°C [44] ^1H NMR (THF-d_8): 2.30 (s, CH_3), 5.00 (d, C_5H_5; J(P, H) = 1.0), 7.0 to 7.5 (m, C_6H_5) [44] IR (THF): 1545 (C_6H_5 vibrational mode), 1575 (ν(C=N)), 1855, 1940 (ν(CO)) [44]

References on pp. 75/7

Table 5 (continued)

No.	compound	method of preparation (yield) properties and remarks
*87	$C_5H_5Mo(CO)_2(P(OCH_3)_3)C(=NC_6H_5)CH_3$	V, in THF (25% by chromatography on Al_2O_3 with C_6H_6/THF) [44] yellow solid, m.p. 82.5 to 83°C (from THF/pentane) [44] ^1H NMR (acetone-d_6): 2.24 (s, CH_3), 3.61 (d, OCH_3; J(P, H) = 5.9), 5.24 (d, C_5H_5; J(P, H) = 1.8), 6.5 to 7.2 (m, C_6H_5) [44] ^{13}C NMR (C_6D_6, 0.06 M $Cr(OC(CH_3)CHC(CH_3)O)_3$): 52.22 (d, CH_3; J(P, C) = 4.5), 94.61 (C_5H_5), 120.44 (d, OCH_3; J(P, C) = 156), 155.49 (C=N), 237.33 (d, CO; J(P, C) = 38.5) [47] IR (hexane): 1570 (ν(C=N)), 1880, 1895, 1955, 1970 (ν(CO)) [44]
	compounds of the type $^5LMo(CO)_2(^2D)C(O)NR_2$	
88	cis-$C_5H_5Mo(CO)_2(NH_3)C(O)NH_2$	VI, at 20°C [17]; similar to VI with $C_5H_5Mo(CO)_3I$ for 10 h at −40°C followed by 1 to 2 h at 20°C [46] IR (KBr): 1235 ($\delta_{sym}(NH_3)$ + $\varrho(NH_2)$), 1502, 1583, and 1610 (ν(C=N) + $\delta(NH_2)$ + $\delta_{asym}(NH_3)$ + ν(C=O)), 1830, 1935 (ν(CO)), other bands between 3100 and 3500 also given; (CH_2Cl_2): 1555, 1585 (ν(C=N) + $\delta(NH_2)$ + $\delta_{asym}(NH_3)$ + ν(C=O)), 1850, 1950 (ν(CO)) [46] well soluble in CH_2Cl_2, acetone, and THF, insoluble in liquid NH_3 between −80 and −60°C stable in the solid state at −20°C, but decomposes above 0°C and in solution even at low temperatures [46]
89	trans-$C_5H_5Mo(CO)_2(P(C_6H_5)_3)C(O)NH_2$	VI, for 24 h at −65°C followed by 5 min at −33°C (58%) [46] IR (KBr): 1215 ($\gamma(NH_2)$), 1432, 1480 (ν(CC)), 1560, 1605 (ν(C=O) + ν(C=N) + $\delta(NH_2)$), 1857, 1946 (ν(CO)); (CH_2Cl_2): 1555, 1590 (ν(C=O) + ν(C=N) + $\delta(NH_2)$), 1865, 1949 (ν(CO)) [46] only stable for a short time in the solid state at room temperature [46] reacts with HCl to $[C_5H_5Mo(CO)_3(P(C_6H_5)_3)]Cl$; with ICN in liquid NH_3 at −60°C to $C_5H_5Mo(CO)_2(P(C_6H_5)_3)NCO$ and $C_5H_5Mo(CO)_2(P(C_6H_5)_3)I$; with NH_4I in liquid NH_3 to $C_5H_5Mo(CO)_2(P(C_6H_5)_3)I$ [46]

References on pp. 75/7

Table 5 (continued)

No.	compound	method of preparation (yield) properties and remarks
	compounds of the type $^5LMo(CO)_2(^2D)C(O)OR$	
90	trans-$C_5H_5Mo(CO)_2(P(C_6H_5)_3)C(O)OH$	VII (96%) [59] yellow powder, m.p. 145 to 153°C (dec.) [59] ^1H NMR ($CDCl_3$): 5.08 (d, C_5H_5; J(P, H) = 1.3), 7.42 (m, C_6H_5) [59] ^{13}C NMR ($CDCl_3$, −7°C): 95.47 (C_5H_5), 128.89, 129.03, 130.91, and 133.23 (C_6H_5), 209.52 (d, C(O)OH; J(P, C) = 11), 238.49 (d, CO; J(P, C) = 26) [59] IR (Nujol): 1616 (ν(C=O)), 1862, 1955 (ν(CO)) [59] reaction with CF_3COOH produces the starting cation; does not decarboxylate upon treatment with $N(C_2H_5)_3$ in acetone at 25°C nor with NaOH in aqueous acetone; no spectral evidence was found for dissociation in polar solvents [59]
91	$C_5H_5Mo(CO)_2(P(C_6H_5)_3)C(O)OCH_3$	VII (ca. 56%) [58] glassy orange solid [58] ^1H NMR (acetone-d_6): 2.57 (CH_3, cis isomer), 3.48 (CH_3, trans isomer), 5.15 (d, C_5H_5, trans isomer; J(P, H) = 1.2), 5.34 (s, C_5H_5, cis isomer), 7.28 to 7.60 (m, C_6H_5) [59] ^{31}P NMR (acetone): 64.6 (cis isomer), 73.4 (trans isomer) [58] IR (CH_2Cl_2): 1610 (ν(C=O)), 1875, 1966 (ν(CO)) [58] thermally unstable [58]
92	trans-$HOCH_2CH_2C_5H_4Mo(CO)_2(P(C_6H_4CH_3\text{-}4)_3)C(O)OCH_3$	formation suggested by the reaction of $O{=}COCH_2CH_2C_5H_4Mo(CO)_2P(C_6H_4CH_3\text{-}4)_3$ (Formula IV on p. 46) with methanol [58]

*Further information:

cis-$(CH_3)_5C_5Mo(CO)_2(P(CH_3)_3)C(O)H$ (Table **5**, No. **1**). The cis/trans mixture was recrystallized from toluene/pentane at −80°C and the crystal structure of the cis isomer was determined at 128 K. It crystallizes in the monoclinic space group $P2_1/c-C_{2h}^5$ (No. 14) with the unit cell parameters a = 14.530 (8), b = 8.244 (12), c = 15.991 (8) Å, β = 94.28 (5)°; Z = 4 molecules per unit cell, and D_{calc} = 1.475 g/cm^3. The molecular structure with the main bond distances and angles is shown in **Fig. 12** [54].

$(CH_3)_5C_5Mo(CO)_2(P(C_6H_5)_3)C(O)H$ (Table **5**, No. **4**). The ^3J(P, H) coupling constant in the ^1H NMR spectrum of the trans isomer increases with temperature: 7.5 Hz at −80°C, 9.3 Hz at

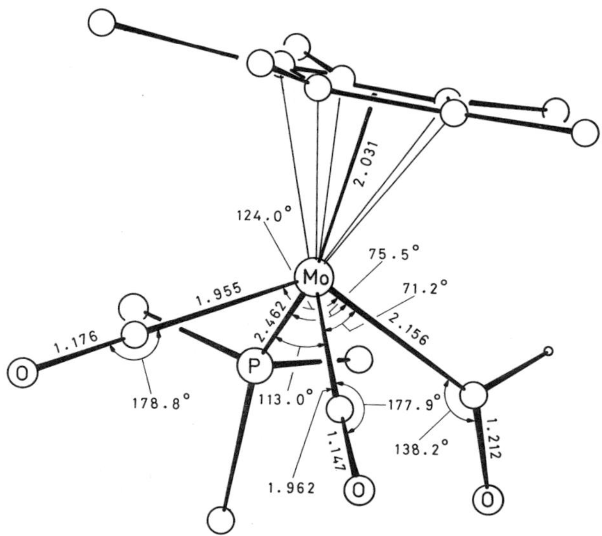

Fig. 12. The molecular structure of cis-$(CH_3)_5C_5Mo(CO)_2(P(CH_3)_3)C(O)H$ [54].

−50°C, 11.9 Hz at −20°C, and 13.3 Hz at 0°C. These changes were attributed to changes in the relative positions of substituent groups in rotamers whose movement about the Mo-C(O)H bond is slightly restricted in rotation. ^1H NMR NOE (Nuclear Overhauser Enhancement) effect reveals a 12% enhancement of the formyl proton by irradiation of the methyl signal which shows that the two groups are close together in the trans isomer [64].

Protonation with 2 equivalents of CF_3COOH in CD_2Cl_2 at −70°C affords $[(CH_3)_5C_5Mo(CO)_2(P(C_6H_5)_3)=C(OH)H]O_2CCF_3$. Methylation with CH_3SO_3F at −90°C in CH_2Cl_2 followed by anion exchange with NH_4PF_6 affords cis/trans-$[(CH_3)_5C_5Mo(CO)_2(P(C_6H_5)_3)=C(OCH_3)H]PF_6$ [54, 64]. Reaction with $[C(C_6H_5)_3]PF_6$ in CH_2Cl_2 at −80°C followed by warming to 25°C gives $[(CH_3)_5C_5Mo(CO)_3P(C_6H_5)_3]PF_6$ [54].

trans-$C_5H_5Mo(CO)_2(P(C_4H_9\text{-}n)_3)C(O)CH_3$ (Table **5**, No. **9**). Kinetic data of the formation in toluene and THF solutions are given in [4, 11, 15]. Decarbonylation is observed in the preparation in CH_3CN at 60°C; $C_5H_5Mo(CO)_2(P(C_4H_9\text{-}n)_3)CH_3$ was first detected after 4 h at a reactant concentration of the order of 0.25 M. The concentration of the decarbonylation product increases with time [12]. Extensive rate determination experiments with respect to the decarbonylation and the dissociation process in No. 9 and related complexes is given in [25, 35]. Decarbonylation activation parameters for the reaction in CH_3CN in the presence of a 5- to 10-fold excess of phosphane are given in the following table [35].

T (°C)	k (× 10^6 s^{-1})	
50.0	0.90	
60.2	3.30	$\Delta H^{\ddagger} = 30.3 \pm 1.0$ kcal/mol
70.2	14.4	$\Delta S^{\ddagger} = 7.4 \pm 3.0$ cal · mol^{-1} · K^{-1}

References on pp. 75/7

The equilibrium constant of the dissociation reaction $C_5H_5Mo(CO)_2(P(C_4H_9\text{-}n)_3)C(O)CH_3 \rightleftharpoons C_5H_5Mo(CO)_3CH_3 + P(C_4H_9\text{-}n)_3$ was obtained by heating a 0.01 M solution in CH_3CN for 1 to 2 h; $K_{equ} = 0.3 \times 10^{-4}$ ($\pm$ 10%) and 5% dissociation [35].

Reaction with HgX_2 (X = Cl, Br, or I) in $CHCl_3$ affords trans-$C_5H_5Mo(CO)_2(P(C_4H_9\text{-}n)_3)HgX$ [45].

trans-$C_5H_5Mo(CO)_2(P(C_6H_5)_3)C(O)CH_3$ (Table **5**, No. **13**) was prepared by treating $C_5H_5Mo(CO)_3CH_3$ with $P(C_6H_5)_3$ (Method II). The conditions are compiled in the following table.

solvent	conditions (yield)
none	at 120°C for 1 h [5]
THF or hexane	refluxing [2] for 3 h under irradiation (6%) or for 30 h using 1.5 equivalents of $P(C_6H_5)_3$ (34%) [3]
THF	at 60°C for 3 h with excess $P(C_6H_5)_3$ (70%) [4]; at 35°C [7]; with 10 equivalents of $P(C_6H_5)_3$ [15]
CH_3CN [12, 41]	for 2 to 3 h (75 to 85%) [25, 35], with two equivalents of $P(C_6H_5)_3$ (93%) [11]

$C_5H_5Mo(CO)_2(P(C_6H_5)_3)C(O)CH_3$ was also obtained by the reaction of trans-$C_5H_5Mo(CO)_2(P(C_6H_5)_3)CH_3$ under 2 atm CO pressure or by treating $Na[C_5H_5Mo(CO)_2(P(C_6H_5)_3)]$ with $(CH_3CO)_2O$ [13]. For the formation in THF at 35°C, no decarbonylation was observed under these conditions; a mechanism with a $C_5H_5Mo(CO)_2C(O)CH_3 \cdot$ solvate intermediate was discussed based on kinetic measurements [7]. A sideproduct of the preparation with Method II in hexane was 13 to 14% of $C_5H_5Mo(CO)_2(P(C_6H_5)_3)CH_3$, decreasing in abundance with decreasing time [3]. The effect of the reaction time on the product ratio for the reaction of $C_5H_5Mo(CO)_3CH_3$ with $P(C_6H_5)_3$ in refluxing THF was determined [3]. Decarbonylation in refluxing THF for 48 h afforded a 14.2% yield of $C_5H_5Mo(CO)_3CH_3$, 34.1% of $C_5H_5Mo(CO)_2(P(C_6H_5)_3)CH_3$, and 25.8% of $C_5H_5Mo(CO)_2(P(C_6H_5)_3)C(O)CH_3$ [2, 3]. Extensive rate determination experiments with respect to the decarbonylation and the dissociation process in No. 13 and related complexes are given in [25, 35]. Decarbonylation activation parameters for the reaction in the presence of a 5- to 10-fold excess of phosphane are given in the following table [25, 35]. Related data have been determined for CD_3CN solutions at 60 $\pm$ 2°C in sealed NMR tubes [12].

solvent	T (°C)	k ($\times 10^6$ s^{-1})	
C_6H_6, hexane, or THF	60.0	14.0	
CH_3CN	50.2	2.8	$\Delta H^{\ddagger} = 29.3 \pm 1.0$ kcal/mol
CH_3CN	60.0	13.22	$\Delta S^{\ddagger} = 4.2 \pm 3.0$ cal $\cdot$ mol^{-1} $\cdot$ K^{-1}
CH_3CN	60.4	13.5	
CH_3CN	70.0	34.1	

The equilibrium constant of the dissociation reaction $C_5H_5Mo(CO)_2(P(C_6H_5)_3)C(O)CH_3 \leftrightharpoons C_5H_5Mo(CO)_3CH_3 + P(C_6H_5)_3$ was obtained by heating a 0.01 M solution in CH_3CN for 1 to 2 h; $K_{equ} = 8.3 \times 10^{-4}$ ($\pm$ 10%) and 25% dissociation [35].

References on pp. 75/7

No. 13 crystallizes in the monoclinic space group $C2/c-C_{2h}^6$ (No. 15) with the unit cell parameters a = 26.13 (3), b = 11.91 (2), c = 17.81 (3) Å, β = 121.91 (3)°; Z = 8 molecules per unit cell, D_{meas} = 1.50 ± 0.04, and D_{calc} = 1.473 g/cm³. The molecular structure with the main bond distances and angles is given in **Fig. 13** [9, 10]. The Mo-C(O) bond is significantly shorter than an Mo-alkyl bond and this observation is interpreted as evidence for dπ-pπ back-donation; a bond order of ca. 1.12 is discussed [10].

Irradiation under decarbonylation affords trans-$C_5H_5Mo(CO)_2(P(C_6H_5)_3)CH_3$ [13]. Reaction with X_2 (X = Br, I) in $CHCl_3$ at 0°C leads to the formation of cis/trans-$C_5H_5Mo(CO)_2(P(C_6H_5)_3)X$ [39]. Reduction with B_2H_6 in C_6D_6 yields a mixture of $C_5H_5Mo(CO)_2(P(C_6H_5)_3)C_2H_5$ and $C_5H_5Mo(CO)_2(P(C_6H_5)_3)H$ [43]. The reaction with t-C_4H_9NC in C_6H_6 substitutes the phosphane ligand with formation of $C_5H_5Mo(CO)_2(CNC_4H_9\text{-t})C(O)CH_3$ [22]. Methylation by CH_3OSO_2F in CH_2Cl_2 followed by anion exchange with NH_4PF_6 affords $[C_5H_5Mo(CO)_2(P(C_6H_5)_3)=C(CH_3)OCH_3]PF_6$ [38, 40]. Treatment with HgX_2 (X = Cl, Br, or I) in $CHCl_3$ affords trans-$C_5H_5Mo(CO)_2(P(C_6H_5)_3)HgX$ [45]. No reaction is observed with HCl in C_6H_6 [5].

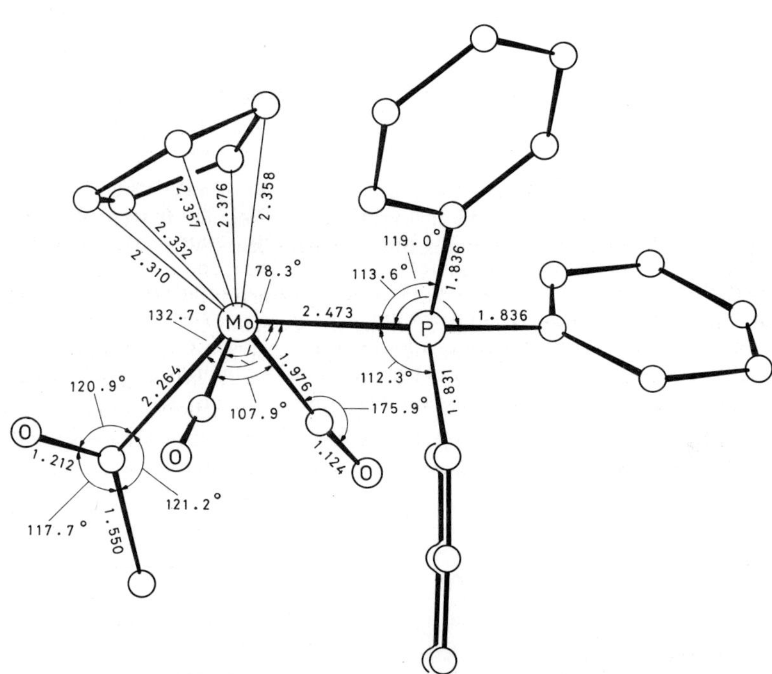

Fig. 13. The molecular structure of trans-$C_5H_5Mo(CO)_2(P(C_6H_5)_3)C(O)CH_3$ [10].

trans-$C_5H_5Mo(CO)_2(P(OR)_3)C(O)CH_3$ (Table **5**, Nos. **19**, **24**; R = CH_3, C_6H_5). The kinetics of the reaction of $C_5H_5Mo(CO)_3CH_3$ with $P(OR)_3$ in CD_3CN at 60 ± 2°C were studied by ^1H NMR spectroscopy. The time-dependent concentrations of $C_5H_5Mo(CO)_2(P(OR)_3)C(O)CH_3$ and its decarbonylation product, $C_5H_5Mo(CO)_2(P(OR)_3)CH_3$, are given [12]. Kinetic parameters of this reaction in CH_3CN at 30.3°C are also given in [11] and in toluene or THF in [4].

References on pp. 75/7

IV

trans-$C_5H_5Mo(CO)_2(P(C_6H_5)_3)C(O)(CH_2)_4X$ (Table **5**, Nos. **63, 64**; X = Br, I). The acyl complexes undergo slow equilibration with the trans isomer of the carbene cation (Formula IV) formed by internal attack of the acyl oxygen atom on the ω-carbon atom, displacing the halogenide. Rate constants (k), activation parameters, and equilibrium constants (K_{equ}) in ca. 0.02 M CH_3CN at 298 K are given in the following table [33].

X	k ($\times 10^5$ s^{-1})	$\Delta H^{\neq}$ (kcal/mol)	$\Delta S^{\neq}$ (e.u.)	ΔH°_{298} (kcal/mol)	ΔS°_{298} (e.u.)	ΔG°_{298} (kcal/mol)	K_{equ}
Br	5.05 ± 2.36	15.0 ± 2.1	−27.6 ± 7.6	−7.68 ± 0.24	−25.8 ± 0.8	0.02 ± 0.07	1.0 ± 0.25
I	15.1 ± 1.6	17.4 ± 2.0	−16.7 ± 6	~0 ± 2	~2.75 ± 0.31		4.0 ± 0.6

trans-$C_5H_5Mo(CO)_2(P(C_6H_5)_3)C(O)CH_2C_6H_5$ (Table **5**, No. **68**). Kinetic data for the formation in CH_3CN and dimethyl sulfoxide were given together with results of related compounds and analyzed in terms of Hammett's sigma substituent parameter [52].

Decarbonylation occurs upon melting at 140°C under vacuum to yield $C_5H_5Mo(CO)_2$-$(P(C_6H_5)_3)CH_2C_6H_5$ [14]. Decarbonylation occurs also in CH_3CN at 60°C with k = 4.7×10^{-6} s^{-1}. In dimethyl sulfoxide-d_6, the initial product is cis-$C_5H_5Mo(CO)_2(P(C_6H_5)_3)CH_2C_6H_5$, which isomerizes to the trans form. The results are given in the following table. Similar results were observed in dimethylformamide with only half the yield of the cis isomer [52].

reaction time (min)	conversion (%)	yield of cis product (%)
1.5	20	13
5	49	14
8	70	5
16	91	1

$(CH_3)_5C_5Mo(CO)_2(P(CH_3)_2C_6H_5)C(O)CH_2C_6H_5$ (Table **5**, No. **73**). The initially formed cis isomer isomerizes in CH_3CN at 30°C into the trans form, as monitored by ^{31}P NMR spectroscopy. After 4 min at 30% reaction the abundance of the cis isomer was 70%. At around 75% reaction (ca. 15 min) the cis and trans forms were present in comparable amounts. An equilibrium between cis and trans isomers with only 17% relative abundance of the cis form was reached after 2 h [60].

$C_5H_5Mo(CO)_2(P(OCH_3)_3)C(=NC_6H_5)CH_3$ (Table **5**, No. **87**) crystallizes in the triclinic space group $P\bar{1}-C_i^1$ (No. 2) with the unit cell parameters a = 8.475 (6), b = 10.571 (8), c =

References on pp. 75/7

11.879 (9) Å, α = 89.46 (5)°, β = 74.95 (5)°, γ = 82.60 (5)°; Z = 2 molecules per unit cell, and D_{calc} = 1.50 g/cm³ at −5°C. The molecular structure with the main bond distances and angles is given in **Fig. 14**. Two of the three methoxy groups of $P(OCH_3)_3$ are disordered; an ORTEP diagram showing this disorder was also given [47].

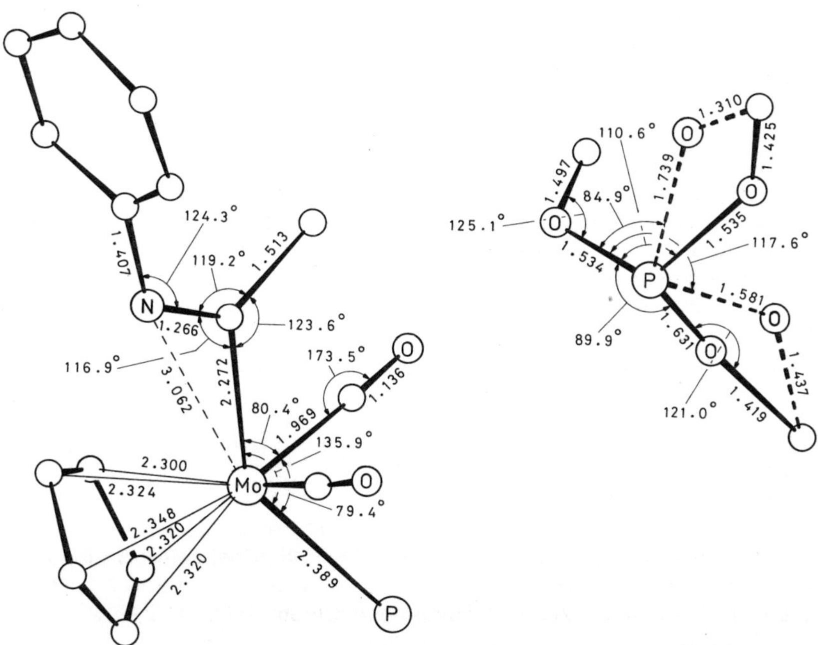

Fig. 14. The molecular structure of $C_5H_5Mo(CO)_2(P(OCH_3)_3)C(=NC_6H_5)CH_3$ [47].

References:

[1] Capron-Cotigny, G.; Poilblanc, R. (Compt. Rend. C **263** [1966] 885/7).
[2] Barnett, K. W. (Diss. Univ. Wisconsin 1967; Diss. Abstr. Intern. B **28** [1968] 3203).
[3] Barnett, K. W.; Treichel, P. M. (Inorg. Chem. **6** [1967] 294/9).
[4] Butler, I. S.; Basolo, F.; Pearson, R. G. (Inorg. Chem. **6** [1967] 2074/9).
[5] Green, M. L. H.; Hurley, C. R. (J. Organometal. Chem. **10** [1967] 188/90).
[6] Craig, P. J.; Green, M. (J. Chem. Soc. Chem. Commun. **1967** 246/7).
[7] Mawby, R. J.; Rowson, C. A. (3rd Intern. Symp. Organometal. Chem., München 1967, pp. 322/3).
[8] Treichel, P. M.; Shubkin, R. L. (Inorg. Chem. **6** [1967] 1328/34).
[9] Churchill, M. R. (Diss. Harvard Univ. 1968; Diss. Abstr. Intern. B **29** [1968] 1594).
[10] Churchill, M. R.; Fennessy, J. P. (Inorg. Chem. **7** [1968] 953/9).
[11] Craig, P. J.; Green, M. (J. Chem. Soc. A **1968** 1978/81).
[12] Barnett, K. W. (Inorg. Chem. **8** [1969] 2009/11).
[13] Bolton, E. S.; Denker, M.; Knox, G. R.; Robertson, C. G. (Chem. Ind. [London] **1969** 327/8).
[14] Craig, P. J.; Green, M. (J. Chem. Soc. A **1969** 157/60).
[15] Hart-Davies, A. J.; Mawby, R. J. (J. Chem. Soc. A **1969** 2403/7).
[16] Makarova, L. G.; Ustynyuk, N. A.; Polovyanyuk, I. V.; Vinogradova, V. N. (4th Intern. Conf. Organometal. Chem., Bristol, Engl., 1969, Abstr. K6).

[17] Behrens, H.; Krohlberger, H.; Lampe, R. J.; Langer, J.; Maertens, D.; Pässler, P. (Proc. 13th Intern. Conf. Coord. Chem., Crawcow-Zakopane 1970, pp. 339/41).
[18] Craig, P. J. (Can. J. Chem. **48** [1970] 3089/94).
[19] King, R. B.; Efraty, A. (Inorg. Chim. Acta **4** [1970] 319/23).
[20] Cotton, J. D.; Kimlin, H. A.; Markwell, R. D. (J. Organometal. Chem. **232** [1982] C 75/C 77).

[21] Nesmeyanov, A. N.; Makarova, L. G.; Ustynyuk, N. A. (J. Organometal. Chem. **23** [1970] 517/23).
[22] Yamamoto, Y.; Yamazaki, H. (Bull. Chem. Soc. Japan **43** [1970] 143/7).
[23] Green, M. L. H.; Mitchard, L. C.; Swanwick, M. G. (J. Chem. Soc. A **1971** 794/7).
[24] Haines, R. J.; Du Preez, A. L.; Mareis, I. L. (J. Organometal. Chem. **28** [1971] 97/104).
[25] Barnett, K. W.; Pollman, T. G.; Solomon, T. W. (J. Organometal. Chem. **36** [1972] C23/C26).
[26] Craig, P. J.; Edwards, J. (J. Organometal. Chem. **46** [1972] 335/7).
[27] King, R. B.; Kapoor, P. N. (Inorg. Chim. Acta **6** [1972] 391/94).
[28] King, R. B.; Zippim, W. C.; Ishaq, M. (Inorg. Chem. **11** [1972] 1361/70).
[29] Merour, J. Y.; Charrier, C.; Benaim, J.; Roustan, J. L. (J. Organometal. Chem. **39** [1972] 321/8).
[30] Nesmeyanov, A. N.; Makarova, L. G.; Ustynyuk, N. A.; Bogatyreva, L. V. (J. Organometal. Chem. **46** [1972] 105/8).

[31] Nesmeyanov, A. N.; Makarova, L. G.; Ustynyuk, N. A.; Kvasov, B. A.; Bogatyreva, L. V. (J. Organometal. Chem. **34** [1972] 185/93).
[32] Alexander, J. J.; Wojcicki, A. (Inorg. Chem. **12** [1973] 74/6).
[33] Cotton, F. A.; Lukehart, C. M. (J. Am. Chem. Soc. **95** [1973] 3552/64).
[34] Craig, P. J.; Edwards, J. (Abstr. Intern. Climax Conf. Chem. Uses Molybdenum, Reading, Engl., 1973, pp. 104/8).
[35] Barnett, K. W.; Pollmann, T. G. (J. Organometal. Chem. **69** [1974] 413/21).
[36] Craig, P. J.; Edwards, J. (J. Less-Common Metals **36** [1974] 193/202).
[37] Harris, A.; Rest, A. J. (J. Organometal. Chem. **78** [1974] C 29/C 30).
[38] Wagner, K. P. (Diss. Univ. Wisconsin 1974; Diss. Abstr. Intern. B **36** [1975] 3235).
[39] Beach, D. L.; Barnett, K. W. (J. Organometal. Chem. **97** [1975] C 27/C 30).
[40] Treichel, P. M.; Wagner, K. P. (J. Organometal. Chem. **88** [1975] 199/206).

[41] Gingell, A. C.; Harris, A.; Rest, A. J.; Turner, R. N. (J. Organometal. Chem. **121** [1976] 205/10).
[42] Lukehart, C. M.; Zeile, J. V. (J. Organometal. Chem. **105** [1976] 231/7).
[43] Van Doom, J. A.; Masters, C.; Volger, H. C. (J. Organometal. Chem. **105** [1976] 245/54).
[44] Adams, R. D.; Chodosh, D. F. (J. Am. Chem. Soc. **99** [1977] 6544/50).
[45] Beach, D. l.; Dattilo, M.; Barnett, K. W. (J. Organometal. Chem. **140** [1977] 47/54).
[46] Pfister, A.; Behrens, H.; Moll, M. (Z. Anorg. Allgem. Chem. **428** [1977] 53/60).
[47] Adams, R. D.; Chodosh, D. F. (Inorg. Chem. **17** [1978] 41/48).
[48] Todd, L. J.; Wilkinson, J. R.; Hickey, J. P.; Beach, D. L.; Barnett, K. W. (J. Organometal. Chem. **154** [1978] 151/7).
[49] Tam, W.; Wong, W.-K.; Gladysz, J. A. (J. Am. Chem. Soc. **101** [1979] 1589/91).
[50] Butts, S. B.; Strauss, S. H.; Holt, E. M.; Stimson, R. E.; Alcock, N. W.; Shriver, D. F. (J. Am. Chem. Soc. **102** [1980] 5093/100).

[51] Alt, H. G.; Eichner, M. E. (J. Organometal. Chem. **212** [1981] 397/403).
[52] Cotton, J. D.; Crisp, G. T.; Daly, V. A. (Inorg. Chim. Acta **47** [1981] 165/9).
[53] Malisch, W.; Blau, H.; Haaf, F. J. (Chem. Ber. **114** [1981] 2956/70).
[54] Asdar, A.; Lapinte, C.; Toupet, L. (Organometallics **8** [1989] 2708/17).

[55] Malisch, W.; Voran, S.; Grötsch, G. (Chemiedozententagung, Kaiserslautern, FRG, 1982, Abstr. B22).
[56] Tam, W.; Lin, G.-Y.; Gladysz, J. A. (Organometallics **1** [1982] 525/9).
[57] Bailey, N. A.; Chell, P. L.; Manuel, C. P.; Mukhopadhyay, A.; Rogers, D.; Tabbron, H. E.; Winter, M. J. (J. Chem. Soc. Dalton Trans. **1983** 2397/403).
[58] Coolbaugh, T. S.; Santarsiero, B. D.; Grubbs, R. H. (J. Am. Chem. Soc. **106** [1984] 6310/8).
[59] Gibson, D. H.; Owens, K.; Ong, T.-S. (J. Am. Chem. Soc. **106** [1984] 1125/27).
[60] Cotton, J. D.; Kimlin, H. A. (J. Organometal. Chem. **294** [1985] 213/17).
[61] Febray, J.; Casabianca, F.; Riess, J. G. (Inorg. Chem. **24** [1985] 3235/9).
[62] Gibson, D. H.; Ong, T.-S.; Owens, K.; Mandal, S.; Sattrich, W. E.; Franco, J. (12th Intern. Conf. Organometal. Chem., Vienna 1985, p. 467).
[63] Akita, M.; Kondoh, A.; Moro-oka, Y. (J. Chem. Soc. Chem. Commun. **1986** 1296/8).
[64] Asdar, A.; Lapinte, C. (J. Organometal. Chem. **327** [1987] C 33/C 36).
[65] Pannell, K. H.; Kapoor, R. N.; Wells, M.; Giasolli, T.; Parkanyi, L. (Organometallics **6** [1987] 663/7).
[66] Akita, M.; Kondoh, A.; Kawaharo, T.; Takugi, T.; Moro-oka, Y. (Organometallics **7** [1988] 366/74).
[67] Leoni, P.; Aquilini, E.; Pasquali, M.; Marchetti, F.; Sabat, M. (J. Chem. Soc. Dalton Trans. **1988** 329/33).
[68] Nolan, S. P.; de la Vega, R. L.; Mukerjce, S. L.; Hoff, C. D. (Inorg. Chem. **25** [1986] 1160/5).

1.5.1.3.2.1.8 Compounds of the Types $^5LMo(CO)_2(^2D)^1L$ and $[^5LMo(CO)_2(^2D)^1L]^+$

This section covers compounds of the type $^5LMo(CO)_2(^2D)^1L$ with a σ-bonded ^1L ligand and some cationic species, $[^5LMo(CO)_2(^2D)^1L]^+$, in which ^1L represents an ylide, $R_3P{=}CR^1R^2$, or $R_3P{=}C{=}CR^1R^2$, bonded to the Mo atom via the α-C carbon atom. Compounds with noncyclic ^1L ligands are listed first in increasing number of carbon atoms and functionality, followed by compounds with cyclic groups and the cationic complexes. Mostly mixtures of cis-trans isomers are obtained in which the trans isomers are more favored and chirality at molybdenum with two enantiomers is expected for the cis isomer; see also General Remarks.

$C_5H_5Mo(CO)_2(^2D)^1L$ complexes have been observed as side products during kinetic investigations on the reaction of $C_5H_5Mo(CO)_3{}^1L$ with P-donor ligands in THF solution at 50°C besides $C_5H_5Mo(CO)_2(^2D)C(O)R$, $(CO)_5Mo^2D$, and $C_5H_5C(O)R$ [82]. Phosphite-containing complexes, $C_5H_5Mo(CO)_2(P(OR)_3)^1L$, were obtained either by the reaction of $Na[C_5H_5Mo(CO)_2P(OR)_3]$ with organic halides [6] or by refluxing $(C_5H_5Mo(CO)_3)_2$ with $P(OR)_3$ in C_6H_6 (R transfer via formation of $P(O)(OR)_2$ side products) [17]. Compounds of the type $C_5H_5Mo(CO)_2(^2D)^1L$ (^1L = aryl, ^2D not specified) may be obtained by one of the following three methods: $Na[C_5H_5Mo(CO)_2{}^2D]$ was allowed to react with $[IR_2]X$ (Method IV), $C_5H_5Mo(CO)_2(^2D)I$ was treated with $Hg(^1L)_2$ (Method V), or $C_5H_5Mo(CO)_2(^2D)C(O)R$ was decarbonylated by $Rh(^2D)_3Cl$ [15]. The compounds listed in Table 6 were prepared in most cases by the following methods. Further information on preparation is given in the table. The complexes were isolated and purified in most cases by column chromatography on silica or alumina.

Method I: $C_5H_5Mo(CO)_3{}^1L$ was irradiated in the presence of the ^2D ligand.

Method II: $C_5H_5Mo(CO)_3{}^1L$ was refluxed with the ^2D ligand in an inert solvent.

Method III: The complexes $C_5H_5Mo(CO)_2(^2D)C(O)R$ were decarbonylated.

Method IV: $Na[^5LMo(CO)_2{}^2D]$ was treated with an excess of RX or $[IR_2]I$ in THF.

References on pp. 103/5

Method V: ($^5LMo(CO)_2{}^2D)_2Hg$ was allowed to react with an excess of ^1LX (X = halogenide) in THF [16].

Method VI: $C_5H_5Mo(CO)_2(^2D)CH_3$ was allowed to react with two equivalents of $(NC)_2C{=}C(CN)_2$ in CH_2Cl_2 at 5 to 10°C ($^2D = P(C_6H_5)_3$) or 25°C ($^2D = P(OC_6H_5)_3$) for 6 to 8 h. $C_5H_5Mo(CO)_2(^2D)N{=}C{=}C(CN)C(CN)_2CH_3$ was also formed in this reaction [40].

Method VII: $C_5H_5Mo(CO)_2(P(OR)_3)H$ (R = CH_3, C_2H_5) was allowed to react with $CH_3C{\equiv}CN(C_2H_5)_2$ in C_6D_6 [61].

General Remarks. On the NMR time scale, two distinct dynamic processes are observed which can be interpreted in terms of a cis-trans isomerization and a cis-cis racemization. As a most likely intermediate is considered a trigonal bipyramid with the ^5L ligand in one axial position and one of the other ligands, CO, ^1L, or ^2D, in the other axial position. If CO occupies this position, the intermediate is chiral but achiral in the other cases indicated by a plane of symmetry. From temperature-dependent NMR studies involving the compounds No. 10, 14, 53, and related compounds, a cis-cis racemization via the achiral intermediate was deduced with lower activation energies than the cis-trans isomerization. Thus, the cis-cis racemization via this process is more likely than the racemization via a cis-trans-cis isomerization; for free energies of activation, see the individual compounds in Table 6 [21]. The dependence of the cis-trans barrier mainly on the nature of ^1L, tending to increase in the order H, D, < CH_3, $CH_2C_6H_5$, and a conversion mechanism via a trigonal-bipyramidal intermediate is studied in [16].

Compounds of the type $C_5H_5Mo(CO)_2(^2D)^1L$ ($^2D = P(C_4H_9\text{-}n)_3$, $P(C_6H_5)_3$, $P(OCH_3)_3$, or $P(OC_6H_5)_3$; $^1L = CH_3$, $CH_2C_6H_5$, or CH_2OCH_3) are air-stable in the solid state for some time but they are noticeably less stable than the corresponding halides, $C_5H_5Mo(CO)_2(^2D)X$ [16].

$C_5H_5Mo(CO)_2(^2D)CH_3$ (^2D not specified) reacts with CO in a variety of solvents, predominantly by the route [25]:

$$C_5H_5Mo(CO)_2(^2D)CH_3 + CO \rightarrow C_5H_5Mo(CO)_3CH_3 + {}^2D \rightarrow C_5H_5Mo(CO)_2(^2D)C(O)CH_3.$$

The propargyl complexes, $C_5H_5Mo(CO)_2(^2D)CH_2C{\equiv}CR$ ($^2D = P(OCH_3)_3$, $P(OC_6H_5)_3$, or $P(C_6H_5)_3$; R = CH_3, C_6H_5), react with $4\text{-}CH_3C_6H_4S(O)_2NCO$ to form cycloaddition products [47].

Table 6
Compounds of the Types $^5LMo(CO)_2(^2D)^1L$ and $[^5LMo(CO)_2(^2D)^1L]^+$.
An asterisk indicates further information at the end of the table.
For explanations, abbreviations, and units see p. X.

No.	compound	method of preparation (yield) properties and remarks
compounds with $^1L = CH_3$		
1	$C_5H_5Mo(CO)_2(N_2)CH_3$	I, with ^{13}CO-enriched $C_5H_5Mo(CO)_3CH_3$ in an N_2 matrix at 12 K (not isolated) [59] IR (N_2 matrix at 12 K): 1913.7 A″, 1969.7 A′ ($\nu(^{12}CO)_2$); 1886.0, 1955.3 ($\nu(^{12}CO, ^{13}CO)$); 1871.4 A″, 1924.5 A′ ($\nu(^{13}CO)_2$); 2190.8 (ν(NN)); the ν(CO) vibrations are in good agreement with the calculated values; spectra given as diagrams [59]

Table 6 (continued)

No.	compound	method of preparation (yield) properties and remarks
		the k_i value of 43.3 N/m (calc. k_{cis} = 43.8, k_{trans} = 49.0) obtained for this complex suggests that it adopts a cis arrangement of the CO groups [59]
*2	$C_5H_5Mo(CO)_2(P(CH_3)_3)CH_3$	I, in pentane for ca. 20 min (23%) [46] yellow solid, m.p. 100 °C (from pentane at −78 °C) [46], yellow crystals, m.p. 102 °C (from C_6H_6/pentane at 0 °C) [60] ^{1}H NMR (acetone-d_6): cis isomer: −0.13 (d, $MoCH_3$; J(P, H) = 11.8), 1.39 (d, PCH_3; J(P, H) = 9.2), 5.16 (s, C_5H_5); trans isomer: 0.16 (d, $MoCH_3$; J(P, H) = 2.8), 1.57 (d, PCH_3; J(P, H) = 9.2), 4.96 (d, C_5H_5; J(P, H) = 1.6) [46]; (C_6D_6): cis isomer: 0.03 (d, $MoCH_3$; J(P, H) = 11.8), 0.93 (d, PCH_3; J(P, H) = 8.3), 4.73 (d, C_5H_5; J(P, H) = 0.4); trans isomer: 0.70 (d, $MoCH_3$; J(P, H) = 2.8), 1.05 (d, PCH_3; J(P, H) = 8.8), 4.61 (d, C_5H_5; J(P, H) = 1.7) [60] ^{31}P NMR (acetone-d_6): cis isomer: 23.4; trans isomer: 25.6 [46]; (C_6D_6): cis isomer: 18.2; trans isomer: 20.2 [60] IR (hexane): 1858, 1938, 1944 (ν(CO)) [46]; (C_6H_6): 1844, 1930 (ν(CO)) [60]
3	$C_9H_7Mo(CO)_2(P(CH_3)_3)CH_3$ (C_9H_7 = indenyl)	I, in pentane for ca. 20 min (17%) [46] orange solid, m.p. 57 °C (from pentane at −78 °C) [46] ^{1}H NMR (acetone-d_6): −0.89 (d, $MoCH_3$; J(P, H) = 11.4), 1.19 (d, CH_3P; J(P, H) = 8.8), 5.19 (t), 5.66 (d), and 7.11 (m, all C_9H_7); −0.20 (d, $MoCH_3$; J(P, H) = 3.3), 1.59 (d, CH_3P; J(P, H) = 9.6), 5.43 (t), 6.66 (d), and 7.11 (m, all C_9H_7); no assignment to a special isomer given [46] ^{31}P NMR (acetone-d_6): 19.77, 28.98 [46] IR (hexane): 1855, 1940 (ν(CO)) [46]
4	$C_5H_5Mo(CO)_2(P(C_4H_9\text{-}n)_3)CH_3$	II, in CH_3CN at 60 °C for 50 to 60 h; time-dependent concentration also given in [8]; II, in refluxing C_6H_6 [16]; III, in CH_3CN at 60 ± 2 °C [34]; V [16] light to dark yellow solid (from methylcyclohexane) [16] ^{1}H NMR ($CDCl_3$): trans isomer: 0.28 (d, $MoCH_3$; J(P, H) = 3), 4.90 (d, C_5H_5; J(P, H) = 1.5 to 2); cis isomer: 0.05 (d, $MoCH_3$; J(P, H) = 9.0), 5.20 (s, C_5H_5) [34]; similar data in [16]; (CD_3CN):

References on pp. 103/5

Table 6 (continued)

No.	compound	method of preparation (yield) properties and remarks
4 (continued)		trans isomer: 0.13 (d, $MoCH_3$; J(P, H) = 2.5), 0.9 and 1.5 (m, n-C_4H_9), 4.97 (d, C_5H_5; J(P, H) = 1.5) [8] cis to trans ratio 0.14 at 25°C (by ^{1}H NMR in $CDCl_3$) [16] IR (cyclohexane): 1869, 1944 (ν(CO)) [16]; similar data in [8] reaction with X_2 (X = Cl, Br, or I) in $CHCl_3$ at 0°C affords cis/trans-$C_5H_5Mo(CO)_2(P(C_4H_9\text{-}n)_3)X$, while with HgX_2 at 25°C cis-$C_5H_5Mo(CO)_2(P(C_4H_9\text{-}n)_3)X$ is obtained [44]
5	trans-$C_5H_5Mo(CO)_2(P(C_4H_9\text{-}i)_3)CH_3$	II, in CH_3CN at 60 ± 2°C [34] ^{1}H NMR ($CDCl_3$): 0.29 (d, $MoCH_3$; J(P, H) = 4), 4.92 (d, C_5H_5; J(P, H) = 1.5 to 2) [34]
6	trans-$C_5H_5Mo(CO)_2(P(C_6H_{11}\text{-cyclo})_3)CH_3$	II, in CH_3CN at 60 ± 2°C [34] ^{1}H NMR ($CDCl_3$): 0.14 (d, CH_3; J(P, H) = 5), 4.87 (d, C_5H_5; J(P, H) = 1.5 to 2) [34]
*7	$C_5H_5Mo(CO)_2(P(C_6H_5)_3)CH_3$	I, with 1 to 1.5 equivalents of $P(C_6H_5)_3$ in refluxing hexane for 3 to 30 h (13%); II, in THF for 48 h (34%) [1, 2], II, in CH_3CN at 60°C (not isolated, time-dependent concentrations given) [1, 8]; III, in CH_3CN at 60°C [34], III, in THF under UV irradiation for 1 h (low) [41] yellow [2, 9], bright yellow solid [16, 41, 77], m.p. 158°C (dec.) [2], 153°C [9], 155°C (dec., from $CH_3C_6H_{11}$-cyclo) [16] ^{1}H NMR: trans isomer: 0.39 (d, CH_3; J(P, H) = 2.5), 4.70 (s, C_5H_5), 7.2 to 7.5 (m, C_6H_5) [2, 34, 41]; 0.58 (d, CH_3; J(P, H) = 2.3), 4.70 (d, C_5H_5; J(P, H) = 1.4) in $CDCl_3$ [16]; 0.34 (d, CH_3; J(P, H) = 3.2), 4.70 (d, C_5H_5; J(P, H) = 1.6), 7 to 8 (m, C_6H_5) in CD_2Cl_2 [64]; 0.39 (d, CH_3; J(P, H) = 3), 4.45 (d, C_5H_5; J(P, H) = 1.5), 7.3 (m, C_6H_5) in CD_3CN [8]; cis isomer: −0.08 (d, CH_3; J(P, H) = 11.0), 5.03 (s, C_5H_5) in $CDCl_3$ [16] ^{13}C NMR ($CDCl_3$ or CD_2Cl_2): trans isomer: −19.2 (d, CH_3; J(P, C) = 9.6), 91.8 (C_5H_5), 235.7 (d, CO; J(P, C) = 23.0) [52] IR ($CHCl_3$): 1850, 1940 (ν(CO)) [2, 41], similar data in [8]; (cyclohexane): 1871, 1947 (ν(CO)) [16]
8	trans-$C_5H_5Mo(CO)_2(P(C_6H_4CH_3\text{-}4)_3)CH_3$	III, in CH_3CN at 60°C [34] ^{1}H NMR ($CDCl_3$): 0.31 (d, $MoCH_3$; J(P, H) = 6), 4.70 (d, C_5H_5; J(P, H) = 1.5 to 2) [34]

References on pp. 103/5

Table 6 (continued)

No.	compound	method of preparation (yield) properties and remarks
9	trans-$C_5H_5Mo(CO)_2(P(C_6H_4OCH_3\text{-}4)_3)CH_3$	III, in CH_3CN at 60°C [34] ^{1}H NMR ($CDCl_3$): 0.40 (d, $MoCH_3$; J(P, H) = 4), 4.73 (d, C_5H_5; J(P, H) = 1.5 to 2); time-dependent spectra of the formation given as diagrams [34]
10	$C_5H_5Mo(CO)_2(P(OCH_3)_2C_6H_5)CH_3$	II, probably cis to trans ratio 0.26 in $CDCl_3$ at 31°C [21] ^{1}H NMR ($CDCl_3$): cis isomer: −0.21 (d, $MoCH_3$; J(P, H) = 11.6), 3.42 (d, CH_3O; J(P, H) = 11.8), 3.46 (d, CH_3O; J(P, H) = 13.0), 5.05 (s, C_5H_5; J(P, H) < 0.2); trans isomer: 0.32 (d, $MoCH_3$; J(P, H) = 3.0), 3.62 (d, CH_3O; J(P, H) = 12.5), 4.84 (d, C_5H_5; J(P, H) = 1.6) [21] IR (cyclohexane): 1882, 1958 (ν(CO)) [21] kinetic data for the interconversion: $\Delta F^{\ddagger}$ = 19.5 (cis to cis) and 19.7 (cis to trans) kcal/mol at 80°C [21]
11	$C_5H_5Mo(CO)_2(P(OCH_3)_3)CH_3$	II, in CH_3CN at 60 ± 2°C (not isolated, only trans isomer) [8], II, in refluxing C_6H_6 with slight excess $P(C_6H_5)_3$ (product contains some impurities) [16]; IV, with CH_3I (81%) [6, 7]; V [16]; $(C_5H_5Mo(CO)_3)_2$ and an excess of $P(OCH_3)_3$ were refluxed in C_6H_6 (>50%) [17] yellow to pale yellow solid, m.p. 67°C (trans isomer) [16] cis to trans ratio 0.43 [7] or 0.17 [16] in $CDCl_3$ and 0.15 in toluene at 25°C [16] conductivity (acetone, 1 to 10 × 10^{-4} M): Λ = 0.4 $cm^2 \cdot \Omega^{-1} \cdot mol^{-1}$ [17] ^{1}H NMR ($CDCl_3$): mixture: 0.33 (d, $MoCH_3$; J(P, H) = 3), 3.55 and 3.65 (d, CH_3O; J(P, H) = 12), 5.10 (d, C_5H_5; J(P, H) = 1 to 2), 5.25 (s, C_5H_5) [7], related values in [17]; cis isomer: −0.01 (d, $MoCH_3$; J(P, H) = 9.7), 5.19 (s, C_5H_5); trans isomer: 0.31 (d, $MoCH_3$; J(P, H) = 3.2), 5.04 (d, C_5H_5; J(P, H) = 1.2) [16]; (CD_3CN): 0.22 (d, $MoCH_3$; J(P, H) = 3), 3.55 (d, CH_3O; J(P, H) = 10), 5.12 (d, C_5H_5; J(P, H) = 1.5) [8] IR ($CHCl_3$): 1867, 1950 (ν(CO)) [7]; similar data in [8] CH_2Cl_2 [17] and cyclohexane [16]
*12	$C_5H_5Mo(CO)_2(P(OC_6H_5)_3)CH_3$	I, in toluene [64]; II, in CD_3CN at 60 ± 2°C (not isolated) [8] or in refluxing C_6H_6 with a slight excess of $P(OC_6H_5)_3$ (product contains some

References on pp. 103/5

Table 6 (continued)

No.	compound	method of preparation (yield) properties and remarks
*12 (continued)		impurities) [16]; IV, with CH_3I (35%) [5, 8]; V [16] yellow [5], light to dark yellow solid (from methylcyclohexane) [16], m.p. 99 to 100°C [5] cis to trans ratio 0.59 [16], 1.22 [5] in $CDCl_3$ [16], 0.63 in toluene at 25°C [11, 16]; $\Delta H = -0.14$ kcal/mol [16] ^1H NMR: cis isomer: 0.27 (d, $MoCH_3$; J(P, H) = 10.3), 4.72 (s, C_5H_5), 7.26, 7.30, and 7.34 (m, C_6H_5) [16] in $CDCl_3$; 0.74 (d, $MoCH_3$; J(P, H) = 10.3), 4.68 (s, C_5H_5; J(P, H) < 0.1), 7.2 (m, C_6H_5) in toluene-d_8 [11, 32, 36]; trans isomer: 0.29 (d, $MoCH_3$; J(P, H) = 2.6), 4.53 (d, C_5H_5; J(P, H) = 1.2), 7.26, 7.30, and 7.34 (m, C_6H_5) in $CDCl_3$ [16]; 0.77 (d, $MoCH_3$; J(P, H) = 2.6), 4.56 (d, C_5H_5; J(P, H) = 1.3), 7.2 (m, C_6H_5) in toluene-d_8 [11, 32, 36]; 0.27 (d, $MoCH_3$; J(P, H) = 3), 4.60 (d, C_5H_5; J(P, H) = 1.5), 7.3 (m, C_6H_5) in CD_3CN [8] the ^1H NMR spectrum in $CDCl_3$ was given in [5] with a reverse assignment; the isomer ratio is stable for up to 48 h at 60 ± 2°C in CD_3CN [8]; at 140°C in toluene, the isomers equilibrate: 0.67 (d, $MoCH_3$; J(P, H) = 5.6), 4.81 (d, C_5H_5; J(P, H) = 0.7), 7.2 (m, C_6H_5) [11, 32, 36] IR (CH_2Cl_2): given from 723 to 3115; 1886, 1968 (ν(CO)) [5]; similar ν(CO) in cyclohexane [8, 16] and i-octane [64], also given as a diagram in [64]
13	$C_5H_5Mo(CO)_2(PF_3)CH_3$	II, in refluxing toluene for 19 h with five equivalents of $Ni(PF_3)_4$, (24% by chromatography in Florisil with hexane) [28] yellow waxy solid, sublimes at 25°C/0.05 Torr [28] IR (Nujol): 818, 828, 842, 857, 886 (ν(PF)), 1013; (pentane or hexane): 1938, 1997 (ν(CO)) [28] air-sensitive in the solid state and in solution; decomposition occurs upon prolonged standing even under vacuum [28]
14	$C_5H_5Mo(CO)_2(P(CH_3)_2C_6H_5)CH_3$	II, in CH_3CN at 60 ± 2°C [34] cis to trans ratio 0.17 in $CDCl_3$ at 31°C [21] ^1H NMR ($CDCl_3$): cis isomer: −0.18 (d, $MoCH_3$; J(P, H) = 11.8), 1.65 (d, CH_3P; J(P, H) = 8.0), 1.68 (d, CH_3P; J(P, H) = 8.2), 5.08 (s, C_5H_5; J(P, H) < 0.3); trans isomer: 0.30 (d, $MoCH_3$; J(P, H) = 2.8), 1.82 (d, PCH_3; J(P, H) = 8.5), 4.71

Table 6 (continued)

No.	compound	method of preparation (yield) properties and remarks
		(d, C_5H_5; J(P, H) = 1.4) [21]; similar data in [34] IR (cyclohexane): 1860, 1941, 1949 (ν(CO)) [21] kinetic data for the interconversion: $\Delta F^{\neq}$ = 19.9 (cis to cis) at 90°C and 20.7 (cis to trans) kcal/mol at 100°C [21] reaction with $CH_3S(O)_2NSO$ in CH_2Cl_2 affords the S-bonded $C_5H_5Mo(CO)_2(P(CH_3)_2C_6H_5)S(O)$-$(CH_3)$=N-$S(O)_2CH_3$; with $(CH_3S(O)_2N)_2S$, the N-bonded $C_5H_5Mo(CO)_2(P(CH_3)_2C_6H_5)N$-$(S(O)_2CH_3)S(CH_3)$=$NS(O)_2CH_3$ is formed [57, 81]
15	$C_5H_5Mo(CO)_2(P(CH_3)(CH_2C_6H_5)C_6H_5)CH_3$	$[P(CH_3)_4][C_5H_5Mo(CO)_2(P(CH_3)CH_2C_6H_5)C_6H_5]$ was allowed to react with CH_3I [76]
16	$C_5H_5Mo(CO)_2(P(C_6H_5)_2CH_3)CH_3$	II, in CH_3CN at 60 ± 2°C [34]; also obtained by the alkylation of $Na_2[C_5H_5Mo(CO)_2P(C_6H_5)_2]$ with $(CH_3O)_2SO_2$ in hexane for 2 h (41% by medium pressure liquid chromatography with hexane/ether 4:1) [72] yellow oil [72] ^{1}H NMR ($CDCl_3$): 0.35 (d, trans-$MoCH_3$; J(P, H) = 5), 4.71 (d, trans-C_5H_5; J(P, H) = 1.5 to 2) [34]; (C_6D_6): 0.81 (d, $MoCH_3$; J(P, H) = 2.6), 1.75 (d, cis-CH_3P; J(P, H) = 7), 1.84 (d, trans-CH_3P; J(P, H) = 7), 4.45 (d, trans-C_5H_5; J(P, H) = 1.5), 4.71 (d, cis-C_5H_5; J(P, H) = 0.7), 7.02 to 7.52 (m, C_6H_5) [72] ^{31}P NMR ($CHCl_3$): 16.1 [72] IR (hexane): 1874, 1949 (ν(CO)) [72] mass spectrum (70 eV): $[M]^+$; other fragments also given [72]
17	trans-$C_5H_5Mo(CO)_2(P(C_6H_5)_2C_3H_7$-i)$CH_3$	II, in CH_3CN at 60 ± 2°C [34] ^{1}H NMR ($CDCl_3$): 0.40 (d, $MoCH_3$; J(P, H) = 4), 4.52 (d, C_5H_5; J(P, H) = 1.5 to 2) [34]
18	$C_5H_5Mo(CO)_2(P(C_6H_5)_2NHC(C_6H_5)(CH_3)H)CH_3$	$[P(CH_3)_4][C_5H_5Mo(CO)_2P(C_6H_5)_2NH$-$C(C_6H_5)(CH_3)H]$ was allowed to react with CH_3I [76]

References on pp. 103/5

Table 6 (continued)

No.	compound	method of preparation (yield) properties and remarks
19	$C_5H_5Mo(CO)_2(P(CH_3)_2C_2H_4P(CH_3)_2)CH_3$ (only one P is coordinated)	$C_5H_5Mo(CO)_3CH_2C(O)OCH_3$ was allowed to react with $(CH_3)_2PC_2H_4P(CH_3)_2$ in CH_3CN (low yield) [13] yellow solid [13] IR: 1882, 1966 (ν(CO)) [13]
20	$C_5H_5Mo(CO)_2(P(C_6H_5)_2CH{=}CHP(C_6H_5)_2$-cis$)CH_3$ (only one P is coordinated)	II, in CH_3CN [6]
21	$C_5H_5Mo(CO)_2(As(C_6H_5)_3)CH_3$	III, with irradiation in THF for 1 h (low yields) [41] bright yellow crystals, m.p. 143°C [41] ^1H NMR ($CDCl_3$): 0.41 (CH_3), 4.79 (C_5H_5), 7.2 to 7.6 (m, C_6H_5) [41] IR ($CHCl_3$): 1848, 1938 (ν(CO)) [41]
22	$C_5H_5Mo(CO)_2(C_4H_8O)CH_3$ (C_4H_8O = THF)	I, in a polyvinyl chloride matrix at 12 K with λ = 330 to 390 nm followed by warming in the presence of THF (not isolated) [70] IR (PVC, 12 K): 1820, 1920 (ν(CO)) [70]
compounds with $^1L = C_2H_5$		
23	$C_5H_5Mo(CO)_2(P(CH_3)_3)C_2H_5$	I, with a large excess of $P(CH_3)_3$ in pentane for 90 min (low yields) [58] ^1H NMR (acetone-d_6): 1.30 (m, C_2H_5), 1.55 (d, CH_3P; J(P, H) = 9.0), 5.00 (d, C_5H_5; J(P, H) = 1.9) [58] ^{13}C NMR (acetone-d_6): 15.9 and 19.0 (CH_3P; two singlets or one doublet?), 84.6 (C_5H_5), 226.3 (d, CO; J(P, C) = 22.2) [58] IR (pentane): 1852, 1928 (ν(CO)) [58]
24	$C_5H_5Mo(CO)_2(P(C_6H_5)_3)C_2H_5$	IV, with C_2H_5Br (40%) [18]; formed by the reaction of $C_5H_5Mo(CO)_2(P(C_6H_5)_3)C(O)CH_3$ with B_2H_6 in C_6D_6 (indicated by NMR) [43] yellow crystalline solid, m.p. 140°C (from heptane/CH_2Cl_2) [18] ^1H NMR ($CDCl_3$): 1.47 to 1.69 (m, C_2H_5), 4.71 (d, C_5H_5; J(P, H) = 1.4), 7.27 to 7.53 (m, C_6H_5) [18]; (C_6D_6): 1.96 (br m, C_2H_5), 4.51 (d, C_5H_5; J(P, H) = 1.4) [43] IR ($CHCl_3$): 1850, 1935 (ν(CO)) [18] air-stable as a solid, air-sensitive in solution [18] hydride abstraction with $[C(C_6H_5)_3]AsF_6$ affords $[C_5H_5Mo(CO)_2(P(C_6H_5)_3)CH_2{=}CH_2]AsF_6$ [64]

References on pp. 103/5

Table 6 (continued)

No.	compound	method of preparation (yield) properties and remarks
25	$C_5H_5Mo(CO)_2(P(OC_2H_5)_3)C_2H_5$	$(C_5H_5Mo(CO)_3)_2$ was allowed to react with an excess of $P(OC_2H_5)_3$ in refluxing C_6H_6 (> 50%) [17] yellow oil [17]
26	$C_5H_5Mo(CO)_2(P(OC_6H_5)_3)C_2H_5$	I, in toluene-d_8 (not isolated) [64]; IV with C_2H_5I (23%) [5] yellow solid, m.p. 92°C [5] cis to trans ratio 7:3 [5] ^{1}H NMR ($CDCl_3$): 1.2 to 1.7 (m, C_2H_5), 4.51 (d, cis-C_5H_5; J(P, H) = 1.2), 4.67 (s, trans-C_5H_5), 7.30 (m, C_6H_5) [5]; (toluene-d_8): 1.46 and 1.56 (m, MoC_2H_5), 4.37 (d, C_5H_5; J(P, H) = 1) [64] IR (CH_2Cl_2): 1880, 1961 (ν(CO)) [5]; similar data in toluene-d_8 [64]
27	$C_5H_5Mo(CO)_2(C_4H_8O)C_2H_5$ (C_4H_8O = THF)	I, in a THF-containing PVC matrix at 12 K with λ = 330 to 390 nm (not isolated) [70] IR (PVC, 12 K): 1818 (ν(CO)), other band is hidden; also given as a diagram [70]
	compounds with other alkyl groups	
28	trans-$C_5H_5Mo(CO)_2(P(C_6H_5)_3)CH_2CH_2CH_2Br$	IV, with $Br(CH_2)_3Br$ for 22 h, extraction with CH_2Cl_2 (60%) [68] yellow, air-stable crystals, m.p. 122 to 123°C (from CH_2Cl_2/light petroleum ether) [68] ^{1}H NMR ($CDCl_3$): 1.47 and 2.24 (m, 2 CH_2), 3.40 (t, $BrCH_2$; J(H, H) = 7), 4.76 (s, C_5H_5), 7.40 (m, C_6H_5) [68] ^{13}C NMR ($CDCl_3$): 0.1 ($MoCH_2$), 37.6 (CH_2), 39.6 (CH_2), 92.7 (C_5H_5), 128.2 (d, C_6H_5, C-4; J(P, C) = 10), 129.9 (C_6H_5, C-3), 133.1 (d, C_6H_5, C-2; J(P, C) = 9), 136.7 (d, C_6H_5, C-1; J(P, C) = 101), 238.0 (d, CO; J(P, C) = 23) [68] IR (THF): 1846, 1923 (ν(CO)) [68] reduction with 3 equivalents of $Li[(C_2H_5)_3BH]$ in THF for 4 h affords $C_5H_5Mo(CO)(P(C_6H_5)_3)$-($\eta^3$-$C_4H_7$) [78]
29	trans-$C_5H_5Mo(CO)_2(P(OCH_3)_3)CH_2CH_2CH_2Br$	IV, with $Br(CH_2)_3Br$ for 19 h, extraction with light petroleum and chromatography on Al_2O_3 with CH_2Cl_2 (53%) [68] yellow crystals, m.p. 40 to 41°C (from CH_2Cl_2/light petroleum ether) [68]

References on pp. 103/5

Table 6 (continued)

No.	compound	method of preparation (yield) properties and remarks
29 (continued)		^{1}H ($CDCl_3$): 1.38 and 2.15 (m, CH_2), 3.37 (t, CH_2Br; J(H, H) = 7), 3.60 (d, CH_3O; J(P, H) = 11), 5.05 (s, C_5H_5) [68] IR (light petroleum ether): 1879, 1952 (ν(CO)) [68]
30	$C_5H_5Mo(CO)_2(P(OC_4H_9\text{-n})_3)C_4H_9\text{-n}$	$(C_5H_5Mo(CO)_3)_2$ was refluxed in C_6H_6 with an excess of $P(OC_4H_9)_3$ (>50%) [17] yellow oil [17] IR (CH_2Cl_2): 1853, 1937 (ν(CO)) [17]
31	$C_5H_5Mo(CO)_2(P(OC_6H_5)_3)CH_2CH{=}CH_2$	IV, with $CH_2{=}CHCH_2Cl$ (38%) [5] yellow solid, m.p. 97 to 98°C (from CH_2Cl_2/hexane) [5] cis to trans ratio 4:1 [5] ^{1}H NMR ($CDCl_3$): 2.07 and 2.15 (d, CH_2; J(H, H) = 3.2), 4.37, 4.53, and 4.72 ($CH_2{=}$), 4.59 (d, cis-C_5H_5; J(P, H) = 1.4), 4.63 (s, trans-C_5H_5), 6.02 (5 lines, CH=), 7.24, 7.31, and 7.34 (m, C_6H_5) [5] IR (CH_2Cl_2): 1883, 1962 (ν(CO)) [5]
*32	$C_5H_5Mo(CO)_2(P(OC_6H_5)_3)CH_2CH{=}CHCH_3$	IV, with $ClCH_2CH{=}CHCH_3$ for 8 to 12 h (moderate to high yields); four isomers present, see "Further information" [51] yellow to orange oil; no attempts were made to crystallize it [51] ^{1}H NMR ($CDCl_3$): 1.65 (d, CH_3; J(H, H) = 6), 2.22 (d, CH_2; J(H, H) = 7), 4.55 (d, C_5H_5; J(P, H) = 1.5), 4.62 (d, C_5H_5; J(P, H) = 1.5), 4.68, 4.78 (s's, C_5H_5), 4.97 to 5.62 (m, CH=CH), 7.55 (m, C_6H_5); the two C_5H_5 doublets are assigned to the two trans (at Mo) isomers [51] IR (CH_2Cl_2): 1880, 1950 (ν(CO)) [51]
*33	$C_5H_5Mo(CO)_2(P(OC_6H_5)_3)CH_2CH{=}C(CH_3)_2$	IV, with $ClCH_2CH{=}C(CH_3)_2$ for 8 to 12 h (moderate to high yields) [51] yellow to orange oil; no attempts were made to crystallize it [51] ^{1}H NMR ($CDCl_3$): 1.53, 1.60, 1.66, and 1.73 (s, CH_3), 2.27 (d of d, CH_2; J(H, H) = 9, J(P, H) = 1.5), 4.62 (d, trans-C_5H_5; J(P, H) = 1.5), 4.80 (s, cis-C_5H_5), 5.66 (t, CH; J(H, H) = 9), 7.55 (m, C_6H_5) [51] IR (CH_2Cl_2): 1877, 1955 (ν(CO)) [51]

References on pp. 103/5

Table 6 (continued)

No.	compound	method of preparation (yield) properties and remarks
*34	$C_5H_5Mo(CO)_2(P(OC_6H_5)_3)CH_2CH{=}CHC_6H_5$	IV, with $ClCH_2CH{=}CHC_6H_5$ for 8 to 12 h (moderate to high yields) [51] yellow to orange oil [51] ^1H NMR ($CDCl_3$): 2.43 (d, CH_2; J(P, H) = 1.5), 4.63 (d, C_5H_5; J(P, H) = 1.5), 4.77 (s, C_5H_5), 6.62 to 7.30 (m, CH=CH), 7.55 (m, C_6H_5); for assignment to cis/trans isomers, see No. 33 [51] IR (CH_2Cl_2): 1878, 1955 (ν(CO)) [51]
35	trans-$C_5H_5Mo(CO)_2(P(C_2H_5)_3)C_5H_7$ (^1L = ^1H_2C–C(H^2)=C(H^3)–C(H^4)=C(H^5)(H^6))	II, with 1.3 equivalents of $P(C_2H_5)_3$ in ether for 4 h (91%) [62] pure yellow-green oil [62] ^1H NMR (C_6D_6): 0.73 (m, CH_3), 1.22 (m, PCH_2), 2.49 (d, H-1), 4.48 (C_5H_5), 4.90 (d of d, H-6), 5.15 (d of d, H-5), 6.21 (d of d, H-3), 6.58 (complex m, H-2, 4); J(H-1, 2) = 7.6, J(H-2, 3) = 16, J(H-3, 4) = 10.4, J(H-4, 5) = 17.4, J(H-4, 6) = 10.4, J(H-5, 6) = 1.5 [62] ^{13}C NMR (C_6D_6): 6.1 (d, $MoCH_2$; J(P, C) = 9.6), 7.9 (d, CH_3; J(P, C) = 8), 22.7 (d, PCH_2; J(P, C) = 26), 91.9 (C_5H_5), 110.7 ($CH_2{=}$), 123.2 and 137.4 (C-2, 3), 146.8 (C-4), 238.4 (d, CO; J(P, C) = 24) [62] IR (pentane): 1617 (ν(C=C)), 1849, 1930 (ν(CO)) [62] mass spectrum (12 eV): $[M-n\,CO]^+$ (n = 0 to 2) and other fragment ions [62] photolysis in ether at −20°C for 18 h affords $C_5H_5Mo(CO)(P(C_2H_5)_3)(\eta^3\text{-}C_5H_7)$ [62]
36	trans-$C_5H_5Mo(CO)_2(P(CH_3)_2C_6H_5)C_5H_7$ (see No. 35 for numbering)	II, with 1.3 equivalents of $P(CH_3)_2C_6H_5$ in ether for 4 h (ca. 85%) [62] pure yellow-green oil [62] ^1H NMR (C_6D_6): 1.37 (d, CH_3; J(P, H) = 8.8), 2.53 (d of d, H-1), 4.35 (C_5H_5), 4.88 (d of d, H-6), 5.12 (d of d, H-5), 6.15 (d of d, H-3), 6.48 to 6.56 (complex m, H-2, 4), 7.01 and 7.28 (m, C_6H_5); J(P, H-1) = 3.2, J(H-1, 2) = 8.2, J(H-2, 3) = 15.2, J(H-3, 4) = 10.2, J(H-4, 5) = 17.3, J(H-4, 6) = 10.2, J(H-5, 6) = 1.8 [62] ^{13}C NMR ($CDCl_3$): 5.6 (d, $MoCH_2$; J(P, C) = 9), 20.3 (d, CH_3; J(P, C) = 32), 91.2 (C_5H_5), 111.6 ($CH_2{=}$), 129.2, 130.4, and 130.9 (C_6H_5), 236.8 (d, CO; J(P, C) = 23) [62] IR (Nujol): 1622 (ν(C=C)), 1845, 1938 (ν(CO)) [62]

References on pp. 103/5

Table 6 (continued)

No.	compound	method of preparation (yield) properties and remarks
36 (continued)		mass spectrum (12 eV): $[M-nCO]^+$ (n = 0 to 2) and other fragment ions [62] photolysis in ether at −20°C for 18 h affords $C_5H_5Mo(CO)(P(CH_3)_2C_6H_5)(\eta^3\text{-}C_5H_7)$ [62]
37	trans-$C_5H_5Mo(CO)_2(P(C_6H_5)_3)CH_2C{\equiv}CH$	IV, with $BrCH_2C{\equiv}CH$ at −78°C for 0.5 h and allowing the reaction mixture to warm up (52%) [35, 48] yellow, air-sensitive crystals, m.p. 115°C (from hexane) [48] ^1H NMR ($CDCl_3$): 1.82 (t, CH_2; J(P, H) = 2.8), 2.03 (t, CH; J(P, H) = 2.8), 4.83 (d, C_5H_5; J(P, H) = 1.4) [35, 48] IR (CH_2Cl_2): 1865, 1945 (ν(CO)) [35, 48] reaction with an excess of HX (X = CH_3O, t-C_4H_9O, or C_6H_5S) in THF affords $C_5H_5Mo(CO)(P(C_6H_5)_3)(\eta^3\text{-}CH_2C(X)CH_2)$ [35, 48]
38	$C_5H_5Mo(CO)_2(P(OCH_3)_3)CH_2C{\equiv}CH$	IV, with $BrCH_2C{\equiv}CH$, similar to No. 37 (34%) [48] yellow, air-sensitive oil [48] IR (THF): 1875, 1955 (ν(CO)) [48]
39	$C_5H_5Mo(CO)_2(P(OC_6H_5)_3)CH_2C{\equiv}CH$	IV, with $BrCH_2C{\equiv}CH$, similar to No. 37 (34%) [48] yellow, air-sensitive oil [48] IR (CH_2Cl_2): 1890, 1970 (ν(CO)) [48]
40	trans-$C_5H_5Mo(CO)_2(P(C_6H_5)_3)CH(CH_3)C{\equiv}CH$	IV, with $BrCH(CH_3)C{\equiv}CH$, similar to No. 37 [35] IR (CH_2Cl_2): 1860, 1940 (ν(CO)) [35] the unstable complex decomposes to give $(C_5H_5Mo(CO)_2P(C_6H_5)_3)_2$ [35]
41	trans-$C_5H_5Mo(CO)_2(P(C_6H_5)_3)CH_2C{\equiv}CCH_3$	IV, with $BrCH_2C{\equiv}CCH_3$, similar to No. 37 (53%) [35, 48] yellow, air-sensitive crystals, m.p. 156°C (from hexane) [48] ^1H NMR ($CDCl_3$): 1.70 (m, CH_2 and CH_3), 4.65 (d, C_5H_5; J(P, H) = 1.4) [35, 48] IR (CH_2Cl_2): 1860, 1940 (ν(CO)) [35, 48] reaction with excess ethanol in THF yields $C_5H_5Mo(CO)(P(C_6H_5)_3)(\eta^3\text{-}CH_2C(C(O)OC_2H_5)\text{-}CHCH_3)$ [35, 48]

Table 6 (continued)

No.	compound	method of preparation (yield) properties and remarks
42	$C_5H_5Mo(CO)_2(P(OCH_3)_3)CH_2C{\equiv}CCH_3$	IV, with $BrCH_2C{\equiv}CCH_3$, similar to No. 37 (52%) [48] yellow, air-sensitive solid, m.p. 74°C (from hexane) cis to trans ratio 0.22 [48] ^1H NMR ($CDCl_3$): 1.82 (m, CH_2 and CH_3), 5.05 (d, trans-C_5H_5; J(P, H) = 1.5), 5.20 (s, cis-C_5H_5) [48] IR (CH_2Cl_2): 1875, 1945 (ν(CO)) [48]
*43	$C_5H_5Mo(CO)_2(P(OC_6H_5)_3)CH_2C{\equiv}CCH_3$	IV, with $BrCH_2C{\equiv}CCH_3$, similar to No. 37 (60%) [23], (37%) [48] yellow solid [23, 48], m.p. 104 to 106°C [23], 120°C from hexane [48] cis to trans ratio 0.59 [48], reverse assignment in [23] ^1H NMR (CCl_4): 2.0 (m, CH_2 and CH_3), 5.01 (d, trans-C_5H_5; J(P, H) = 1.5), 5.12 (s, cis-C_5H_5), 7.5 (m, C_6H_5) [23]; ($CDCl_3$): 1.82 (m, CH_2 and CH_3), 4.74 (d, trans-C_5H_5; J(P, H) = 1.4), 4.86 (cis-C_5H_5) [48] IR ($CHCl_3$): 1883, 1961 (ν(CO)), 2200 (ν(C≡C)) [23]; (CH_2Cl_2): 1885, 1965 (ν(CO)) [48]
44	trans-$C_5H_5Mo(CO)_2(P(C_6H_5)_3)CH_2C{\equiv}CC_6H_5$	IV, with $BrCH_2C{\equiv}CC_6H_5$, similar to No. 37 (35%) [35, 48] yellow, air-sensitive solid, m.p. 135°C (from hexane) [48] ^1H NMR ($CDCl_3$): 2.08 (d, CH_2; J(P, H) = 2.5), 4.86 (d, C_5H_5; J(P, H) = 1.4) [35, 48] IR (CH_2Cl_2): 1860, 1940 (ν(CO)) [35, 48]
45	$C_5H_5Mo(CO)_2(P(OCH_3)_3)CH_2C{\equiv}CC_6H_5$	IV, with $BrCH_2C{\equiv}CC_6H_5$, similar to No. 37 (38%) [48] yellow, air-sensitive solid (from hexane) [48] cis to trans ratio 0.16 [48] ^1H NMR ($CDCl_3$): 2.08 (d, CH_2; J(P, H) = 3.1), 5.18 (d, trans-C_5H_5; J(P, H) = 1.5), 5.36 (s, cis-C_5H_5) [48] IR (CH_2Cl_2): 1875, 1945 (ν(CO)) [48]
46	$C_5H_5Mo(CO)_2(P(OC_6H_5)_3)CH_2C{\equiv}CC_6H_5$	IV, with $BrCH_2C{\equiv}CC_6H_5$, similar to No. 37 (35%) [48] yellow, air-sensitive solid, m.p. 114°C (from hexane) [48] cis to trans ratio 0.61 [48]

References on pp. 103/5

Table 6 (continued)

No.	compound	method of preparation (yield) properties and remarks
46 (continued)		^{1}H NMR ($CDCl_3$): 1.90 (d, CH_2; J(P, H) = 3), 4.75 (d, trans-C_5H_5; J(P, H) = 1.4), 4.88 (s, cis-C_5H_5) [48] IR (CH_2Cl_2): 1885, 1965 (ν(CO)) [48]
*47	$C_5H_5Mo(CO)_2(P(C_6H_5)_3)(CH_2)_2CH{=}C{=}CH_2$	IV, with $BrCH_2CH_2CH{=}C{=}CH_2$ (not isolated) [53] IR (THF): 1855, 1940 (ν(CO)) [53]
*48	$C_5H_5Mo(CO)_2(P(OCH_3)_3)(CH_2)_2CH{=}C{=}CH_2$	IV, with $BrCH_2CH_2CH{=}C{=}CH_2$ (not isolated) [53] IR (THF): 1870, 1950 (ν(CO)) [53]
*49	$C_5H_5Mo(CO)_2(P(OC_6H_5)_3)(CH_2)_2CH{=}C{=}CH_2$	IV, with $BrCH_2CH_2CH{=}C{=}CH_2$ (not isolated) [53] IR (THF): 1880, 1970 (ν(CO)) [53]
*50	$C_5H_5Mo(CO)_2(P(C_6H_5)_3)(CH_2)_2C(CH_3){=}C{=}CH_2$	IV, with $BrCH_2CH_2C(CH_3){=}C{=}CH_2$ (not isolated) [53] IR (THF): 1850, 1935 (ν(CO)) [53]
51	$C_5H_5Mo(CO)_2(P(C_6H_5)_3)CH_2C_6H_5$	III, without solvent for 0.5 h at 140°C (39%) [10]; IV, V [16] light to dark yellow [16], yellow-orange solid, m.p. 119°C (twice from CH_2Cl_2/hexane) [10] cis to trans ratio <0.02 [16] ^{1}H NMR ($CDCl_3$): 1.25 (s, CH_2), 5.27 (s, C_5H_5), 7.1 to 7.6 (m, C_6H_5) [10]; 2.19 (d, CH_2; J(P, H) = 2.0), 4.68 (d, trans-C_5H_5; J(P, H) = 1.4) [16] IR (cyclohexane): 1896, 1973 (ν(CO)) [10]; 1868, 1943 (ν(CO)) [16]
52	$C_5H_5Mo(CO)_2(P(OC_6H_5)_3)CH_2C_6H_5$	I [11, 16]; IV, with $C_6H_5CH_2Cl$ (36%) [5]; V [11, 16] light to dark yellow [16], yellow solid [5], m.p. 113 to 115°C [5], 115°C [16] cis to trans ratio 4.0 in $CDCl_3$ [5], 0.18 in $CDCl_3$, 0.15 in toluene at 25°C; ΔH = 0.83 kcal/mol in toluene [16] ^{1}H NMR ($CDCl_3$): cis isomer: 2.60 (d of d, CH_2; J(P, H) = 11, J(H, H) = 10.3), 2.96 (d of d, CH_2; J(P, H) = 10.20, J(H, H) = 10.3), 4.63 (C_5H_5); trans isomer: 2.76 (d, CH_2; J(P, H) = 2.7), 4.45 (d, C_5H_5; J(P, H) = 1.3) [16]; similar spectra with reverse assignment of cis and trans isomers in [5]; diagram given in [11] IR (CH_2Cl_2): 1882, 1963 (ν(CO)) [5]; (cyclohexane): 1889, 1962 (ν(CO)) [16]

References on pp. 103/5

Table 6 (continued)

No.	compound	method of preparation (yield) properties and remarks
		kinetic data for cis to trans and trans to cis (in parentheses) interconversion in toluene by ^{1}H NMR: $E_a = 19.6 \pm 0.2$ (20.6 ± 0.2) kcal/mol, log A = 13.1 ± 0.2 (13.0 ± 0.2), $\Delta H^{\neq} = 19.0 \pm 0.2$ (20.0 ± 0.2) kcal/mol, $\Delta S^{\neq} = -0.8 \pm 0.7$ (-1.2 ± 0.8) cal · mol^{-1} · K^{-1}, $\Delta F^{\neq}$ (at 25°C) = 19.2 (20.4) kcal/mol [16]
53	$C_5H_5Mo(CO)_2(P(OCH_3)_2C_6H_5)CH_2C_6H_5$	cis to trans ratio 0.08 in $CDCl_3$ at 31°C [21] ^{1}H NMR ($CDCl_3$): trans isomer: 2.78 (d, CH_2; J(P, H) = 2.9), 3.53 (d, CH_3O; J(P, H) = 11.9), 4.69 (d, C_5H_5; J(P, H) = 1.6); cis isomer: 4.90 (s, C_5H_5; J(P, H) < 0.2) [21] IR (cyclohexane): 1877, 1953 (ν(CO)) [21]
*54	trans-$C_5H_5Mo(CO)_2(P(C_6H_5)_3)CH_2OCH_3$	II, in refluxing C_6H_6 (product often contains impurities); V [16] light to dark yellow solid (from methylcyclohexane) [16] ^{1}H NMR ($CDCl_3$): 3.38 (s, CH_3), 4.74 (d, CH_2; J(P, H) = 3.6), 4.84 (d, C_5H_5; J(P, H) = 1.2), 7 to 8 (m, C_6H_5) [16, 65]
55	$C_5H_5Mo(CO)_2(P(OC_6H_5)_3)CH_2OCH_3$	II, in refluxing C_6H_6 (product often contains impurities); V [16] light to dark yellow solid (from methylcyclohexane) [16] ^{1}H NMR ($CDCl_3$): cis isomer: 3.31 (s, CH_3), 4.78 (s, C_5H_5); trans isomer: 3.56 (s, CH_3), 4.60 (d, CH_2; J(P, H) = 3.6), 4.62 (d, C_5H_5; J(P, H) = 1.2) [16]
56	$C_5H_5Mo(CO)_2(P(OC_6H_5)_3)CH_2SCH_3$	II, in refluxing C_6H_6 (product often contains impurities); V [16] light to dark yellow solid (from methylcyclohexane) [16] ^{1}H NMR (C_6D_6): cis isomer: 2.49 (d, CH_3; J(P, H) = 0.5), 2.91 (d of d, CH_2: J(P, H) = 11.5), 3.02 (d of d, CH_2; J(P, H) = 8.6, J(H, H) = 8.2); trans isomer: 2.35 (s, CH_3), 2.79 (d, CH_2; J(P, H) = 4.1) [16] decomposes in solution to $C_5H_5Mo(CO)_2$-(η^2-CH_2SCH_3) [16]

References on pp. 103/5

Table 6 (continued)

No.	compound	method of preparation (yield) properties and remarks
*57	trans-$C_5H_5Mo(CO)_2(P(C_6H_5)_3)CH_2OCH_2C_6H_5$	^1H NMR ($CDCl_3$): 4.54 (d, $MoCH_2$; J(P, H) = 3.7), 4.83 (d, C_5H_5; J(P, H) = 1.8), 5.29 (s, $CH_2C_6H_5$), 7 to 8 (m, C_6H_5) [65]
*58	trans-$C_5H_5Mo(CO)_2(P(C_6H_5)_3)CH_2OC(O)C_4H_9$-t	^1H NMR ($CDCl_3$): 1.21 (t-C_4H_9), 4.87 (d, C_5H_5; J(P, H) = 1.8), 5.55 (d, CH_2; J(P, H) = 3.8), 7 to 8 (m, C_6H_5) [65]
*59	trans-$C_5H_5Mo(CO)_2(P(C_6H_5)_3)C(CH_3)(OCH_3)H$	^1H NMR ($CDCl_3$): 5.15 (d of q, CH; J(H, H) = 6.5, J(P, H) = 3) [79]
*60	trans-$C_5H_5Mo(CO)_2(P(C_6H_5)_3)C(CN)_2C(CN)_2CH_3$	VI (40%) [40] yellow crystals [38, 40], m.p. 111 to 112°C (dec.) [40] ^1H NMR ($CDCl_3$): 2.37 (s, CH_3), 5.53 (d, C_5H_5; J(P, H) = 1.5), 7.30 to 7.44 (m, C_6H_5) [40] IR (CH_2Cl_2): 1901, 1984 (ν(CO)), 2212 (ν(CN)) [40] $C_5H_5Mo(CO)_2(P(C_6H_5)_3)N{=}C{=}C(CN)C(CN)_2CH_3$ was formed in 16% yield after standing in CH_2Cl_2 for 60 h [40]
61	trans-$C_5H_5Mo(CO)_2(P(OC_6H_5)_3)C(CN)_2C(CN)_2CH_3$	VI (15%) [40] yellow solid, m.p. 117°C (dec.) [40] ^1H NMR ($CDCl_3$): 1.77 (s, CH_3), 5.45 (d, C_5H_5; J(P, H) = 2.0), 7.30 (m, C_6H_5) [40] IR (CH_2Cl_2): 1918, 2006 (ν(CO)), 2212 (ν(CN)) [40] $C_5H_5Mo(CO)_2(P(OC_6H_5)_3)N{=}C{=}C(CN)C(CN)_2CH_3$ was formed in 24% yield on standing in CH_2Cl_2 for 72 h at 25°C [40]
compounds with ^1L = alkenyl		
62	trans-$C_5H_5Mo(CO)_2(P(CH_3)_3)C({=}C{=}O)CH_3$	$C_5H_5Mo(CO)(P(CH_3)_3)O{\cdots}C{\cdots}CCH_3$ was allowed to react with CO under pressure; no further information given [55]
63	trans-$C_5H_5Mo(CO)_2(P(CH_3)_3)C({=}C{=}O)C_6H_4CH_3$-4	$C_5H_5Mo(CO)(P(CH_3)_3)O{\cdots}C{\cdots}C_6H_4CH_3$-4 was allowed to react in CH_2Cl_2 at −30°C with CO (30 bar) for 24 h (95%) [55]; also obtained by treating $C_5H_5Mo(CO)(P(CH_3)_3){\equiv}CC_6H_4CH_3$-4 in ether at −30°C with 1 atm CO for 5 h (69%) [50, 55, 75]

References on pp. 103/5

Table 6 (continued)

No.	compound	method of preparation (yield) properties and remarks
		yellow crystals (from CH_2Cl_2/ether/pentane) [55] ^{1}H NMR (CD_2Cl_2, −20°C): 1.67 (d, CH_3P; J(P, H) = 9.2), 2.31 (CH_3-4), 5.26 (d, C_5H_5; J(P, H) = 2.3), 7.13 (m, C_6H_4) [55] ^{13}C NMR (CD_2Cl_2, −40°C): −5.12 (d, MoC; J(P, C) = 7.3), 19.90 (d, CH_3P; J(P, C) = 31.7), 21.04 (CH_3-4), 92.78 (C_5H_5), 129.03, 129.24, 131.29, and 137.01 (C_6H_4), 155.03 (=C=O), 232.00 (d, CO; J(P, C) = 26.9) [55] ^{31}P NMR (CD_2Cl_2, −60°C): 19.2 [55] IR (CH_2Cl_2): 1855, 1945 (ν(CO)), 2029 (ν(C=C=O)) [50, 55] mass spectrum: decomposes in source to $C_5H_5Mo(CO)(P(CH_3)_3)$≡$CC_6H_4CH_3$-4 at 125°C [55] soluble in CH_2Cl_2 or acetone, insoluble in ether or pentane [50] reaction with 80 atm CO pressure at 60°C in CH_2Cl_2 for 24 h affords $C_5H_5Mo(CO)_2(P(CH_3)_3)C$≡$CC_6H_4CH_3$-4 and CO_2 [63]
64	$C_5H_5Mo(CO)_2(P(C_6H_5)_3)CH=C(CN)_2$	II, in refluxing C_6H_6 with 3 equivalents of $P(C_6H_5)_3$ for 15 h (32% after chromatography) [27, 33] yellow crystals, m.p. 179 to 181°C (from CH_2Cl_2/hexane) [33] ^{1}H NMR (CH_2Cl_2): ca. 5.2 (C_5H_5, under solvent), 7.40 and 7.48 (br m, C_6H_5), 10.6 (CH=) [33] IR (CH_2Cl_2): 1904, 1981 (ν(CO)), 2210, 2221, 2233 (ν(CN)) [33]
65	trans-$C_5H_5Mo(CO)_2(P(C_6H_5)_3)C(CN)=CHCN$	I, in THF; other product is $C_5H_5Mo(CO)_2(\eta^2$-$CH(CN)=C(CN)P(C_6H_5)_3)$ [69] ^{1}H NMR ($CDCl_3$): 5.00 (d, C_5H_5; J(P, H) = 1.5), 6.72 (CH=), 7.30 (m, C_6H_5) [69] IR (CH_2Cl_2): 1510 (ν(C=C)), 1880, 1960 (ν(CO)), 2180, 2210 (ν(CN)) [69]
66	$C_5H_5Mo(CO)_2(P(C_6H_5)_3)C(CN)=C(CN)_2$	II, in refluxing C_6H_6 with 22 equivalents of $P(C_6H_5)_3$ (28%) [39] brown solid (from CH_2Cl_2/hexane) [39]

References on pp. 103/5

Table 6 (continued)

No.	compound	method of preparation (yield) properties and remarks
67	trans-$C_5H_5Mo(CO)_2(P(OCH_3)_3)CCl{=}C(CN)_2$	II, in refluxing C_6H_6 with 3 equivalents of $P(OCH_3)_3$ (60%) [54] yellow solid, m.p. 134 °C (from CH_2Cl_2/hexane) [54] ^1H NMR ($CDCl_3$): 3.76 (d, CH_3; J(P, H) = 12), 5.43 (s, C_5H_5) [54] ^{13}C NMR ($CDCl_3$): 54.0 (d, CH_3; J(P, C) = 8), 95.2 (C_5H_5), 99.1 (C=, not specified), 113.0, 116.4 (CN), 226.8 (C=, not specified), 232.0 (d, CO; J(P, C) = 38) [54] IR (CH_2Cl_2): 1913, 1987 (ν(CO)), 2222 (ν(CN)) [54]
68	$C_5H_5Mo(CO)_2(P(C_6H_5)_3)C(=C(CF_3)_2)C(CF_3){=}CF_2$	IV, with $(F_3C)_2C{=}C{=}C(CF_3)_2$ at −70 °C for 1h and 3 h at ambient temperature (67%) [49] orange crystals, m.p. 130 to 131 °C (from pentane at −70 °C) [49] IR (KBr): 1000 to 1350 (ν(CF)), 1540, 1740 (ν(C=C)), 1875, 1960 (ν(CO)) [49] soluble in polar solvents; air-stable as a solid and in solution [49] refluxing in n-heptane for 4 h under Ar affords $C_5H_5Mo(CO)_2(\eta^3\text{-}CF_2 \cdots C(CF_3) \cdots C{=}C(CF_3)_2)$ [49]
69	trans-$C_5H_5Mo(CO)_2(P(OCH_3)_3)C(N(C_2H_5)_2){=}CHCH_3$ (N cis to CH_3)	VII, (not isolated) [61] ^1H NMR (C_6D_6): 0.9 (t, **CH_3**CH_2; J(H, H) = 7.0), 1.67 (d, CH_3C=; J(H, H) = 6.0), 2.63 (q, CH_2), 3.03 (d, CH_3O; J(P, H) = 11.0), 4.52 (d, C_5H_5; J(P, H) = 4.0) [61] IR (ether): 1640 (ν(C=C)), 1850, 1920 (ν(CO)) [61]
70	$C_5H_5Mo(CO)_2(P(OC_2H_5)_3)C(N(C_2H_5)_2){=}CHCH_3$ (N cis to CH_3)	VII (not isolated) [61] IR (ether): 1640 (ν(C=C)), 1850, 1920 (ν(CO)) [61]
	compounds with ^1L = alkynyl	
71	$C_5H_5Mo(CO)_2(P(C_6H_5)_3)C{\equiv}CC_6H_5$	II, in octane at 120 °C for 2.5 h (10%) [73]; also obtained by the reaction of $(C_5H_5Mo(CO)_2P(C_6H_5)_3)_2$ with 2 equivalents of $NaC{\equiv}CC_6H_5$ in THF for 5 h (42%) [66]

Table 6 (continued)

No.	compound	method of preparation (yield) properties and remarks
		yellow crystals, m.p. 199 to 200°C (from C_6H_6/ hexane 1:1) [66] IR (THF): 1892, 1970 (ν(CO)) [66, 73] treatment with $HBF_4/P(C_6H_5)_3$ in CH_2Cl_2 between −30 to −20°C leads to the formation of $[C_5H_5Mo(CO)_2(P(C_6H_5)_3)C(=CHC_6H_5)$-$P(C_6H_5)_3]BF_4$ [66]
72	$C_5H_5Mo(CO)_2(P(CH_3)_3)C{\equiv}CC_6H_4CH_3$-4	$C_5H_5Mo(CO)_2(P(CH_3)_3)C(=C=O)C_6H_4CH_3$-4 was treated in CH_2Cl_2 at 60°C with 80 atm CO for 24 h (69%) [63] yellow crystals (from pentane) [63] ^1H NMR (acetone-d_6): 1.73 (d, CH_3P; J(P, H) = 9.8), 2.41 (s, CH_3-4), 5.60 (s, C_5H_5), 7.17 (m, C_6H_4) [63] ^{13}C NMR (acetone-d_6): 19.75 (d, CH_3P; J(P, C) = 33.8), 21.67 (CH_3-4), 93.35 (C_5H_5), 109.60 (d, CMo; J(P, C) = 47.1), 127.08 (d, ≡C; J(P, C) = 2.9), 127.47 (d, C_6H_4; J(P, C) = 2.9), 129.88 (s, C_6H_4), 131.18 (d, C_6H_4; J(P, C) = 2.9), 135.47 (C_6H_4), 241.43 (d, cis-CO; J(P, C) = 1.5), 254.92 (d, trans-CO; J(P, C) = 35.3) [63] ^{31}P NMR (acetone-d_6): 20.61 [63] IR (CH_2Cl_2): 1863, 1956 (ν(CO)), 2088 (ν(C≡C)) [63] mass spectrum: $[M-n\,CO]^+$ (n = 0 to 2) and fragment ions [63] well soluble in acetone and CH_2Cl_2 but nearly insoluble in pentane or ether stable for longer times under N_2 in the solid state and in solution [63]

compound with ^1L = perfluorinated alkyl group

No.	compound	method of preparation (yield) properties and remarks
73	$C_5H_5Mo(CO)_2(P(C_6H_5)_3)CF_3$	I, in hexane for 12 h (63%) [14] orange solid, dec. 154 to 156°C (from CH_2Cl_2/ hexane) [14] ^{19}F NMR (CH_2Cl_2): −11.3 (d, CF_3; J(P, F) = 1.5) [14] IR: 1885, 1979 (ν(CO)) [14] treatment with SbF_5 in SO_2 affords $[C_5H_5Mo(CO)_2(P(C_6H_5)_3)=CF_2]SbF_6$; with BF_3 in CH_2Cl_2, $[C_5H_5Mo(CO)_3P(C_6H_5)_3]BF_4$ was observed [45]

References on pp. 103/5

Table 6 (continued)

No.	compound	method of preparation (yield) properties and remarks
compounds with ^{1}L = cyclic group		
74	$C_5H_5Mo(CO)_2(P(OC_6H_5)_3)C{=}C(CH_3)C(CN)_2C(CN)_2CH_2$ (^{1}L 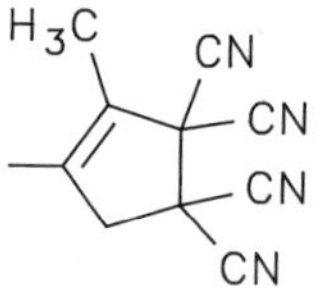 = H$_3$C CN CN CN CN)	$C_5H_5Mo(CO)_2(P(OC_6H_5)_3)CH_2C{\equiv}CCH_3$ was allowed to react in CH_3CN, THF, or C_6H_6 with $(CN)_2C{=}C(CN)_2$ (65 to 70%); purification by chromatography on Al_2O_3 (10% H_2O) with CH_2Cl_2 or $CHCl_3$ reduced the yield to 25 to 30% [37] orange-yellow solid (from CH_2Cl_2/pentane) [37] ^{1}H NMR ($CDCl_3$): 1.98 (t, CH_3; J = 1.5), 3.28 (q, CH_2; J = 1.5), 5.82 (d, C_5H_5; J(P, H) = 1.0), 7.28 (m, C_6H_5) [37] IR (Nujol): 2256 (ν(CN)); (CH_2Cl_2): 1900, 1980 (ν(CO)) [37]
75	trans-$C_5H_5Mo(CO)_2(P(C_6H_5)_3)C_5H_4Mn(CO)_3$	$C_5H_5Mo(CO)_3C(O)C_5H_4Mn(CO)_3$ and a slight excess of $P(C_6H_5)_3$ were irradiated at 10 to 15°C in THF for 8 h (60%) [71] yellow crystals, m.p. 204 to 205°C (from THF/pentane) [71] ^{1}H NMR ($CDCl_3$): 4.59 and 4.79 (t, C_5H_4 each 2H), 4.88 (d, C_5H_5; J(P, H) = 1.4) [71] IR ($CHCl_3$): 1882, 1930, 1970, 2016 (ν(CO)) [71] reaction with $HgCl_2$ in C_6H_6 affords $C_5H_5Mo(CO)_2(P(C_6H_5)_3)Cl$ and $(CO)_3MnC_5H_4HgCl$ [71]
76	trans-$C_5H_5Mo(CO)_2(P(C_6H_5)_3)C_5F_4N$-4 ($C_5F_4N$ = tetrafluoropyridyl)	IV with C_5F_5N for 1 h (40%) [26] bright yellow crystals, m.p. 220 to 223°C (from ether) [26] ^{1}H NMR: 4.84 (d, C_5H_5; J(P, H) = 0.5), 7.49 (m, C_6H_5) [26] ^{19}F NMR: 100.0, 103.9 [26] IR (Nujol): 1870, 1965 (ν(CO)); ($CHCl_3$): 1886, 1971 (ν(CO)) [26]
77	$C_5H_5Mo(CO)_2(P(C_6H_5)_3)CH(CH_2)_3O$ (^{1}L 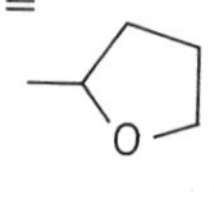 = O)	$C_5H_5Mo(CO)_2CH(CH_2)_3O$ (No. 46, p. 126) was allowed to react with $P(C_6H_5)_3$ in THF (55%); $[C_5H_5Mo(CO)_2(P(C_6H_5)_3){=}C(CH_2)_3O\text{-cyclo}]Br$ was treated with $Li[(C_2H_5)_3BH]$ in THF (21%) [67, 78, 83] yellow crystals, m.p. 152 to 154°C (from CH_2Cl_2/petroleum ether) [78] ^{1}H NMR (C_6D_6): 1.73 (d of d of q, central CH_2, 1H; J = 11.5, 9.5, 8.0), 1.86 (d of d of d of d of d, central CH_2, 1H; J = 11.5, 9, 7.5, 4.5, 2.5), 2.53

Table 6 (continued)

No.	compound	method of preparation (yield) properties and remarks
		(d of d of d of d, C**H**$_2$CH, 1H; J = 12.5, 10.5, 9.5, 9.2), 2.85 (d of d of d of d, C**H**$_2$CH, 1H; J = 12.5, 8, 5.5, 2.5), 3.76 (t of d, OCH_2, 1H; J = 8.0, 4.5), 3.08 (q, OCH_2, 1H; J = 7.5), 4.83 (s, C_5H_5), 5.70 (d of d, MoCH; J = 10.5, 5.5) [78] ^{13}C NMR ($CDCl_3$): 19.0 (central CH_2), 39.0 (**C**H_2CH), 69.7 (OCH_2), 75.3 (d, MoCH; J(P, C) = 11), 93.0 (C_5H_5), 128.1 (d, C_6H_5, C-3; J(P, C) = 9), 129.8 (d, C_6H_5, C-4; J(P, C) = 2), 133.1 (d, C_6H_5, C-2; J(P, C) = 11), 136.5 (d, C_6H_5, C-1; J(P, C) = 43), 240.1 (d, CO; J(P, C) = 25) [78] IR (CH_2Cl_2): 1850, 1934 (ν(CO)) [78]
*78	trans-$C_5H_5Mo(CO)_2(P(C_6H_5)_3)$C=CHCH=CHO (^{1}L =)	$Na[C_5H_5Mo(CO)_3]$ was allowed to react with 1.5 equivalents of $P(C_6H_5)_3$ and 5 equivalents of epibromohydrin in THF for 12 h (20%) [85] orange-red crystals, m.p. 178°C (from CH_2Cl_2/hexane 7:3) [85] ^{1}H NMR (C_6D_6): 4.69 (d, C_5H_5; J(P, H) = 0.7), 7.12 and 7.61 (m, C_6H_5), 6.70, 7.23, 7.30, and 8.13 (m, C_4H_3O) [85] ^{13}C NMR (C_6D_6): 92.4 (C_5H_5), 111.7, 126.7, 128.3, 128.7, 130.3, 133.6, and 148.6 (C_6H_5 and C_4H_3O), 199.3 (CO) [85] IR (CH_2Cl_2): 1870, 1955 (ν(CO)) [85]
79	$C_5H_5Mo(CO)_2(P(OC_6H_5)_3)C{=}C(CH_3)S(O)OCH_2$ (^{1}L =)	$C_5H_5Mo(CO)_2(P(OC_6H_5)_3)CH_2C{\equiv}CCH_3$ was allowed to react with SO_2 in pentane or CH_2Cl_2 (55%) [19, 23] unstable yellow solid [23] IR (KBr): 900 to 915 (ν(SO)); ($CHCl_3$): 1900, 1977 (ν(CO)) [23]

compounds with ^{1}L = aryl

No.	compound	method of preparation (yield) properties and remarks
*80	trans-$C_5H_5Mo(CO)_2(P(C_6H_5)_3)C_6H_5$	II, in C_6H_6 with excess $P(C_6H_5)_3$ (traces) [30]; also obtained by irradiation of $C_5H_5Mo(CO)_2(P(C_6H_5)_3)Sn(C_6H_5)_3$ (60%) [56, 74]; see "Further information" yellow crystals, m.p. 174°C (from CH_2Cl_2/hexane) [56] ^{1}H NMR ($CDCl_3$): 4.84 (d, C_5H_5; J(P, H) = 1.3), 7.29 to 7.49 (m, C_6H_5) [18]; similar data in [56] ^{31}P NMR (CH_2Cl_2): 69.8 (s) [54]

Table 6 (continued)

No.	compound	method of preparation (yield) properties and remarks
*80 (continued)		IR ($CHCl_3$): 1864, 1950 (ν(CO)) [56]; 1859, 1935 (ν(CO)) [18] mass spectrum: $[M]^+$ [56] air-stable in the solid state but air-sensitive in solution reaction with $HgCl_2$ in C_6H_6 yields $C_5H_5Mo(CO)_2(P(C_6H_5)_3)Cl$ besides C_6H_5HgCl; treatment with I_2 in CH_2Cl_2 at −50°C gives $C_5H_5Mo(CO)_2(P(C_6H_5)_3)I$ [18]
*81	trans-$C_5H_5Mo(CO)_2(P(OC_6H_5)_3)C_6H_5$	II, in C_6H_6 with an excess of $P(OC_6H_5)_3$ (20%) [30]; see "Further information" yellow solid, m.p. 115 to 116°C (from $CHCl_3$/CH_3OH 1:1) [30] ^1H NMR ($CDCl_3$): 4.90 (d, C_5H_5; J(P, H) = 1.3), 7.18 to 7.40 (m, C_6H_5) [29] IR ($CHCl_3$): 1891, 1970 (ν(CO)) [29]
82	trans-$C_5H_5Mo(CO)_2(P(C_6H_5)_3)C_6H_4F$-3	IV, with $[I(C_6H_4F\text{-}3)_2]I$ (12%) [31] yellow solid (from $CHCl_3$/CH_3OH 1:1), m.p. 174 to 175°C (dec.) [31] ^1H NMR ($CDCl_3$): 4.84 (d; J(P, H) = 1.4), 7.34 to 7.56 (m, C_6H_5) [31] ^{19}F NMR ($CHCl_3$, vs. C_6H_5F): 3.10 [31]; mentioned in [22] IR ($CHCl_3$): 1863, 1947 (ν(CO)) [31] air-stable in the solid state but decomposes under air in solution [31]
83	trans-$C_5H_5Mo(CO)_2(P(C_6H_5)_3)C_6H_4F$-4	IV, with $[I(C_6H_4F\text{-}4)_2]I$ (4%); also obtained by decarbonylation of $C_5H_5Mo(CO)_2(P(C_6H_5)_3)$-$C(O)C_6H_4F$-4 with one equivalent of $Rh(P(C_6H_5)_3)_3Cl$ in C_6H_6 at 70°C for 4 h (75%) [31] yellow crystals, m.p. 174°C (from heptane/C_6H_6 1:1) [31] ^1H NMR ($CDCl_3$): 4.80 (d, C_5H_5; J(P, H) = 1.4), 7.34 to 7.56 (m, C_6H_5) [31] ^{19}F NMR ($CHCl_3$ vs. C_6H_5F): 9.9 [31]; mentioned in [22] IR ($CHCl_3$): 1860, 1944 (ν(CO)) [31] air-stable in the solid state but decomposes in solution [31]
84	trans-$C_5H_5Mo(CO)_2(P(C_6H_5)_3)C_6F_3(CN)_2$-3,4	IV, with $C_6F_4(CN)_2$-1, 2 (30%) [26] bright yellow crystals, m.p. 215°C (from acetone/hexane) [26]

References on pp. 103/5

Table 6 (continued)

No.	compound	method of preparation (yield) properties and remarks
		^{1}H NMR ($CDCl_3$): 4.83 (d, C_5H_5; J(P, H) = 0.8), 7.40 (m, C_6H_5) [26] ^{19}F NMR: 59.8 (d of q, F-2), 76.6 (d of q, F-5), 135.2 (q, F-6); J(P, F-2) = 1.6, J(P, F-5) = 1.8, J(F-2, 5) = 5.1, J(F-2, 6) = 12, J(F-5, 6) = 28 [26] IR ($CHCl_3$): 1870, 1970 (ν(CO)), 2240 (ν(CN)) [26]
$[^5LMo(CO)_2(^2D)^1L]^+$ compounds with ^1L = ylide		
85	trans-$[C_5H_5Mo(CO)_2(P(CH_3)_3)CH(Si(CH_3)_3)P(CH_3)_3]O_3SCF_3$	$C_5H_5Mo(CO)_2(P(CH_3)_3)OSO_2CF_3$ was allowed to react with $(CH_3)_3P{=}CHSi(CH_3)_3$ in ether at −60 °C (79%) [80] yellow crystals, m.p. 148 °C (dec.) [80]
86	$[C_5H_5Mo(CO)_2(P(C_6H_5)_3)C({=}CHC_6H_5)P(C_6H_5)_3]BF_4$	$C_5H_5Mo(CO)_2(P(C_6H_5)_3)C{\equiv}CC_6H_5$ was allowed to react in CH_2Cl_2 with 2 equivalents of $P(C_6H_5)_3$ and 4 equivalents of HBF_4 at −20 to −30 °C; the mixture was allowed to warm up and stirred for 4 h (43%) [3, 66] yellow crystalline solid (from CH_2Cl_2/hexane 2:1), m.p. 119 to 120 °C (dec.) [66], 105 to 107 °C (dec.) [3] ^{1}H NMR (CH_3CN): 4.0 (m, C_5H_5), 7.39 (m, C_6H_5) [66] ^{31}P NMR (CH_2Cl_2): 33.0 (s, $CP(C_6H_5)_3$), 59.0 ($MoP(C_6H_5)_3$) [66] IR (CH_2Cl_2): 1872, 1957 (ν(CO)) [3, 66]
87	$[C_5H_5Mo(CO)_2(P(C_6H_5)_3)C(CH_3)HP(C_6H_5)_3]BF_4$	from No. 59 and $P(C_6H_5)_3$; see "Further information" of No. 59 [79]

*Further information:

$C_5H_5Mo(CO)_2(P(CH_3)_3)CH_3$ (Table **6**, No. **2**) was obtained in nearly quantitative yield by the reaction of $[P(CH_3)_4][C_5H_5Mo(CO)_2P(CH_3)_3]$ with CH_3I in C_6H_6 or pentane suspension. The slow decomposition of $[P(CH_3)_4][C_5H_5Mo(CO)_2P(CH_3)_3]$ in CH_3CN or dimethyl sulfoxide also led to the formation of complex No. 2 (42% yield) [60].

$C_5H_5Mo(CO)_2(P(C_6H_5)_3)CH_3$ (Table **6**, No. **7**) was formed by the reaction of $Na[BH_4]$ with $[C_5H_5Mo(CO)_3P(C_6H_5)_3]PF_6$ [4]. Alkylation of cis-$Na[C_5H_5Mo(CO)_2P(C_6H_5)_3]$ with $(CH_3O)_2SO_2$, cis-$C_5H_5Mo(CO)_2(P(C_6H_5)_3)H$ with CH_2N_2, or cis-$C_5H_5Mo(CO)_2(P(C_6H_5)_3)Cl$ with $LiCH_3$ also led to the formation of complex No. 7 [9]. It was formed as a side product in the reaction of $C_5H_5Mo(CO)_2(P(C_6H_5)_3)CH_2OR$ with $(CH_3)_3SiOSO_2CF_3$ in CD_2Cl_2 at −90 °C and obtained by treating $C_5H_5Mo(CO)_2(P(C_6H_5)_3)CH_2OR$ (R = CH_3, $CH_2C_6H_5$, or $C(O)C_4H_9$-t) with $[C(C_6H_5)_3]AsF_6$ at −90 °C in CH_2Cl_2, together with equal amounts of $[C_5H_5Mo(CO)_2(P(C_6H_5)_3){=}C(OR)H]AsF_6$ [64]. $Li[C_5H_5Mo(CO)_2(\eta^2\text{-}O{=}CHCH_3)]$ was allowed to react with CH_3I and two equivalents of $P(C_6H_5)_3$

in THF for 4 h; a 52% yield of the title complex was obtained by chromatography on Al_2O_3 with CH_2Cl_2/petroleum ether (1:1) [77, 79].

The cis to trans ratio is 0.06 in 1,2-$C_6H_4Cl_2$ and 0.08 in toluene at 25°C; $\Delta H_{cis/trans} = 1.6$ kcal/mol. Kinetic data were obtained by several NMR techniques for the cis to trans and trans to cis (in parentheses) interconversion in 1,2-$C_6H_4Cl_2$: $E_a = 21.6 \pm 0.2$ (23.3 ± 0.2) kcal/mol, log A = 13.3 ± 0.1 (13.3 ± 0.1), $\Delta H^{\ddagger} = 20.9 \pm 0.4$ (20.9 ± 0.2) kcal/mol, $\Delta S^{\ddagger} = -0.2 \pm 0.6$ (-0.2 ± 0.6) cal · $mol^{-1} \cdot K^{-1}$, $\Delta F^{\ddagger} = 20.9$ (22.6) kcal/mol [16]; see also General Remarks.

Reaction with X_2 (X = Cl, Br, or I) in $CHCl_3$ at 0°C or HgX_2 (X = Cl, Br, I, SCN, or CN) in CH_2Cl_2 at ambient temperature gives cis/trans-$C_5H_5Mo(CO)_2(P(C_6H_5)_3)X$ [44]. trans-$C_5H_5Mo(CO)_2(P(C_6H_5)_3)C(O)CH_3$ is obtained by treatment with 2 atm CO pressure [9]. Insertion of SO_2 in $CHCl_3$ or of neat SO_2 at room temperature into the Mo-CH_3 bond leads to the formation of $C_5H_5Mo(CO)_2(P(C_6H_5)_3)SO_2CH_3$ [12]. Treatment with $(CN)_2C{=}C(CN)_2$ in CH_2Cl_2 at 25°C [20] or for 6 to 8 h at 5 to 10°C [40] gives a mixture of $C_5H_5Mo(CO)_2(P(C_6H_5)_3)C(CN)_2C(CN)_2CH_3$ and $C_5H_5Mo(CO)_2(P(C_6H_5)_3)N{=}C{=}C(CN)C(CN)_2CH_3$ [20, 40]. Reaction in CH_2Cl_2 solution with $CH_3S(O)_2NSO$ at 25°C affords $C_5H_5Mo(CO)_2(P(C_6H_5)_3)S(O)(CH_3){=}NS(O)_2CH_3$ [57, 81]. Oxidation with t-$C_4H_9OOC_4H_9$-t under irradiation yields detectable amounts of a radical cation [84].

$C_5H_5Mo(CO)_2(P(OC_6H_5)_3)CH_3$ (Table **6**, No. **12**). Kinetic data for the cis to trans and trans to cis (in parentheses) interconversion in toluene were obtained by NMR methods: $E_a = 18.4 \pm 0.3$ (18.3 ± 0.3) kcal/mol, log A = 12.3 ± 0.2 (12.0 ± 0.2), $\Delta H^{\ddagger} = 17.7 \pm 0.3$ (17.6 ± 0.3) kcal/mol, $\Delta S^{\ddagger} = -4.7 \pm 0.7$ (-5.9 ± 0.7) cal · $mol^{-1} \cdot K^{-1}$, $\Delta F^{\ddagger}$ (at 25°C) = 19.1 (19.4) kcal/mol [16].

The reaction with four equivalents of $(CN)_2C{=}C(CN)_2$ in CH_2Cl_2 for 6 to 8 h affords a mixture of $C_5H_5Mo(CO)_2(P(OC_6H_5)_3)C(CN)_2C(CN)_2CH_3$, $C_5H_5Mo(CO)_2(P(OC_6H_5)_3)N{=}C{=}C(CN)C(CN)_2CH_3$, and $C_5H_5Mo(CO)_2(P(OC_6H_5)_3)CN$ [40].

$C_5H_5Mo(CO)_2(P(OC_6H_5)_3)CH_2CH{=}C(R)R'$ (Table **6**, Nos. **32** to **34**). No. 32 (R = H, R' = CH_3) is obtained as a mixture of cis/trans isomers (major) concerning the position of the CO groups and each of these isomers is a mixture of two isomers which arise from cis-trans orientations of the CH_3 group at the C=C bond. No. 34 (R = H, R' = C_6H_5) forms only two isomers (cis/trans at Mo). The compounds react with SO_2 by insertion into the Mo-C bond and formation of $C_5H_5Mo(CO)_2(P(OC_6H_5)_3)S(O)_2R''$; SO_2 insertion produces mainly the trans (at Mo) isomers in all cases. Conditions and products obtained are given in the following table [19, 51].

R	R'	conditions	R'' (yield)
H	CH_3	in refluxing SO_2	$C(CH_3)HCH{=}CH_2$ (20%)/$CH_2CH{=}CHCH_3$ (80%)
		in SO_2 at −45°C	$C(CH_3)HCH{=}CH_2$ (40%)/$CH_2CH{=}CHCH_3$ (60%)
		in C_6H_6 for 10 h or CH_2Cl_2 for 1 h at 27°C	$C(CH_3)HCH{=}CH_2$ (20%)/$CH_2CH{=}CHCH_3$ (80%)
CH_3	CH_3	in refluxing SO_2	only $CH_2CH{=}C(CH_3)_2$
H	C_6H_5	in refluxing SO_2	only $CH_2CH{=}CHC_6H_5$
		in SO_2 at −45°C for 6 h	only $CH_2CH{=}CHC_6H_5$

No. 33 reacts with one equivalent of $(CN)_2C{=}C(CN)_2$ in CH_2Cl_2 for ca. 50 h to yield $C_5H_5Mo(CO)_2(P(OC_6H_5)_3)N{=}C{=}C(CN)C(CN)_2CH_2CH{=}C(CH_3)_2$ and cis-$C_5H_5Mo(CO)_2(P(OC_6H_5)_3)CN$ [24, 37].

$C_5H_5Mo(CO)_2(P(OC_6H_5)_3)CH_2C{\equiv}CCH_3$ (Table **6**, No. **43**) reacts with SO_2 in pentane or CH_2Cl_2 to give $C_5H_5Mo(CO)_2(P(OC_6H_5)_3)C{=}C(CH_3)S(O)OCH_2$-cyclo [19, 23]. Reaction with one equivalent of $(CN)_2C{=}C(CN)_2$ in CH_3CN, THF, or C_6H_6 yields $C_5H_5Mo(CO)_2(P(OC_6H_5)_3)C{=}CCH_3C(CN)_2$-

References on pp. 103/5

$C(CN)_2CH_2$-cyclo [37]. Reaction with an excess of CH_3OH in THF affords $C_5H_5Mo(CO)(P(OC_6H_5)_3)(\eta^3-CH_2C(C(O)OCH_3)CHCH_3)$ [48].

$C_5H_5Mo(CO)_2(PR_3)CH_2CH_2C(R)=C=CH_2$ (Table **6**, Nos. **47** to **50**). Heating in THF for 2 h at 65°C leads to the formation of the allyl complexes shown in Formula I [53].

$C_5H_5Mo(CO)PR_3$

R O

I

trans-$C_5H_5Mo(CO)_2(P(C_6H_5)_3)CH_2OR$ (Table **6**, Nos. **54**, **57**, and **58**). The molybdenum methylene complex, $[C_5H_5Mo(CO)_2(P(C_6H_5)_3)=CH_2]^+$, is generated by the reaction with $(CH_3)_3SiOSO_2CF_3$ at -90°C in CH_2Cl_2. When $[C(C_6H_5)_3]AsF_6$ is added dropwise to a solution of $C_5H_5Mo(CO)_2(P(C_6H_5)_3)CH_2OR$ in CD_2Cl_2 equimolar amounts of $C_5H_5Mo(CO)_2(P(C_6H_5)_3)CH_3$ and $[C_5H_5Mo(CO)_2(P(C_6H_5)_3)=CHOR]^+$ (R = CH_3, $CH_2C_6H_5$, or $C(O)C_4H_9$-t) are obtained [65].

trans-$C_5H_5Mo(CO)_2(P(C_6H_5)_3)C(CH_3)(OCH_3)H$ (Table **6**, No. **59**) was obtained by treatment of $[C_5H_5Mo(CO)_2(\eta^2-O=CHCH_3)]^-$ at -78°C in THF with $[O(CH_3)_3]BF_4$ and $P(C_6H_5)_3$. On warming up, at first alkylation occurs at the oxygen atom to give the intermediate $C_5H_5Mo(CO)_2(\eta^2-CH_3OCHCH_3)$, followed by $P(C_6H_5)_3$ addition to give No. 59. Isolation of the complex is possible if the reaction mixture is filtered through alumina before removal of the solvent to avoid excess of $[O(CH_3)_3]BF_4$. However, if excess $[O(CH_3)_3]BF_4$ is not removed, proceeding reaction occurs to generate the ylide complex No. 87 [79].

trans-$C_5H_5Mo(CO)_2(P(C_6H_5)_3)C(CN)_2C(CN)_2CH_3$ (Table **6**, No. **60**) crystallizes in the triclinic space group $P\overline{1}-C_i^1$ (No. 2) with the unit cell parameters a = 8.606 (1), b = 12.158 (2), c = 13.987 (2) Å, α = 85.60 (2)°, β = 75.95 (1)°, γ = 88.37 (1)°; Z = 2 molecules per unit cell, D_{calc} = 1.460, and D_{meas} = 1.468 g/cm^3. The molecular structure with the main bond distances and angles is shown in **Fig. 15**; another view is given in [38].

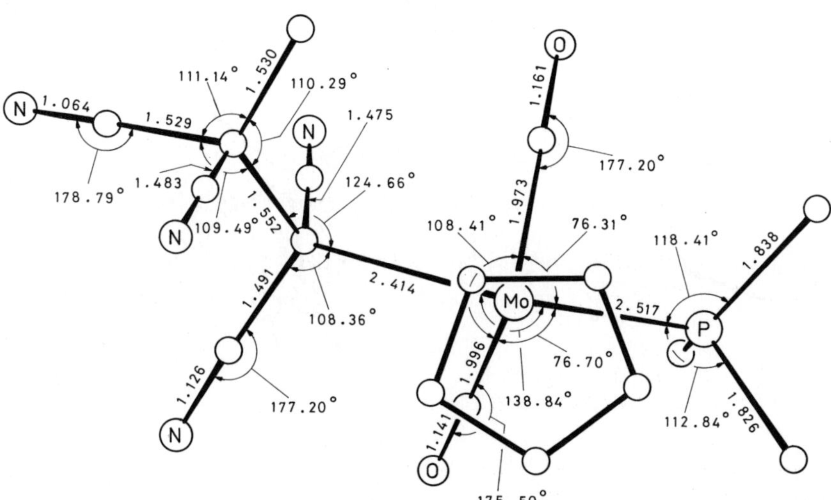

Fig. 15. Molecular structure of trans-$C_5H_5Mo(CO)_2(P(C_6H_5)_3)C(CN)_2C(CN)_2CH_3$ [38].

References on pp. 103/5

trans-$C_5H_5Mo(CO)_2(P(C_6H_5)_3)C_4H_3O$-cyclo (Table **6**, No. **78**) crystallizes in the triclinic space group $P\bar{1}-C_i^1$ (No. 2) with the unit cell parameters a = 9.150 (5), b = 12.028 (5), c = 12.716 (8) Å, α = 112.67 (3)°, β = 100.97 (3)°, γ = 98.83 (8)°; Z = 2 molecules per unit cell. The molecular structure with the main bond distances is shown in **Fig. 16** [85].

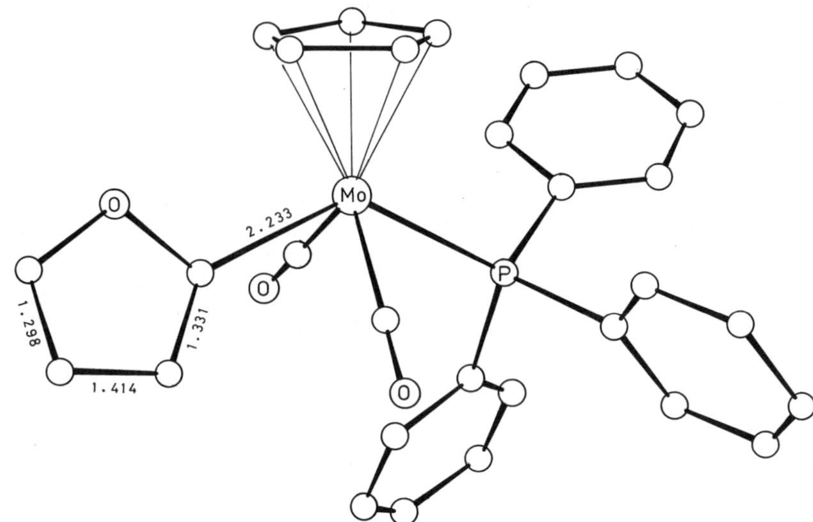

Fig. 16. Molecular structure of trans-$C_5H_5Mo(CO)_2(P(C_6H_5)_3)C{=}CHCH{=}CHO$ [85].

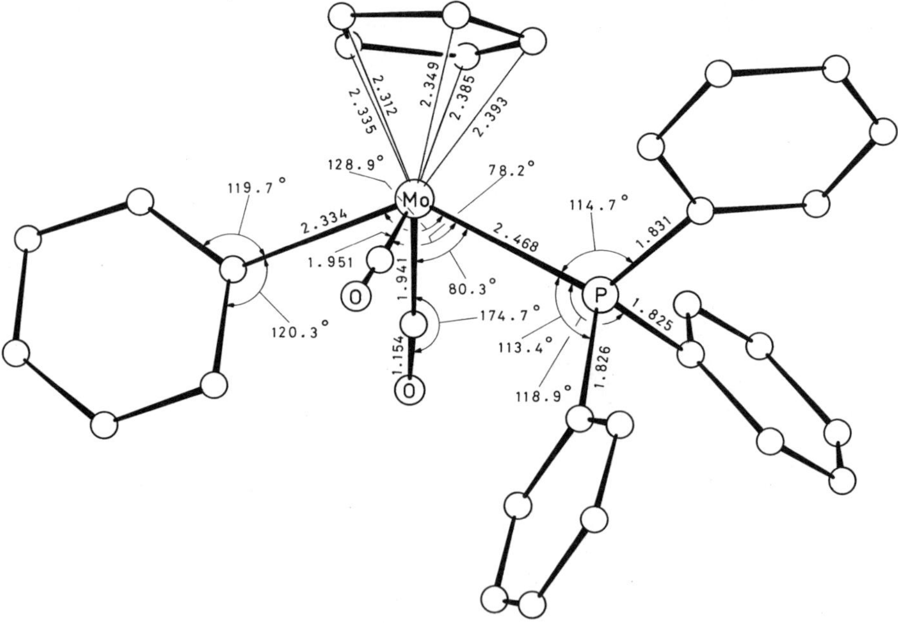

Fig. 17. Molecular structure of trans-$C_5H_5Mo(CO)_2(P(C_6H_5)_3)C_6H_5$ [74].

trans-$C_5H_5Mo(CO)_2(PR_3)C_6H_5$ (Table **6**, Nos. **80** and **81**). No. 80 (R = C_6H_5) was also obtained by the following reactions: $C_5H_5Mo(CO)_2(P(C_6H_5)_3)C(O)C_6H_5$ was allowed to react with one equivalent of $Rh(P(C_6H_5)_3)_3Cl$ in CH_2Cl_2 at 30°C for 8 h to give a 10% yield [18]. $C_5H_5Mo(CO)_2(P(C_6H_5)_3)I$ and $Hg(C_6H_5)_2$ were refluxed in C_6H_6 for 12 h to give an 8% yield besides $C_5H_5Mo(CO)_2(P(C_6H_5)_3)HgI$ and $C_5H_5Mo(CO)_2(P(C_6H_5)_3)I$. Treatment of $Na[C_5H_5Mo(CO)_2P(C_6H_5)_3]$ with $[I(C_6H_5)_2]I$ in THF at −70°C gave a 6% yield; the main products were $(C_5H_5Mo(CO)_2P(C_6H_5)_3)_2$ and $C_5H_5Mo(CO)_2(P(C_6H_5)_3)I$ [18]. $C_5H_5Mo(CO)_2(P(OC_6H_5)_3)C_6H_5$ and five equivalents of $P(C_6H_5)_3$ were refluxed in C_6H_6 for 10 h. Chromatography on Al_2O_3 with $CHCl_3/CH_3OH$ (1:1) gave a 30% yield besides 40% of starting material. This reaction is reversible. Treatment with three equivalents of $P(OC_6H_5)_3$ in refluxing C_6H_6 for 14 h gives a 40% yield of $C_5H_5Mo(CO)_2(P(OC_6H_5)_3)C_6H_5$ (No. 81) [29].

No. 80 crystallizes in the triclinic space group $P\bar{1}-C_i^1$ (No. 2) with the unit cell parameters a = 9.126 (4), b = 12.417 (4), c = 12.666 (4) Å, α = 112.77 (3)°, β = 100.81 (3)°, γ = 100.00 (3)°; Z = 2 molecules per unit cell, and D_{calc} = 1.470 (1) g/cm^3. The molecular structure with important bond distances and angles is shown in **Fig. 17** [74]; mentioned without details in [56].

References:

[1] Barnett, K. W. (Diss. Univ. Wisconsin 1967; Diss. Abstr. Intern. B **28** [1968] 3203).
[2] Barnett, K. W.; Treichel, P. M. (Inorg. Chem. **6** [1967] 294/9).
[3] Kolobova, N. E.; Shripkin, V. V.; Rozantseva, T. V. (Izv. Akad. Nauk SSSR Ser. Khim. **1979** 2393/4; Bull. Acad. Sci. USSR Div. Chem. Sci. **1979** 2218/9).
[4] Shubkin, R. L. (Diss. Univ. Wisconsin 1967; Diss. Abstr. Intern. B **28** [1968] 561).
[5] King, R. B.; Pannell, K. H. (Inorg. Chem. **7** [1968] 2356/61).
[6] King, R. B.; Pannell, K. H.; Houk, L. W.; Kapoor, R. N. (Proc. 1st Intern. Symp. New Aspects Chem. Metal Carbonyl Deriv., Padua, Italy, 1968, Abstract E 8, pp. 1/6).
[7] Mays, M. J.; Pearson, S. M. (J. Chem. Soc. A **1968** 2291/4).
[8] Barnett, K. W. (Inorg. Chem. **8** [1969] 2009/11).
[9] Bolton, E. S.; Denker, M.; Knox, G. R.; Robertson, C. G. (Chem. Ind. London **1969** 327/8).
[10] Craig, P. J.; Green, M. (J. Chem. Soc. A **1969** 157/60).

[11] Faller, J. W.; Anderson, A. S. (J. Am. Chem. Soc. **91** [1969] 1550/1).
[12] Graziani, M.; Bibler, J. P.; Montesano, R. M.; Wojcicki, A. (J. Organometal. Chem. **16** [1969] 507/11).
[13] King, R. B.; Houk, L. W.; Kapoor, R. N. (Inorg. Chem. **8** [1969] 1792/4).
[14] King, R. B.; Kapoor, R. N.; Pannell, K. H. (J. Organometal. Chem. **20** [1969] 187/93).
[15] Makarova, L. G.; Ustynyuk, N. A.; Polovyanyuk, I. V.; Vinogradova, V. N. (4th Intern. Conf. Organometal. Chem., Bristol, Engl., 1969, K 6).
[16] Faller, J. W.; Anderson, A. S. (J. Am. Chem. Soc. **92** [1970] 5852/60).
[17] Haines, R. J.; Nolte, C. R. (J. Organometal. Chem. **24** [1970] 725/36).
[18] Nesmeyanov, A. N.; Makarova, L. G.; Ustynyuk, N. A. (J. Organometal. Chem. **23** [1970] 517/23).
[19] Ross, D. A. (Diss. Ohio State Univ. 1970; Diss. Abstr. Intern. B **41** [1971] 3905).
[20] Su, S. R.; Hanna, J. A.; Wojcicki, A. (J. Organometal. Chem. **21** [1970] P 21/P 22).

[21] Faller, J. W.; Anderson, A. S.; Jakubowski, A. (J. Organometal. Chem. **27** [1971] C 47/C 52).
[22] Makarov, L. G.; Ustynyuk, N. A.; Polovyanyuk, I. V. (5th Intern. Conf. Organometal. Chem., Moscow 1971, pp. 229, 614/5).

[23] Thomason, J. E.; Robinson, P. W.; Ross, D. A.; Wojcicki, A. (Inorg. Chem. **10** [1971] 2130/7).
[24] Su, S. R.; Wojcicki, A. (J. Organometal. Chem. **31** [1971] C 34/C 36).
[25] Barnett, K. W.; Pollmann, T. G. (unpublished results from Barnett, K. W.; Slocum, D. W.; J. Organometal. Chem. **44** [1972] 1/37).
[26] Bruce, M. I.; Goodall, B. L.; Shamrocks, D. N.; Stone, F. G. A. (J. Organometal. Chem. **39** [1972] 139/43).
[27] King, R. B. (AD-746829 [1972] 1/27).
[28] King, R. B.; Efraty, A. (J. Am. Chem. Soc. **94** [1972] 3768/73).
[29] Nesmeyanov, A. N.; Ustynyuk, N. A.; Bogatyreva, L. V.; Makarova, L. G. (Izv. Akad. Nauk SSSR Ser. Khim. **1973** 62/7; Bull. Acad. Sci. USSR Div. Chem. Sci. **1973** 55/60).
[30] Nesmeyanov, A. N.; Makarova, L. G.; Ustynyuk, N. A.; Bogatyreva, L. V. (J. Organometal. Chem. **46** [1972] 105/8).

[31] Nesmeyanov, A. N.; Makarova, L. G.; Ustynyuk, N. A.; Kvasov, B. A.; Bogatyreva, L. V. (J. Organometal. Chem. **34** [1972] 185/93).
[32] Craig, P. J.; Edwards, J. (Intern. Climax Conf. Chem. Uses Molybdenum, Reading, Engl., 1973, pp. 104/8).
[33] King, R. B.; Saran, M. S. (J. Am. Chem. Soc. **95** [1973] 1817/24).
[34] Barnett, K. W.; Pollmann, T. G. (J. Organometal. Chem. **69** [1974] 413/21).
[35] Collin, J.; Savignac, M.; Lambert, P. (J. Organometal. Chem. **82** [1974] C 19/C 22).
[36] Craig, P. J.; Edwards, J. (J. Less-Common Metals **36** [1974] 193/202).
[37] Su, S. R.; Wojcicki, A. (Inorg. Chim. Acta **8** [1974] 55/60).
[38] Churchill, M. R.; Chang, S. W.-Y. (Inorg. Chem. **14** [1975] 98/105).
[39] King, R. B.; Saran, M. S. (Inorg. Chem. **14** [1975] 1018/26).
[40] Su, S. R.; Wojcicki, A. (Inorg. Chem. **14** [1975] 89/98).

[41] Gingell, A. C.; Rest, A. J.; Turner, R. N. (J. Organometal. Chem. **121** [1976] 205/10).
[42] Lukehart, C. M.; Zeile, J. V. (J. Organometal. Chem. **105** [1976] 231/7).
[43] Van Doom, J. A.; Masters, C.; Volger, H. C. (J. Organometal. Chem. **105** [1976] 245/54).
[44] Beach, D. L.; Dattilo, M.; Barnett, K. W. (J. Organometal. Chem. **140** [1977] 47/54).
[45] Regler, D. L.; Dukes, M. D. (J. Organometal. Chem. **153** [1978] 67/72).
[46] Alt, H. G.; Schwärzle, J. A. (J. Organometal. Chem. **162** [1978] 45/56).
[47] Bell, P. B. (Diss. Ohio State Univ. 1978; Diss. Abstr. Intern. B **39** [1979] 4878).
[48] Collin, J.; Roustan, J. L.; Cadiot, P. (J. Organometal. Chem. **162** [1978] 67/78).
[49] Kolobova, N. E.; Zlotina, I. B.; Yudin, E. N. (Izv. Akad. Nauk SSSR Ser. Khim. **1978** 681/3; Bull. Acad. Sci. USSR Div. Chem. Sci. **1978** 588/90).
[50] Kreißl, F. R.; Uedelhoven, W.; Eberl, K. (Angew. Chem. **90** [1978] 908; Angew. Chem. Intern. Ed. Engl. **90** [1978] 860).

[51] Ross, D. A.; Wojcicki, A. (Inorg. Chim. Acta **28** [1978] 59/67).
[52] Todd, L. J.; Wilkinson, J. R.; Hickey, J. P.; Beach, D. L.; Barnett, K. W. (J. Organometal. Chem. **154** [1978] 151/7).
[53] Collin, J.; Roustan, J. L.; Cadiot, P. (J. Organometal. Chem. **169** [1979] 53/62).
[54] King, R. B.; Saran, M. D.; McDonald, D. P.; Diefenbach, S. P. (J. Am. Chem. Soc. **101** [1979] 1138/42).
[55] Uedelhoven, W.; Eberl, K.; Kreißl, F. R. (Chem. Ber. **112** [1979] 3376/89).
[56] Kunze, U.; Sastrawan, S. B. (Z. Naturforsch. **35 b** [1980] 421/7).
[57] Leung, T. W. (Diss. Ohio State Univ. 1980; Diss. Abstr. Intern. B **41** [1981] 3779).
[58] Alt, H. G.; Eichner, M. E. (J. Organometal. Chem. **212** [1981] 397/403).

[59] Mahmoud, K. A.; Narayanaswamy, R.; Rest, A. J. (J. Chem. Soc. Dalton Trans. **1981** 2199/204).
[60] Malisch, W.; Blau, H.; Haaf, F. J. (Chem. Ber. **114** [1981] 2956/70).

[61] Brix, H.; Beck, W. (J. Organometal. Chem. **234** [1982] 151/74).
[62] Lee, T.-W.; Liu, R.-S. (Organometallics **7** [1988] 878/83).
[63] Eberl, K.; Uedelhoven, W.; Wolfsgruber, M.; Kreißl, F. R. (Chem. Ber. **115** [1982] 504/12).
[64] Kazlauskas, R. J.; Wrighton, M. S. (J. Am. Chem. Soc. **104** [1982] 6005/15).
[65] Kegley, S. E.; Brookhart, M.; Husk, G. R. (Organometallics **1** [1982] 760/2).
[66] Kolobova, N. E.; Shripkin, V. V.; Rozantseva, T. V.; Struchkov, Yu. T.; Aleksandrov, G. G.; Antonovich, V. A.; Bakhmutov, V. I. (Koord. Khim. **8** [1982] 1655/63; Soviet J. Coord. Chem. **8** [1982] 910/8).
[67] Adams, H.; Bailey, N. A.; Cahill, P.; Rogers, D.; Winter, M. J. (J. Chem. Soc. Chem. Commun. **1983** 831/3).
[68] Bailey, N. A.; Chell, P. L.; Manuel, C. P.; Mukhopadhyay, A.; Rogers, D.; Tabbron, H. E.; Winter, M. J. (J. Chem. Soc. Dalton Trans. **1983** 2397/403).
[69] Scordia, H.; Kergoat, R.; Kubicki, M. M.; Guerchais, J. E.; L'Haridon, P. (Organometallics **2** [1983] 1681/7).
[70] Hooker, R. H.; Rest, A. J. (J. Chem. Soc. Dalton Trans. **1984** 761/70).

[71] Leont'eva, L. I.; Perevalova, E. G. (Izv. Akad. Nauk SSSR Ser. Khim. **1984** 2352/8; Bull. Acad. Sci. USSR Div. Chem. Sci. **1984** 2149/54).
[72] Lindner, E.; Kuester, E. U.; Hiller, W.; Fawzi, R. (Chem. Ber. **117** [1984] 127/41).
[73] Ustynyuk, N. A.; Vinogradova, V. N.; Korneva, V. N.; Kratsov, D. N.; Andrianov, V. G.; Struchkov, Y. T. (J. Organometal. Chem. **277** [1984] 285/96).
[74] Butters, T.; Winter, W.; Kunze, U.; Sastrawan, S. B. (Acta Crystallogr. C **41** [1985] 23/4).
[75] Kreißl, F. R.; Uedelhoven, W.; Neugebauer, D. (J. Organometal. Chem. **344** [1988] C 27/C 30).
[76] Hofmockel, U.; Greißinger, D.; Malisch, W. (12th Intern. Conf. Organometal. Chem., Vienna 1985, p. 441).
[77] Gauntlett, J. T.; Taylor, B. F.; Winter, M. J. (J. Chem. Soc. Dalton Trans. **1985** 1815/20).
[78] Adams, H.; Bailey, N. A.; Cahill, P.; Rogers, D.; Winter, M. J. (J. Chem. Soc. Dalton Trans. **1986** 2119/26).
[79] Gauntlett, J. T.; Winter, M. J. (Polyhedron **5** [1986] 451/9).
[80] Griessmann, K.-H.; Stasunik, A.; Angerer, W.; Malisch, W. (J. Organometal. Chem. **303** [1986] C 29/C 32).

[81] Leung, T. W.; Christoph, G. G.; Gallucc, J.; Wojcicki, A. (Organometallics **5** [1986] 366/74).
[82] Nolan, S. P.; de la Vega, R. L.; Mukerju, S. L.; Hoff, C. D. (Inorg. Chem. **25** [1986] 1160/5).
[83] Osborne, V. A.; Winter, M. J. (Polyhedron **5** [1986] 435/7).
[84] Solodovnikov, S. P.; Jumanskii, B. L.; Bubnov, N. N.; Kabachnik, M. I. (Izv. Akad. Nauk SSSR Ser. Khim **1986** 2147/50; Bull. Acad. Sci. USSR Div. Chem. Sci. **1986** 1960/2).
[85] Pannell, K. H.; Cea-Olivares, R.; Toscano, R. A.; Kapoor, R. N. (Organometallics **6** [1987] 1821/2).

1.5.1.3.2.1.9 Compounds of the Type [^{5}LMo(CO)$_2$(^{2}D)=C(R)R′]X

This chapter covers cationic carbene complexes of the general type [^{5}LMo(CO)$_2$-(^{2}D)=C(R)R′]X. For cis-trans isomers, orientation of the plane of the carbene ligand with respect to the basal plane of the ligands at Mo (upright or perpendicular), different conformers with respect to the position of R and R′ according to a rotation barrier around the Mo=C bond, and chirality of the cis isomers, see the neutral compounds ^{5}LMo(CO)$_2$(X)=C(R)R′ in Section 1.5.1.3.2.1.4.

[C_5H_5Mo(CO)$_2$(P(C_6H_5)$_3$)=CH_2]O_3SCF_3 was formed by treating C_5H_5Mo(CO)$_2$(P(C_6H_5)$_3$)CH_2OR (R = CH_3, $CH_2C_6H_5$, or C(O)C_4H_9-t) with $(CH_3)_3SiOSO_2CF_3$ in CH_2Cl_2 at −90°C. The reaction was monitored by ^{1}H NMR spectroscopy. Side products were small amounts of [C_5H_5Mo(CO)$_2$(P(C_6H_5)$_3$)=C(OR)H]O_3SCF_3 and C_5H_5Mo(CO)$_2$(P(C_6H_5)$_3$)CH_3 [8].

^{1}H NMR spectrum (CD_2Cl_2, −90°C): δ = 5.84 (d, C_5H_5; J(P, H) = 1.2 Hz), 15.4 (d, =CH_2; J(P, H) = 12.5 Hz) ppm. An upright conformation for the CH_2=Mo unit with a barrier of rotation of about 6.7 kcal/mol was concluded based on NMR data [8].

Disproportionation occurs at −70°C in solution to [C_5H_5Mo(CO)$_2$(P(C_6H_5)$_3$)CH_2=CH_2]$^+$ and [C_5H_5Mo(CO)$_2$P(C_6H_5)$_3$]$^+$, the latter stabilized by the solvent or triflate anion. Reaction with styrene in CH_2Cl_2 at −70°C affords phenylcyclopropane in > 50% yield [8].

[$(CH_3)_5C_5$Mo(CO)$_2$(P(C_6H_5)$_3$)=C(OH)H]O_2CCF_3 was formed by the protonation of trans/cis-$(CH_3)_5C_5$Mo(CO)$_2$(P(C_6H_5)$_3$)C(O)H in CD_2Cl_2 at −70°C with two equivalents of CF_3COOH. Solutions of the complex are stable up to +20°C. The cis/trans isomer ratio of 1/9 was identical with that of the starting acyl complex [11, 12].

^{1}H NMR spectrum (CD_2Cl_2, 0°C): δ = 1.74 (s, CH_3), 12.99 (d, =CH; J(P, H) = 14.6 Hz) ppm [11]; ($CDCl_3$, −60°C): δ = 1.74 (s, CH_3), 7.50 (m, C_6H_5), 9.27 (br, OH), 12.99 (d, =CH, trans isomer; J(P, H) = 12.6 Hz), 13.20 (m, =CH, cis isomer) ppm [12]. ^{13}C NMR spectrum ($CDCl_3$, −60°C): trans isomer: δ = 10.23 (CH_3), 106.18 (C_5), 131.0 (m, C_6H_5), 155.23 (q, CF_3; J(F, C) = 287.0 Hz), 244.80 (d, CO; J(P, C) = 27 Hz), 307.78 (d of d, =C; J(P, C) = 22.0, J(H, C) = 134.8 Hz) ppm [12]. IR spectrum (CH_2Cl_2): 1930, 1990 (ν(CO)) cm^{-1} [11].

[C_5H_5Mo(CO)$_2$(P(C_6H_5)$_3$)=C(OCH_3)H]X (X = O_3SCF_3 or AsF_6). The cation was obtained as a side product of the reaction of C_5H_5Mo(CO)$_2$(P(C_6H_5)$_3$)CH_2OCH_3 with $(CH_3)_3SiOSO_2CF_3$ in CD_2Cl_2 at −90°C or obtained in a 1:1 ratio with C_5H_5Mo(CO)$_2$(P(C_6H_5)$_3$)CH_3 by treatment with [C(C_6H_5)$_3$]AsF_6 in CD_2Cl_2 at −90°C [8].

^{1}H NMR spectrum (CD_2Cl_2): δ = 4.58 (s, CH_3), 5.56 (d, C_5H_5; J(P, H) = 1.2 Hz), 7 to 8 (m, C_6H_5), 12.10 (s, =CH) ppm [8].

[$(CH_3)_5C_5$Mo(CO)$_2$(P(C_6H_5)$_3$)=C(OCH_3)H]PF_6. The formyl complex, $(CH_3)_5C_5$Mo(CO)$_2$(P(C_6H_5)$_3$)-CHO (90% trans and 10% cis), was allowed to react with CH_3SO_3F at −90°C in CH_2Cl_2. Anion exchange with NH_4PF_6 and recrystallization from $CHCl_3$ gave microcrystals in 60% yield. The cis to trans ratio of the product depended on the mode of workup. Monitoring of the reaction at −60°C showed that the ratio was still 90% trans and 10% cis, but after recrystallization the ratio was 70:30 [11, 12].

^{1}H NMR spectrum ($CDCl_3$): trans isomer: δ = 1.95 (s, $(CH_3)_5C_5$), 4.02 (s, CH_3O), 7.60 (m, C_6H_5), 12.45 (d, =CH; J(P, H) = 5.5 Hz); cis isomer: δ = 1.96 (s, $(CH_3)_5C_5$), 4.80 (s, CH_3O), 7.60 (m, C_6H_5), 12.52 (d, =CH; J(P, H) = 1.5 Hz) ppm. ^{13}C NMR spectrum ($CDCl_3$, −50°C): trans isomer: δ = 10.70 ((**C**H_3)$_5C_5$), 66.38 (CH_3O), 108.81 (**C**$_5$(CH_3)$_5$), 118.49 (d, C_6H_5, C-1; J(P, C) = 87.5 Hz), 131.64 (m, C_6H_5), 227.57 (d, CO; J(P, C) = 30.5 Hz), 339.33 (d, =CH; J(P, C) = 25.4 Hz); cis isomer: δ = 10.66 ((**C**H_3)$_5C_5$), 64.94 (CH_3O), 109.37 (**C**$_5$(CH_3)$_5$), 118.49 (d, C_6H_5, C-1; J(P, C) = 87.5 Hz), 131.64 (m, C_6H_5), 233.62 (d, CO; J(P, C) = 4 Hz), 235.81 (d, CO; J(P, C) =

29.7 Hz), 330.58 (d, =CH; J(P, C) = 6.8 Hz) ppm. ^{31}P NMR spectrum ($CDCl_3$, −50°C): δ = −143.81 (PF_6; J(F, P) = 713 Hz), 52.79 (cis isomer), 54.62 (trans isomer) ppm. IR spectrum (Nujol): 1130 (ν(PF_6)), 1935, 1995, 2030 (ν(CO)) cm^{-1} [11, 12].

The complex is air-sensitive but thermally stable [11, 12].

[$C_5H_5Mo(CO)_2(P(C_6H_5)_3)$=C(OR)H]X (X = O_3SCF_3 or AsF_6; R = $CH_2C_6H_5$ or $C(O)C_4H_9$-t) was observed in small amounts together with $C_5H_5Mo(CO)_2(P(C_6H_5)_3)CH_3$ in the reaction of $C_5H_5Mo(CO)_2(P(C_6H_5)_3)CH_2OR$ with $(CH_3)_3SiOSO_2CF_3$ in CD_2Cl_2 at −90°C or $[C(C_6H_5)_3]AsF_6$ in CH_2Cl_2 at −70°C. The complex was not isolated [8].

[$(CH_3)_5C_5Mo(CO)_2(P(C_6H_5)_2CH_3)$=C(OSi($CH_3)_3$)H]$O_3SCF_3$ was obtained by treating $(CH_3)_5C_5$-$Mo(CO)_2(P(C_6H_5)_2CH_3)CHO$ with 1.2 equivalents of $CF_3S(O)_2OSi(CH_3)_3$ in CH_2Cl_2 at 0°C for a few minutes. The thermally unstable red oily compound was a mixture of 70% trans and 30% cis isomer [12].

^{1}H NMR spectrum (CD_2Cl_2, 27°C): cis isomer: δ = 0.02 (s, CH_3Si), 1.79 (s, $(CH_3)_5C_5$), 2.01 (d, CH_3P; J(P, H) = 8.5 Hz), 7.50 (m, C_6H_5), 13.15 (d, =CH; J(P, H) = 6.0 Hz); trans isomer: δ = 0.01 (s, CH_3Si), 1.77 (s, $(CH_3)_5C_5$), 2.01 (d, CH_3P; J(P, H) = 8.5 Hz), 7.50 (m, C_6H_5), 12.80 (d, =CH; J(P, H) = 8.2 Hz) ppm [12]. ^{13}C NMR spectrum (CD_2Cl_2, −60°C): cis isomer: δ = 2.16 (CH_3Si), 10.87 ((**C**$H_3)_5C_5$), 19.8 (d, CH_3P; J(P, C) = 31.8 Hz), 109.66 (**C**$_5$ $(CH_3)_5$), 131.0 (m, C_6H_5), 178.64 (q, CF_3; J(F, C) = 318 Hz), 248.50 (d, CO; J(P, C) = 24.2 Hz), 341.69 (br, =C); trans isomer: δ = 1.02 (CH_3Si), 10.87 ((**C**$H_3)_5C_5$), 18.0 (d, CH_3P; J(P, C) = 32.0 Hz), 109.25 (**C**$_5(CH_3)_5$), 131.0 (m, C_6H_5), 178.64 (q, CF_3; J(F, C) = 318 Hz), 238.10 (d, CO; J(P, C) = 24.2 Hz), 341.69 (br, =C) ppm [12]. ^{31}P NMR spectrum (CD_2Cl_2, −60°C): δ = 37.09 (trans isomer), 54.50 (cis isomer) ppm [12].

The complex decomposes slowly at room temperature over 24 h but is indefinitely stable at −20°C [12].

[$C_5H_5Mo(CO)_2(P(C_6H_5)_3)$=C($CH_3$)$OCH_3$]$PF_6$ · $CH_3C(O)CH_3$. $C_5H_5Mo(CO)_2(P(C_6H_5)_3)C(O)CH_3$ was allowed to react with CH_3OSO_2F in C_6H_6. Anion exchange with NH_4PF_6 was carried out in acetone. Bright yellow needles were obtained from acetone/hexane melting at 126 to 129°C. The acetone of crystallization could not be removed by drying under vacuum at 39°C for 10 h [4, 5].

Molar conductivity (CH_2Cl_2): Λ = 54.7 $cm^2 \cdot \Omega^{-1} \cdot mol^{-1}$. ^{1}H NMR spectrum ($CD_3CN$): δ = 3.11 (s, CH_3C), 4.33 (s, CH_3O), 5.43 (s, C_5H_5), 7.48 (m, C_6H_5) ppm. IR spectrum (CH_2Cl_2): 1712 (ν(C=O)), 1916, 1997 (ν(CO)) cm^{-1} [5].

The complex is air-sensitive; surface oxidation affords a green oxidation product [4, 5]. Reaction with PR_3 (R not specified) affords $[CH_3PR_3]^+$ and $C_5H_5Mo(CO)_2(P(C_6H_5)_3)C(O)CH_3$ [4].

[$C_5H_5Mo(CO)_2(P(C_6H_{11}$-cyclo$)_3$)=C($CH_3$)$OC_2H_5$]$BF_4$ · 0.5 CH_2Cl_2 was obtained in 80% yield as a yellow-green crystalline solid by treating $C_5H_5Mo(CO)_2(P(C_6H_{11}$-cyclo$)_3)C(O)CH_3$ with $[O(C_2H_5)_3]BF_4$ in CH_2Cl_2 [2].

^{1}H NMR spectrum (acetone-d_6): δ = 1.62 (m, C_6H_{11}-cyclo), 3.25 (=CCH_3), 4.80 (q, CH_2), 5.87 (d, C_5H_5; J(P, H) = 0.8 Hz) ppm. IR spectrum: 1280 (OC_2H_5), 1900, 1990 (ν(CO)) cm^{-1} [2].

[$C_5H_5Mo(CO)_2(P(C_6H_5)_3)$=$CF_2$]$SbF_6$ was formed by the reaction of trans-$C_5H_5Mo(CO)_2$-$(P(C_6H_5)_3)CF_3$ with a 10% excess of SbF_5 in liquid SO_2. Attempts to isolate the complex afforded only the disproportionation product $[C_5H_5Mo(CO)_3P(C_6H_5)_3]SbF_6$ [6, 7].

^{1}H NMR spectrum (liquid SO_2): δ = 5.2 (s, C_5H_5), 6.9 (m, C_6H_5) ppm. ^{13}C NMR spectrum (liquid SO_2): δ = 98.3 (C_5H_5), 132.7 (m, C_6H_5), 264.1 (d of t, =CF_2; J(F, C) = 392, J(P, C) = 100 Hz) ppm. ^{19}F NMR spectrum (liquid SO_2): δ = 239.04 (s, =CF_2) ppm [6, 7].

References on pp. 110/1

cis-[$C_5H_5Mo(CO)_2(P(C_6H_5)_3)$=$CO(CH_2)_3$][$B(C_6H_5)_4$] (Formula I; R = H) was obtained in quantitative yield by stirring the corresponding Br^- salt in a concentrated solution of $Na[B(C_6H_5)_4]$. The yellow precipitate melts at 152 to 153°C [1].

^1H NMR spectrum ($CDCl_3$): δ = 2.33 (complex t, 2 CH_2), 3.46 and 4.10 (br m, CH_2), 5.16 (s, C_5H_5), 7.16 (m, C_6H_5) ppm. IR spectrum (CH_2Cl_2): 1930, 1990 (ν(CO)) cm^{-1} [1].

Kinetic data for cis to trans conversion in CH_3CN at 298 K are: $k_i = 3.43 \pm 0.42 \times 10^{-6}$ s^{-1}, $\Delta H^{\ddagger} = 24.1 \pm 1.1$ kcal/mol, $\Delta S^{\ddagger} = -3 \pm 3$ cal · mol^{-1} · K^{-1}. The reaction follows first-order kinetics [3].

trans-[$C_5H_5Mo(CO)_2(P(C_6H_5)_3)$=$CO(CH_2)_3$][$B(C_6H_5)_4$] (Formula I; R = H) was obtained in quantitative yield by stirring the corresponding Br^- salt in a concentrated solution of $Na[B(C_6H_5)_4]$. The yellow precipitate melts at 159 to 161°C [1].

^1H NMR spectrum ($CDCl_3$): δ = 1.46 (br qui, 2 CH_2), 3.06 and 4.35 (br t, CH_2), 5.10 (d, C_5H_5; J(P, H) ≈ 1.0 Hz), 7.20 (m, C_6H_5) ppm. IR spectrum ($CHCl_3$): 1910, 1985 (ν(CO)) cm^{-1} [3].

For kinetic data concerning cis to trans interconversion compare the cis isomer [3].

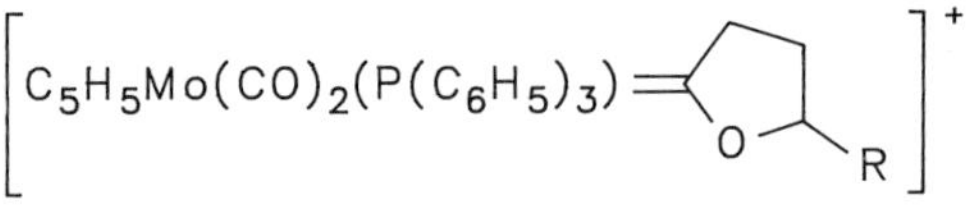

$$\left[C_5H_5Mo(CO)_2(P(C_6H_5)_3)=\text{C}\langle\text{CH}_2\text{CH}_2\text{CH(R)O}\rangle\right]^+ \qquad \left[C_5H_5Mo(CO)_2(P(C_6H_5)_3)=\text{C}\langle\text{CH}_2\text{CH}_2\text{CH}_2\text{CH}_2\text{O}\rangle\right]^+$$

I 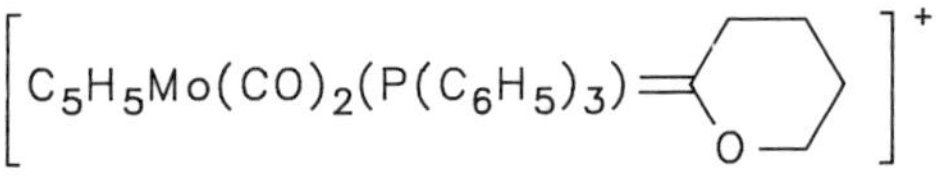 II

cis-[$C_5H_5Mo(CO)_2(P(C_6H_5)_3)$=$CO(CH_2)_3$]Br (Formula I; R = H). $C_5H_5Mo(CO)_3(CH_2)_3Br$ was allowed to react with three equivalents of $P(C_6H_5)_3$ in CH_3CN for 15 min. A bright yellow precipitate was obtained in 71% yield; m.p. 140 to 143°C [1].

^1H NMR spectrum ($CDCl_3$): δ = 1.60 (br m, CH_2), 3.67 (complex sext, CH_2), 4.61 (complex sext, CH_2), 5.83 (s, C_5H_5), 7.46 (m, C_6H_5) ppm. IR spectrum (CH_2Cl_2): 1910, 1980 (ν(CO)) cm^{-1}. For conductivity information compare the trans isomer [1].

Kinetic data for the isomerization into the trans isomer of a ca. 0.3 M solution in $CDCl_3$ at 27 ± 1°C were obtained by ^1H NMR: $k = 1.3 \pm 0.1 \times 10^{-5}$ s^{-1}, $\tau_{1/2}$ = 14.8 h, and a limiting cis to trans ratio of 0.025 [1]; in CH_3CN at 298 K: $k = 1.17 \pm 0.17 \times 10^{-5}$ s^{-1}, $\Delta H^{\ddagger} = 22.4 \pm 1.0$ kcal/mol, $\Delta S^{\ddagger} = 5 \pm 4$ cal · mol^{-1} · K^{-1} [3].

Reduction of the carbene function with $Li[(C_2H_5)_3BH]$ in THF affords trans-C_5H_5-$Mo(CO)_2(P(C_6H_5)_3)CH(CH_2)_3O$-cyclo (No. 46, p. 126) [9, 10].

trans-[$C_5H_5Mo(CO)_2(P(C_6H_5)_3)$=$CO(CH_2)_3$]Br (Formula I; R = H) $C_5H_5Mo(CO)_3(CH_2)_3Br$ was allowed to react with 2.3 equivalents of $P(C_6H_5)_3$ in CH_3CN for 62 h. The solvent was removed and the residue was washed with CH_2Cl_2/hexane 1:1 and pentane to give a pale yellow powder; m.p. 150 to 160°C [1].

The conductivity of Λ = 82.7 $cm^2 \cdot \Omega^{-1} \cdot mol^{-1}$ in methanol indicates a 1:1 electrolyte. ^1H NMR spectrum ($CDCl_3$): δ = 2.16 (br qui, CH_2), 4.06 (br t, CH_2), 5.23 (br t, CH_2), 5.56 (d, C_5H_5; J(P, H) ≈ 1.1 Hz), 7.45 (m, C_6H_5) ppm. IR spectrum (CH_2Cl_2): 1910, 1985 (ν(CO)) cm^{-1} [1].

Reaction with two equivalents of pyridine N-oxide in CH_2Cl_2 affords trans-C_5H_5-$Mo(CO)_2(P(C_6H_5)_3)Br$ and γ-butyrolactone [1].

[$C_5H_5Mo(CO)_2(P(C_6H_5)_3)$=$COCH(CH_3)(CH_2)_2$]Br (Formula I; R = CH_3). $C_5H_5Mo(CO)_3CH_2$-$CH_2CH(CH_3)Br$ and two equivalents of $P(C_6H_5)_3$ were stirred in CH_3CN for 20 min. The solvent

References on pp. 110/1

was removed and the residue was washed with pentane/CH_2Cl_2 6:1 to give 98.5% yield as a bright yellow solid which was a 97:3 mixture of cis and trans isomers; m.p. 105 to 110°C [3].

^1H NMR ($CDCl_3$): δ = 0.60 and 1.26 (d, ca. 1.5 CH_3; J(H, H) $\approx$ 7 Hz), 0.83 (br m, CH_2 and CH), 4.00 (br m, CH_2), 5.50 (d, ca. 0.15 C_5H_5, trans isomer; J(P, H) $\approx$ 1 Hz), 5.80 (s, C_5H_5, cis isomer), 7.43 (br m, C_6H_5) ppm. The cis isomer possesses two chiral centers, one at molybdenum and another at $CHCH_3$. A diastereotopic splitting of ca. 1.2 Hz of the C_5H_5 resonance is only observable in polar solvents such as CH_3CN, CH_3OD, dimethylformamide-d_7, or dimethyl sulfoxide-d_6. The optical center at the molybdenum atom racemized slowly on the NMR time scale up to 90°C in dimethyl sulfoxide-d_6, at which point the isomerization became rapid enough to prevent measurement of the C_5H_5 resonance of the cis isomer [3].

trans-[$C_5H_5Mo(CO)_2(P(C_6H_5)_3)$=$CO(CH_2)_4$][$B(C_6H_5)_4$] (Formula II). trans-$C_5H_5Mo(CO)_2$-$(P(C_6H_5)_3)C(O)(CH_2)_4X$ (X = Br, I) and $AgBF_4$ were stirred for 6 h in CH_3NO_2. After filtration and removal of the solvent the residue was treated with methanolic Na[$B(C_6H_5)_4$]. Purification by column chromatography on Florisil gave a 60% (X = Br) to 74% (X = I) yield of a yellow solid, m.p. 139.5 to 141°C. The kinetics of the formation were studied in CH_3CN. It was found that the rate depends markedly on the concentration of $AgBF_4$ [3].

^1H NMR spectrum ($CDCl_3$): δ = 1.67 (br m, 2 CH_2), 3.34 (br m, 1 CH_2), 4.40 (br m, 1 CH_2), 5.26 (d, C_5H_5; J(P, H) $\approx$ 1 Hz), 7.20 (br m, C_6H_5) ppm. IR spectrum (CH_2Cl_2): 1900, 1975 (ν(CO)) cm^{-1} [3].

trans-[$C_5H_5Mo(CO)_2(P(C_6H_5)_3)$=$COC(R)HC_2H_4$]X (Formula I; R = OCH_3, OC_2H_5, SC_2H_5, OC_3H_7-n, OC_3H_7-i, OC_4H_9-t, X = BF_4; R = OC_2H_5, X = $B(C_6H_5)_4$). trans-$C_5H_5Mo(CO)_2(P(C_6H_5)_3)$-$C(O)(CH_2)_2CH(OC_2H_5)_2$ was allowed to react with 1.33 equivalents of $BF_3 \cdot O(C_2H_5)_2$ in ether at 25°C for 15 min. The product (R = OC_2H_5, X = BF_4) precipitated. The obtained complex was stirred with an excess of R′EH (R′ = n-C_3H_7, i-C_3H_7, t-C_4H_9, E = O; R′ = C_2H_5, E = S) in acetone for 2 h or with methanol as solvent for 5 min. The product precipitated from the reaction mixture after the addition of ether. Treating with Na[$B(C_6H_5)_4$] in ethanol leads to the formation of the corresponding $[B(C_6H_5)_4]^-$ salt. Yields and properties are given in the following table [13].

R/X	yield and properties
OCH_3/BF_4	56% of a bright yellow solid, dec. 188 to 190°C ^1H NMR (CD_3CN): 2.41 (m, diastereomeric CH_2), 3.68 (s, CH_3), 3.82 (t, CH_2C), 5.49 (d, C_5H_5; J(P, H) $\approx$ 1.3), 6.10 (t, CH), 7.48 (m, C_6H_5) IR (CH_2Cl_2): 1914, 1918 (ν(CO))
OC_2H_5/BF_4	91% yield of a pale lemon-yellow solid, dec. 164 to 166°C ^1H NMR ($CDCl_3$): 1.30 (t, CH_3), 1.75 and 2.26 (m, each one diastereomeric H), 3.86 (t, CH_2C), 4.05 (m, CH_2O), 5.43 (d, C_5H_5; J(P, H) $\approx$ 1.3), 6.35 (t, CH), 7.49 (m, C_6H_5) IR (CH_2Cl_2): 1915, 1994 (ν(CO))
$OC_2H_5/B(C_6H_5)_4$	high yield of a yellow solid ^1H NMR ($CDCl_3$): 1.19 (t, CH_3), 1.45 (br m, diastereomeric CH_2), 3.65 (t, CH_2C), 3.86 (m, CH_2O), 5.01 (d, C_5H_5; J(P, H) $\approx$ 1.3), 5.49 (t, CH), 6.8 to 7.4 (m, C_6H_5) IR (CH_2Cl_2): 1915, 1995 (ν(CO))

References on pp. 110/1

R/X	yield and properties
SC_2H_5/BF_4	15% yield of a bright yellow needles, dec. 142 to 144°C ^{1}H NMR ($CDCl_3$): 1.41 (t, CH_3), 1.66 and 2.47 (m, each one diastereomeric H), 2.85 (q, CH_2S), 3.86 (t, CH_2C), 5.4 (d, C_5H_5; J(P, H) ≈ 1.3), 6.46 (t, CH), 7.39 (m, C_6H_5) IR (CH_2Cl_2): 1913, 1992 (ν(CO))
OC_3H_7-n/BF_4	55% yield of a lemon-yellow solid, dec. 165 to 166.5°C ^{1}H NMR ($CDCl_3$): 0.92 (t, $C\mathbf{H}_3CH_2$), 1.66 (m, $C\mathbf{H}_2CH_3$ and one diastereomeric H), 2.26 (m, one diastereomeric H), 3.83 (m, CH_2C and CH_2O), 5.40 (d, C_5H_5; J(P, H) ≈ 1.3), 6.31 (t, CH), 7.42 (m, C_6H_5) IR (CH_2Cl_2): 1913, 1988 (ν(CO))
OC_3H_7-i/BF_4	76% yield of a light yellow solid, dec. 161 to 163°C ^{1}H NMR ($CDCl_3$): 1.30 (d of d, 2 CH_3), 1.69 and 2.29 (m, each one diastereomeric H), 3.83 (t, CH_2C), 4.23 (m, i-C_3H_7, CH), 5.4 (d, C_5H_5; J(P, H) ≈ 1.3), 6.40 (t, CH), 7.42 (m, C_6H_5) IR (CH_2Cl_2): 1914, 1987 (ν(CO))
OC_4H_9-t/BF_4	40% yield of a light yellow solid, dec. 153 to 155°C ^{1}H NMR ($CDCl_3$): 1.39 (t-C_4H_9), 1.63 and 2.17 (m, each one diastereomeric H), 3.82 (t, CH_2C), 5.36 (d, C_5H_5; J(P, H) ≈ 1.3), 6.58 (t, CH), 7.42 (m, C_6H_5) IR (CH_2Cl_2): 1909, 1988 (ν(CO))

trans-[$C_5H_5Mo(CO)_2(P(C_6H_5)_3)$=$COC(OC_2H_5)HC_2H_4$]$BF_4$ is oxidized with five equivalents of pyridine N-oxide in CH_2Cl_2 for 2 h to afford 4-ethoxy-γ-butyrolactone. The reaction with $NaOCH_3$ gives trans-$C_5H_5Mo(CO)_2(P(C_6H_5)_3)C(O)(CH_2)_2C(OCH_3)(OC_2H_5)H$ in CH_2Cl_2 and trans-C_5H_5-$Mo(CO)_2(P(C_6H_5)_3)C(O)(CH_2)_2C(O)H$ in dimethyl sulfoxide [13].

[$C_5H_5Mo(CO)_2(P(OCH_3)_3)$=$C(CH_3)N(C_2H_4)_2O$]$BF_4$ (Formula IV). $C_5H_5Mo(CO)(P(OCH_3)_3)C(O)$-$C(=CH_2)N(C_2H_4)_2O$ (Formula III) was protonated in CH_2Cl_2 with HBF_4. The obtained yellow solid melts at 103 to 105°C [14].

^{1}H NMR spectrum (acetone-d_6): δ = 2.97 (CH_3C), 3.88 (d, OCH_3; J(P, H) = 12 Hz), 5.85 (d, C_5H_5; J(P, H) = 1.5 Hz) ppm. IR spectrum (KBr): 1510 (ν(C=N)); (CH_2Cl_2): 1910, 1988 (ν(CO)) cm^{-1} [14].

References:

[1] Cotton, F. A.; Lukehart, C. M. (J. Am. Chem. Soc. **93** [1971] 2672/6).

[2] Green, M. L. H.; Mitchard, L. C.; Swanwick, M. G. (J. Chem. Soc. A **1971** 794/7).

[3] Cotton, F. A.; Lukehart, C. M. (J. Am. Chem. Soc. **95** [1973] 3552/64).
[4] Wagner, K. P. (Diss. Univ. Wisconsin 1974; Diss. Abstr. Intern. B **35** [1975] 3235).
[5] Treichel, P. M.; Wagner, K. P. (J. Organometal. Chem. **88** [1975] 199/206).
[6] Dukes, M. D. (Diss. Univ. South Carolina 1977; Diss. Abstr. Intern. B **38** [1978] 5925).
[7] Reger, D. L.; Dukes, M. D. (J. Organometal. Chem. **153** [1978] 67/72).
[8] Kegley, S. E.; Bookhart, M.; Husk, G. R. (Organometallics **1** [1982] 760/2).
[9] Adams, H.; Bailey, N. A.; Cahill, P.; Rogers, D.; Winter, M. J. (J. Chem. Soc. Chem. Commun. **1983** 831/3).
[10] Adams, H.; Bailey, N. A.; Cahill, P.; Rogers, D.; Winter, M. J. (J. Chem. Soc. Dalton Trans. **1986** 2119/26).

[11] Asdar, A.; Lapinte, C. (J. Organometal. Chem. **327** [1987] C 33/C 36).
[12] Asdar, A.; Lapinte, C.; Toupet, L. (Organometallics **8** [1989] 2708/17).
[13] Lukehart, C. M.; Zeile, J. V. (J. Organometal. Chem. **105** [1976] 231/7).
[14] Brix, H.; Beck, W. (J. Organometal. Chem. **234** [1982] 151/74).

1.5.1.3.2.1.10 Compounds with One Chelate-Bonded ^{1}L Ligand (^{1}L-^{2}D)

The compounds described in this section contain one ^{1}L ligand chelate-bonded to molybdenum with formation of a three- to six-membered metallacycle. The ^{5}L ligand is in the most cases C_5H_5, some complexes with $^5L = CH_3C_5H_4$, $(CH_3)_5C_5$, and indenyl are also known. The complexes are arranged first on order of increasing ring size followed by increasing group number of the donor element. The complexes were obtained in many cases by the following methods.

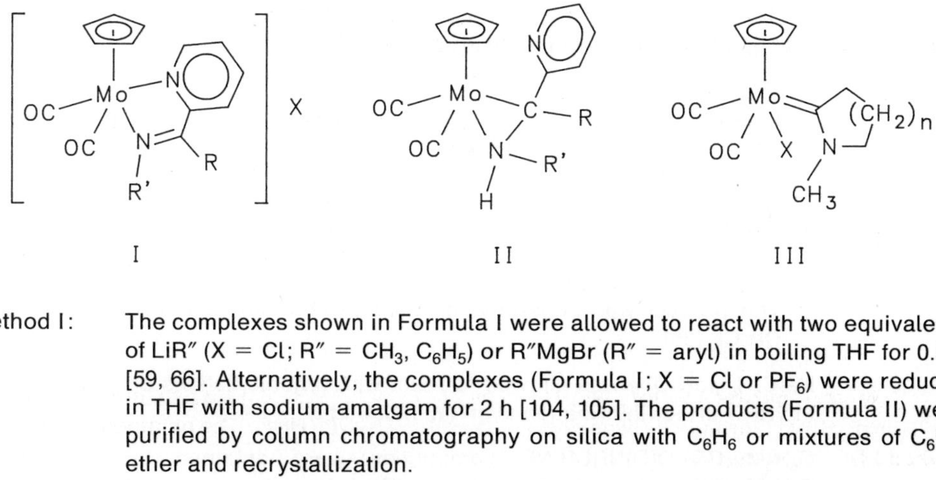

I II III

Method I: The complexes shown in Formula I were allowed to react with two equivalents of LiR″ (X = Cl; R″ = CH_3, C_6H_5) or R″MgBr (R″ = aryl) in boiling THF for 0.5 h [59, 66]. Alternatively, the complexes (Formula I; X = Cl or PF_6) were reduced in THF with sodium amalgam for 2 h [104, 105]. The products (Formula II) were purified by column chromatography on silica with C_6H_6 or mixtures of C_6H_6/ether and recrystallization.

Method II:

a. A 20-molar excess of an aziridine, $N(R^1)C(R^2)(R^3)C(R^4)R^5$-cyclo, was added to a solution of $C_5H_5Mo(CO)_3H$ in THF cooled at −78°C. The mixture was allowed to warm up, stirred for 4 h, and then cooled at −20°C overnight [87]. The mechanism of the reaction was discussed [41].
b. NH_3 was passed through a solution of $[C_5H_5Mo(CO)_3CH_2{=}CHR]PF_6$ in C_6H_6. After 30 min the precipitate was isolated and dissolved in acetone. The solution was treated with 10% aqueous NaOH [37, 87].

References on pp. 173/6

Method III: $^5LMo(CO)_3CH_3$ was irradiated in pentane in the presence of an excess of alkyne for 20 to 60 min followed by chromatography and recrystallization [49, 50, 92, 112]. Thermal reaction of $^5LMo(CO)_3CH_3$ with alkynes for prolonged times was also possible [16, 64].

Method IV: A solution of $Li[(C_2H_5)_3BH]$ in THF was added to a solution cis-$C_5H_5Mo(CO)_2$-(=$CN(CH_3)CH_2(CH_2)_nCH_2)I$ (Formula III; n = 1, 2; X = I) in THF. The resulting solution was stirred for 2 h [101].

Method V: CH_3I was added to a solution of $Na[C_5H_5Mo(CO)_2C{\equiv}NR]$ (R = CH_3, C_6H_5) in THF cooled to −78 °C. The mixture was allowed to warm to ambient temperature and stirred for an additional 2 h [38, 48, 53].

Method VI: $[(CH_3)_2C{=}C{=}NR_2]Cl$ in THF was added dropwise to a solution of $Na[C_5H_5Mo(CO)_3]$ in THF. The resulting mixture was stirred for 24 h. After removal of the solvent the residue was extracted with CH_2Cl_2 [35].

Method VII: Reactions of $[C_5H_5Mo(CO)_2P(C_6H_5)_2]^{2-}$.

a. A solution of $Na_2[C_5H_5Mo(CO)_2P(C_6H_5)_2]$ in THF was added dropwise during 6 h to a solution of a 3- to 4-molar excess of $RCHCl_2$ in hexane. The products were separated by medium pressure liquid chromatography on silica with n-hexane/ether (7:1) [91].
b. $RCHCl_2$ in THF was added during 2 h to a solution of $Li_2[C_5H_5Mo(CO)_2$-$P(C_6H_5)_2]$ in THF cooled to 0 °C. The reaction mixture was allowed to warm up. After removal of the solvent the residue was extracted with ether/hexane (1:5) and the extracts chromatographed on Al_2O_3 [17].

Method VIII: $^5LMo(CO)_2{=}PR_2$ was treated with CH_2N_2 to yield $^5LMo(CO)_2CH_2PR_2$ [89, 98, 99, 103, 108].

Method IX: $C_5H_5Mo(CO)_3R$ or $C_5H_5Mo(CO)_2(I){=}C(CH_2)_3O$-cyclo was reduced with 1.2 to 1.5 equivalents of $Li[(C_2H_5)_3BH]$ in THF [78, 88, 95, 96, 100, 106].

Method X: $C_5H_5Mo(CO)_3H$ was allowed to react with $(t\text{-}C_4H_9)_2EN{=}S{=}NE(C_4H_9\text{-}t)_2$ in pentane for 2 h at −30 °C (E = P) or in refluxing hexane for 3 h (E = As) [83].

Method XI: Hydrolysis or alcoholysis of $C_5H_5Mo(CO)_3C(Cl){=}C(CN)_2$.

a. The complex was stirred in CH_2Cl_2 in the presence of alumina for 10 to 15 min. The resulting slurry was chromatographed [36].
b. The complex was treated in ROH (R = CH_3, C_2H_5) with NaOH for 5 to 10 min. The product was separated by column chromatography on Florisil with CH_2Cl_2 [36].

Method XII: $Na[C_5H_5Mo(CO)_3]$ was allowed to react with organic halides or diazoacetates in THF. Further information is given in the table.

Method XIII: $C_5H_5Mo(CO)_2C(O)NRCR'NR$ compounds were obtained by addition of $C_5H_5Mo(CO)_3Cl$ to a solution of $K[CR(NR')_2]$ (from the appropriate amidine and KOC_4H_9-t) in benzene [55, 60, 63] or $Li[CR(NR')_2]$ (from the appropriate amidine and LiC_4H_9) in ether [55]. They were also formed as major compounds when the reaction was carried out with $M[CR(NR')_2]$ (M = Li, K) in THF at −35 °C in the presence of CS_2; CS_2 insertion products were formed in minor amounts [70]. The products were isolated in many cases by column chromatography on silica with ether or ether/C_6H_6 [60] or extraction with $CHCl_3$ [55]. Further information is given in the table.

References on pp. 173/6

Method XIV: A solution of $C_5H_5Mo(CO)_3CH_2C{\equiv}CH$ in THF containing a slow excess of ROH ($R = CH_3$, C_4H_9-t, $CH_2C_6H_5$, or C_6H_5) cooled at −20°C was allowed to warm to +10°C and stirred for 4 to 12 h. The products were purified by chromatography on alumina with C_6H_6/pentane (1:1) [20, 62].

Method XV: $[C_5H_5Mo(CO)(CH_3C{\equiv}CR)_2]BF_4$ ($R = CH_3$, C_2H_5) was allowed to react with $K[(s\text{-}C_4H_9)_3BH]$ in THF at −78°C. After 5 min CO was passed through the solution and the mixture was allowed to warm. The products were isolated by column chromatography on alumina with ether/hexane (1:4) [79].

Method XVI: $[N(P(C_6H_5)_3)_2][C_5H_5Mo(CO)_3]$ and $CH_3C{\equiv}C(CH_2)_nI$ (n = 3, 4) were refluxed for 1 to 3 h in THF. The products were purified by column chromatography on alumina with C_6H_6 [69].

Method XVII: The complexes, $^5LMo(CO)_2(ER_2Y)X$ (E = P, As; Y has a halogen in 4-position), were treated with an excess of sodium amalgam in THF or ether for several hours [47, 90, 91]. After removal of the solvents the products were often extracted with hexane [90].

Table 7
Compounds with One Chelate-Bonded ^1L Ligand.
An asterisk indicates further information at the end of the table.
For explanations, abbreviations, and units see p. X.

No.	compound	method of preparation (yield) properties and remarks
compounds with a three-membered Mo-C-N ring		
1	$C_5H_5\overline{Mo(CO)_2CH_2N}(CH_3)_2$	$(CO)_3Mo(NCCH_3)(I)CH_2N(CH_3)_2$ was allowed to react with 1 equivalent of NaC_5H_5 in THF (43% by sublimation) [32] yellow crystals, subl. at 40 to 50°C/0.05 Torr [32] ^1H NMR ($CDCl_3$): 2.52 (s, CH_2), 2.69 (s, CH_3), 5.36 (s, C_5H_5) [32] IR (cyclohexane): 1844, 1936 (ν(CO)) [32] mass spectrum (70 eV, 50°C): $[M]^+$ (34%), $[C_5H_5Mo(CO)_2C_3H_5]^+$ (19%), $[C_5H_5Mo(CO)CH_2N(CH_3)_2]^+$ (42%), $[C_5H_5Mo(CO)C_3H_5]^+$ (23%), $[C_5H_5Mo(CO)_2]^+$ (100%), $[C_5H_5MoC_3H_5]^+$ (28%), $[C_5H_5MoC_3H_3]^+$ (46%), $[C_5H_5MoCO]^+$ (42%), $[C_5H_5Mo]^+$ (47%), $[MoC_3H_3]^+$ (21%) [32]
2	$C_5H_5\overline{Mo(CO)_2CH(C_5H_4N\text{-}2)N}HCH_3$	I, with $LiCH_3$ (23% by chromatography on silica with C_6H_6/ether 30:1) [59, 66] wine-red prisms (from ether/pentane 2:1), m.p. 78°C (dec.) [66] ^1H NMR ($CDCl_3$): 2.83 (d, CH_3; J(H, H) = 5.5), 3.75 (d, CH; J(H, H) = 7), 4.98 (s, C_5H_5), 6.81 (m, C_5H_4N, H-5), 7.51 (m, C_5H_4N, H-3), 8.16

References on pp. 173/6

Table 7 (continued)

No.	compound	method of preparation (yield) properties and remarks
2 (continued)		(m, C_5H_4N, H-6); H-4 of C_5H_4N covered by other resonances [66] $^{13}C\{^1H\}$ NMR ($CDCl_3$, 0.05 M $Cr(OC(CH_3)CHC(CH_3)O)_3$): 46.6 (CH_3), 48.0 (MoC), 93.1 (C_5H_5), 119.1 to 148.1 (C_5H_4N), 165.3 (C_5H_4N, C-2), 247.9 and 251.4 (CO) [66] IR (KBr): 1590 to 1600 (ν(CN)), 1800 to 1825, 1900 to 1925 (ν(CO)), 3145 to 3230 (ν(NH)); after exchange with D_2O in CCl_4, ND bands were found at 1220 to 1270 and 2380 [66] mass spectrum: $[M-n\,CO]^+$ (n = 0 to 2), $[C_5H_5MoC_5H_4N]^+$ [66]
*3	$C_5H_5\overline{Mo(CO)_2CH(C_5H_4N\text{-}2)N}HC_{11}H_{19}$ ($C_{11}H_{19}$ 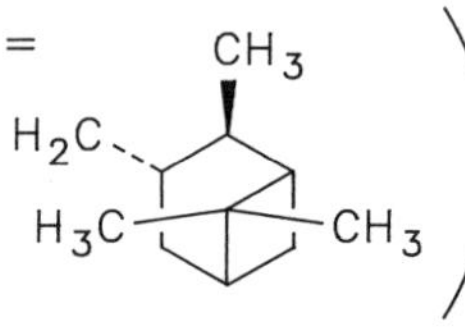 = [structure: CH_3, H_2C, H_3C, CH_3])	I, with sodium amalgam (46%, diastereomer ratio 34:66); see "Further information" [105] red prisms, m.p. 165 to 167 °C [105] for the absolute configuration see Formulas IVa, b and Va, b on p. 158 isomers IV(a/b = 97:3): optical rotation (toluene, 0.05 M): $[\alpha]^{20}(\lambda)$ = +2500° (365), −1400° (436), +1140° (546), +760° (578); given as a diagram in [105] ^1H NMR (toluene, 90 °C): 0.5 to 2.7 (m, $C_{11}H_{19}$), 3.65 (d, CH; J(H, H) = 6.95), 4.80 (s, C_5H_5), 5.18 (m, 1H), 6.3 to 7.0 (m, 3H), 7.92 (d, 1H; J(H, H) = 4.93); (toluene, −70 °C): 0.24 to 2.6 (m, $C_{11}H_{19}$), 3.72 (d, CH), 4.76 (s, C_5H_5, isomer IVa), 4.95 (s, C_5H_5, isomer IVb), 5.32 (m, 1H), 6.2 to 7.1 (m, 3H), 7.87 (d, 1H); a to b ratio 97:3 at −70 °C [105] isomers V (a/b = 94:6): optical rotation (toluene, 0.05 M): $[\alpha]^{20}(\lambda)$ = −2595° (365), +1508° (436), −1215° (546), −830° (578); given as a diagram in [105] ^1H NMR (toluene, 90 °C): 0.5 to 2.7 (m, $C_{11}H_{19}$), 3.63 (d, CH; J(H, H) = 6.79), 4.81 (s, C_5H_5), 5.09 (m, 1H), 6.3 to 7.1 (m, 3H), 7.91 (d, 1H; J(H, H) = 5.01); (toluene, −70 °C): 0.4 to 2.4 (m, $C_{11}H_{19}$), 3.36 (d, CH, isomer Vb), 3.54 (d, CH, isomer Va), 4.73 (s, C_5H_5, isomer Va), 4.88 (s, C_5H_5, isomer Vb), 4.98 (m, 1H, Vb), 5.17 (m, 1H, Va), 6.2 to 7.1 (m, 3H), 7.87 (d, 1H, Va), 8.11 (d, 1H, Vb); a to b ratio 94:6 at −70 °C [105] both: IR (KBr): 1590 (ν(CN)), 1800, 1910 (ν(CO)) [105] mass spectrum (field desorption, toluene): $[M]^+$ [105]

Table 7 (continued)

No.	compound	method of preparation (yield) properties and remarks
4	$C_5H_5Mo(CO)_2CH(C_5H_4N\text{-}2)NHCH_2C_6H_5$	I, with $LiCH_3$ (5%) [59, 66] red-brown rhombohedra, m.p. 123 to 123.5°C (from ether/pentane/C_6H_6 4:2:1) [66] ^{1}H NMR ($CDCl_3$): 3.68 and 4.30 (d, CH_2,1H each; J(H, H) = 15), 3.91 (s, MoCH), 5.00 (s, C_5H_5), 6.75 (m, C_6H_5N, H-5), 8.04 (m, C_6H_5N, H-6); H-3, 4 of C_5H_4N are covered [66] IR (KBr): 1590 to 1600 (ν(CN)), 1800 to 1825, 1900 to 1925 (ν(CO)), 3145 to 3230 (ν(NH)) [66] mass spectrum: $[M-n\,CO]^+$ (n = 0 to 2), $[C_5H_5MoC_5H_4N]^+$ [66]
*5	$C_5H_5Mo(CO)_2CH(C_5H_4N\text{-}2)NHCH_2C(C_2H_5)(CH_3)H$ ($CH_2C(C_2H_5)(CH_3)H$ 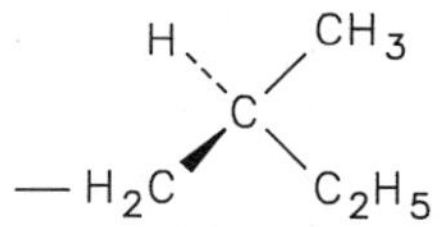 = H, CH_3, C, $—H_2C$, C_2H_5)	I, with sodium amalgam (43%); see "Further information" [105] orange needles, m.p. 100 to 101°C (from ether/pentane 1:1 at −20°C) [105] optical rotation (toluene, 0.05 M): $[\alpha]^{20}(\lambda)$ = −2621° (365), +2086° (436), −1671° (546), −1138° (578); given as a diagram in [105] ^{1}H NMR (toluene-d_8, 90°C): 0.75 (t, 3H; J(H, H) = 7.44), 0.77 (d, 3H; J(H, H) = 6.71), 1.0 to 1.3 (m, 2H), 1.7 to 2.6 (m, 3H), 3.64 (d, 3H; J(H, H) = 6.58), 4.86 (s, C_5H_5), 5.05 (m, 1H), 6.4 to 7.1 (m, 3H), 7.96 (d, 1H; J(H, H) = 4.98); (toluene-d_8, −70°C): 0.53 (d, 3H), 0.66 (t, 3H), 0.8 to 1.2 (m, 3H), 1.4 to 2.6 (m, 3H), 3.55 (d, 1H), 4.71 and 4.84 (s, C_5H_5), 4.99 (m, 1H), 6.2 to 7.1 (m, 3H), 7.83 (d, 1H); diastereomer ratio of Va to b 96:4 at −70°C [105] IR (KBr): 1590 (ν(CN)), 1800, 1910 (ν(CO)) [105] mass spectrum (field desorption, toluene): $[M]^+$ [105]
6	$C_5H_5Mo(CO)_2CH(C_5H_4N\text{-}2)NHC_3H_7\text{-}i$	I, with $LiCH_3$ (18%) [59, 66] orange needles (from pentane/ether 1:2), m.p. 120°C (dec.) [66] ^{1}H NMR ($CDCl_3$): 1.26 and 1.41 (d, CH_3; J(H, H) = 6), 2.08 (m, **CH**$(CH_3)_2$), 3.78 (d, MoCH; J(H, H) = 7), 4.98 (C_5H_5), 6.81 (m, C_5H_4N, H-5), 7.14 (m, C_5H_4N, H-4), 7.51 (m, C_5H_4N, H-3), 8.15 (m, C_5H_4N, H-6) [66] $^{13}C\{^1H\}$ NMR ($CDCl_3$, 0.05 M $Cr(OC(CH_3)CHC(CH_3)O)_3$): 22.2 and 25.0 (CH_3), 47.0 (MoC), 60.4 (**CH**$(CH_3)_2$), 119.1 to 148.1

Table 7 (continued)

No.	compound	method of preparation (yield) properties and remarks
6 (continued)		(C_5H_4N), 166.1 (C_5H_4N, C-2), 248.5 and 250.7 (CO) [66] IR (KBr): 1590 to 1600 (ν(CN)), 1800 to 1825, 1900 to 1925 (ν(CO)), 3145 to 3230 (ν(NH)); after treating with D_2O in CCl_4, ND bands were found at 1220 to 1270 and 2380 [66] mass spectrum: $[M-n\ CO]^+$ (n = 0 to 2), $[C_5H_5MoC_5H_4N]^+$ [66]
*7	$C_5H_5Mo(CO)_2CH(C_5H_4N\text{-}2)NHC(C_6H_{11}\text{-cyclo})(CH_3)H$ ($C(C_6H_{11}\text{-cyclo})(CH_3)H$ = [structure: H---C(CH$_3$) with wedge bond, attached to cyclohexyl])	I, with sodium amalgam (41%, diastereomer ratio 51:59); see "Further information" [105] orange oil [105] for the absolute configuration, see Formulas IVa, b and Va, b isomers IVa, b: optical rotation (toluene, 0.05 M): $[\alpha]^{20}$ (λ) = −2440° (365), +2300° (436), −1483° (546), −970° (578); given as a diagram in [105] ^{1}H NMR (toluene-d_8, 100°C): 0.9 to 1.7 (m, 11H), 0.89 (d, 3H; J(H, H) = 6.61), 2.11 (m, 1H), 3.65 (d, 1H; J(H, H) = 7.16), 4.87 (s, C_5H_5), 5.10 (m, 1H), 6.47 to 7.1 (m, 3H), 7.95 (d, 1H; J(H, H) = 4.62); (toluene, −70°C): 0.70 (d, 3H), 0.8 to 1.9 (m, 11H), 2.34 (m, 1H), 3.53 (d, 1H), 4.79, 4.92 (s, C_5H_5), 5.13 (m, 1H), 6.3 to 7.2 (m, 3H), 7.92 and 8.12 (d, 1H); a to b ratio 87:13 at −70°C [105] isomers Va, b: optical rotation (toluene, 0.05 M): $[\alpha]^{20}$ (λ) = +2280° (365), −2100° (436), +1345° (546), +812° (578); given as a diagram in [105] ^{1}H NMR (toluene-d_8, 100°C): 0.8 to 1.8 (m, 10H), 1.19 (d, 3H; J(H, H) = 6.71), 1.88 (m, 1H), 3.77 (d, 1H; J(H, H) = 7.27), 4.86 (m, 5H), 4.90 (s, 1H), 6.4 to 7.1 (m, 3H), 7.93 (d, 1H; J(H, H) = 4.50); (toluene, −70°C): 0.6 to 1.7 (m, 10H), 1.26 (d, 3H), 1.86 (m, 1H), 3.51 and 3.83 (d, 1H), 4.73 and 4.86 (s, C_5H_5), 5.05 (m, 1H), 6.2 to 7.2 (m, 3H), 7.83 and 8.15 (d, 1H); a to b ratio 87:13 at −70°C [105] all isomers: IR (KBr): 1590 (ν(CN)), 1810, 1910 (ν(CO)) [105] mass spectrum (field desorption, toluene): $[M]^+$ [105]

Table 7 (continued)

No.	compound	method of preparation (yield) properties and remarks
*8	$C_5H_5Mo(CO)_2CH(C_5H_4N\text{-}2)NHC(C_6H_5)(CH_3)H$	I, with $LiCH_3$ (3%) [59, 66]; I with sodium amalgam (46%) [104, 105] orange-yellow needles, dec. 115°C ($(-)_{365}$ diastereomer pair, from ether/pentane 2:1) [66]; orange prisms, m.p. 123 to 124°C (from ether/pentane at −20°C; diastereomer ratio 45:55) [104, 105] isomers IVa, b: optical rotation (toluene, 0.05 M): $[\alpha]^{20}$ (λ) = +2178° (365), −2467° (436), +1933° (546), +1267° (578); given as a diagram in [105] ^1H NMR (toluene-d_8, 100°C): 1.59 (d, CH_3; J(H, H) = 6.8), 3.11 (m, CH), 3.72 (d, MoCH; J(H, H) = 6.9), 4.89 (s, C_5H_5), 5.29 (m, NH), 6.28 to 7.96 (m, C_6H_5 and C_6H_4N); (toluene-d_8, −70°C): 1.51 (d, 3H), 3.04 (m, 1H), 3.67 (d, 1H), 4.70 (s, C_5H_5, isomer IVa), 4.91 (s, C_5H_5, isomer IVb), 5.38 (m, 1H), 6.18 to 7.86 (m, 9H); a to b ratio is 96:4 at −70°C [104]; related data in [66] IR (cyclohexane): 1850, 1940 (ν(CO)) [104] isomers Va, b: optical rotation (toluene, 0.05 M): $[\alpha]^{20}$ (λ) = −2003° (365), +1910° (436), −1685° (546) −1190° (578) [105] ^1H NMR (toluene, 100°C): 1.21 (d, CH_3; J(H, H) = 6.7), 3.06 (m, CH), 3.76 (d, MoCH; J(H, H) = 7.2), 4.75 (s, C_5H_5), 5.37 (m, NH), 6.38 to 8.02 (m, C_6H_5 and C_5H_4N); (toluene, −70°C): 1.05 (d, 3H), 2.51 (m, 1H), 3.61 (d, 1H), 4.59 (s, C_5H_5, isomer a), 4.67 (s, C_5H_5, isomer b), 5.30 (m, 1H), 6.27 to 7.93 (m, 9H); a to b ratio 90:10 at −70°C [104]; related data in [66] IR (cyclohexane): 1850, 1940 (ν(CO)) [104] all isomers: IR (KBr): 1600 (ν(CN)), 1785, 1865 (ν(CO)) [105] mass spectrum (field desorption, toluene): $[M]^+$ [105]
9	$C_5H_5Mo(CO)_2CH(C_5H_4N\text{-}2)NHC_6H_5$	I, with $LiCH_3$(15%) [59, 66] red-brown needles, dec. 127°C (from ether/pentane 2:1) [66] ^1H NMR ($CDCl_3$): 5.08 (s, C_5H_5), 7.18 (m, C_6H_5), 7.53 (m, C_5H_4N, H-3), 8.27 (m, C_5H_4N, H-6); H-4, 5 of C_5H_4N are covered by other resonances [66]

References on pp. 173/6

Table 7 (continued)

No.	compound	method of preparation (yield) properties and remarks
9 (continued)		^{13}C NMR ($CDCl_3$, 0.05 M $Cr(OC(CH_3)CHC(CH_3)O)_3$): 48.6 (MoCH), 93.5 (C_5H_5), 119.4 to 148.2 (C_5H_4N), 119.9 to 129.2 (C_6H_5), 149.3 (C_6H_5, C-1), 165.2 (C_5H_4N, C-2), 246.3 and 249.3 (CO) [66] IR (KBr): 1590 to 1600 (ν(CN)), 1800 to 1825, 1900 to 1925 (ν(CO)), 3145 to 3230 (ν(NH)) [66] mass spectrum: $[M-nCO]^+$ (n = 0 to 2), $[C_5H_5MoC_5H_4N]^+$ [66]
10	$C_5H_5Mo(CO)_2C(CH_3)(C_5H_4N\text{-}2)NHCH_3$	I, with $LiCH_3$ (traces); main product was compound No. 2 [59, 66] ^1H NMR ($CDCl_3$): 2.04 (s, CCH_3), 2.95 (d, NCH_3; J(H, H) = 5.5), 4.91 (s, C_5H_5), 6.82 (m, C_5H_4N, H-5), 7.36 (m, C_5H_4N, H-3, 4), 8.10 (m, C_5H_4N, H-6) [66] IR (KBr): 1590 to 1600 (ν(CN)), 1800 to 1825, 1900 to 1925 (ν(CO), not specified), 3145 to 3230 (ν(NH)) [66] mass spectrum: $[M-n\,CO]^+$ (n = 0 to 2), $[C_5H_5MoC_5H_4N]^+$ [66]
11	$C_5H_5Mo(CO)_2C(CH_3)(C_5H_4N\text{-}2)NHC_3H_7\text{-}i$	I, with $LiCH_3$ (10%); main product of the reaction is compound No. 6 [59, 66] yellow rhombohedra, dec. 80.5°C (from ether/pentane 2:1) [66] ^1H NMR ($CDCl_3$): 1.30 and 1.51 (d, $C(CH_3)_2$; J(H, H) = 6), 2.10 (s, CH_3), 2.95 (m, CH), 4.95 (s, C_5H_5), 6.82 (m, C_5H_4N, H-5), 7.43 (m, C_5H_4N, H-3, 4), 8.12 (m, C_5H_4N, H-6) [66] ^{13}C NMR ($CDCl_3$, 0.05 M $Cr(OC(CH_3)CHC(CH_3)O)_3$): 22.6 and 23.0 ((**CH_3**)$_2$C), 23.6 (CH_3), 53.5 (MoC), 54.3 (CH), 93.4 (C_5H_5), 119.2 to 147.0 (C_5H_4N), 168.0 (C_5H_4N, C-2), 248.0 and 249.8 (CO) [66] IR (KBr): 1590 to 1600 (ν(CN)), 1800 to 1825, 1900 to 1925 (ν(CO)), 3145 to 3230 (ν(NH)), after treating with D_2O in CCl_4; ND bands appear at 1220 to 1270 and 2380 [66] mass spectrum: $[M-n\,CO]^+$ (n = 0 to 2), $[C_5H_5MoC_5H_4N]^+$ [66]
*12	$C_5H_5Mo(CO)_2C(CH_3)(C_5H_4N\text{-}2)NHC(C_6H_5)(CH_3)H$	I, with $LiCH_3$ (7%) [59, 66, 75] orange-red prisms, dec. 116°C (from pentane/ether 1:2 at −35°C) [59, 66]

References on pp. 173/6

Table 7 (continued)

No.	compound	method of preparation (yield) properties and remarks
		^{1}H NMR ($CDCl_3$): 1.74 (s, $MoCCH_3$), 1.79 (d, $NCCH_3$; J(H, H) = 6.6), 3.67 (m, CH), 4.98 (s, C_5H_5), 7.20 (m, C_6H_5), 8.16 (m, C_5H_4N, H-6); other resonances of C_5H_4N are covered [66] IR (KBr): 1590 to 1600 (ν(CN)), 1800 to 1825, 1900 to 1925 (ν(CO)), 3145 to 3230 (ν(NH)) [66] mass spectrum: $[M-n\,CO]^+$ (n = 0 to 2), $[C_5H_5MoC_5H_4N]^+$ [66]
13	$C_5H_5\overline{Mo(CO)_2C(CH_3)(C_5H_4N\text{-}2)N}HC_6H_5$	I, with $LiCH_3$ (traces); main product is compound No. 9 [59, 66] IR (KBr): 1590 to 1600 (ν(CN)), 1800 to 1825, 1900 to 1925 (ν(CO)), 3145 to 3230 (ν(NH)) [66] mass spectrum: $[M-n\,CO]^+$ (n = 0 to 2), $[C_5H_5MoC_5H_4N]^+$ [66]
*14	$C_5H_5\overline{Mo(CO)_2C(C_6H_5)(C_5H_4N\text{-}2)N}HC_3H_7\text{-i}$	I, with C_6H_5MgBr (X = Cl; 27 to 36%) [66], with LiC_6H_5 (X = PF_6) [75] red-orange needles, dec. 150°C [66] ^{1}H NMR ($CDCl_3$): 0.73 and 1.43 (d, CH_3; J(H, H) = 6), 2.49 (m, CH), 4.93 (s, C_5H_5), 6.82 (m, C_5H_4N, H-5), 7.21 (m, C_6H_5), 8.09 (m, C_5H_4N, H-6); H-3, 4 of C_5H_4N are covered [66] ^{13}C NMR ($CDCl_3$, 0.05 M $Cr(OC(CH_3)CHC(CH_3)O)_3$): 22.5 and 23.9 (CH_3), 53.9 (MoC), 65.4 (CH), 93.9 (C_5H_5), 119.5 to 147.0 (C_5H_4N), 124.4 to 127.7 (C_6H_5), 139.4 (C_6H_5, C-1), 166.7 (C_5H_4N, C-2), 247.7 and 248.5 (CO) [66] IR (KBr): 1590 to 1600 (ν(CN)), 1800 to 1825, 1900 to 1925 (ν(CO)), 3145 to 3230 (ν(NH)) [66] mass spectrum: $[M-n\,CO]^+$ (n = 0 to 2), $[C_5H_5MoC_5H_4N]^+$ [66]
15	$C_5H_5\overline{Mo(CO)_2C(C_6H_5)(C_5H_4N\text{-}2)N}HC(C_6H_5)(CH_3)H$	I, with C_6H_5MgBr (X = Cl) [66] red-orange needles, m.p. 149°C ($(-)_{365}$ isomer), 141°C ($(+)_{365}$ isomer) [66] ^{1}H NMR ($CDCl_3$): $(+)_{365}$ isomer: 1.71 (d, CH_3; J(H, H) = 7), 3.27 (m, CH), 5.07 (s, C_5H_5), 6.94 (m, C_6H_5), 8.23 (m, C_5H_4N, H-6), other resonances of C_5H_4N are covered; $(-)_{365}$ isomer: 1.09 (d, CH_3; J(H, H) = 7), 4.66 (C_5H_5), 6.87 (m, C_5H_4N, H-5), 7.40 (m, C_6H_5), 8.21 (m, C_5H_4N, H-6); other resonances of C_5H_4N are covered [66]

References on pp. 173/6

Table 7 (continued)

No.	compound	method of preparation (yield) properties and remarks
15 (continued)		IR (KBr): 1590 to 1600 (ν(CN)), 1800 to 1825, 1900 to 1925 (ν(CO)), 3145 to 3230 (ν(NH)) [66] mass spectrum: $[M - n\ CO]^+$ (n = 0 to 2), $[C_5H_5MoC_5H_4N]^+$ [66]
16	$C_5H_5Mo(CO)_2C(C_6H_4CH_3\text{-}4)(C_5H_4N\text{-}2)NHC_3H_7\text{-}i$	I, with $4\text{-}CH_3C_6H_4MgBr$ (X = Cl; 27 to 36%) [66] red-orange needles, m.p. 164 to 166°C [66] ^1H NMR ($CDCl_3$): 0.82 and 1.44 (d, $(CH_3)_2C$; J(H, H) = 6), 2.31 (s, CH_3-4), 4.97 (s, C_5H_5), 6.87 (m, C_5H_4N, H-5), 8.14 (m, C_5H_4N, H-6); other resonances of C_5H_4N and C_6H_5 are covered [66] IR (KBr): 1590 to 1600 (ν(CN)), 1800 to 1825, 1900 to 1925 (ν(CO)), 3145 to 3230 (ν(NH)) [66] mass spectrum: $[M-n\ CO]^+$ (n = 0 to 2), $[C_5H_5MoC_5H_4N]^+$ [66]
17	$C_5H_5Mo(CO)_2C(C_6H_4OCH_3\text{-}4)(C_5H_4N\text{-}2)NHC_3H_7\text{-}i$	I, with $4\text{-}CH_3OC_6H_4MgBr$ (X = Cl; 27 to 36%) [66] red-orange needles, m.p. 181 to 183°C [66] ^1H NMR ($CDCl_3$): 0.83 and 1.44 (d, $(CH_3)_2C$; J(H, H) = 6), 2.43 (m, CH), 3.77 (s, CH_3O), 4.97 (s, C_5H_5), 6.80 (m, C_5H_4N, H-5), 8.12 (m, C_5H_4N, H-6); other resonances of C_5H_4N and C_6H_4 are covered [66] IR (KBr): 1590 to 1600 (ν(CN)), 1800 to 1825, 1900 to 1925 (ν(CO)), 3145 to 3230 (ν(NH)) [66] mass spectrum: $[M-n\ CO]^+$ (n = 0 to 2), $[C_5H_5MoC_5H_4N]^+$ [66]
18	$C_5H_5Mo(CO)_2C(C_6H_4Cl\text{-}4)(C_5H_4N\text{-}2)NHC_3H_7\text{-}i$	I, with $4\text{-}ClC_6H_4MgBr$ (X = Cl; 4%) [66] red-orange needles, m.p. 176 to 177°C [66] ^1H NMR ($CDCl_3$): 0.84 and 1.46 (d, CH_3; J(H, H) = 6), 2.46 (m, CH), 4.97 (s, C_5H_5), 6.86 (m, C_5H_4N, H-5), 8.15 (m, C_5H_4N, H-6) [66] IR (KBr): 1590 to 1600 (ν(CN)), 1800 to 1825, 1900 to 1925 (ν(CO)), 3145 to 3230 (ν(NH)) [66] mass spectrum: $[M - n\ CO]^+$ (n = 0 to 2), $[C_5H_5MoC_5H_4N]^+$ [66]
19	$C_5H_5Mo(CO)_2C(CH_2C_6H_5){=}NC_6H_{11}\text{-cyclo}$	$C_5H_5Mo(CO)_3CH_2C_6H_5$ was allowed to react with 1.2 equivalents of cyclo-$C_6H_{11}NC$ in C_6H_6 at 25°C for 24 h (40% by chromatography on Al_2O_3 with $CHCl_3$); formation is dramatically accelerated by light [86] yellow powder [86]

Table 7 (continued)

No.	compound	method of preparation (yield) properties and remarks
		^{1}H NMR ($CDCl_3$): 4.17 and 4.27 (CH_2), 5.27 (C_5H_5) [86] IR: 1595 (ν(CN)), 1895, 1967 (ν(CO)) [86]
20	$C_5H_5\overline{Mo(CO)_2CHN}(CH_3)(CH_2)_3$ [Mo] (pyrrolidine ring, N–CH$_3$) 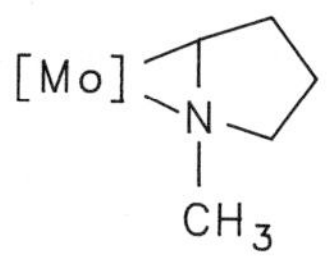	IV (36%) [101, 109, 110]; also formed by the rearrangement of $C_5H_5Mo(CO)_2({=}CN(CH_3)(CH_2)_3)H$ (Formula III, n = 1, X = H) [109] yellow crystals, m.p. 69 to 71 °C (from light petroleum) [101] ^{1}H NMR (C_6D_6): 1.10 (d of t, central CH_2, 1H; J(H, H) = 12.0, 7.0), 1.52 (q of t, central CH_2, 1H; J(H, H) = 12.0, 7.0), 1.88 (s, CH_3), 1.92 (d of d of d, NCH_2, 1H; J(H, H) = 12.0, 10.0, 7.0), 2.00 (d of d of d of d, MoCHC**H**$_2$, 1H; J(H, H) = 13.0, 11.5, 7.0, 6.0), 2.41 (d of d, MoCHC**H**$_2$, 1H; J(H, H) = 13.0, 7.0), 2.69 (d, MoCH; J(H, H) = 6.0), 2.86 (d of d, NCH_2, 1H; J(H, H) = 10.0, 7.5), 4.89 (s, C_5H_5) [101] ^{13}C NMR (toluene-d_8, −50 °C): 23.0 (central CH_2), 30.1 (MoCH**C**H$_2$), 52.9 (CH_3), 54.6 (MoCH), 63.2 (NCH_2), 92.6 (C_5H_5), 245.3 and 253.1 (CO) [101] IR (light petroleum): 1840, 1932 (ν(CO)) [101] mass spectrum: $[M]^+$ [101] protonation with H_2SO_4 affords N-methylpyrrolidine [101]
21	$C_5H_5\overline{Mo(CO)_2CHN}(CH_3)(CH_2)_4$ 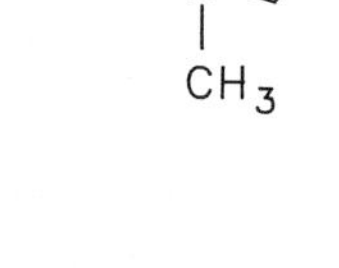 [Mo] (piperidine ring, N–CH$_3$)	IV (27%) [101, 110] yellow crystals, m.p. 70 to 72 °C (from light petroleum) [101] ^{1}H NMR (C_6D_6): 0.78 (d of d of d of d of d, NCH_2C**H**$_2$, 1H; J(H, H) = 14.0, 10.0, 7.4, 5.5, 2.6), 1.05 (d of d of d of d of d, MoCHCH$_2$C**H**$_2$, 1H; J(H, H) = 13.3, 12.3, 10.1, 9.9, 4.8), 1.20 (d of d of d of d of d, NCH_2C**H**$_2$, 1H; J(H, H) = 14.0, 11.8, 9.9, 5.6, 3.2), 1.46 (d of d of d of d of d, MoCHCH$_2$C**H**$_2$, 1H; J(H, H) = 13.3, 12.3, 7.4, 4.8, 3.2), 1.79 (m, MoCHC**H**$_2$, 1H), 1.87 (s, CH_3), 2.14 (d of d of d, MoCHC**H**$_2$, 1H; J(H, H) = 13.4, 11.8, 5.4), 2.27 (d of d of d of d, NCH_2, 1H; J(H, H) = 14.4, 6.0, 4.8, 3.7), 2.36 (d of d, MoCH, 1H; J(H, H) = 8.8, 6.0), 2.54 (d of d of d of d, MoCHC**H**$_2$, 1H; J(H, H) = 13.4, 5.6, 2.6, 0.7), 4.90 (s, C_5H_5) [101] ^{13}C NMR (C_6D_6): 19.3 (MoCHCH$_2$**C**H$_2$), 20.8 (NCH_2**C**H$_2$), 30.4 (MoCH**C**H$_2$), 51.4 (CH_3), 56.2 (MoCH), 60.4 (NCH_2), 92.4 (C_5H_5), 244.9 and 252.6 (CO) [101]

References on pp. 173/6

Table 7 (continued)

No.	compound	method of preparation (yield) properties and remarks
21 (continued)		IR (light petroleum): 1841, 1937 (ν(CO)) [101] mass spectrum: $[M]^+$ [101] protonation with H_2SO_4 affords N-methylpiperidine [101]
*22	$C_5H_5Mo(CO)_2C(CH_3)=NCH_3$	V (51%) [38, 48, 53] red solid, m.p. 94 to 101 °C (from hexane at −78 °C) [48] ^{1}H NMR (C_6H_6): 1.98 and 2.44 (CH_3), 5.03 (C_5H_5) [48] ^{13}C NMR (toluene-d_8, −10 °C): 18.88 (**C**CH_3), 36.65 (NCH_3), 93.91 (C_5H_5), 195.15 (C=N), 249.97 and 252.63 (CO) [57] IR (THF): 1720 (ν(C=N)), 1825, 1930 (ν(CO)) [38, 48]; similar (ν(C=N)) in [57] reacts with $(NC)_2C=C(CN)_2$ in THF giving $C_5H_5Mo(CO)_2((NC)_2C=C(CN)_2)$-($\eta^1$-$C(CH_3)=NCH_3$) [38, 48]
*23	$CH_3C_5H_4Mo(CO)_2C(CH_3)=NCH_3$	similar to V [57] ^{13}C NMR (toluene-d_8, −10 °C): 13.70 (**C**$H_3C_5H_4$), 18.74 (**C**H_3C=), 35.50 (NCH_3); 90.28, 90.64, 94.12, and 94.81 (C_5H_4, CH), 114.55 (C_5H_4, **C**CH_3), 196.99 (C=N), 250.90 and 253.06 (CO); spectra given as a diagram between 0 and 55 °C with the related simulated spectra; sharp signals observed at ca. 60 °C [57]
*24	$C_5H_5Mo(CO)_2C(CH_3)=NC_6H_5$	V (79%) [39, 48, 53] red crystals, m.p. 84 to 87.5 °C (from hexane at −78 °C) [48] ^{1}H NMR (acetone-d_6): 3.09 (CH_3), 5.59 (C_5H_5), 7.25 to 7.50 (C_6H_5) [48] IR (pentane): 1585 (ν(C_6H_5)), 1670 (ν(C=N)), 1850, 1945 (ν(CO)) [48]; similar data in [57] reaction with D ligands in THF at 25 °C affords $C_5H_5Mo(CO)_2(D)C(CH_3)=NC_6H_5$ (D = $(NC)_2$-$C=C(CN)_2$, $P(OCH_3)_3$, or $P(C_6H_5)_3$) [48, 53]
25	$C_5H_5Mo(CO)_2C(C_6H_5)=NC_6H_5$	$Na[C_5H_5Mo(CO)_3]$ was allowed to react with $ClC(C_6H_5)=NC_6H_5$ [53]
26	$C_5H_5Mo(CO)_2C(=C(CH_3)_2)N(CH_3)_2$	VI (38%) [31, 35] yellow solid, m.p. 120 to 122 °C (from CH_2Cl_2/hexane) [31, 35] ^{1}H NMR ($CDCl_3$): 1.96 and 1.99 (CH_3C), 2.71 (CH_3N), 5.34 (C_5H_5) [31, 35]; (acetone): 1.95 (CH_3C), 2.69 (CH_3N), 5.30 (C_5H_5); (C_6D_6): 1.66

References on pp. 173/6

Table 7 (continued)

No.	compound	method of preparation (yield) properties and remarks
		and 1.97 (CH_3C), 2.18 (CH_3N), 4.90 (C_5H_5); (C_6F_6): 1.91, 2.02 (CH_3C), 2.89 (CH_3N), 5.21 (C_5H_5) [35] ^{13}C NMR ($CDCl_3$): 20.0 and 22.3 (**C**H_3C), 55.6 (CH_3N), 92.1 (C_5H_5), 123.2 and 156.5 (C=C) [35] IR (KBr): 1591 (ν(C=N or C=C)); (hexane): 1832, 1930 (ν(CO)) [35], similar data in [31] mass spectrum (70 eV, 60 to 210°C): $[M-n\ CO]^+$ (n = 0 to 2), $[C_5H_5Mo]^+$ [35] reaction with I_2 in CH_2Cl_2 affords $C_5H_5Mo(I)_2C(CH_3)_2{=}C{=}N(CH_3)_2$; decomposition occurs on irradiation in hexane [35]; protonation with strong acids such as HPF_6 affords $[C_5H_5Mo(CO)_2(\eta^3\text{-}C(CH_3)_2CHN(CH_3)_2)]^+$ [28, 35]
27	$C_5H_5\overline{Mo(CO)_2C(=C(CH_3)_2)=N}(CH_2)_5$ 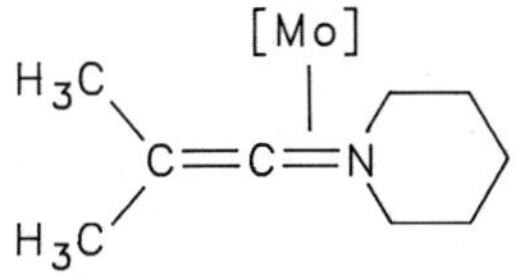 [Mo] H_3C, H_3C C=C=N	VI (15%) [35] burnt orange crystals, m.p. 89 to 90°C (from CH_2Cl_2/hexane) [35] ^1H NMR ($CDCl_3$): ca. 1.6 (br, CH_2), 1.93 and 1.98 (CH_3C), ca. 3.2 (br, CH_2), 5.16 (C_5H_5) [35] ^{13}C NMR ($CDCl_3$): 18.7 and 23.2 (CH_3), 23.8, 24.9, and 62.5 (CH_2), 92.4 (C_5H_5), 124.7 and 155.3 (C=C); the CO resonances were not detected [35] IR (hexane): 1853, 1937 (ν(CO)) [35] mass spectrum (70 eV, 60 to 210°C): $[M-n\ CO]^+$ (n = 0 to 2), $[C_5H_5Mo]^+$ [35] treatment with strong acids like HPF_6 affords $[C_5H_5Mo(CO)_2(\eta^3\text{-}C(CH_3)_2CHN(CH_2)_5)]^+$ [28]
28	$C_5H_5Mo(CO)_2C(CH_3)_2{=}N{=}NC_6H_5$	$C_5H_5Mo(CO)_3Cl$ was allowed to react with 1.5 equivalents of $Li[C(CH_3)_2{=}NNC_6H_5]$ in ether at −78°C followed by 4 h at 25°C (85%); different structures were suggested [56] deep red-orange crystals, m.p. 133 to 135°C (from CH_2Cl_2/hexane) [56] ^1H NMR ($CDCl_3$): 2.05 and 2.32 (CH_3), 5.65 (C_5H_5), 6.54 (d, C_6H_5, H-3; J(H, H) = 8), 6.83 (d, C_6H_5, H-4; J(H, H) = 8), 7.12 (t, C_6H_5, H-2; J(H, H) = 8) [56] ^{13}C NMR ($CDCl_3$, 0.15% $Cr(OC(CH_3)CHC(CH_3)O)_3$): 23.3 (CH_3), 93.5 (C_5H_5), 117.0 (C_6H_5, C-2), 120.4, 128.5, and 161.6 (C_6H_5), 159.5 (C=N), 224.2 and 232.6 (CO) [56] IR (cyclohexane): 1868, 1953 (ν(CO)) [56]

References on pp. 173/6

Table 7 (continued)

No.	compound	method of preparation (yield) properties and remarks
28 (continued)		air-stable for a short time in the solid state but is rapidly oxidized in solution; decomposes on attempts to chromatograph [56]
compounds with a three-membered Mo-C-P ring		
29	$C_5H_5Mo(CO)_2C(Si(CH_3)_3)_2P(CH_3)H$	$C_5H_5Mo(CO)_2$=P=C(Si(CH$_3$)$_3$)$_2$ was allowed to react with Li[(C_2H_5)$_3$BH]; the reaction mixture was treated with CH_3I (not isolated) [29] ^{31}P NMR (C_6D_6): −23.0 [29] reacts further with CH_3I to No. 30 [29]
*30	$C_5H_5Mo(CO)_2C(Si(CH_3)_3)_2P(CH_3)_2$	see No. 29 [29] ^{31}P NMR (C_6D_6): 14.0 [29]
*31	$C_5H_5Mo(CO)_2CH_2P(C_6H_5)_2$	VIIa (17.1%) [91]; VIIb (65%) [17] yellow crystals [17, 91], m.p. 125°C [91], 133°C [17] ^1H NMR (C_6D_6): 0.79 (d, CH_2; J(P, H) = 1.3), 4.76 (s, C_5H_5), 6.91 to 7.51 (m, C_6H_5) [17], similar data in [91]; (CD_2Cl_2, −80°C): 0.62 (ABX system, CH_2; J(H-A, B) = 8.265, J(P, H-A) = −2.25, J(P, H-B) = 0.02), 5.13 (s, C_5H_5), 6.84 to 7.87 (C_6H_5) [91]; CH_2 region given as a diagram between 193 and 310 K in [91], activation data for the rotation of $CH_2P(C_6H_5)_2$ are: T_c = 263 K, ν = 44 Hz, $\Delta G^{\ddagger}$ = 53.9 ± 3 kJ/mol [17, 91] ^{13}C NMR (C_6D_6): 30.3 ($MoCH_2$?), 90.51 (C_5H_5), 203.65 (CO) [91]; in [17] the CH_2 resonance was found at −21.7 ($MoCH_2$) ^{31}P NMR ($CHCl_3$): 1.33 [91]; (C_6D_6): −2.7 [17] IR (n-hexane): 1873, 1956 (ν(CO)) [91], similar data in [17]; (KBr): 1838, 1920 (ν(CO)) [17] mass spectrum (70 eV, 200°C): $[M]^+$ (18), $[M-CO]^+$ (16), $[M-2CO]^+$ (17), $[C_5H_5MoP(C_6H_5)_2]^+$ (3), $[C_6H_5MoPC_6H_5]^+$ (5), $[C_5H_5MoPC_6H_5]^+$ (100), and other fragment ions [91]
32	$C_5H_5Mo(CO)_2C(CH_3)HP(C_6H_5)_2$	VIIa (7.5%) [91]; VIIb (88%) [17] yellow crystals, m.p. 68°C [17] ^1H NMR (C_6D_6): 1.3 to 1.9 (m, A_3BX system, CH_3 and CH), 4.74 (s, C_5H_5); computed values for A_3BX system: 1.55 (CH_3), 1.70 (CH) with J(H, H) = 7.1, ^2J(P, H) = 11.8, ^3J(P, H) = 2.4; activation data for rotation of $C(CH_3)HP(C_6H_5)_2$: $\Delta G^{\ddagger}$ = 50.2 ± 2 kJ/mol [17]

Table 7 (continued)

No.	compound	method of preparation (yield) properties and remarks
		^{13}C NMR (C_6D_6): −9.7 (CH), 18.9 (d, CH_3; J(P, C) = 3.7), 90.8 (C_5H_5) [17] ^{31}P NMR (hexane): 8.1; (hexane, −60 °C): 7.1, 10.0 (two diastereomers) [17] IR (KBr): 1850, 1923 (ν(CO)) [17]; (n-hexane): 1879, 1951 (ν(CO)) [91] mass spectrum (70 eV, 200 °C): $[M]^+$ (4), $[M-CO]^+$ (4), $[M-2CO]^+$ (1), $[C_5H_5MoP(C_6H_5)_2]^+$ (13), $[M-CO-C_6H_8]^+$ (4), $[C_5H_5MoPC_6H_5]^+$ (8), and other fragment ions; base peak $[P(C_6H_4)_2]^+$ [17, 91]
33	$C_5H_5\overline{Mo(CO)_2C(C_6H_5)HP}(C_6H_5)_2$	VIIb (27%) [17] yellow solid, m.p. 62 °C [17] ^1H NMR (C_6D_6): 3.03 (d, CH; J(P, H) = 1.7), 4.73 (s, C_5H_5) [17] ^{13}C NMR (C_6D_6): 2.5 (MoCH), 91.7 (C_5H_5) [17] ^{31}P NMR (hexane): 6.1; (hexane, −60 °C): 1.4, 9.0 (two diastereomers) [17] IR (KBr): 1852, 1935 (ν(CO)) [17] mass spectrum (field desorption, 50 °C): $[M]^+$ [17]
34	$C_5H_5\overline{Mo(CO)_2C(Si(CH_3)_3)_2P}(OCH_3)H$	$C_5H_5Mo(CO)_2{=}P{=}C(Si(CH_3)_3)_2$ was allowed to react with $[OCH_3]^-$ in CH_3OH; other minor products were $C_5H_5Mo(CO)_2P(OCH_3)C(Si(CH_3)_3)_2H$ and $C_5H_5Mo(CO)_2(H)P(OCH_3)_2C(Si(CH_3)_3)_2H$ [29] ^{31}P NMR (CD_2Cl_2): 42.0 [29]
35	$C_5H_5\overline{Mo(CO)_2CH_2P}(N(CH_3)_2)_2$	VIII [89, 98, 99, 103] undergoes olefin-like rotation of the $CH_2P(N(CH_3)_2)_2$ unit with $\Delta G^{\ddagger}$ = 13.6 kcal/mol and T_c = 251 K [89, 99, 103]
36	$(CH_3)_5C_5\overline{Mo(CO)_2CH_2P}(N(CH_3)_2)_2$	VIII [99, 103] undergoes olefin-like rotation of the $CH_2P(N(CH_3)_2)_2$ unit [99, 103]
37	$C_5H_5\overline{Mo(CO)_2CH_2P}OC(CH_3)_2C(CH_3)_2O$ 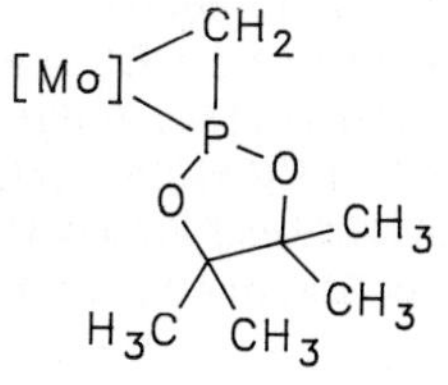 [Mo] CH2 P O O CH3 CH3 H3C CH3	VIII [108]

References on pp. 173/6

Table 7 (continued)

No.	compound	method of preparation (yield) properties and remarks
	compounds with a three-membered Mo-C-O ring	
38	$[Li \cdot C_8H_{16}O_4][C_5H_5\overline{Mo(CO)_2CH(CH_3)O}]$ ($C_8H_{16}O_4$ = [12]crown-4)	IX, in the presence of [12]crown-4 [106] for reactions, see No. 39
*39	$Li[C_5H_5\overline{Mo(CO)_2CH(CH_3)O}]$	IX ($R = CH_3$), for 5 min (high yields, not isolated) [88, 95, 96] orange-brown solution in THF [95] ^1H NMR (THF): 5.06 (s, C_5H_5) [88, 95] ^{13}C NMR (THF, −50°C): 56.2 (CHO; J(H, C) = 156), 92.5 (C_5H_5), 254.4 and 255.5 (CO) [88, 95] IR (THF, −10°C): 1766, 1870 (ν(CO)); also given as a diagram in [88, 95]
40	$Li[C_5H_5\overline{Mo(CO)_2CD(CH_3)O}]$	IX ($R = CH_3$), with $Li[(C_2H_5)_3BD]$ [95] ^2H NMR (THF, −30°C): 3.5 (s, CD) [95]
41	$Li[C_5H_5\overline{Mo(CO)_2CH(CD_3)O}]$	IX ($R = CD_3$), with $C_5H_5Mo(CO)_3CD_3$ [95] ^2H NMR (THF, −30°C): 1.6 (s, CD_3) [95]
42	$C_5H_5\overline{Mo(CO)_2CH(CH_3)O}CH_3$	No. 39 was alkylated with $[O(CH_3)_3]BF_4$ in THF at −78°C; the IR spectrum indicated the formation of No. 42; in the presence of $P(C_6H_5)_3$, trans-$C_5H_5Mo(CO)_2(P(C_6H_5)_3)C(CH_3)(OCH_3)H$ is obtained upon warm-up [106] extremely air-sensitive, decomposes upon attempted isolation by chromatography [106]
43	$Li[C_5H_5\overline{Mo(CO)_2CH(R)O}]$ ($R = C_2H_5$, $CH_2C_5H_5$)	IX [106] oxidation with $[(C_5H_5)_2Fe]^+$ affords $(C_5H_5)_2(Mo)_2(CO)_3(CO\text{-}\mu)CH(CH_2C_6H_5)O\text{-}\mu$ (Formula VII, $R = CH_2C_6H_5$) [106]
44	$Li[C_5H_5\overline{Mo(CO)_2CH(C_3H_6Br)O}]$	IX ($R = C_3H_6Br$), between −78 to −5°C; an intermediate in the preparation of No. 46 [100] IR (THF, −5°C): 1773, 1877 (ν(CO)) [100, 106]
45	$Li[C_5H_5\overline{Mo(CO)_2CH(C_4H_8Br)O}]$	IX ($R = C_4H_8Br$) [100] IR (THF): 1774, 1878 (ν(CO)) [100] decomposes on heating in solution; IR spectrum shows the formation of $[C_5H_5Mo(CO)_3]^-$ [100]
46	$C_5H_5\overline{Mo(CO)_2CH(CH_2)_3O}$ [Mo]—O; ^1H, ^2H, ^3H, ^4H, H^5, ^6H, ^7H	IX ($R = C_3H_6Br$), at −78°C followed by warm-up (20% by chromatography on Al_2O_3 with light petroleum/CH_2Cl_2 85:15) [100, 106]; from trans-$C_5H_5Mo(CO)_2(I)$=$C(CH_2)_3O$-cyclo under similar conditions (31%) [78, 79, 100] orange crystals, m.p. 69.5 to 70.5°C (from pentane) [100]

Table 7 (continued)

No.	compound	method of preparation (yield) properties and remarks
		^{1}H NMR (C_6D_6): 1.07 (H-7; J(H-1, 7) = 0.5, J(H-2, 7) = 0.9, J(H-3, 7) = 5.7, J(H-4, 7) = 0.7, J(H-5, 7) = 7.5, J(H-6, 7) = 12.5), 1.59 (H-6; J(H-2, 6) = 8.4, J(H-3, 6) = 11.9, J(H-4, 6) = 7.4), 1.94 (H-5; J(H-1, 5) = 6.5, J(H-2, 5) = 0.5, J(H-3, 5) = 0.3, J(H-4, 5) = 13.1, J(H-6, 5) = 11.9), 2.60 (H-4; J(H-1, 4) = J(H-2, 4) = J(H-3, 4) = 0.5), 3.27 (H-3; J(H-1, 3) = 0.5, J(H-2, 3) = 8.4), 3.33 (H-2; J(H-1, 2) = 0.5), 4.22 (H-1), 5.10 (C_5H_5) [100] ^{13}C NMR (C_6D_6): 21.9 (t, central CH_2), 30.2 (t, CH**C**H_2), 75.6 (d, MoCH), 79.9 (t, OCH_2), 94.6 (d, C_5H_5), 246.7, 256.7 (s, CO) [100] IR (light petroleum): 1845, 1937 (ν(CO)) [100]; (THF): 1829, 1925 (ν(CO)) [79] mass spectrum: $[M]^+$ [100] air-stable in the solid state, somewhat air-sensitive in solution [79, 100] reaction with $P(C_6H_5)_3$ in THF for 2 h yields trans-$C_5H_5Mo(CO)_2(P(C_6H_5)_3)CH(CH_2)_3O$-cyclo [79, 100]
47	$C_5H_5\overline{Mo(CO)_2CD(CH_2)_3O}$	IX (62:38 mixture with No. 46) [78, 100] ^{1}H NMR ($CDCl_3$): 4.22 (H-1; see No. 46); other signals are more complex than in No. 46 and no coupling less than 1 Hz is resolved [100]
*48	$CH_3C_5H_4\overline{Mo(CO)_2CH(CH_2)_3O}$ 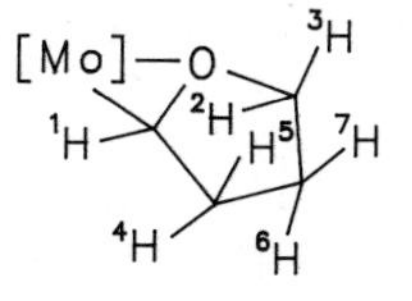 [Mo]—O, ^{1}H, ^{2}H, ^{3}H, ^{4}H, H^5, ^{6}H, ^{7}H	IX (ca. 30% by column chromatography on Al_2O_3 with light petroleum) [78, 106] orange solid [100], irregular red crystals [78], m.p. 63 to 64°C (from pentane) [100] ^{1}H NMR (C_6D_6): 0.97 (H-7; J(H-2, 7) = J(H-3, 7) = J(H-4, 7) = 0.8, J(H-5, 7) = 7.4, J(H-6, 7) = 12.3), 1.52 (d, CH_3; J(H, H) = 0.5), 1.59 (H-6; J(H-1, 6) = 0.5, J(H-2, 6) = 8.2, J(H-3, 6) = 11.8, J(H-4, 6) = 7.4, J(H-5, 6) = 12.3), 1.92 (H-5; J(H-1, 5) = 6.5, J(H-4, 5) = 12.9), 2.58 (H-4), 3.23 (H-3; J(H-2, 3) = 8.3), 3.32 (H-2), 4.05 (H-1), 4.80 (m, C_5H_4, 2H), 5.14 (t of d, C_5H_4, 1H; J(H, H) = 2.6, 1.9), 5.20 (m, C_5H_4, 1H) [100] ^{13}C NMR (C_6D_6): 13.2 (CH_3), 21.8 (central CH_2), 30.1 (CH**C**H_2), 77.6 (MoCH), 79.7 (OCH_2), 89.9, 90.6, 95.1, and 96.1 (CH of C_5H_4), 113.4 (**C**CH_3), 246.9, 256.4 (CO) [100] IR (hexane): 1844, 1939 (ν(CO)) [100] mass spectrum: $[M]^+$ [100]

References on pp. 173/6

Table 7 (continued)

No.	compound	method of preparation (yield) properties and remarks
compounds with a three-membered Mo-C-S ring		
*49	$C_5H_5Mo(CO)_2CH_2SCH_3$	$C_5H_5Mo(CO)_3CH_2SCH_3$ was irradiated for 112 h in C_6H_6 (17.4%) or heated at 70 to 80°C in vacuum (0.5 Torr) for 2 h (45% after sublimation) [1, 3, 8]; formed by the decomposition of $C_5H_5Mo(CO)_2(P(OC_6H_5)_3)CH_2SCH_3$ in solution [15] yellow-orange crystals, m.p. 65 to 67°C (from pentane at −78°C), subl. at 60 to 80°C/0.1 Torr [1, 3] ^1H NMR (CS_2): 1.88 (CH_3), 1.89 and 2.67 (AB system, CH_2S; J(A, B) = 6.0), 5.15 (C_5H_5) [3, 10] IR (halocarbon oil): given from 686 to 3060, 1838, 1922 (ν(CO)), 3060 (ν(CH)) [3, 10]; (cyclohexane): 1869, 1952 (ν(CO)) [5] UV (cyclohexane): λ_{max} (ε) = 220 (18100), 254 (11750), 419 (705) [3] mass spectrum: $[M-n\ CO]^+$ (n = 0 to 2), $[MoC_6H_6S]^+$ most abundant metal-containing ion [4, 11] decomposes slowly at 25°C under N_2 [1, 3]
50	$C_9H_7Mo(CO)_2CH_2SCH_3$	$C_9H_7Mo(CO)_3CH_2SCH_3$ loses CO either on irradiation or on sublimation at 70 to 100°C/0.1 Torr (8%) [2] yellow crystals, m.p. 87 to 89°C [2] ^1H NMR (CS_2): 1.64 (s, CH_3), 5.96 (C_9H_7, five-membered ring, 1H), 5.62 (C_9H_7, five-membered ring, 2H; J(H, H) = 2), 6.89 (C_9H_7, six-membered ring, 4H) [2] IR (halocarbon oil): given from 745 to 1940; 1827, 1858, 1940 (ν(CO)) [2] UV (cyclohexane): λ_{max} (ε) = 220 (41200), 287 (7960) [2]
51	$C_5H_5Mo(CO)_2C(SC(C_6H_5){=}NCH_3)SO$ [Mo] O S C S N—CH3	CS_2 was dropped into a solution of $Li[(NCH_3)_2CC_6H_5]$ in ether cooled to −20°C; the resulting mixture was added to a solution of $C_5H_5Mo(CO)_3Cl$ cooled to −78°C; the mixture was stirred for 2 h at −78°C and 20 h at room temperature (12% by column chromatography on SiO_2 with ether/C_6H_6 14:1) [67]

References on pp. 173/6

Table 7 (continued)

No.	compound	method of preparation (yield) properties and remarks
		other possible coordinations of the chelating ligand were discussed; the one shown is the most probable; ether is proposed as O-source [67] orange powder, m.p. 168 to 170°C (from CH_2Cl_2/ ether 1:1) [67] ^{1}H NMR ($CDCl_3$): 3.53 (s, CH_3), 5.42 (s, C_5H_5), 7.49 (m, C_6H_5) [67] ^{13}C NMR ($CDCl_3$): 35.9 (CH_3), 93.2 (C_5H_5), 128.4 to 135.2 (C_6H_5), 170.2 (SCN), 211.7 (SCS), 252.0 (CO) [67] IR (KBr): 1084 (ν(SO)), 1314, 1467 (ν(NCS and NCN)), 1654 (ν(C=N)), 1860, 1940 (ν(CO)) [67] mass spectrum: $[M]^+$ (28), $[M-CO]^+$ (<1), $[M-2CO]^+$ (100), $[M-2CO-CH_3CN]^+$ (31), $[M-2CO-C_6H_5CN]^+$, $[C_5H_5MoS_2OCH]^+$ (6), $[C_5H_5MoS_2]^+$ (42) [67]
52	$C_5H_5Mo(CO)_2C(N(CH_3)_2)S$	$ClC(S)N(CH_3)_2$ was added to a solution of $Na[C_5H_5Mo(CO)_3]$ in THF cooled to −78°C and the mixture was allowed to warm up overnight (75.5% by column chromatography on Al_2O_3 with $CHCl_3$/hexane 1:3); $Na[C_5H_5Mo(CO)_2P(C_6H_5)_3]$ reacted similarly (67.5%) [21, 26]; also prepared by ligand exchange between $Co_3(CO)_7(\mu_3\text{-}S)C(S)(Cl)$-$N(CH_3)_2$ and $Na[C_5H_5Mo(CO)_3]$ in THF (34%) [84] red-brown crystals [21, 26], m.p. 110 to 111°C (from $CHCl_3$/hexane) [21, 26], 112°C [84] ^{1}H NMR ($CDCl_3$): 3.60 and 3.73 (s, CH_3), 5.42 (s, C_5H_5) [21, 26] IR ($CHCl_3$): given from 615 to 2920; 1841, 1934 (ν(CO)) [21, 26] high resolution mass spectrum: $[M]^+$ [84] alkylation with $[O(CH_3)_3]BF_4$ in CH_2Cl_2 affords $[C_5H_5Mo(CO)_2C(SCH_3)N(CH_3)_2]^+$ [21, 26]
53	$[C_5H_5Mo(CO)_2C(N(CH_3)_2)SCH_3]BF_4$	formed by the alkylation of No. 52 with $[O(CH_3)_3]BF_4$ in CH_2Cl_2 for 18 h [21, 26] converted by anion exchange with aqueous NH_4PF_6 into No. 54 [21, 26]

Table 7 (continued)

No.	compound	method of preparation (yield) properties and remarks
54	$[C_5H_5Mo(CO)_2C(N(CH_3)_2)SCH_3]PF_6$	obtained by anion exchange of No. 53 with aqueous NH_4PF_6 (73%) [21, 26] orange crystals, m.p. 168 to 169 °C (from acetone/ether) [26] ^1H NMR (acetone-d_6): 2.35 (s, SCH_3), 3.93 and 4.03 (s, NCH_3), 6.02 (s, C_5H_5) [21, 26] IR (acetone): given from 715 to 3122; 1935, 2006 (ν(CO)) [21, 26] irradiation in acetone in the presence of $P(C_6H_5)_3$ affords $[C_5H_5Mo(CO)(P(C_6H_5)_3)C(N(CH_3)_2)$-$SCH_3]PF_6$ [26]

compounds with a four-membered Mo-^{1}L-^{2}D ring

No.	compound	method of preparation (yield) properties and remarks
55	$C_5H_5Mo(CO)_2C(C(O)N(C_2H_5)_2)(CH{=}C(CH_3)C_6H_5)CH_2N(C_2H_5)_2$ 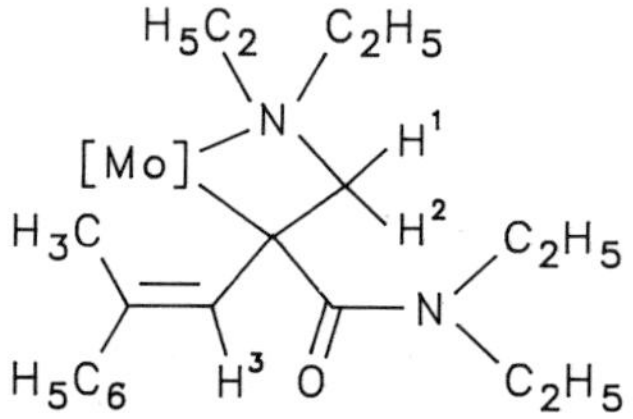	$C_5H_5Mo(CO)_3(\eta^2$-$CH_2{=}C{=}C{=}C(CH_3)C_6H_5)$ was treated with 2 equivalents of $(C_2H_5)_2NH$ [97] ^1H NMR ($CDCl_3$): 1 to 1.2 (s, CH_3), 1.4 (t, CH_3), 1.85 (s, CH_3), 3.0 (d, H-1; J(H, H) = 16), 3.1 to 3.6 (m, CH_2), 3.25 (q, CH_2), 3.47 (d, H-2; J(H, H) = 16), 4.0 (s, H-3), 4.81 (s, C_5H_5), 7.0 to 7.6 (m, C_6H_5) [97] IR (CH_2Cl_2): 1650 (ν(C=O)), 1790, 1890 (ν(CO)) [97] mass spectrum: $[M-nCO]^+$ (n = 0, 2), $[M-2CO-C_2H_5NH]^+$, $[M-3CO-C_2H_5NH]^+$ [97]
56	$C_5H_5Mo(CO)_2(CH_2)_2P(C_6H_5)_2$	$C_5H_5Mo(CO)_2(P(C_6H_5)_2(CH_2)_2Cl)Br$ was treated with an excess of sodium amalgam in THF for 2 h (36%) [91] yellow solid, m.p. 98 °C [91] ^1H NMR (C_6D_6): 0.88 to 1.87 (m, $MoCH_2$; J(H, H) = 8.6), 3.26 to 3.56 (m, part of an ABMNX system, CH_2; J(H, H) = 8.5), 4.71 (s, C_5H_5), 6.88 to 7.66 (m, C_6H_5) [91] ^{31}P NMR ($CHCl_3$): 57.6 [91] IR (n-hexane): 1879, 1951 (ν(CO)) [91] mass spectrum (70 eV, 200 °C): $[M-n\,CO]^+$ (n = 0 to 2) [91] less soluble than the related five-membered ring compound (No. 140) in unpolar solvents; decomposes in solution even at 0 °C reaction with liquid SO_2 at −40 °C affords $C_5H_5Mo(CO)_2S(O)_2(CH_2)_2P(C_6H_5)_2$ [91]

Structure of No. 55: [Mo] bonded to N(C_2H_5)$_2$ (with H_5C_2, C_2H_5) and to C bearing CH^1H^2, C(=O)N(C_2H_5)$_2$, and CH^3=C(CH_3)(C_6H_5).

References on pp. 173/6

Table 7 (continued)

No.	compound	method of preparation (yield) properties and remarks
*57	$(CH_3)_5C_5\overline{Mo(CO)_2C(O)N(C_4H_9\text{-t})P}(PC_5(CH_3)_5)NC_4H_9\text{-t}$ 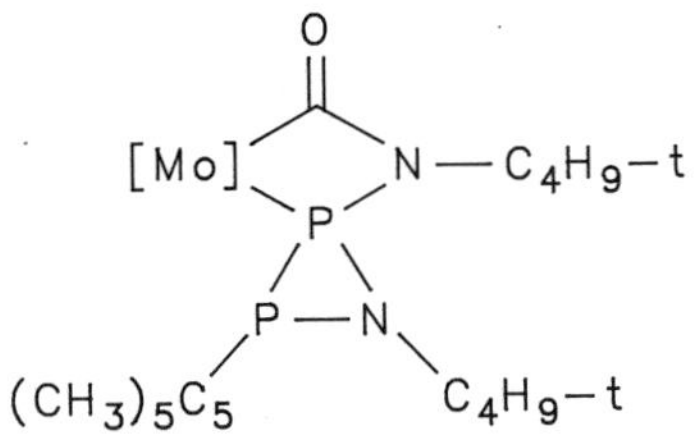	a solution of two equivalents of $(CH_3)_5C_5P{=}N\text{-}C_4H_9\text{-t}$ in CH_2Cl_2 was added to $(CO)_3Mo(NCCH_3)_3$ cooled at 0°C; the mixture was stirred for 2 h at room temperature (8 to 10%) [107] orange crystals, m.p. 142 to 144°C (from hexane) [107] ^1H NMR (C_6D_6): 1.29 and 1.51 (s, $t\text{-}C_4H_9$), 1.74 (d, $(CH_3)_5C_5$; J(P, H) = 1.7), 1.94 (d, $(CH_3)_5C_5$; J(P, H) = 1.1) [107] ^{31}P NMR: −42.5 and −20.3 (AB system; J(A, B) = 163.4) [107] IR (Nujol): 1618 (ν(C=O)), 1862, 1934 (ν(CO)) [107] thermally stable and only moderately air-sensitive in the solid state; decomposes in solution within several hours at 25°C [107]
58	$C_5H_5\overline{Mo(CO)_2C(O)N(SNHP(C_4H_9\text{-t})_2)P}(C_4H_9\text{-t})_2$	X (80%) [83] yellow crystals, dec. ca. 140°C (from pentane below −30°C) [83] ^1H NMR ($CDCl_3$): 1.05 (d, $t\text{-}C_4H_9$ exocyclic; J(P, H) = 11.5), 1.38 (d, $t\text{-}C_4H_9$ endocyclic; J(P, H) = 14.4), 1.52 (d, $t\text{-}C_4H_9$ endocyclic; J(P, H) = 14.2), 4.11 (d, NH; J(P, H) = 10.7), 5.29 (s, C_5H_5) [83] ^{13}C NMR ($CDCl_3$, −20°C): 27.0 to 30.0 (four d, $(\mathbf{C}H_3)_3C$), 32.6 to 39.5 (four d, $(CH_3)_3\mathbf{C}$), 92.3 (C_5H_5), 202.7 (d, C=O; J(P, C) = 24.8), 236.9 (CO), 249.0 (d, CO; J(P, C) = 23.9) [83] ^{31}P NMR ($CDCl_3$, −20°C): 95.6 (d, exocyclic P), 132.9 (d, endocyclic P; J(P, P′) = 3.7) [83] IR (KBr): 1628, 1657 (ν(C=O)), 3358 (ν(NH)); (CH_2Cl_2): 1872, 1952 (ν(CO)) [83]
59	$C_5H_5\overline{Mo(CO)_2C(O)N(SNHAs(C_4H_9\text{-t})_2)As}(C_4H_9\text{-t})_2$	X (40%) [83] brown-yellow crystals, dec. ca. 143°C (from hexane at −30°C) [83] ^1H NMR ($CDCl_3$): 1.12 (s, $t\text{-}C_4H_9$ exocyclic), 1.42 and 1.58 (s, $t\text{-}C_4H_9$ endocyclic), 3.84 (s, NH), 5.32 (s, C_5H_5) [83] ^{13}C NMR ($CDCl_3$, −20°C): 27.8, 28.6, and 29.5 ($(\mathbf{C}H_3)_3C$), 36.3, 36.7, 42.3, and 44.0 ($(CH_3)_3\mathbf{C}$),

References on pp. 173/6

Table 7 (continued)

No.	compound	method of preparation (yield) properties and remarks
59 (continued)		91.3 (C_5H_5), 206.2 (C=O), 237.8 and 250.5 (CO) [83] IR (KBr): 1611, 1639 (ν(C=O)), 3355 (ν(NH)); (CH_2Cl_2): 1871, 1930 (ν(CO)) [83]
compounds with a five-membered Mo-^{1}L-^{2}D ring (^{2}D bonded through N)		
60	$C_5H_5\overline{Mo(CO)_2C(O)CH_2C_5H_4N}$-2 ($C_5H_4N$ = pyridyl)	XII, with 2 equivalents of $ClCH_2C_5H_4N$-2 [12] mass spectrum (70 eV, 200 °C): main process is elimination of HCN and formation of $[(C_5H_5)_2Mo]^+$; given as diagram and discussed in [12]
61	$C_5H_5\overline{Mo(CO)_2C(O)NHC_5H_4N}$-2 ($C_5H_4N$ = pyridyl)	a mixture of $C_5H_5Mo(CO)_3Cl$ and $LiNHC_5H_4N$ in ether was allowed to warm from −196 to 25 °C and stirred for 1 h; isolated in small quantities contaminated by decomposition products [55] yellow crystals (from $CHCl_3$/hexane at −10 °C) [55] IR (KBr): 1873, 1959 (ν(CO)) [55] mass spectrum (70 eV, 80 to 220 °C): $[M-n\,CO]^+$ (n = 0 to 3) [55]
62	$C_5H_5\overline{Mo(CO)_2CH=C(CN)C(OH)=N}H$	XIa (20%); also formed by hydrolysis of $C_5H_5Mo(CO)_3C(Cl)=C(CN)_2$ in CH_2Cl_2 in the presence of $N(C_2H_5)_3$ for 16 h (25%, without addition of water 3%) [36] red-orange crystals, decomposes > 150 °C, m.p. 215 to 126 °C (from CH_2Cl_2/hexane) [36] ^{1}H NMR (acetone-d_6): 5.50 (s, C_5H_5), 7.54 (br s, NH, width at half-height ca. 10), 11.71 (s, =CH) [36] ^{13}C NMR (acetone-d_6): 94.3 (d, C_5H_5), 101.2 (s, C=), 115.6 (CN), 178.0 (s, C=), 247.1 (d, CH=), 249.2 (s, CO) [36] IR (CH_2Cl_2): 1540, 1620 (ν(C=C)), 1891, 1975 (ν(CO)), 2212 (ν(CN)), 2950, 3110, 3190, 3220, and 3393 (ν(CH, NH, and OH)) [36] stirring in THF/D_2O leads to H-D exchange of the NH group (compound No. 63) [36]
63	$C_5H_5\overline{Mo(CO)_2CH=C(CN)C(OD)=N}D$	obtained from No. 62 by H-D exchange in THF/D_2O (3 times) for 1 h followed by crystallization (50% recovery, 70% deuteration) or from the reaction of $C_5H_5Mo(CO)_3C(Cl)=C(CN)_2$ with $N(C_2H_5)_3/D_2O$ in CH_2Cl_2 (17%, 90% D by ^{1}H) [36] orange-red solid (from CH_2Cl_2/hexane) [36]

References on pp. 173/6

Table 7 (continued)

No.	compound	method of preparation (yield) properties and remarks
		IR (CH_2Cl_2): 1565 (ν(C=C)), 1888, 1972 (ν(CO)), 2210 (ν(CN)), 2388, 2549, 2955, and 3100 (ν(ND, OD, CH)) [36] prolonged treatment with D_2O in CH_2Cl_2 (ca. 16 h) led to complete decomposition [36]
64	$C_5H_5\overline{Mo(CO)_2C(CO_2CH_3){=}C(CN)C(OCH_3){=}N}H$	XIb (60 to 62%) [36] red solid, dec. > 145 °C, m.p. 205 to 206 °C (from CH_2Cl_2/hexane) [36] ^1H NMR (acetone-d_6): 3.82 and 4.03 (s, CH_3), 5.42 (s, C_5H_5), 9.0 (br s, NH) [36] ^{13}C NMR (acetone-d_6): 51.9 and 56.0 (CH_3), 94.1 (C_5H_5), 102.1 (C=), 114.1 (CN), 168.6 (CO_2), 176.1 and 233.6 (C=), 247.5 (CO) [36] IR (CH_2Cl_2): 1570 (ν(C=C, N=C)), 1700 (ν(CO_2)), 1902, 1976 (ν(CO)), 2224 (ν(CN)), 2950, 2991, 3006, 3116, and 3268 (ν(CN, NH)) [36] no deuteration in THF/D_2O is observed after 12 h reaction time [36]
65	$C_5H_5\overline{Mo(CO)_2C(CO_2C_2H_5){=}C(CN)C(OC_2H_5){=}N}H$	XIb (29%) [36] red solid, m.p. 169 to 171 °C (from CH_2Cl_2/hexane) [36] ^1H NMR (acetone-d_6): 1.32 and 1.35 (t, **CH_3**CH_2; J(H, H) ≈ 7), 4.28 and 4.35 (q, **CH_2**CH_3), 5.41 (s, C_5H_5), 9.0 (br, NH) [36] ^{13}C NMR (acetone-d_6): 14.4 and 14.8 (CH_3), 64.1 and 64.7 (CH_2), 94.1 (C_5H_5), 102.5 (C=), 114.0 (CN), 168.3 (CO_2), 175.8 and 233.5 (C=), 247.7 (CO) [36] IR (KBr): 1570 (ν(C=C)), 1693 (ν(CO_2)), 1900, 1979 (ν(CO)), 2222 (ν(CN)), 2871, 2905, 2935, 2976, 2986, 3114, and 3305 (ν(CH, NH)) [36]
66	$C_5H_5\overline{Mo(CO)_2C_6H_4CH{=}N}CH_2C_6H_5$ [Mo] bonded to ring position 1; ring positions 3, 4, 5, 6; N=C(H) with N–$C_6H_5CH_2$	$C_5H_5Mo(CO)_3Cl$ and $C_6H_5CH{=}NCH_2C_6H_5$ were heated in pyridine at 120 °C for 2 h (4% by column chromatography on silica) [44] yellow-orange crystals (from ether/pentane), m.p. 114 °C (dec.) [44] ^1H NMR ($CDCl_3$): 4.86 (s, C_5H_5), 5.20 (s, CH_2), 7.30 (m, C_6H_5), 7.77 (m, H-3), 8.10 (m, H-6), 8.41 (CH=); resonances of H-4, 5 covered [44]

References on pp. 173/6

Table 7 (continued)

No.	compound	method of preparation (yield) properties and remarks
66 (continued)		IR (KBr): 1578 (ν(C=N)), 1835, 1925 (ν(CO)) [44] mass spectrum: $[M-n\,CO]^+$ (n = 0 to 2) [44]
67	$C_5H_5\overline{Mo(CO)_2C_6H_4CH=N}C_3H_7$-i 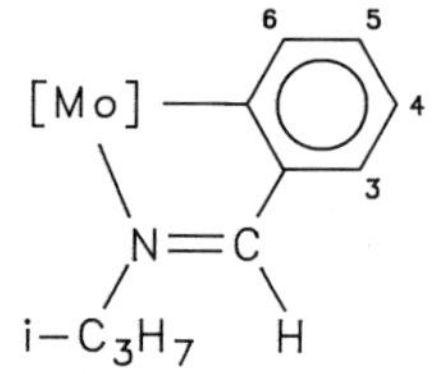	XII, with 2-$BrC_6H_4CH=NC_3H_7$-i for 2 h at 80 °C; the product was isolated by column chromatography on silica with pentane/C_6H_6 1:2 (20%) [44] m.p. 147 °C (from pentane) [44] ^{1}H NMR ($CDCl_3$): 1.30 and 1.41 (d, CH_3), 4.34 (sept, CH on i-C_3H_7), 5.23 (s, C_5H_5), 7.09 (m, H-4 and H-5), 7.69 (m, H-3), 8.04 (m, H-6), 8.37 (s, CH=); (C_6D_6): 0.79 and 1.07 (d, CH_3), 3.98 (sept, CH on i-C_3H_7), 4.73 (s, C_5H_5), 7.54 (m, H-3), 7.94 (s, CH=), 8.22 (m, H-6), H-4, 5 covered [44] IR (CH_2Cl_2): 1580 (ν(C=N)), 1855, 1943 (ν(CO)) [44] mass spectrum: $[M-n\,CO]^+$ (n = 0 to 2) [44]
*68	$C_5H_5\overline{Mo(CO)_2C_6H_4CH=N}C(C_6H_5)(CH_3)H$ 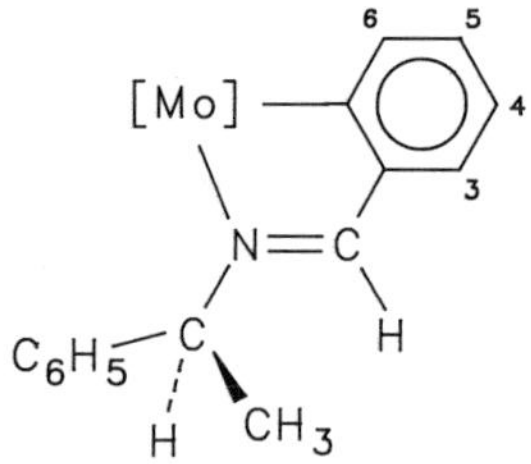	XII, with 2-$BrC_6H_4C=NC(C_6H_5)(CH_3)H$ for 2 h at 80 °C; (19%); the product was isolated by column chromatography on silica with pentane/C_6H_6 (1:2); a mixture of diastereomers [44] m.p. 134 °C (from ether/pentane 1:1) optical rotation (toluene, 3 mg/mL): $[\alpha]^{20}(\lambda)$ = +5020° (365), −1100° (436), +120° (546), small value (579) $(+)_{365}$ isomer; −2400° (365), +625° (436), −525° (546), −360° (579) eperimized mixture; figures of the spectra in [44] ^{1}H NMR ($CDCl_3$): (+) diastereomer: 1.72 (d, CH_3; J(H, H) = 6 to 7), 5.25 (s, C_5H_5), 5.48 (q, CH), 7.09 (m, H-4 and H-5), 7.34 (m, C_6H_5), 7.75 (m, H-3), 8.14 (m, H-6), 8.41 (s, CH=); (−) diastereomer: 1.76 (d, CH_3; J(H, H) = 6 to 7), 4.87 (s, C_5H_5), 5.51 (q, CH), 7.08 (m, H-4 and H-5), 7.34 (m, C_6H_5), 7.75 (m, H-3), 8.06 (m, H-6), 8.61 (s, CH=) [44]; in parts mentioned in [43, 94] ^{95}Mo NMR (acetone-d_6, vs. to 2 M Na_2MoO_4 in D_2O at pH 11): −383 ($H_{1/2}$ = 50), −396 ($H_{1/2}$ = 150) [94] IR (KBr): 1580 (ν(CN)), 1845, 1933 (ν(CO)) [44] mass spectrum: $[M-n\,CO]^+$ (n = 0 to 2) [44]

Table 7 (continued)

No.	compound	method of preparation (yield) properties and remarks
69	$C_5H_5Mo(CO)_2C_6H_4N{=}NC_6H_5$ 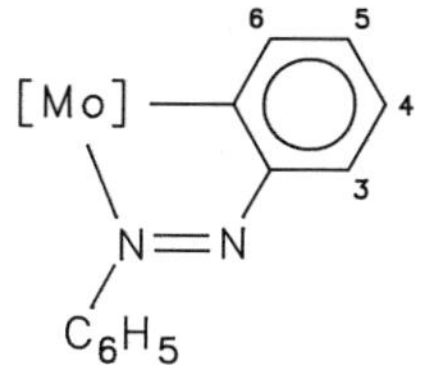	$(C_5H_5Mo(CO)_3)_2$ was allowed to react with 2.5 equivalents of $C_6H_5N{=}NC_6H_5$ in refluxing petroleum ether (b.p. 100 to 120°C) for 64 h (1.7%); $C_5H_5Mo(CO)_3CH_3$ and $C_6H_5N{=}NC_6H_5$ were refluxed in petroleum ether (b.p. 40 to 60°C) for 5 h (2.9%); in both cases the complex was isolated by column chromatography on Florisil with ether/light petroleum (1:1) [14, 18] deep violet (almost black) needles, m.p. 120°C (dec.) [14] ^1H NMR (CS_2): 5.15 (s, C_5H_5), 6.77 (t of d, H-3), 7.02 to 7.44 (m, H-4 and C_6H_5), 8.09 (d of d, H-5), 8.35 (d of d, H-6) [14] IR (cyclohexane): given from 697 to 1978; 1919, 1978 (ν(CO)) [14] reduction with $LiAlH_4$ affords small amounts of $C_6H_5NHNHC_6H_5$ [14]
*70	$C_5H_5Mo(CO)_2C(O)CH_2CH_2NH_2$ 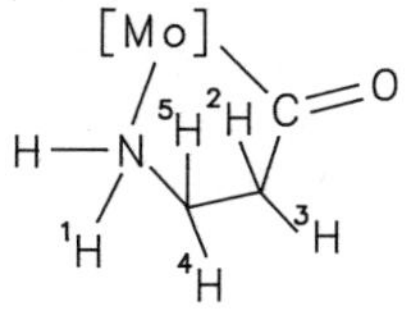	IIa (70%) [23, 87]; IIb (R = H, 40%) [37] yellow crystals, m.p. 187 to 191°C (dec.), sublimes at 140°C in vacuum with partial decomposition [87] ^1H NMR (CD_3CN): 2.0 to 2.70 (m, 2CH_2), 5.20 (s, C_5H_6) [37] IR (KBr): 1545 (δ(NH_2)), 1598 (ν(CO)), 3240, 3312 (ν(NH)); (CH_2Cl_2): 1847, 1935 (ν(CO)) [23]; (Nujol): 1540, 1590 (C=O, NH_2), 1820, 1950 (ν(CO)) [37]; ($CHCl_3$): 1545 (ν(C=O)), 1845, 1935 (ν(CO)) [87] mass spectrum: $[M-n\,CO]^+$ (n = 0 to 3), $[M-2CO-H_2]^+$ [87] stable as solid over weeks, in solution for a few days [87] moderately soluble in polar solvents like CH_2Cl_2, THF, nearly insoluble in ether and petroleum ether [87]
71	$C_5H_5Mo(CO)_2C(O)CH_2CH(CH_3)NH_2$ 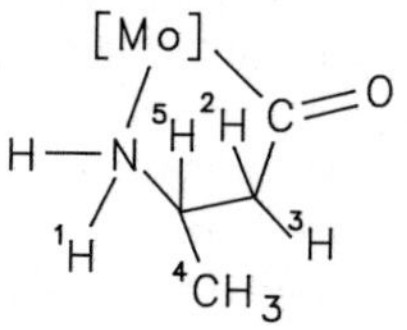	IIa (70 to 95%) [23, 87]; IIb (R = CH_3, 75%) [87] orange-yellow crystals, m.p. 200°C [87] ^1H NMR (dimethyl sulfoxide-d_6): 1.14 (CH_3), 1.97 (CH_2), 2.67 (CH), 4.4 (br, NH), 5.22 (C_5H_5) [87] IR ($CHCl_3$): 1600 (ν(C=O)), 1854, 1945 (ν(CO)) [87]

References on pp. 173/6

Table 7 (continued)

No.	compound	method of preparation (yield) properties and remarks
72	$C_5H_5\overline{Mo(CO)_2C(O)CH_2CH(C_2H_5)N}H_2$ 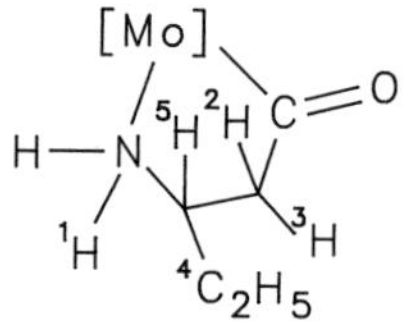	IIa (70 to 95%) [87] yellow-orange crystalline solid [41, 87] optical rotation (acetone, 1 mg/mL): $[\alpha]^{23}_{578}$ = +51.3° [41, 87]; spectrum given as diagram in [87] ^{1}H NMR (CD_3OD): 0.99 (CH_3-4), 1.55 (CH_2-4), 5.30 (C_5H_5); the spectrum shows at −60 °C a splitting of the C_5H_5 resonance of 1 Hz (T_c = −28 °C) [87] IR ($CHCl_3$): 1598 (ν(C=O)), 1845, 1945 (ν(CO)) [87]
73	$C_5H_5\overline{Mo(CO)_2C(O)C(C_6H_5)HCH_2N}H_2$ 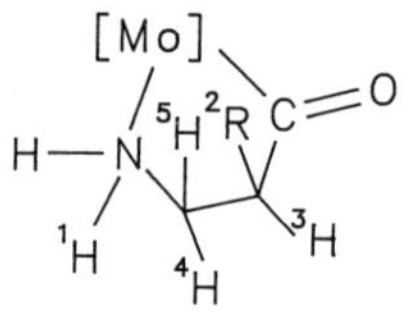 R = C_6H_5	IIa (70 to 95%) [87] yellow to orange crystalline solid [87] ^{1}H NMR (dimethyl sulfoxide-d_6): 5.38 (C_5H_5), 7.20 (m, C_6H_5) [87] IR ($CHCl_3$): 1855, 1945 (ν(CO)) [87]
74	$C_5H_5\overline{Mo(CO)_2C(O)CH_2C(CH_3)_2N}H_2$ 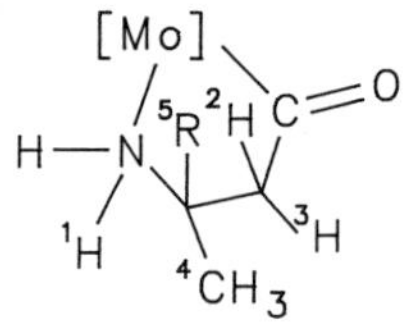 R = CH_3	IIa (70 to 95%) [87] yellow to orange crystalline solid, m.p. 150 °C [87] ^{1}H NMR (CD_3OD): 1.16 (H-4,5), 2.00 (H-2,3), 5.23 (C_5H_5) [87] IR ($CHCl_3$): 1598 (ν(C=O)), 1853, 1945 (ν(CO)) [87]
75	$C_5H_5\overline{Mo(CO)_2C(O)CH_2CH_2N}HC_4H_9$-t 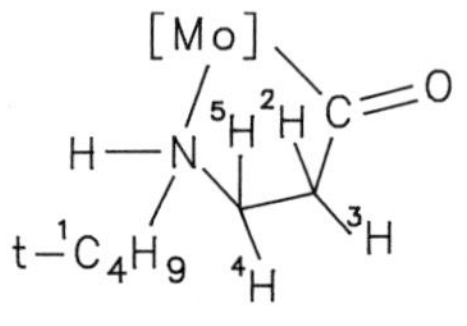	IIa (70 to 95%) [41, 87] yellow to orange crystalline solid, m.p. 150 °C [87] ^{1}H NMR (dimethyl sulfoxide-d_6): 1.20 (t-C_4H_9), 2.17 to 2.67 (m, H-2 to 5), 5.19 (C_5H_5) [87]
76	$C_5H_5\overline{Mo(CO)_2C(O)CH_2CH_2N}HC(C_6H_5)(CH_3)H$ 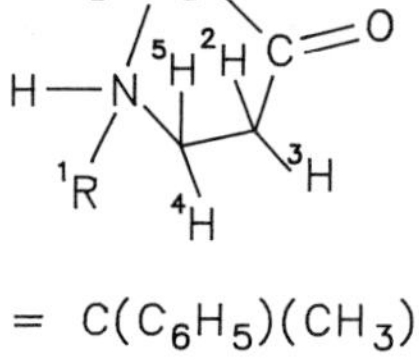 R = $C(C_6H_5)(CH_3)H$	L- or D-$N(C_6H_5)(CH_3)H$ was added to a solution of $[C_5H_5Mo(CO)_3CH_2{=}CH_2]PF_6$; the product precipitated in a few minutes (90 to 91%); two diastereomers in a ratio of 74:26 for both amines [87] orange-yellow, transparent sticks (from CH_2Cl_2/pentane) [87] CD spectra are given as diagram in [87]

References on pp. 173/6

Table 7 (continued)

No.	compound	method of preparation (yield) properties and remarks
		^{1}H NMR: 4.63 and 5.34 (C_5H_5) [87] IR (KBr): 1610 (ν(C=O)), 1846, 1936 (ν(CO)) [87]
77	$C_5H_5Mo(CO)_2C(O)CH(C_6H_5)C(CH_3)HNHCH_3$ 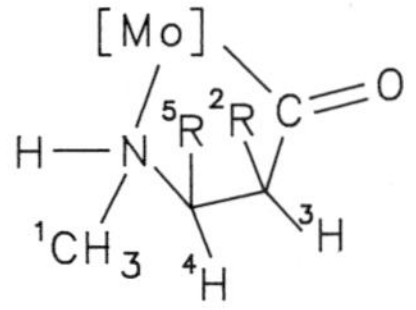 $R^5 = CH_3$; $R^2 = C_6H_5$	IIa, for 15 min (70 to 95%) [87] yellow to orange solid [87] optical rotation (acetone, 1 mg/mL): $[\alpha]^{23}_{578}$ = +100° [41, 87]; the CD spectrum is given as figure in [87] no isomerization to No. 78 is observed in solution [87]
*78	$C_5H_5Mo(CO)_2C(O)C(C_6H_5)HC(CH_3)HNHCH_3$ 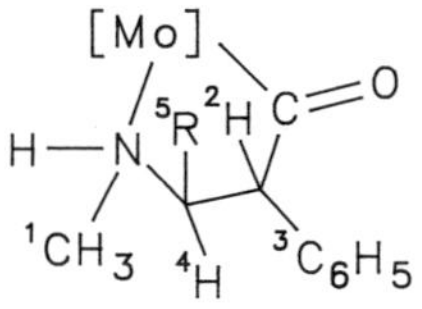 R = CH_3	IIa, for 7 h (70 to 95% yield) [41, 87] orange to yellow crystalline solid, m.p. 162°C [87] optical rotation (acetone, 1 mg/mL): $[\alpha]^{23}_{578}$ = −687° [41, 87]; the CD-spectrum is given as diagram in [87] ^{1}H NMR (dimethyl sulfoxide-d_6): 1.00 (H-5), 2.69 (H-1), 4.4 (br, NH), 5.35 (C_5H_5), 7.37 (C_6H_5) [87] ^{13}C NMR (dimethyl sulfoxide-d_6): 15.2 (**C**H_3C), 44.1 (NCH_3), 66.2 (N**C**CH_3), 79.7 (**C**C_6H_5), 93.5 and 94.4 (C_5H_5), 126.1, 128.1, and 140.1 (C_6H_5), 218.9 (C=O), 238.4 and 245.2 (CO) [87] IR ($CHCl_3$): 1852, 1943 (ν(CO)) [87] no isomerization to No. 77 is found in solution [87]
79	$C_5H_5Mo(CO)_2C(O)C(CH{=}CH_2)HCH_2N(CH_3)_2$ 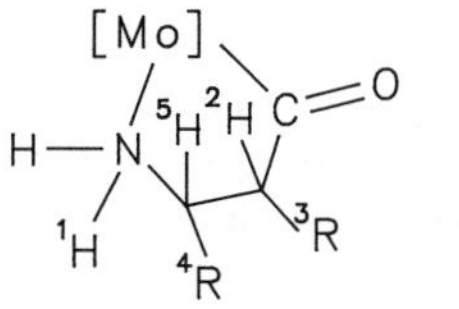 R^3 or R^4 = CH=CH_2	$(CH_3)_2NH$ was passed through a solution of $[C_5H_5Mo(CO)_3(\eta^2\text{-}CH_2{=}CHCH{=}CH_2)]BF_4$ cooled at −40°C in CH_2Cl_2; two isomers were obtained [76] yellow microcrystalline powder [76] ^{1}H NMR (acetone-d_6): 2.80 and 2.85 (m, CH_3), 5.33 and 5.39 (s, C_5H_5) [76] IR (Nujol): 1582, 1590, 1602 (ν(CC)), 1816, 1824, 1933, 1945 (ν(CO)) [76]
80	$C_5H_5Mo(CO)_2C(O)CH_2C_5H_4N$-2	XII, with $ClCH_2C_5H_4N$-2 (C_5H_4N = pyridine) for 12 h (56%) yellow crystals, m.p. 146 to 150°C (from CH_2Cl_2/hexane) [6] ^{1}H NMR (acetone-d_6): 4.00 (s, CH_2), 5.38 (s, C_5H_5), 7.29 (t of d, H-4; J(H, H) = 7, 2), 7.71 (H-3; J(H, H) ≈ 7), 8.96 (d, H-2; J(H, H) = 7) [6] IR (halocarbon oil): given from 758 to 3060; 1624 (ν(C=O)), 1854, 1934, 1946 (ν(CO)) [6]

References on pp. 173/6

Table 7 (continued)

No.	compound	method of preparation (yield) properties and remarks
80 (continued)		UV (dioxane): without maxima in proceeding from 200 to 400 nm [6]
81	$C_5H_5Mo(CO)_2C_{10}H_8N$ 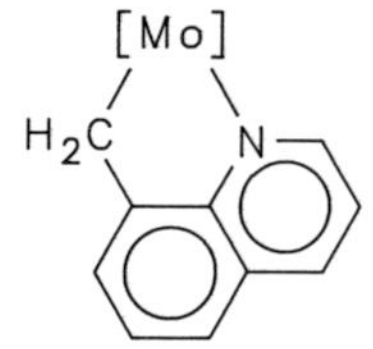	XII, with 8-bromomethylquinoline for 8 h; isolated by column chromatography on Al_2O_3 with ether/hexane (1:1) (16%) [33] orange crystals, m.p. 83 to 85°C [33] ^{13}C NMR ($CDCl_3$ at 34°C): −1.0 ($MoCH_2$), 93.2 (C_5H_5), 120.6, 123.21, 126.4, 127.8, 128.4, 137.5, 145.2, 147.9, and 150.9 (all aromatic C) [33]
82	$C_5H_5Mo(CO)_2C_{13}H_8N$ 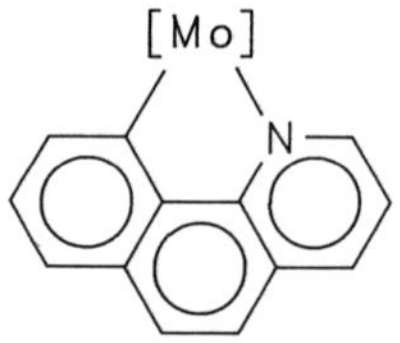	$C_5H_5Mo(CO)_3CH_3$ was allowed to react with benzo[h]quinoline; the main product was $(C_5H_5Mo(CO)_3)_2$ [24] IR ($CHCl_3$): 1917, 1961 (ν(CO)) [24] air-sensitive [24]
83	$Na[C_5H_5\overline{Mo(CO)_2C(O)C(C(O)OC_2H_5){=}N}NH]$	$C_5H_5Mo(CO)_3NHN{=}CHC(O)OC_2H_5$ was deprotonated by alkaline in aqueous solution in a reversible reaction; maybe the complex is identical with No. 84 [7] alkylation with CH_3I in THF affords $C_5H_5Mo(CO)_3NCH_3N{=}CHC(O)OC_2H_5$ [7]
84	$Na[C_5H_5\overline{Mo(CO)_2C(OH)C(C(O)OC_2H_5)N}N]$	No. 86 was reduced with an excess of sodium amalgam (2% Na) in THF for 10 min at 25°C (high yield); directly converted into the $[N(C_4H_9\text{-}n)_4]^+$ salt (No. 85) by treating with aqueous $[N(C_4H_9\text{-}n)_4]OH$ [16]
85	$[N(C_4H_9\text{-}n)_4][C_5H_5\overline{Mo(CO)_2C(OH)C(C(O)OC_2H_5)N}N]$	anion exchange of No. 84 with excess aqueous $[N(C_4H_9\text{-}n)_4]OH$ (ca. 70%) [16] yellow solid (from CH_2Cl_2/ether) [16] ^1H NMR ($CDCl_3$): 0.93 (m, C_4H_9-n), 1.28 (t, CH_3; J(H, H) = 7), 1.45 and 3.17 (m, C_4H_9-n), 4.19 (q, OCH_2), 5.16 (s, C_5H_5), 10.84 (s, OH) [16] IR (mull): given from 1665 to 3220; 1822, 1920 (ν(CO)) [16] reaction with excess of CH_3I in THF affords No. 87 [16]

Table 7 (continued)

No.	compound	method of preparation (yield) properties and remarks
*86	$C_5H_5Mo(CO)_2C(OH)C(C(O)OC_2H_5)NHN$	XII, with $N_2CHC(O)OC_2H_5$ at −40°C followed by warming; the product of this reaction was neutralized in CH_2Cl_2/H_2O with 2 M HCl (20% by chromatography on Al_2O_3 with acetone/H_2O 4:12) [16] yellow solid (from acetone/water or SO_2/ether) [16] ^1H NMR ($CDCl_3$): 2.43 (t, CH_3; J(H, H) = 7), 4.39 (q, CH_2), 5.66 (s, C_5H_5), 12.31 and 12.57 (s, OH and NH) [16] IR (mull): given from 1643 to 3210; 1926, 1998 (ν(CO)) [16] mass spectrum: $[M]^+$ and fragments such as $[C_5H_5Mo(CO)_nNH]^+$ [16] protonation affords after anion exchange Nos. 88 and 89; reduction with excess sodium amalgam affords No. 84 [16]
87	$C_5H_5Mo(CO)_2C(OH)C(C(O)OC_2H_5)N(CH_3)N$	No. 85 and excess of CH_3I were stirred in THF for 6 h (60% by chromatography) [16] red solid (from CH_2Cl_2/ether) [16] ^1H NMR ($CDCl_3$): 1.45 (t, CH_3; J(H, H) = 7), 4.55 (q, CH_2), 4.68 (s, NCH_3), 5.63 (s, C_5H_5), 10.03 (s, OH) [16] IR (mull): given from 1585 to 3115; 1895, 2000 (ν(CO)) [16] protonation with aqueous HCl followed by anion exchange affords No. 90 [16]
88	$[C_5H_5Mo(CO)_2C(OH)C(C(O)OC_2H_5)NHNH]BF_4$	No. 85 was dissolved in aqueous HCl followed by anion exchange with $NaBF_4$ (high yield) [16] red solid (from liquid SO_2) [16]
89	$[C_5H_5Mo(CO)_2C(OH)C(C(O)OC_2H_5)NHNH]PF_6$	No. 85 was dissolved in aqueous HCl and NH_4PF_6 was added to the resulting red solution (high yield) [16] red solid, decomposes at ca. 200°C (from liquid SO_2) [16] ^1H NMR ($CDCl_3$): 1.39 (t, CH_3; J(H, H) = 7), 4.48 (q, CH_2), 5.59 (s, C_5H_5), 11.1 and 12.79 (s, OH and NH; not visible after H/D exchange) [16] IR (mull): given from 1706 to 3400; 1971, 2040 (ν(CO)) [16]

References on pp. 173/6

Table 7 (continued)

No.	compound	method of preparation (yield) properties and remarks
89 (continued)		soluble in acetone and liquid SO_2 [16] 4% aqueous dioxane solution show the cation is a strong acid, having a pK_a value of 1.7 [16]
*90	$[C_5H_5\overline{Mo(CO)_2C(OH)C(C(O)OC_2H_5)N(CH_3)N}H]PF_6$	No. 87 was dissolved in aqueous HCl and NH_4PF_6 was added to the obtained red solution to precipitate (80%) [16] red solid (from CH_2Cl_2/ether) [16] ^1H NMR ($CDCl_3$): 1.43 (t, CH_3; J(H, H) = 7), 4.59 (q, CH_2), 4.59 (s, NCH_3), 5.65 (s, C_5H_5), 11.91 and 12.18 (NH and OH) [16] IR (mull): given from 1668 to 3305; 1975, 2040 (ν(CO)) [16]
91	$C_5H_5\overline{Mo(CO)_2C(O)NHC(CH_3)N}H$	XIII, M = Li, in ether (−196°C, 1 h at 20°C; 40%) [55] for 11 h (20%) [70] yellow crystals (from $CHCl_3$/hexane at −10°C) [55], brown solid (from acetone/ether) [70], m.p. 155 to 156°C [55], 212°C (dec.) [70] ^1H NMR (acetone-d_6): 2.31 (d, CH_3; J(H, H) = 0.75), 5.29 (s, C_5H_5), 8.25 and 9.59 (m, NH) [70] IR (KBr): 1569 (ν(C=O)), 1628 (ν(CN)), 3157, 3380 (ν(NH)); (CH_2Cl_2): 1868, 1960 (ν(CO)) [55, 70]; (KBr): 1630 (ν(C=O)), 1827, 1853, 1928, 1953 (ν(CO)) [55] mass spectrum (70 to 220°C, 70 eV): $[M-n\,CO]^+$ (n = 0 to 3) [55, 70] air-stable as a solid, but sensitive in solution [55]
92	$C_5H_5\overline{Mo(CO)_2C(O)NHC(C_6H_5)N}H$	XIII, M = Li, in ether with CS_2 for 11 h (32%) [70] yellow-brown solid (from acetone/ether), m.p. 206°C (dec.) [70] ^1H NMR (acetone-d_6): 5.37 (s, C_5H_5), 7.65 and 7.92 (m, C_6H_5), 9.09 and 10.39 (m, NH) [70] IR (KBr): 1578 (ν(C=O)), 1610 (ν(CN)), 3155, 3375 (ν(NH)); (CH_2Cl_2): 1872, 1960 (ν(CO)) [70] mass spectrum: $[M]^+$ [70]
93	$C_5H_5\overline{Mo(CO)_2C(O)NHC((CH_2)_5)N}H_2$ [Mo]—C(=O)—NH—C(cyclohexane)—H_2N—[Mo]	a solution of pentamethylene diaziridine was added to a solution of $C_5H_5Mo(CO)_3H$ in THF; the product precipitated [25] (55%) [45] yellow microcrystals, dec. 140°C [45] ^1H NMR (dimethyl sulfoxide-d_6): 1.40 (m, $3CH_2$), 1.60 (m, $2CH_2$), 5.4 (s, C_5H_5) [45] IR (KBr): 1533 (ν(C=O)), 1588 (δ(NH_2)), 1807, 1933 (ν(CO)), 3159, 3220 (ν(NH_2)), 3345 (ν(NH)) [45], similar data in [25]; after H/D exchange: 2312, 2380 (ν(ND_2)), 2470 (ν(ND)) [45]

Table 7 (continued)

No.	compound	method of preparation (yield) properties and remarks
		mass spectrum (70 eV): compound decomposes; $[M-C(O)HNH_2]^+$ is the base peak [45] reaction with HCl in C_6H_6 gives $C_5H_5Mo(CO)_3Cl$ [25, 45]; $C_5H_5Mo(CO)_3H$ produces $(C_5H_5Mo(CO)_3)_2$ [45]
94	$C_5H_5\overline{Mo(CO)_2C(O)N(CH_3)CHN}CH_3$	XIII, M = K, in C_6H_6 in the dark for 18 h (22%); also obtained in low yields from the reaction of $C_5H_5Mo(CO)_2(NCH_3)_2CH$ with CO [54] yellow solid, decomposes at 180°C (from acetone/hexane); sublimes at ca. 100°C in vacuum [54] ^1H NMR ($CDCl_3$): 3.10 (s, CH_3), 3.45 (CH_3; J(H, H) ≈ 1), 5.27 (s, C_5H_5), 7.60 (CH; J(H, H) ≈ 1) [54] ^{13}C NMR ($CDCl_3$): 27.8, 49.8 (CH_3), 92.5 (C_5H_5), 156.9 (CH), 247.4, 251.9 (CO) [54] IR (CH_2Cl_2): 1642 (ν(C=O)), 1867, 1956 (ν(CO)) [54] air-stable in the solid state [54]
95	$C_5H_5\overline{Mo(CO)_2C(O)N(C_4H_9\text{-}t)CHN}C_6H_5$	XIII, M = K, in C_6H_6 (ca. 20% by chromatography on SiO_2 with toluene/ether 1:1) [63] yellow solid (from acetone/hexane) [63] ^1H NMR ($CDCl_3$): 1.54 (s, t-C_4H_9), 5.27 (s, C_5H_5), 7.01 and 7.21 (m, C_6H_5), 7.91 (s, CH) [63] ^{13}C NMR ($CDCl_3$): 29.8 (C(**C**H_3)$_3$), 57.2 (**C**(CH_3)$_3$), 93.3 (C_5H_5), 123.3 (C-2), 125.8 (C-4), 129.0 (C-3), 153.9 (C-1, all C_6H_5), 247.0 (C=O), 231.9 and 251.1 (CO); three CO resonances at −30°C and one at +30°C indicate fluxional behavior [63] IR (CH_2Cl_2): 1609 (ν(C=O)), 1864, 1952 (ν(CO)) [63] air-stable solid; UV irradiation in benzene affords $C_5H_5Mo(CO)_2(N(C_4H_9\text{-}t)CHN(C_6H_5))$ [63]
96	$C_5H_5\overline{Mo(CO)_2C(O)N(C_6H_5)CHN}C_6H_5$	XIII, M = Li, in ether (40%) [55]; M = K, in C_6H_6 (19%) [63] yellow crystalline solid (from $CHCl_3$/hexane at −10°C [55] or acetone/hexane [63]) ^1H NMR ($CDCl_3$): 5.37 (s, C_5H_5), 7.29 (m, C_6H_5), 7.96 (s, CH) [63] ^{13}C NMR ($CDCl_3$, −55°C): 92.9 (C_5H_5), 122.8, 125.8 (C-2), 125.8, 127.5 (C-4), 128.8, 128.9 (C-3), 134.4, 150.0 (C-1), 155.5 (CH), 247.3 (C=O), 229.4, 249.5 (CO); three CO resonances at −30°C and one at +30°C indicate fluxional behavior [63]

References on pp. 173/6

Table 7 (continued)

No.	compound	method of preparation (yield) properties and remarks
96 (continued)		IR (CH_2Cl_2): 1617 (ν(C=O)), 1924, 1961 (ν(CO)) [63]; (KBr): 1628 (ν(C=O)), 1869, 1956 (ν(CO)); ($CHCl_3$): 1886, 1969 (ν(CO)) [55] air-stable in the solid state, decomposes above 40°C in $CHCl_3$ [55, 63] other product of the preparation is $C_5H_5Mo(CO)_2(NC_6H_5)_2CH$, which is also isolated after irradiation of No. 96 in C_6H_6 [55, 63]
97	$C_5H_5\overline{Mo(CO)_2C(O)N(CH_3)C(C_6H_5)N}CH_3$	XIII, M = Li, in ether (25%) [60], (40%) [55] yellow crystals (from $CHCl_3$/hexane at −10°C) [55], yellow prisms (from ether/pentane) [60], m.p. 171°C [60], 174 to 175°C [55] ^1H NMR ($CDCl_3$): 3.29 and 3.71 (CH_3), 5.90 (C_5H_5), 7.64 and 8.03 (C_6H_5) [55]; 2.79 and 3.20 (s, CH_3), 5.36 (s, C_5H_5), 7.31 (m, C_6H_5) [60] IR (KBr): 1612 (ν(C=O)), 1853, 1861, 1949 (ν(CO)); ($CHCl_3$): 1877, 1963 (ν(CO)) [55]; (KBr): 1434, 1495, 1537 (ν(NCN)), 1594 (ν(CN)), 1608 (ν(C=O)); (CH_2Cl_2): 1868, 1956 (ν(CO)) [60] mass spectrum (70 eV, 80 to 220°C): $[M-n\ CO]^+$ (n = 0 to 3) [55, 60]; other fragment ions are also given in [60] stable as solid but sensitive in solution [55] decomposes during photolysis in benzene to $(C_5H_5Mo(CO)_3)_2$; the carbamoyl group can be reduced with $LiAlH_4$ [60]
98	$C_5H_5\overline{Mo(CO)_2C(O)N(CH_2C_6H_5)C(C_6H_5)N}CH_2C_6H_5$	XIII, M = Li, in ether (41%) [60] yellow needles (from ether), m.p. 136°C (dec.) [60] ^1H NMR ($CDCl_3$): 4.58 and 4.74 (AB system, CH_2; J(A, B) = 14), 4.93 (s, C_5H_5), 7.19 (m, C_6H_5) [60] IR (KBr): 1444, 1495 (ν(NCN)), 1595 (ν(CN)), 1608 (ν(C=O)); (CH_2Cl_2): 1860, 1950 (ν(CO)) [60] mass spectrum (70 eV): $[M-n\ CO]^+$ (n = 0 to 3) and other fragments ions [60]
99	$C_5H_5\overline{Mo(CO)_2C(O)N(C_3H_7\text{-}i)C(C_6H_5)N}C_3H_7\text{-}i$	XIII, M = Li, in ether (34%) [60] yellow prisms (from ether/pentane), m.p. 161°C (dec.) [60] ^1H NMR ($CDCl_3$): 1.03, 1.27, and 1.30 (d, CH_3C; J(H, H) ≈ 6.5 to 7), 3.31 and 3.87 (m, CH), 5.29 (s, C_5H_5), 7.32 (m, C_6H_5) [60]

References on pp. 173/6

Table 7 (continued)

No.	compound	method of preparation (yield) properties and remarks
		IR (KBr): 1413, 1490, 1548 (ν(NCN)), 1593 (ν(CN)), 1615 (ν(C=O)); (CH_2Cl_2): 1860, 1950 (ν(CO)) [60] mass spectrum (70 eV): $[M-n\,CO]^+$ (n = 0 to 3) and other fragment ions [60]
100	$C_5H_5\overline{Mo(CO)_2C(O)N(C_6H_5)C(CH_3)}NC_6H_5$	XIII, M = Li, in ether (21%) [60], (40%) [55]; also obtained in low yields by the reaction of $C_5H_5Mo(CO)_2(P(C_6H_5)_3)Cl$ with $Li(NC_6H_5)_2CCH_3$ in $CH_3OC_2H_4OCH_3$ for 2 h or by refluxing of $(C_5H_5Mo(CO)_3)_2$ with $C_6H_5NHC(CH_3)NC_6H_5$ in toluene [55] yellow crystals ($CHCl_3$/hexane at −10°C) [55], yellow prisms (from ether/pentane) [60], m.p. 166°C [55], 176°C (dec.) [60] ^1H NMR ($CDCl_3$): 2.30 (s, CH_3), 5.87 (s, C_5H_5), 7.82 (m, C_6H_5) [55]; 1.81 (s, CH_3), 5.61 (s, C_5H_5), 7.00 (m, C_6H_5) [60] IR (KBr): 1408, 1499 (ν(NCN)), 1594 (ν(CN)), 1617 (ν(C=O)), 1862, 1949 (ν(CO)) [60], similar data in [55]; (toluene): 1878, 1960 (ν(CO)) [55] mass spectrum (70 eV): $[M-n\,CO]^+$ (n = 0 to 3) [55, 60], other fragment ions are given in [60] stable as solid but sensitive in solution [55] irradiation results in decomposition products and a new unidentified complex with ν(CO) at 2026; in the presence of CH_3I, the following products have been identified by IR and mass spectrum: $C_5H_5Mo(CO)_2(\eta^3\text{-}N(C_6H_5)C(CH_3)NC_6H_5)$, $C_5H_5Mo(CO)_3I$, and $[C_5H_5Mo(CO)_3N(C_6H_5)C(CH_3)N(C_6H_5)CH_3]I$ [55]
101	$C_5H_5\overline{Mo(CO)_2C(O)N(C_6H_4CH_3\text{-}4)C(CH_3)}NC_6H_4CH_3\text{-}4$	XIII, M = Li, in ether (40%) [55] yellow crystals, m.p. 158 to 159°C (from $CHCl_3$/hexane at −10°C) [55] ^1H NMR ($CDCl_3$): 2.24 (NCH_3), 2.84 (CH_3-4), 5.80 (C_5H_5), 7.76 (m, C_6H_4) [55] IR (KBr): 1865, 1952 (ν(CO)) [55] mass spectrum (70 eV, 80 to 220°C): $[M-n\,CO]^+$ (n = 0 to 3) [55] stable as a solid but sensitive in solution [55]

References on pp. 173/6

Table 7 (continued)

No.	compound	method of preparation (yield) properties and remarks
102	$C_5H_5Mo(CO)_2C(O)N(C_6H_5)C(C_6H_5)NC_6H_5$	XIII, M = Li, in ether (33%) [60] dark yellow needles, m.p. 150°C (dec., from ether) [60] ^1H NMR ($CDCl_3$): 5.37 (C_5H_5), 6.92 (m, C_6H_5) [60] IR (KBr): 1417, 1495 (ν(CNC)), 1593 (ν(CN)), 1626 (ν(C=O)), 1862, 1952 (ν(CO)) [60] mass spectrum (70 eV): $[M-n\ CO]^+$ (n = 0 to 3) and other fragment ions [60]
103	$C_5H_5Mo(CO)_2C(O)N(C_6H_5)NNC_6H_5$	similar to XIII, with $K[(NC_6H_5)_2N]$ in C_6H_6 for 2 h (10%) [63] orange-yellow solid (from acetone/ether) [63] ^1H NMR ($CDCl_3$): 5.43 (C_5H_5), 7.34 (m, C_6H_5) [63] ^{13}C NMR ($CDCl_3$): 93.2 (C_5H_5), 122.3, 124.9 (C-2), 127.8 (C-4), 128.6 (C-3), 135.6, 154.5 (C-1), 238.1 (C=O), 231.0 and 241.9 (CO); three CO resonances at −30°C and one at +30°C indicate fluxional behavior [63] IR (CH_2Cl_2): 1618 (ν(C=O)), 1904, 1980 (ν(CO)) [63] irradiation in C_6H_6 affords $C_5H_5Mo(CO)_2(NC_6H_5)_2N$; decomposes above 40°C in $CHCl_3$ [63]
104	$[C_5H_5Mo(CO)_2C(C_6H_5)N(CH_3)C(C_6H_5)=NCH_3]Cl$	XII, with an excess of $ClC(C_6H_5)=NCH_3$ (27%) [61, 81] dark violet prisms (from CH_2Cl_2/ether 1:1) [61, 81] IR (KBr): 1851, 1954 (ν(CO)) [61, 81] anion exchange with NH_4PF_6 in CH_3OH/CH_2Cl_2 (8:1) [61] or C_2H_5OH [81] affords the corresponding PF_6 salt (No. 105) [61]
105	$[C_5H_5Mo(CO)_2C(C_6H_5)N(CH_3)C(C_6H_5)=NCH_3]PF_6$	anion exchange of the corresponding chloride (No. 104) with NH_4PF_6 in CH_3OH/CH_2Cl_2 (8:1) [61] or C_2H_5OH [81] (35%) [61] yellow-brown prisms, m.p. 157°C (from acetone/ether 2:1 at −35°C) [61] ^1H NMR (acetone-d_6): 3.22 and 3.83 (s, CH_3), 5.93 (s, C_5H_5), 7.15 to 7.65, and 7.69 (m, C_6H_5) [61] ^{13}C NMR (acetone-d_6): 42.0 and 52.6 (CH_3), 96.1 (C_5H_5), 124 to 132 (C_6H_5), 149.7 (C_6H_5, C-1), 168.7 (NCN), 280.1 (CO) [61] IR (KBr): 850 (ν(PF_6)), 1580 (ν(CN)), 1950, 2005 (ν(CO)) [61], similar data in [81]

Table 7 (continued)

No.	compound	method of preparation (yield) properties and remarks
		sensitive to decomposition [81] compound No. 106 is obtained by the reduction of No. 105 with $NaBH_4$ [61, 71]; reaction with 2 equivalents of C_6H_5MgBr in THF for 17 h affords compound No. 107 [82]
106	$C_5H_5\overline{Mo(CO)_2CH(C_6H_5)N(CH_3)C(C_6H_5)=N}CH_3$ 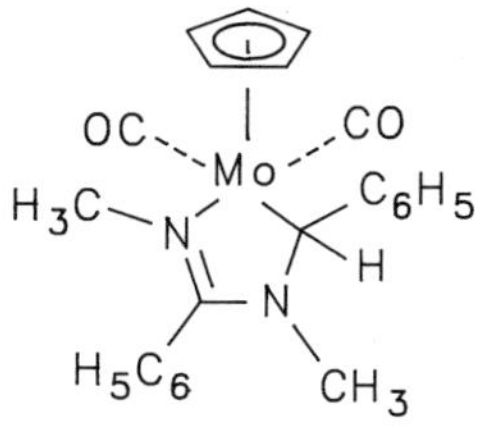 a 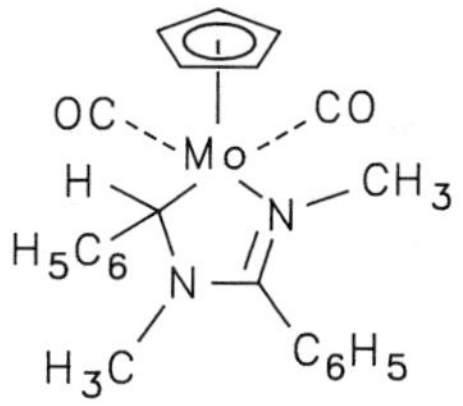 b	side product of the preparation of No. 105 (5%) [61]; compound No. 105 was allowed to react with $NaBH_4$ in THF at 60°C; two diastereomers were separated by chromatography on silica with toluene at 0°C (33% of isomer a and 11% of isomer b) [71] the a to b isomer ratio in equilibrium is 85:15 [71] ^{1}H NMR spectroscopically investigations of mixtures by Nuclear-Overhauser-Effect Difference Spectroscopy in [82] isomer a: orange-brown needles (from ether/pentane 2:1 at −35°C), m.p. 141°C (dec.) [71] ^{1}H NMR ($CDCl_3$): 2.56 and 2.80 (s, CH_3), 4.89 (s, C_5H_5), 5.65 (s, CH), 7.16 and 7.43 (m, C_6H_5) [61, 71] ^{13}C NMR ($CDCl_3$, 0.06 M $Cr(OC(CH_3)CHC(CH_3)O)_3$): 38.7 and 47.3 (CH_3), 69.2 (CH), 96.6 (C_5H_5), 121.5 to 133.2 (C_6H_5), 155.4 (C_6H_5, C-1), 170.1 (NCN), 250.9 and 260.4 (CO) [71] IR (KBr): 1588 (ν(CN)), 1805, 1911 (ν(CO)) [71] isomer b: orange-brown needles (from ether/pentane 2:1 at −35°C), m.p. 130°C [71] ^{1}H NMR ($CDCl_3$): 2.53 and 2.89 (s, CH_3), 5.36 (s, C_5H_5), 5.48 (s, CH), 7.07 and 7.43 (m, C_6H_5) [71] ^{13}C NMR ($CDCl_3$, 0.06 M $Cr(OC(CH_3)CHC(CH_3)O)_3$): 38.3 and 47.3 (CH_3), 65.4 (CH), 94.7 (C_5H_5), 123.2 to 133.1 (C_6H_5), 153.7 (C_6H_5, C-1), 170.1 (CH), 255.4 and 263.9 (CO) [71], similar data in [61] IR (KBr): 1586 (ν(CN)), 1804, 1907 (ν(CO)) [71], similar data in [61] the configurations are not stable in solution; activation parameters for the isomerization of a

Table 7 (continued)

No.	compound	method of preparation (yield) properties and remarks
106 (continued)		to b in $CDCl_3$ at 55°C: k = 4.80 × 10^{-6} ± 5.38 × 10^{-7} s^{-1}, $\Delta G^{\neq}$ = 27.2 ± 0.2 kcal/mol; for the isomerization of b to a: k = 2.80 × 10^{-5} ± 1.41 × 10^{-6} s^{-1}, $\Delta G^{\neq}$ = 26.1 ± 0.2 kcal/mol, $\tau_{1/2}$ = 351 min [71]
*107	$C_5H_5\overline{Mo(CO)_2C(C_6H_5)_2N(CH_3)C(C_6H_5)=N}CH_3$	compound No. 105 was allowed to react with 2 equivalents of C_6H_5MgBr in refluxing THF for 17 h (20% by chromatography on silica with C_6H_6) [82] dark red prisms (from ether/pentane) [82] ^{1}H NMR ($CDCl_3$): 2.13 and 2.87 (s, CH_3), 4.81 (s, C_5H_5), 7.36 (m, C_6H_5) [82] IR (CH_2Cl_2): 1827, 1932 (ν(CO)) [82]

compounds with a five-membered Mo-^{1}L-^{2}D ring of the type MoC(R)=C(R′)C(R″)O

No.	compound	method of preparation (yield) properties and remarks
108	$C_5H_5\overline{Mo(CO)_2CH=CHC(CH_3)O}$	III, under irradiation with HC≡CH (19%) [92, 102], (9%) [40, 49, 50] red to dark red crystals, m.p. 65°C (from pentane at −78°C) [40, 49] ^{1}H NMR (acetone-d_6, −20°C): 2.36 (s, CH_3), 5.42 (s, C_5H_5), 6.98 (d, MoCH; J(H, H) = 8.3), 11.64 (d, =CH; J(H, H) = 8.3) [49, 92]; similar data in [40] ^{13}C NMR (acetone-d_6): 24.38 (CH_3), 93.75 (C_5H_5), 131.75 (MoCH=**C**H), 201.96 (C=O), 247.05 (MoCH), 248.35 (CO) [40]; similar data in [92] at −20°C IR (KBr): 1488 (ν(C=O)); (pentane): 1905, 1980 (ν(CO)) [40]; nearly identical data in [49, 92] mass spectrum: $[M-n\,CO]^+$ (n = 0 to 2) [49, 92] reaction with $P(CH_3)_3$ leads to the formation of No. 112 [58]; no constant product is obtained on reaction with NO due to decomposition, while the corresponding W compound affords an oxo complex [93]
109	$CH_3C_5H_4\overline{Mo(CO)_2CH=CHC(CH_3)O}$	III, under irradiation with HC≡CH (26%) [92] red solid [92]

References on pp. 173/6

Table 7 (continued)

No.	compound	method of preparation (yield) properties and remarks
		^{1}H NMR (acetone-d_6, −20 °C): 1.63 (s, **C**$H_3C_5H_4$), 2.39 (s, CH_3CO), 5.08 and 5.40 (m, C_5H_4), 6.97 and 11.57 (d, CH=CH; J(H, H) = 8.5) [92] ^{13}C NMR (acetone-d_6, −20 °C): 12.9 (**C**$H_3C_5H_4$; J(H, C) = 128.2), 24.4 (**C**H_3C=O; J(H, C) = 123.4), 86.6 (C_5H_4; J(H, C) = 177.7) and 94.4 (C_5H_4; J(H, C) = 177.0), 119.0 (C_5H_4, C-1), 131.3 (CH; J(H, C) = 160.7), 248.1 (MoCH; J(H, C) = 143.6), 248.2 (CO) [92] IR (pentane): 1905, 1975 (ν(CO)) [92] mass spectrum: $[M]^+$ [92]
110	$C_9H_7Mo(CO)_2CH{=}CHC(CH_3)O$ (C_9H_7 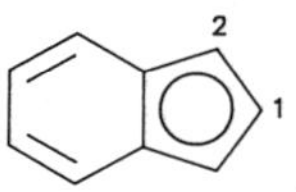 = indenyl, positions 1, 2)	III, under irradiation with CH≡CH (20%) [50] orange-red crystals, m.p. 84 °C (from pentane at −78 °C) [50] ^{1}H NMR (acetone-d_6): 2.20 (s, CH_3), 5.34 (t, H-1; J(H, H) = 2.8), 5.84 (d, H-2; J(H, H) = 2.8), 7.10 (m, C_9H_7, six-membered ring), 6.58 (d, CH=; J(H, H) = 8.5), 11.29 (MoCH; J(H, H) = 8.5) [50] IR (KBr): 1503 (ν(C=O)); (hexane): 1903, 1973, 1980 (ν(CO)) [50]
111	$(CH_3)_5C_5Mo(CO)_2CH{=}CHC(CH_3)O$	III, under irradiation with HC≡CH (32%) [92] red solid [92] ^{1}H NMR (acetone-d_6, −20 °C): 1.77 (s, $(CH_3)_5C_5$), 2.42 (s, CH_3CO), 6.93 and 11.30 (d, HC=CH; J(H, H) = 8.5) [92] ^{13}C NMR (acetone-d_6, −20 °C): 10.4 ((**C**$H_3)_5C_5$; J(H, C) = 127.4), 24.6 (CH_3C=O; J(H, C) = 127.4), 105.5 ($(CH_3)_5$**C**$_5$), 129.8 (CH=; J(H, C) = 159.0), 198.8 (C=O), 249.8 (CO), 253.0 (MoCH; J(H, C) = 141.0) [92] IR (pentane): 1890, 1963 (ν(CO)) [92] mass spectrum: $[M]^+$ [92]
112	$C_5H_5Mo(CO)_2CH(P(CH_3)_3)CH{=}C(CH_3)O$ 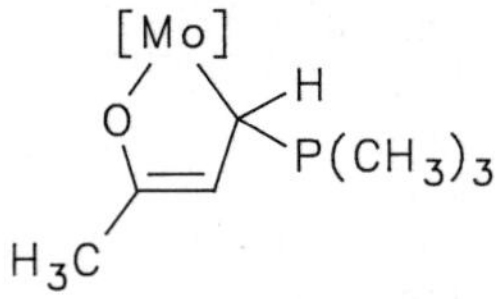 ([Mo]–O–C(H_3C)=CH–CH(H)(P(CH_3)$_3$)–[Mo] ring)	compound No. 108 was allowed to react with 2 equivalents of $P(CH_3)_3$ in pentane at −30 °C for 24 h (87%) [58] yellow solid, decomposes at 35 °C [58] ^{1}H NMR (acetone-d_6): 1.69 (d, PCH_3; J(P, H) = 13.6), 1.97 and 2.47 (CHCH=; J(H, H) = 2.1, J(P, H) = 9.7), 2.13 (s, CH_3C), 4.91 (s, C_5H_5) [58] IR (THF): 1635 (ν(C=O)), 1778, 1885 (ν(CO)) [58]

References on pp. 173/6

Table 7 (continued)

No.	compound	method of preparation (yield) properties and remarks
113	$C_5H_5\overline{Mo(CO)_2CH{=}C(CH_3)C(CH_3)O}$	III, under irradiation with $HC{\equiv}CCH_3$; ca. 15% in a mixture with No. 114 [49] ^1H NMR (acetone-d_6): 11.15 (q, MoCH); the other part of the spectrum is identical with No. 114 [49]
114	$C_5H_5\overline{Mo(CO)_2C(CH_3){=}CHC(CH_3)O}$	III, under irradiation with $HC{\equiv}CCH_3$ (9%); product contains ca. 15% of No. 113 [49] red to dark red crystals, m.p. 77 °C (from pentane at −78 °C) [49] ^1H NMR (acetone-d_6): 2.26 (s, CH_3CO), 2.93 (d, $CH_3C{=}$; J(H, H) = 0.9), 5.42 (s, C_5H_5), 6.78 (q, CH=; J(H, H) = 0.9) [49] IR (KBr): 1510 (ν(C=O)); (hexane): 1890, 1976 (ν(CO)) [49] mass spectrum: $[M-nCO]^+$ (n = 0 to 2) [49]
115	$C_9H_7\overline{Mo(CO)_2CH{=}C(CH_3)C(CH_3)O}$ (C_9H_7 = indenyl, positions 1 and 2 numbered on the five-membered ring)	III, under irradiation with $HC{\equiv}CCH_3$; ca. 15% in a mixture with No. 116 [50] ^1H NMR (acetone-d_6): 1.87 (d, $CH_3C{=}$; J(H, H) = 0.8), 2.17 (s, CH_3CO), 5.38 (t, H-1; J(H, H) = 2.8), 5.87 (d, H-2), 7.23 (m, C_9H_7, six-membered ring), 10.77 (q, CH=; J(H, H) = 0.8) [50] IR (KBr): 1495 (ν(C=O)), 1898, 1978 (ν(CO)) [50]
116	$C_9H_7\overline{Mo(CO)_2C(CH_3){=}CHC(CH_3)O}$	III, under irradiation with $HC{\equiv}CCH_3$ (52%); the product contains 15% of No. 115 [50] ^1H NMR (acetone-d_6): 2.11 (s, CH_3CO), 2.92 (d, $CH_3C{=}$; J(H, H) = 1), 5.32 (t, H-1; J(H, H) = 2.8), 5.83 (d, H-2), 6.43 (q, CH=; J(H, H) = 1), 7.23 (m, C_9H_7) [50] IR (KBr): 1478 (ν(C=O)), 1897, 1970 (ν(CO)) [50]
117	$C_9H_7\overline{Mo(CO)_2C(C_6H_5){=}CHC(CH_3)O}$	III, under irradiation with $HC{\equiv}CC_6H_5$ (75%) [50] orange-red crystals, m.p. 156 °C (from pentane at −78 °C) [50] ^1H NMR (acetone-d_6): 2.21 (s, CH_3), 5.27 (t, H-1; J(H, H) = 2.8), 6.03 (d, H-2), 6.56 (s, CH=), 7.20 (m, C_6H_5), 7.37 (m, C_9H_7) [50] IR (KBr): 1495 (ν(C=O)); (hexane): 1878, 1903, 1972 (ν(CO)) [50]
118	$C_5H_5\overline{Mo(CO)_2C(C_6H_5){=}CHC(CH_3)O}$	III, under irradiation with $HC{\equiv}CC_6H_5$ (10%) [49] red to dark red crystals, m.p. 147 °C (from pentane at −78 °C) [49]

References on pp. 173/6

Table 7 (continued)

No.	compound	method of preparation (yield) properties and remarks
		^{1}H NMR (acetone-d_6): 2.36 (s, CH_3), 5.46 (s, C_5H_5), 6.89 (s, CH=), 7.34 (m, C_6H_5) [49] IR (KBr): 1480 (ν(C=O)); (hexane): 1887, 1910, 1982 (ν(CO)) [49] mass spectrum: $[M-n\ CO]^+$ (n = 0 to 2) [49]
119	$(CH_3)_5C_5\overline{Mo(CO)_2CH{=}C(C_6H_5)C(CH_3)O}$	III, under irradiation with $HC{\equiv}CC_6H_5$ (49% of a 1:2 mixture with No. 120); the complex was separated by fractional crystallization from pentane, isomer No. 119 crystallizing better [112] crystalline solid, m.p. 115°C [112] ^{1}H NMR (acetone-d_6, −20°C): 1.82 (s, $(CH_3)_5C_5$), 2.46 (s, CH_3), 7.29 (m, C_6H_5), 11.20 (s, CH=) [112] ^{13}C NMR (acetone-d_6, −20°C): 10.1 ((**C**$H_3)_5C_5$), 25.5 (CH_3), 105.4 ($(CH_3)_5$**C**$_5$), 127.5, 128.8, 129.3, and 151.4 (C_6H_5), 143.5 (**C**C_6H_5), 195.9 (C=O), 248.8 (CO), 251.5 (MoC) [112] IR (pentane): 1475 (ν(C=O)), 1913, 1968 (ν(CO)) [112] mass spectrum: $[M]^+$ [112]
120	$(CH_3)_5C_5\overline{Mo(CO)_2C(C_6H_5){=}CHC(CH_3)O}$	III, under irradiation with $HC{\equiv}CC_6H_5$ (49% of a 2:1 mixture with No. 119); the complex was separated by fractional crystallization from pentane, isomer No. 119 crystallizing better [112] crystalline solid, m.p. 115°C [112] ^{1}H NMR (acetone-d_6, −20°C): 1.69 (s, $(CH_3)_5C_5$), 2.42 (s, CH_3), 6.77 (s, CH=), 7.38 (m, C_6H_5) [112] ^{13}C NMR (acetone-d_6, −20°C): 10.2 ((**C**$H_3)_5C_5$), 24.9 (CH_3), 106.1 ($(CH_3)_5$**C**$_5$), 126.7, 128.4, 129.4, and 152.9 (C_6H_5), 127.9 (CH=), 199.9 (C=O), 254.3 (CO), 265.1 (MoC) [112] IR (pentane): 1475 (ν(C=O)), 1875, 1897, 1962 (ν(CO)); the third ν(CO) band was interpreted as evidence for the presence of a noncyclic isomer in solution [112] mass spectrum: $[M]^+$ [112]
*121	$C_5H_5\overline{Mo(CO)_2C(CH_3){=}CHC(OCH_3)O}$	XIV (59%) [20, 62] m.p. 108 [20], 109°C [62] ^{1}H NMR ($CDCl_3$): 2.82 (d, CH_3; J(H, H) = 1.1), 3.79 (s, OCH_3), 5.28 (s, C_5H_5), 6.09 (q, CH=; J(H, H) = 1.1) [20, 62]

References on pp. 173/6

Table 7 (continued)

No.	compound	method of preparation (yield) properties and remarks
*121	(continued)	^{13}C NMR (CD_2Cl_2, −70°C): 35.4 (CH_3), 53.9 (OCH_3), 93.8 (C_5H_5; J(H, C) = 175), 115.6 (CH=; J(H, C) = 175), 178.3 (**C**(O)OCH_3), 246.1 (MoC), 254.0 and 256.1 (CO) [62] IR ($CHCl_3$): 1550 (ν(C=O)), 1860, 1960 (ν(CO)) [20, 62]
122	$C_5H_5\overline{Mo(CO)_2C(CH_3)=CHC(OCH_2C_6H_5)O}$	XIV (26%) [62] m.p. 99°C [62] ^{1}H NMR ($CDCl_3$): 2.83 (d, CH_3), 5.19 (s, CH_2), 5.22 (s, C_5H_5), 6.12 (q, CH=), 7.33 (s, C_6H_5) [62] IR (CH_2Cl_2): 1555 (ν(C=O)), 1860, 1960 (ν(CO)) [62]
123	$C_5H_5\overline{Mo(CO)_2C(CH_3)=CHC(OC_4H_9\text{-t})O}$	XIV (30%) [62] m.p. 91°C [62] ^{1}H NMR ($CDCl_3$): 1.42 (s, t-C_4H_9), 2.79 (d, CH_3), 5.27 (s, C_5H_5), 6.00 (q, CH=) [62] IR (CH_2Cl_2): 1555 (ν(C=O)), 1860, 1960 (ν(CO)) [62]
124	$C_5H_5\overline{Mo(CO)_2C(CH_3)=CHC(OC_6H_5)O}$	XIV (25%) [62] m.p. 94°C [62] ^{1}H NMR ($CDCl_3$): 2.90 (d, CH_3), 5.24 (s, C_5H_5), 6.23 (q, CH=), 7.16 (m, C_6H_5) [62] IR (CH_2Cl_2): 1550 (ν(C=O)), 1860, 1960 (ν(CO)) [62]
*125	$C_5H_5\overline{Mo(CO)_2C(CH_3)=C(CH_3)C(C(CH_3)=CHCH_3)O}$ [Mo], $^1CH_3$, $^2CH_3$, O, H, $H_3\overset{3}{C}$, $^4CH_3$	XV (42%) [79] red needles (from hexane at −30°C) [79] ^{1}H NMR (C_6D_6): 1.46 (d of m, CH_3-4; J(H, H) = 6.9), 1.80 (m, CH_3-3), 1.98 (s, CH_3-2), 2.88 (s, CH_3-1), 4.76 (s, C_5H_5), 5.68 (q of q, CH; ^{3}J(H, H) = 6.9, ^{4}J(H, H) = 1.4) [79] ^{13}C NMR (C_6D_6): 13.6 (CH_3-4), 14.2 (CH_3-3), 15.1 (CH_3-2), 34.6 (CH_3-1), 93.2 (C_5H_5), 129.0 (**C**CH_3-4), 134.3 (**C**CH_3-3), 136.0 (**C**CH_3-2), 199.0 (C=O), 248.7 (CO), 261.0 (**C**CH_3-1) [79] IR (hexane): 1890, 1973 (ν(CO)) [79] mass spectrum: $[M-n\ CO]^+$ (n = 0 to 2) [79]
126	$C_5H_5\overline{Mo(CO)_2C(CH_3)=C(CH_3)C(C(CH_3)=C(CH_3)_2)O}$ [Mo], $^1CH_3$, $^2CH_3$, O, $^5CH_3$, $H_3\overset{3}{C}$, $^4CH_3$	similar to XV with $LiCu(CH_3)_2$ (42%) [79] orange powder (from hexane at −30°C) [79] ^{1}H NMR (C_6D_6): 1.47 (m, CH_3-4), 1.67 (m, CH_3-5), 1.75 (br s, CH_3-3), 1.86 (s, CH_3-2), 2.89 (s, CH_3-1), 4.77 (s, C_5H_5) [79] IR (hexane): 1886, 1975 (ν(CO)) [79] mass spectrum: $[M-n\ CO]^+$ (n = 1, 2) [79]

References on pp. 173/6

Table 7 (continued)

No.	compound	method of preparation (yield) properties and remarks
127	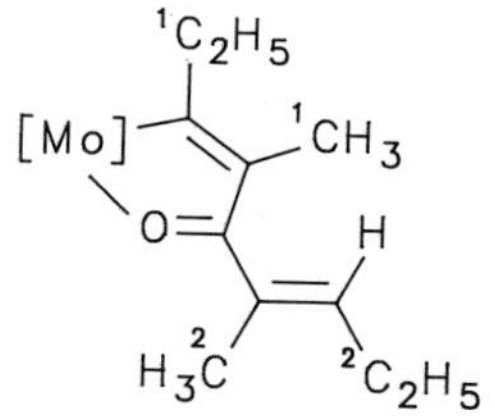 $C_5H_5Mo(CO)_2C(R)$=$C(R')C(CCH_3$=$CHC_2H_5)O$ (R = C_2H_5, R′ = CH_3; R = CH_3, R′ = C_2H_5) 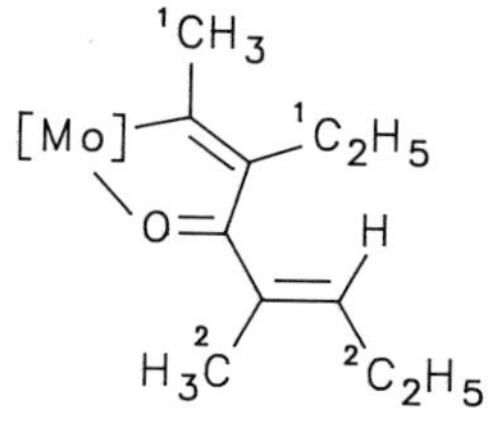	XV (60% of a mixture), a to b ratio 1.5 [79] isomer a: ^{1}H NMR (C_6D_6): 0.84 (t, **CH_3**CH_2-2; J(H, H) = 7), 1.04 (t, **CH_3**CH_2-1; J(H, H) = 7.0), 1.84 (m, CH_3-2), 1.98 (q, CH_2-2), 2.54 (q, CH_2-1), 2.98 (s, CH_3-1), 4.76 (s, C_5H_5), 5.64 (t of q, CH=; ^{3}J(H, H) = 7.0, ^{4}J(H, H) = 1.5) [79] isomer b: ^{1}H NMR (C_6D_6): 1.02 (t, **CH_3**CH_2-2; J(H, H) = 7.0), 1.26 (t, **CH_3**CH_2-1; J(H, H) = 7.0), 1.84 (m, CH_3-2), 2.04 (s, CH_3-1), 2.44 (q, CH_2-2), 3.38 (q, CH_2-1), 4.82 (s, C_5H_5), 5.64 (t of q, CH=; ^{3}J(H, H) = 7.0, ^{4}J(H, H) = 1.5) [79] IR (hexane): 1896, 1897, 1977 (ν(CO)) [79] mass spectrum: $[M-n\ CO]^+$ (n = 0 to 2) [79]
128	$C_5H_5Mo(CO)_2C(CH_3)$=$C(CH_3)C(CH_3)O$	III, with ca. 8 equivalents of $CH_3C{\equiv}CCH_3$ in hexane at 60°C for 48 h (60%) [16, 64] orange-red crystals, m.p. 158 to 159°C (from hexane at −20°C) [64] ^{1}H NMR ($CDCl_3$): 1.93, 2.28, and 2.84 (s, CH_3), 5.24 (s, C_5H_5) [64] IR (CCl_4): 1514 (ν(C=O)), 1878, 1970 (ν(CO)) [16, 64] mass spectrum: $[M]^+$ [64] reaction with D (D = $P(C_6H_5)_3$ or CNC_4H_9-t) in CH_2Cl_2 affords $C_5H_5Mo(CO)(D)(\eta^3$-$C(CH_3){\cdots}C(CH_3){\cdots}C(CH_3)C(O)O$-cyclo) [16, 64]
129	$(CH_3)_5C_5Mo(CO)_2C(CH_3)$=$C(CH_3)C(CH_3)O$	reacts with 2 equivalents of NOCl in CH_2Cl_2 at −78°C with formation of $(CH_3)_5C_5Mo(CO)(Cl)_2$-$C(CH_3)$=$C(CH_3)C(CH_3)O$ and $((CH_3)_5C_5Mo(CO)_3)_2$ [111]

References on pp. 173/6

Table 7 (continued)

No.	compound	method of preparation (yield) properties and remarks
130	$C_9H_7Mo(CO)_2C(CH_3){=}C(CH_3)C(CH_3)O$ (structure with H-a to H-f and C-1 to C-10 numbering)	III, with excess $CH_3C{\equiv}CCH_3$ in a sealed tube for 2 d (44%) [65] red crystals (from hexane at −78°C) [65] ^1H NMR (C_6D_6): 1.55 (s, H-e), 1.80 (s, H-f), 2.80 (s, H-d), 4.7 (t, H-c; J(H, H) = 3.0), 5.2 (d, H-b; J(H, H) = 3.0), 6.5 to 6.9 (m, H-a) [65] ^{13}C NMR (C_6D_6): 13.7 (C-8), 25.0 (C-10), 33.3 (C-6), 79.2 (C-3), 88.9 (C-4), 115.1 (C-2), 125.1 and 125.7 (C-1), 128.9 (C-7), 134.1 (C-5), 261.1 (C-9) [65] IR (Nujol): 1530 (ν(C=O)), 1892, 1970 (ν(CO)) [65] mass spectrum: $[M-n\ CO]^+$ (n = 0 to 3) [65]
131	$C_5H_5Mo(CO)_2C(CH_3){=}C(CH_3)C(CH_2C_6H_5)O$	III, with $C_5H_5Mo(CO)_3CH_2C_6H_5$ and ca. 12 equivalents of $CH_3C{\equiv}CCH_3$ in hexane at 60°C for 69 h (20%) [46, 64] bright red crystals (from hexane at −10°C), m.p. 145 to 146°C (dec.) [46, 64] ^1H NMR ($CDCl_3$): 1.90 and 2.84 (s, CH_3), 3.90 (s, CH_2), 5.22 (s, C_5H_5), 7.16 (m, C_6H_5) [64] IR (CCl_4): 1498 (ν(C=O)), 1885, 1972 (ν(CO)) [64] mass spectrum: $[M]^+$ [64]
*132	$C_5H_5Mo(CO)_2C(CH_3){=}C(CH_3)C(CF_3)O$	III, with $C_5H_5Mo(CO)_3C(O)CF_3$ and ca. 6 equivalents of $CH_3C{\equiv}CCH_3$ in hexane at 60°C for 24 h (40% by chromatography on Florisil with hexane) [46, 64] bright red crystals (from hexane at −20°C), m.p. 82 to 83°C (dec.) [64] ^1H NMR ($CDCl_3$): 2.10 (q, CH_3; J(F, H) = 2.0), 3.04 (s, CH_3), 5.34 (s, C_5H_5) [64] ^{19}F NMR (acetone-d_6): 67.7 (q, CF_3; J(H, F) = 2.0) [64] IR (cyclohexane): 1505 (ν(C=O)), 1915, 2000 (ν(CO)) [64] mass spectrum: $[M]^+$ [64]
*133	$C_9H_7Mo(CO)_2CH{=}C(C_4H_9\text{-}t)C(CH_3)O$ (C_9H_7 = indenyl, positions 1 and 2 numbered)	III, with $HC{\equiv}CC_4H_9$-t for 4 d in a sealed tube (16% by chromatography on alumina with hexane/ether 4:1) [65] red crystals (from hexane at −78°C) [65] ^1H NMR (C_6D_6): 1.4 (s, t-C_4H_9), 1.8 (s, CH_3), 4.6 (t, H-1; J(H, H) = 3), 5.45 (br m, H-2), 6.4 (s, CH=), 6.6 to 7.1 (m, C_9H_7) [65] IR (Nujol): 1530 (ν(C=O)); (hexane): 1888, 1970 (ν(CO)) [65] mass spectrum: $[M-n\ CO]^+$ (n = 0 to 2) [65]

References on pp. 173/6

Table 7 (continued)

No.	compound	method of preparation (yield) properties and remarks
134	$C_5H_5Mo(CO)_2C(C_6H_5)=C(C_6H_5)C(CH_3)O$	III, with an excess of $C_6H_5C{\equiv}CC_6H_5$ under irradiation (4%) [49] red to dark red crystals, m.p. 134°C (from pentane at −78°C) [49] ^{1}H NMR (acetone-d_6): 2.21 (s, CH_3), 5.47 (s, C_5H_5), 7.02 (m, C_6H_5) [49] IR (KBr): 1490 (ν(C=O)); (hexane): 1887, 1905, 1978 (ν(CO)) [49] mass spectrum: $[M]^+$ [49]
135	$C_5H_5Mo(CO)_2C(CH_3)=C((CH_2)_3)CO$ 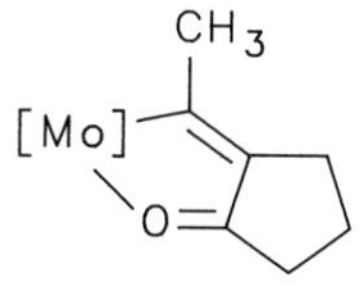	XVI (72%); the reaction is independent of the presence of CO [69] orange powder, m.p. 124 to 126°C (from C_6H_6/petroleum ether) [69] ^{1}H NMR (C_6D_6): 1.71 to 2.41 (m, CH_2), 2.81 (s, CH_3), 4.82 (s, C_5H_5) [69] ^{13}C NMR (C_6D_6): 34.3, 36.8, 43.6, and 43.8 (CH_2 and CH_3), 93.0 (C_5H_5), 127.2 (C=), 145.4 (C=O), 212.6 (MoC), 251.8 and 251.9 (CO) [69] IR (C_6H_6): 1870, 1960 (ν(CO)) [69] hydrogenation results completely analogous in products to those of No. 136 [69]
*136	$C_5H_5Mo(CO)_2C(CH_3)=C((CH_2)_4)CO$ 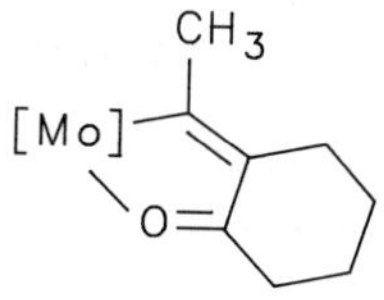	XVI (35%); also obtained by the cyclization of $C_5H_5Mo(CO)_3(CH_2)_4C{\equiv}CCH_3$ in C_6H_6 for 14 h (35%) [69] orange powder, m.p. 163 to 164°C [69] ^{1}H NMR (C_6D_6): 1.44 and 2.44 (m, 2CH_2), 2.92 (s, CH_3), 4.84 (s, C_5H_5) [69] IR (THF): 1875, 1965 (ν(CO)) [69]

compounds with a five-membered Mo-^{1}L-^{2}D ring (^{2}D bonded through S)

No.	compound	method of preparation (yield) properties and remarks
137	$C_5H_5Mo(CO)_2C(CF_3)=C(CF_3)C(O)SCH_3$	$(C_5H_5Mo(CO)_3)_2$ was irradiated in the presence of $S_2(CH_3)_2$ and $CF_3C{\equiv}CCF_3$ in pentane (5 to 30% by chromatography on silica with CH_2Cl_2/C_6H_6); the other product was $C_5H_5Mo(CO)S(CH_3)C(O)C(CF_3)=C(CF_3)CO$ ("Organomolybdenum Compounds" 6, p. 343, Formula VI) [73] m.p. 97°C [73] ^{1}H NMR ($CDCl_3$): 1.68 (s, CH_3), 5.58 (s, C_5H_5) [73] ^{19}F NMR ($CDCl_3$): 51.6 and 59.4 (q, CF_3; J(C, F) = 14.1) [73], partly listed in [74]

References on pp. 173/6

Table 7 (continued)

No.	compound	method of preparation (yield) properties and remarks
137 (continued)		^{95}Mo NMR ($CDCl_3$, vs. 2 M Na_2MoO_4 at pH 11): −845 ($\tau_{1/2}$ = 114) [74] IR (KBr): 1430 (ν(C=C), C_5H_5), 1520 (ν(C=C)), 1735 (ν(C=O)); (CCl_4): 1910, 1990 (ν(CO)) [73] mass spectrum: $[M]^+$ [73]
138	$C_5H_5\overline{Mo(CO)_2C(O)N(CH_3)C(C_6H_5)S}$	$LiN(CH_3)C(S)C_6H_5$ was added to a solution of $C_5H_5Mo(CO)_3Cl$ in THF cooled to −78°C; the mixture was allowed to warm and stirred for 1 h (30% by chromatography on silica with C_6H_6); the other product was $C_5H_5Mo(CO)_2S{\cdots}C(C_6H_5){\cdots}NCH_3$ [52] red needles (from ether/pentane), decompose at 134°C [52] ^{1}H NMR ($CDCl_3$): 3.23 (s, CH_3), 5.31 (s, C_5H_5), 7.50 (m, C_6H_5) [52] IR (CH_2Cl_2): 1621 (ν(C=O)), 1894, 1971 (ν(CO)) [52] mass spectrum (70 eV): $[M]^+$ (1), $[M-CO]^+$ (31), $[M-2CO]^+$ (11), $[M-3CO]^+$ (100), and other organic fragment ions [52]
139	$C_5H_5\overline{Mo(CO)_2C(O)C(N(C_2H_5)_2)=CHS}C_6H_5$	$C_5H_5Mo(CO)_3H$ was allowed to react with $C_6H_5SC{\equiv}CN(C_2H_5)_2$ in THF [77] yellow crystals, m.p. 93 to 95°C (from THF/pentane) [77] ^{1}H NMR ($CDCl_3$): 1.11 (t, CH_3; J(H, H) = 7.0), 3.21 (q, CH_2), 5.16 (s, C_5H_5), 5.70 (s, CH=), 7.33 (s, C_6H_5); the low-temperature NMR spectrum in $CDCl_3$ at −65°C shows the presence of two conformers in 4:1 ratio; main conformer: 1.15 (t, CH_3; J(H, H) = 7.0), 3.20 (q, CH_2), 5.45 (s, C_5H_5), 5.98 (s, CH=), 7.45 (s, C_6H_5); minor conformer: 1.15 (t, CH_3; J(H, H) = 7.0), 3.20 (q, CH_2), 4.90 (s, C_5H_5), 5.65 (s, CH=), 7.37 (s, C_6H_5); coalescence temperature is +5°C [77] IR (CH_2Cl_2): 1582 (C_6H_5), 1560 (ν(C=C)), 1597 (ν(C=O)), 1870, 1953 (ν(CO)) [77] mass spectrum: $[M-n\,CO]^+$ (n = 0 to 3)

compounds with a five-membered Mo-^{1}L-^{2}D ring (^{2}D bonded through P or As)

No.	compound	method of preparation (yield) properties and remarks
140	$C_5H_5\overline{Mo(CO)_2(CH_2)_3P}(C_6H_5)_2$	XVII, in THF for 3 h, R = C_6H_5, Y = $(CH_2)_2Cl$, X = Cl (62%) [91] yellow solid, m.p. 121°C [91] ^{1}H NMR (C_6D_6): 1.25 to 2.50 (m, CH_2), 4.49 (s, C_5H_5), 6.93 to 7.55 (m, C_6H_5) [91]

References on pp. 173/6

Table 7 (continued)

No.	compound	method of preparation (yield) properties and remarks
		^{31}P NMR ($CHCl_3$): 92.3 [91] IR (hexane): 1877, 1950 (ν(CO)) [91] mass spectrum (70 eV, 200°C): $[M-n\,CO]^+$ (n = 0 to 2) [91] reaction in liquid SO_2 at −40°C results in insertion of SO_2 into the MoC bond with formation of $C_5H_5Mo(CO)_2S(O)_2(CH_2)_3P(C_6H_5)_2$ [91]
141	$C_5H_5Mo(CO)_2(CH_2)_2OP(C_6H_5)_2$	XVII, in ether for 2 h, R = C_6H_5, Y = $O(CH_2)_2Cl$, X = Br (43%) [90] pale yellow, amorphous crystals (from pentane), decompose at 93 to 95°C [90] ^{1}H NMR (C_6D_6): 2.2 to 2.8 (m, $MoCH_2$), 3.3 to 3.7 (m, OCH_2), 4.9 (s, C_5H_5), 7.2 to 7.8 (m, C_6H_5) [90] ^{13}C NMR (C_6D_6): 12.5 (d, $MoCH_2$; J(P, C) = 17.1), 69.7 (d, OCH_2; J(P, C) = 11.4), 92.3 (C_5H_5) [90] ^{31}P NMR (toluene): 188.3 [90] IR (KBr): 1022 (ν(P-O)); (hexane): 1884, 1955 (ν(CO)) [90] mass spectrum (field desorption, 8 kV): $[M]^+$; (70 eV, 120°C): $[C_5H_5MoOP(C_6H_5)_2]^+$; (160°C): $[(Mo(CO)_2OP(C_6H_5)_2-H)_2]$ [90] heating in n-heptane at 70°C for 2 h leads to decomposition with formation of $H_2C{=}CH_2$ [90]
*142	$C_5H_5Mo(CO)_2(CH_2)_3As(CH_3)_2$	XVII, in THF for 0.5 h, R = CH_3, Y = $(CH_2)_3Cl$, X = I (94%) [22, 47] yellow, air-stable crystals, m.p. 107 to 108°C (from C_6H_6), sublimes at 80°C/0.1 Torr [22, 47] ^{1}H NMR ($CDCl_3$): 1.25 and 1.35 (s, CH_3), 4.95 (s, C_5H_5); ($C_6H_4Cl_2$-1,2, 20°C): 1.29 and 1.34 (CH_3); ($C_6H_4Cl_2$-1,2, 90°C): 1.28 (CH_3), interpreted as an equilibrium between the approximately square-pyramidal isomer (determined molecular structure) and a trigonal-bipyramidal isomer; $\Delta G^{\ddagger} \approx 82.2$ kJ/mol [47] ^{13}C NMR ($CDCl_3$): 41.9 and 42.9 (CH_3) [47] IR (heptane): 1860, 1940 (ν(CO)) [22, 47] mass spectrum: $[M]^+$ [22] reacts with $[C(C_6H_5)_3]BF_4$ with formation of $[C_5H_5Mo(CO)_2(\eta^3\text{-}CH_2{=}CHCH_2As(CH_3)_2)]^+$ [22]
143	$C_5H_5Mo(CO)_2CH_2C_6H_4As(CH_3)_2$-2	XVII, in THF for 12 h, R = CH_3, Y = C_6H_4-CH_2Br-2, X = Br or I (72%) [47] pale yellow plates, m.p. 114 to 114°C (from toluene/petroleum ether) [47]

References on pp. 173/6

Table 7 (continued)

No.	compound	method of preparation (yield) properties and remarks
143 (continued)		^{1}H NMR ($CDCl_3$): 1.52 and 1.61 (s, CH_3), 4.94 (s, C_5H_5); ($C_6H_4Cl_2$-1,2, 20 °C): 1.39 and 1.41 (CH_3); ($C_6H_4Cl_2$-1,2, 149 °C): 1.39 (CH_3); coalescence at 149 °C, $\Delta G^{\neq} \approx 99.6$ kJ/mol; compare No. 142 for interpretation [47] IR ($CHCl_3$): 1855, 1925 (ν(CO)) [47]
144	$C_5H_5\overline{Mo(CO)_2C_6H_4CH_2As}(CH_3)_2$-2	XVII, in THF for 16 h, R = CH_3, Y = $CH_2C_6H_4Br$-2, X = I (70%) [47] yellow plates, m.p. 126 to 128 °C (from toluene/petroleum ether) [47] ^{1}H NMR ($CDCl_3$): 1.49 (s, CH_3), 3.41 (s, CH_2), 5.04 (s, C_5H_5); no splitting of the $AsCH_3$ resonance was detected in $C_6H_4Cl_2$-1,2 in the range from −98 to +140 °C; compare Nos. 142, 143 [47] IR ($CHCl_3$): 1850, 1922 (ν(CO)) [47]
145	$C_5H_5\overline{Mo(CO)_2(CH_2)_3As}(C_6H_5)_2$	XVII, in THF for 0.5 h, R = C_6H_5, Y = $(CH_2)_3Cl$, X = I (82%) [47] yellow prisms, m.p. 134 to 135 °C (from toluene/petroleum ether) [47] ^{1}H NMR ($CDCl_3$): 5.04 (C_5H_5) [47] IR ($CHCl_3$): 1870, 1955 (ν(CO)) [47]
146	$C_5H_5\overline{Mo(CO)_2(CH_2)_3As}(C_6H_4CH_3\text{-}4)_2$	XVII, in THF for 0.5 h, R = $C_6H_4CH_3$-4, Y = $(CH_2)_3Cl$, X = I (75%) [47] yellow prisms, m.p. 139 to 140 °C (from toluene/petroleum ether) [47] ^{1}H NMR ($CDCl_3$): 2.29, 2.38, and 2.43 (CH_3-4), 4.94 (C_5H_5); ($C_6H_4Cl_2$-1,2, 20 °C): 2.16 and 2.29 (CH_3-4 of square-pyramidal isomer), 2.24 (CH_3-4 of trigonal-bipyramidal isomer); ($C_6H_4Cl_2$-1,2, 115 °C): 2.24 (CH_3-4); coalescence at 115 °C, $\Delta G^{\neq} \approx 85.3$ kJ/mol, CH_3 region given as a diagram for 4, 10, 20, 94, and 115 °C [47] IR ($CHCl_3$): 1845, 1930 (ν(CO)) [47]

compound with a five-membered Mo-^{1}L-^{2}D ring (^{2}D bonded through Br)

No.	compound	method of preparation (yield) properties and remarks
147	$C_5H_5\overline{Mo(CO)_2C(CH_3)OAl(Br)_2Br}$ [Mo]–C(CH3)=O–Al(Br)2–Br (ring: [Mo], CH3, O, Al, Br, Br, Br)	$C_5H_5Mo(CO)_3CH_3$ was treated with $AlBr_3$ in toluene (not isolated) [68, 72] ^{1}H NMR (toluene-d_8): 2.28 (CH_3), 4.56 (C_5H_5) [68, 72] IR (toluene): 1363 (ν(C=O)), 1915, 2004 (ν(CO)) [68, 72] reacts smoothly with CO to yield $C_5H_5Mo(CO)_3C(OAlBr_3)CH_3$ [68, 72]; rate for CO uptake for a 1.0 M complex solution in

References on pp. 173/6

Table 7 (continued)

No.	compound	method of preparation (yield) properties and remarks
		toluene is 35 × 10^6 mol · L^{-1} · s^{-1} at 20°C and p_{CO} = 383 ± 17 Torr [72] hydrolysis under mild condition produces the corresponding simple acyl complex [68]
	compound with a six-membered ring	
148	$C_5H_5\overline{Mo(CO)_2(CH_2)_3OP}(C_6H_5)_2$	XVII, in ether for 2 h, R = C_6H_5, Y = $O(CH_2)_3Cl$, X = Br (46%) [90] pale yellow powder (from pentane), decomposes > 85°C [90] ^{1}H NMR (C_6D_6): 1.3 to 2.1 (m, $MoCH_2CH_2$), 3.2 to 3.6 (m, OCH_2), 4.62 (s, C_5H_5), 6.7 to 7.8 (m, C_6H_5) [90] ^{31}P NMR (toluene): 161.0 [90] IR (KBr): 1018 (ν(P-O)); (hexane): 1877, 1950 (ν(CO)) [90] mass spectrum (field desorption, 8 kV): $[M]^+$ [90]

*Further information:

$C_5H_5Mo(CO)_2CH(C_5H_4N\text{-}2)NHR^*$ (Table **7**, Nos. **3, 5, 7, 8**). Samples of the complexes were chromatographed on two connected columns on LiChroprep Si60 (40 to 63 μm) with exclusion of light in toluene/ether (50:1). With the exception of No. 5, there was complete separation into two red zones. With Nos. 3 and 8, the first zone contained the complexes IVa, b and the second zone the complexes Va, b. With No. 7, the first zone contained Va, b and the second zone IVa, b. The complexes from each zone were recrystallized from ether/pentane at −20°C. The single zone of No. 5 contained isomer V [105].

The chirality of R* at the appropriate ligands arises from the fact that they were initially derived from (1S,2S,3S)-(+)-(aminomethyl)pinane (No. 3), (S)-(−)-2-methylbuthylamine (No. 5), (S)-(+)-1-cyclohexylethylamine (No. 7), and (R)-(+)-1-phenylethylamine (No. 8). The compounds contain three chiral centers (Mo, N, and C). The isomers a ($S_{Mo}S_NS_C$) and b ($R_{Mo}S_NS_C$) of IV differ by opposite configuration at Mo. Similary, V contains the stereoisomers a ($R_{Mo}R_NR_C$) and b ($S_{Mo}R_NR_C$) [105].

At ambient temperature a fast exchange between IVa and IVb is observed. At −70°C the ^{1}H NMR spectrum shows the separated resonances of IVa and b. The Va, b pair shows a similar behavior. For the a to b interconversion an intramolecular rotation was assumed similar to the rotation of an olefin in a π-complex. A IV to V interconversion takes place only in strong alkaline solutions; thus starting from pure IV or V of No. 3, with KOH/18-crown-6 in toluene for 10 min a 35:65 mixture of IV and V is obtained. A mechanism via an anionic intermediate is assumed [104, 105].

The optical rotations were determined at room temperature and obtained from isomer mixtures a and b with unknown a/b ratios. The CD spectra were dominated by the major isomers a (both for IV and V). The corresponding CD spectra of IV and V are mirror images, but with intensity differences due to slightly different a/b ratios [105].

References on pp. 173/6

a b

IV

a b

V

One isomer of No. 3 (Formula IVa; R* = $C_{11}H_{19}$) crystallizes in the orthorhombic space group $P2_12_12_1-D_2^4$ (No. 19) with the unit cell parameters a = 6.876 (2), b = 11.726 (2), c = 28.487 (5) Å; Z = 4 molecules per unit cell, and D_{calc} = 1.37 g/cm³. The molecular structure with the main bond distances and angles is shown in **Fig. 18**. One isomer of No. 5 (Formula Va; R* = $C(C_2H_5)(CH_3)H$) crystallizes in the tetragonal space group $P4_1-C_4^2$ (No. 76) with the unit cell parameters a = 12.933 (2), b = 12.933 (2), c = 10.858 (2) Å; Z = 4 molecules per

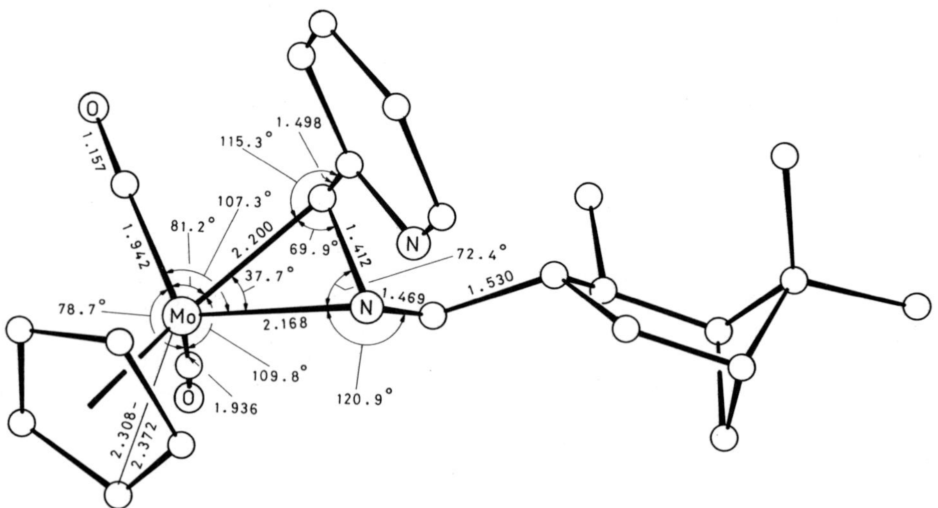

Fig. 18. Molecular structure of $C_5H_5Mo(CO)_2CH(C_5H_4N\text{-}2)NHC_{11}H_{19}$ [105].

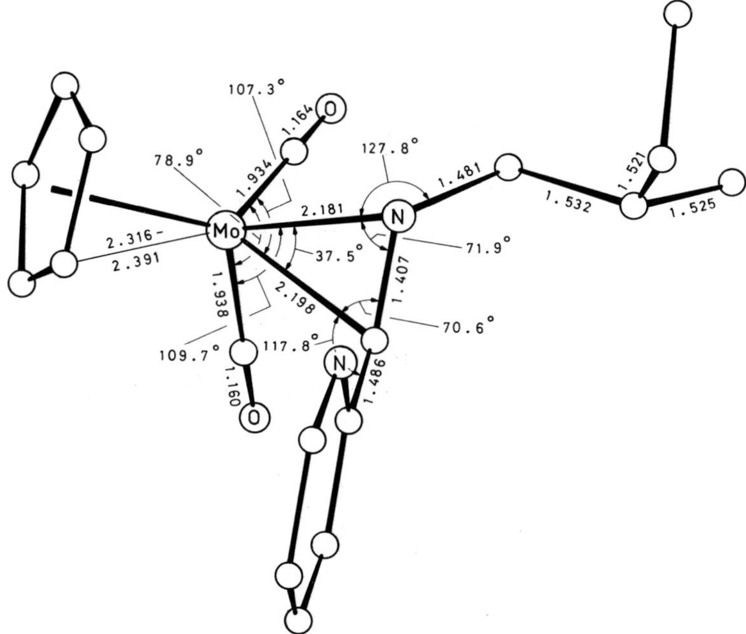

Fig. 19. Molecular structure of $C_5H_5Mo(CO)_2CH(C_5H_4N\text{-}2)NHC(C_2H_5)(CH_3)H$ [105].

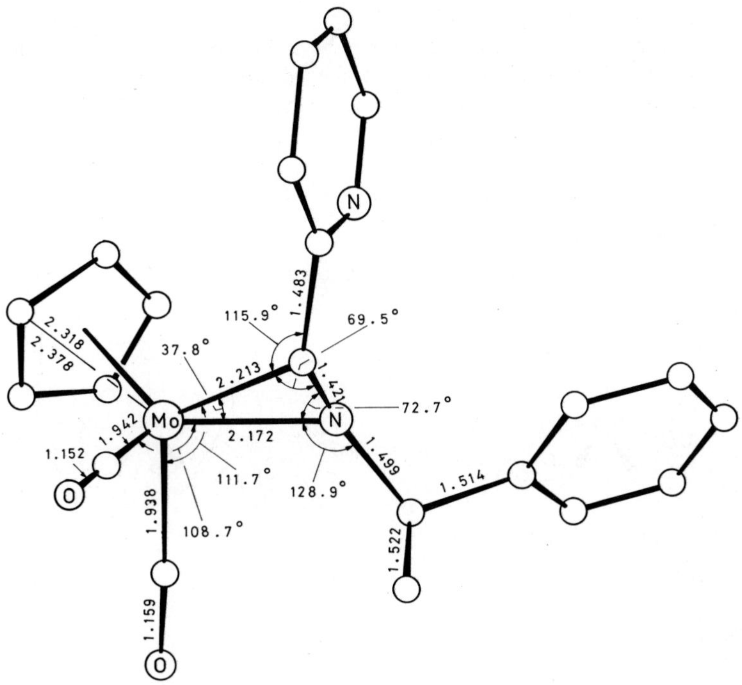

Fig. 20. Molecular structure of $C_5H_5Mo(CO)_2CH(C_5H_4N\text{-}2)NHC(C_6H_5)(CH_3)H$ (isomer IVa) [104].

References on pp. 173/6

unit cell, and D_{calc} = 1.44 g/cm^3. The molecular structure with the main bond distances and angles is shown in **Fig. 19**, p. 159 [105]. The isomer IVa (R^* = $C(C_6H_5)(CH_3)H$) of No. 8 crystallizes in the orthorhombic space group $P2_12_12_1-D_2^4$ (No. 76) with the unit cell parameters a = 10.084 (2), b = 13.516 (2), c = 29.203 (5) Å; Z = 8 molecules per unit. The molecular structure with the main bond distances and angles is shown in **Fig. 20**, p. 159. The second, not shown, independent molecule in the unit cell has a ca. 30° rotated C_6H_5 ring. The major Va (R^* = $C(C_6H_5)(CH_3)H$) isomer of No. 8 crystallizes in the orthorhombic space group $P2_12_12_1-D_4^2$ (No. 76) with the unit cell parameters a = 6.795 (1), b = 11.744 (2), c = 24.110 (3) Å; Z = 4 molecules per unit cell. The molecular structure with the main bond distances and angles is shown in **Fig. 21** [104].

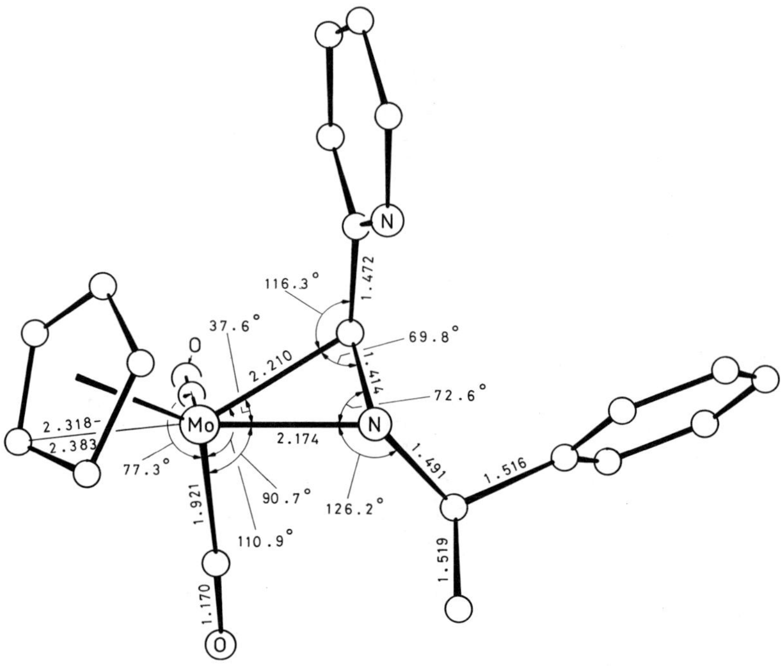

Fig. 21. Molecular structure of $C_5H_5Mo(CO)_2CH(C_5H_4N\text{-}2)NHC(C_6H_5)(CH_3)H$ (isomer Va) [104].

OC, OC, Mo, H, N—R*, N, Rh — H, Mo, CO, R*—N, CO, N, Rh

VI

References on pp. 173/6

Reaction of Nos. 3, 5, 7, or 8 (IV or V) with 0.5 equivalents $(Rh(C_7H_8)Cl)_2$ and excess of KOH in the presence of [18]crown-6 in toluene gives the heteronuclear compounds shown in Formula VI [104, 105].

$C_5H_5Mo(CO)_2C(CH_3)(C_5H_4N\text{-}2)NHC(C_6H_5)(CH_3)H$ (Table **7**, No. **12**) crystallizes in the monoclinic space group P $2_1/n - C_{2h}^5$ (No. 14) with the unit cell parameters a = 17.053 (9), b = 11.185 (5), c = 11.011 (3) Å, β = 104.48 (3)°; Z = 4 molecules per unit cell, and D_{calc} = 1.44 g/cm^3. The molecular structure with the main bond distances and angles is shown in **Fig. 22** [59, 75]; a stereoview and a packing diagram are also given in [75].

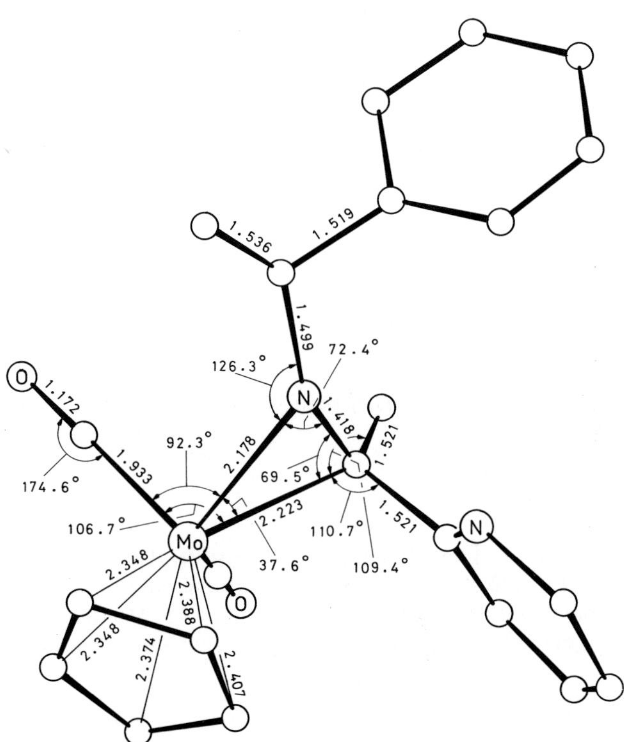

Fig. 22. Molecular structure of $C_5H_5Mo(CO)_2C(CH_3)(C_5H_4N\text{-}2)NHC(C_6H_5)(CH_3)H$ [59, 75].

$C_5H_5Mo(CO)_2C(C_6H_5)(C_5H_4N\text{-}2)NHC_3H_7\text{-}i$ (Table **7**, No. **14**) crystallizes in the orthorhombic space group $P2_12_12_1 - D_2^4$ (No. 19) with the unit cell parameters a = 8.477 (3), b = 14.132 (7), c = 16.670 (8) Å; Z = 4 molecules per unit cell, and D_{calc} = 1.471 g/cm^3. The molecular structure with the main bond distances and angles is shown in **Fig. 23**, p. 162; a stereoview and packing diagram are also given in [75].

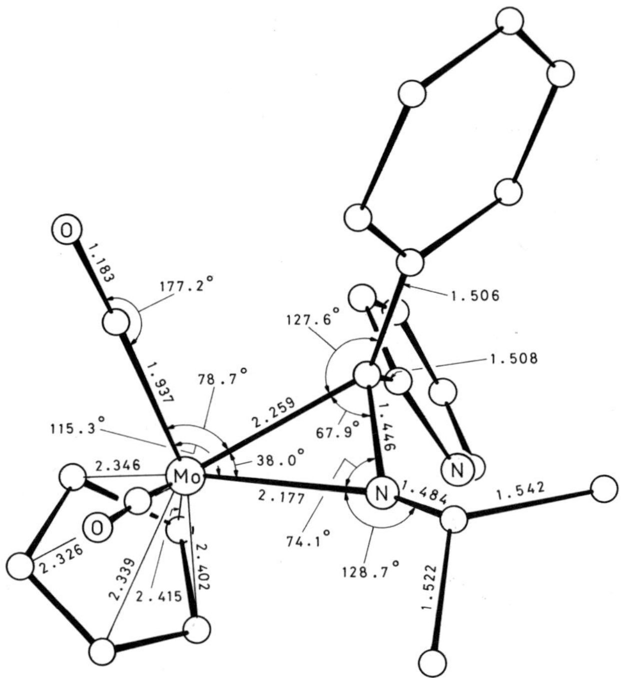

Fig. 23. Molecular structure of $C_5H_5Mo(CO)_2C(C_6H_5)(C_5H_4N\text{-}2)NHC_3H_7\text{-}i$ [75].

$C_5H_5Mo(CO)_2C(CH_3)$=NCH_3 and **$CH_3C_5H_4Mo(CO)_2C(CH_3)$=$NCH_3$** (Table **7**, Nos. **22, 23**) both show fluxional behavior. The most likely mechanism for the fluxional process can be described as a rotation of the iminoacyl ligand about an axis defined by the Mo atom and the center of the CN multiple bond. The activation parameters for this process are given in the following table [53, 57].

^5L	$\Delta G^{\ddagger}_{298}$ (kcal/mol)	$\Delta H^{\ddagger}$ (kcal/mol)	$\Delta S^{\ddagger}$ (cal · mol^{-1} · K^{-1})
C_5H_5	15.3 ± 0.6	14.7 ± 0.5	−1.7 ± 1.5
$C_5H_4CH_3$	15.3 ± 0.4	13.8 ± 0.3	−5.1 ± 0.9

$C_5H_5Mo(CO)_2C(CH_3)$=NC_6H_5 (Table **7**, No. **24**) crystallizes in the monoclinic space group $P2_1/n - C^5_{2h}$ (No. 14) with the unit cell parameters a = 7.789 (4), b = 6.810 (3), c = 26.754 (15) Å, β = 91.91 (4)°; Z = 4 molecules per unit cell, and D_{calc} = 1.57 g/cm^3. The molecular structure with the main bond distances and angles is shown in **Fig. 24** [39, 57], and was mentioned in [53].

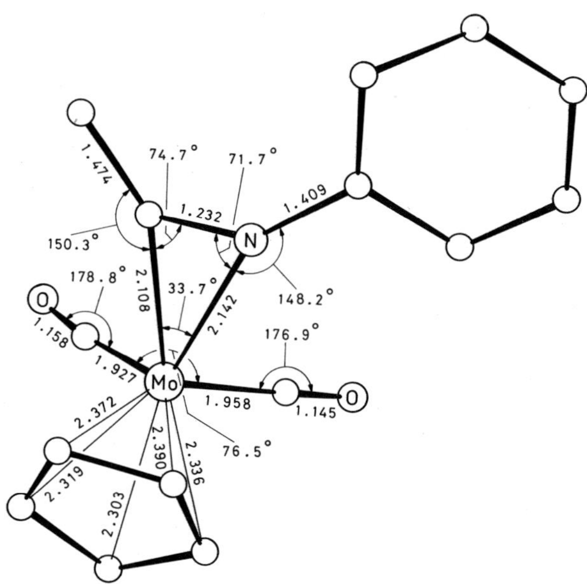

Fig. 24. Molecular structure of $C_5H_5Mo(CO)_2C(CH_3)=NC_6H_5$ [57].

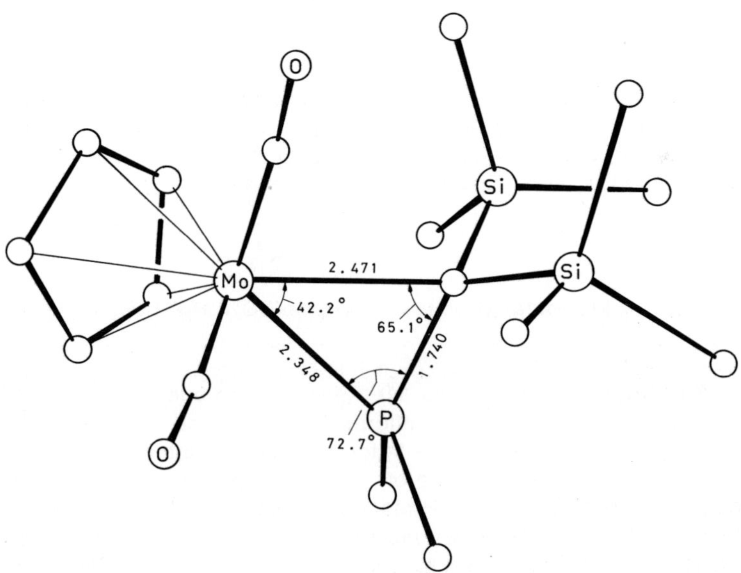

Fig. 25. Molecular structure of $C_5H_5Mo(CO)_2C(Si(CH_3)_3)_2P(CH_3)_2$ [29].

$C_5H_5Mo(CO)_2C(Si(CH_3)_3)_2P(CH_3)_2$ (Table **7**, No. **30**) crystallizes in the monoclinic space group $P2_1/n-C^5_{2h}$ (No. 14) with the unit cell parameters a = 14.295 (3), b = 9.144 (2), c = 15.751 (3) Å, β = 92.60 (2)°; Z = 4 molecules per unit cell, and D_{calc} = 1.410 g/cm³. The molecular structure with the main bond distances and angles is shown in **Fig. 25**, p. 163 [29].

$C_5H_5Mo(CO)_2CH_2P(C_6H_5)_2$ (Table **7**, No. **31**) crystallizes in the monoclinic space group $P2_1/n-C^5_{2h}$ (No. 14) with the unit cell parameters a = 7.944 (4), b = 16.388 (7), c = 14.239 (6) Å, β = 94.99 (4)°; Z = 4 molecules per unit cell, and D_{calc} = 1.497 g/cm³. The molecular structure with the main bond distances and angles is shown in **Fig. 26** [91].

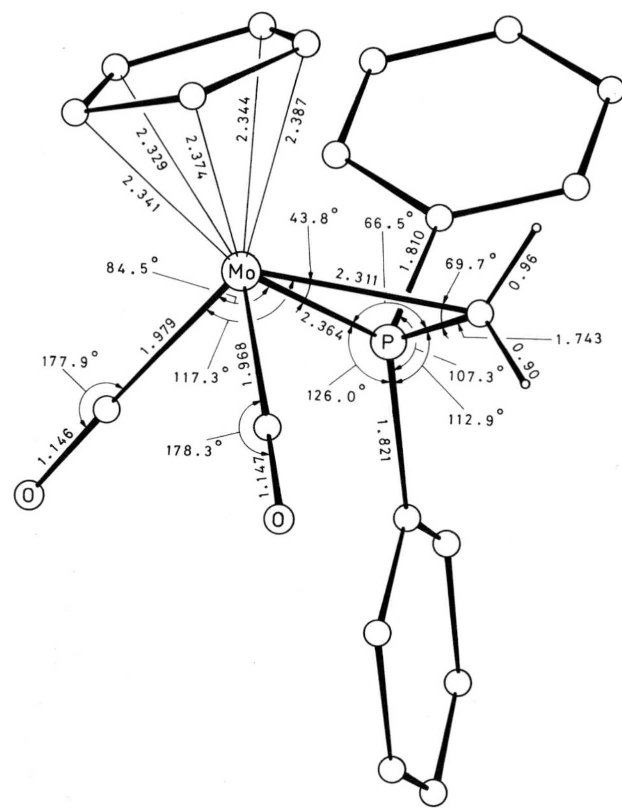

Fig. 26. Molecular structure of $C_5H_5Mo(CO)_2CH_2P(C_6H_5)_2$ [91].

$Li[C_5H_5Mo(CO)_2CH(CH_3)O]$ (Table **7**, No. **39**). Reaction with one equivalent of $P(C_6H_5)_3$ and two equivalents of CH_3I in THF at 25°C affords $CH_3C(O)H$ and $C_5H_5Mo(CO)_2(P(C_6H_5)_3)CH_3$, besides traces of $C_5H_5Mo(CO)_3CH_3$. The latter complex is obtained in high yields when the reaction is conducted in the absence of $P(C_6H_5)_3$ under 1 atm CO pressure. With C_6H_5-CH_2Br at 25°C for 4 h, $CH_3C(O)H$, $C_5H_5Mo(CO)_3CH_2C_6H_5$, $C_5H_5Mo(CO)_2(\eta^3$-$CH_2C_6H_5)$, and $(C_5H_5)_2Mo_2(CO)_3(\mu$-$CO)(\mu$-$OCHCH_3)$ (Formula VII) were obtained [85, 88, 95, 96]. Addition of $C_5H_5Mo(CO)_3C_2H_5$ to a solution of No. 39 affords CH_3CHO and C_2H_5CHO after addition of $BrCH_2C_6H_5$ [95]. Reaction with $[C_7H_7]BF_4$ or CH_2Cl_2 affords $(C_5H_5)_2Mo_2(CO)_3(\mu$-$CO)(\mu$-$OCHCH_3)$ (Formula VII; R = CH_3) besides $C_5H_5Mo(CO)_2C_7H_7$ (C_7H_7 bonded as ³L ligand) and

$(C_5H_5Mo(CO)_3)_2$ [85, 106]. The dinuclear complex (Formula VII, R = CH_3) is also obtained by reaction with $[C_7H_7Mo(CO)_3]BF_4$, $[(C_5H_5)_2Fe]PF_6$, HBF_4, or $(CF_3CO)_2O$ [96]. Alkylation with $[O(CH_3)_3]BF_4$ affords $C_5H_5Mo(CO)_2CH(CH_3)OCH_3$ (No. 42) [106].

VII VIII

$CH_3C_5H_4Mo(CO)_2CH(CH_2)_3O$ (Table **7**, No. **48**) crystallizes in the triclinic space group $P\bar{1}-C_i^1$ (No. 2) with the unit cell parameters a = 8.416 (9), b = 11.541 (12), c = 7.063 (3) Å, α = 105.23 (3)°, β = 105.71 (3)°, γ = 104.938 (12)°; Z = 2 molecules per unit cell, and D_{calc} = 1.684 g/cm^3. The molecular structure with the main bond distances and angles is shown in **Fig. 27** [78, 100].

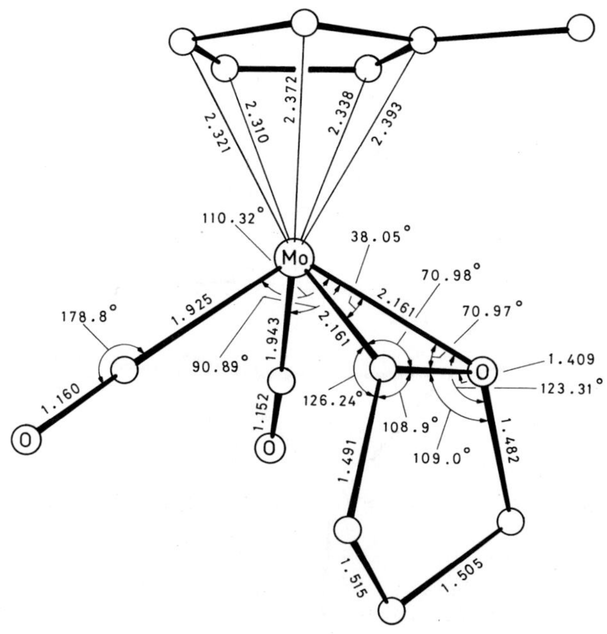

Fig. 27. Molecular structure of $CH_3C_5H_4Mo(CO)_2CH(CH_2)_3O$ [78, 100].

$C_5H_5Mo(CO)_2CH_2SCH_3$ (Table **7**, No. **49**) crystallizes in monoclinic space group $P2_1/n-C_{2h}^5$ (No. 14) with the unit cell parameters a = 13.57 (2), b = 7.97 (1), c = 9.70 (1) Å, β = 97.0 (2)°; Z = 4 molecules per unit cell, D_{calc} = 1.77, and D_{meas} = 1.76 g/cm^3. The molecular structure

References on pp. 173/6

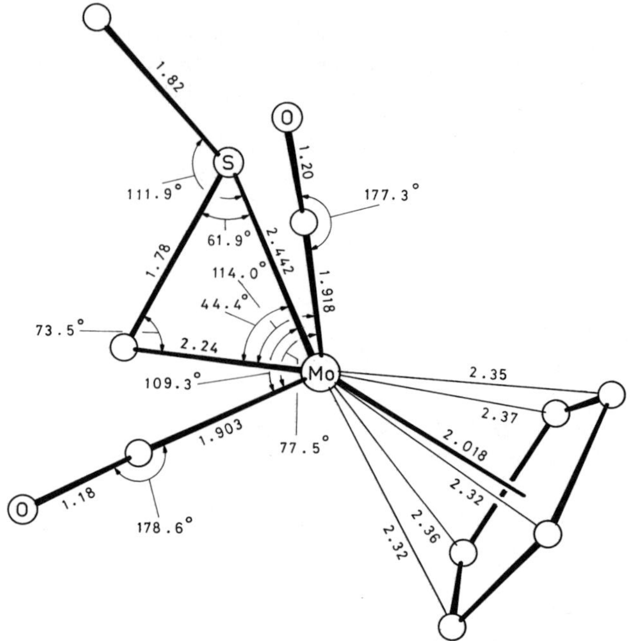

Fig. 28. Molecular structure of $C_5H_5Mo(CO)_2CH_2SCH_3$ [10].

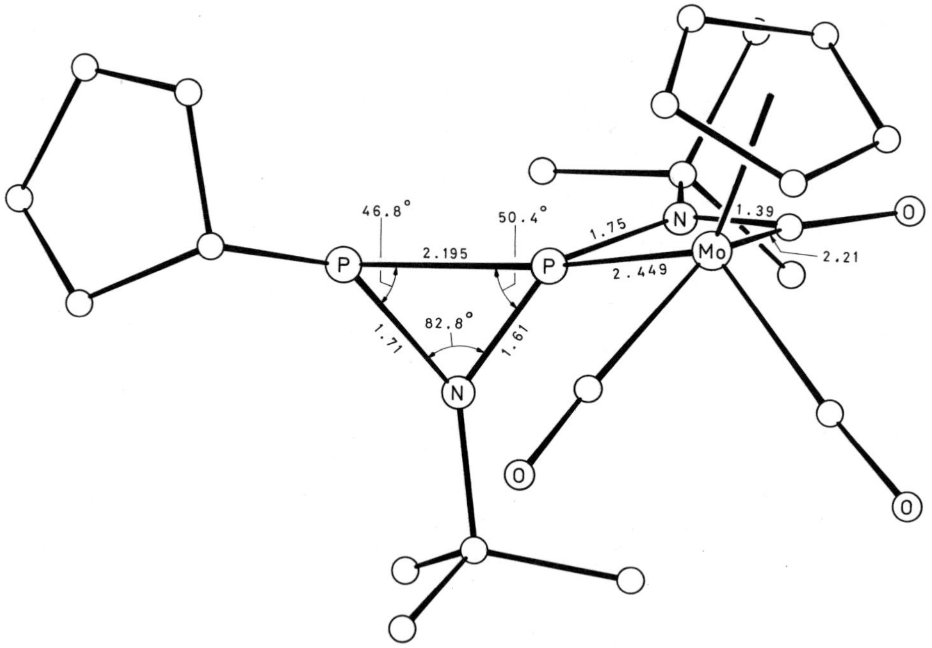

Fig. 29. Molecular structure of $(CH_3)_5C_5Mo(CO)_2C(O)N(C_4H_9\text{-t})P(PC_5(CH_3)_5)NC_4H_9\text{-t}$. Methyl groups of the $(CH_3)_5C_5$ rings are omitted for clarity [107].

References on pp. 173/6

with the main bond distances and angles is shown in **Fig. 28**. The observed CH_2-S bond length of 1.78 Å is not indicative of any significant degree of multiple bond character. The CH_2SCH_3 ligand acts as 3-electron donor [10]; mentioned in part in [9].

$(CH_3)_5C_5Mo(CO)_2C(O)N(C_4H_9\text{-t})P(PC_5(CH_3)_5)NC_4H_9\text{-t}$ (Table **7**, No. **57**) crystallizes (measured at −133°C) in the orthorhombic space group $P2_12_12_1-D_2^4$ (No. 19) with the unit cell parameters a = 12.219 (7), b = 16.891 (10), c = 15.768 (10) Å; Z = 4 molecules per unit cell, and D_{calc} = 1.33 g/cm^3. The molecular structure with the main bond distances and angles is shown in **Fig. 29** [107].

$C_5H_5Mo(CO)_2C_6H_4CH{=}NC(C_6H_5)(CH_3)H$ (Table **7**, No. **68**). Mixtures of diastereomers were formed, from which the $(+)_{365}$ compound can be obtained optically pure by fractional crystallization from ether/pentane (1:1) at −25°C, in the less soluble fraction. In the solid state the configurations are stable, but in solution the diastereomers epimerize even at room temperature. The $(+)_{365}/(-)_{365}$ equilibrium ratios are 71:29 in $CHCl_3$ [43, 44], 67:33 in acetone at 33°C, and 72:38 in toluene at 33°C [43]. The epimerization follows first-order kinetics in toluene at 20°C whose parameters are: $\tau_{1/2}$ = 11.5 min, E_A = 19.5 ± 0.5 kcal/mol, $\Delta H^{\neq}$ = 18.9 ± 0.5 kcal/mol, $\Delta G^{\neq}$ = 21.4 ± 0.6 kcal/mol, and $\Delta S^{\neq}$ = −8.5 ± 0.8 cal · mol^{-1} · K^{-1} [44].

$C_5H_5Mo(CO)_2C(O)CH_2CH_2NH_2$ (Table **7**, No. **70**) crystallizes in the orthorhombic space group $Pna2_1-C_{2v}^9$ (No. 33) with the unit cell parameters a = 18.932 (35), b = 8.734 (12), c = 6.472 (18) Å; Z = 4 molecules per unit cell, D_{calc} = 1.81, and D_{meas} = 1.79 g/cm^3. The molecular structure with the main bond distances and angles is shown in **Fig. 30**. The molecules are associated through C=O···H-N intermolecular hydrogen bonds with the distances O···H 2.090, N-H 1.04, and O···N 3.087 Å; the O···H-N angle is 159° [34].

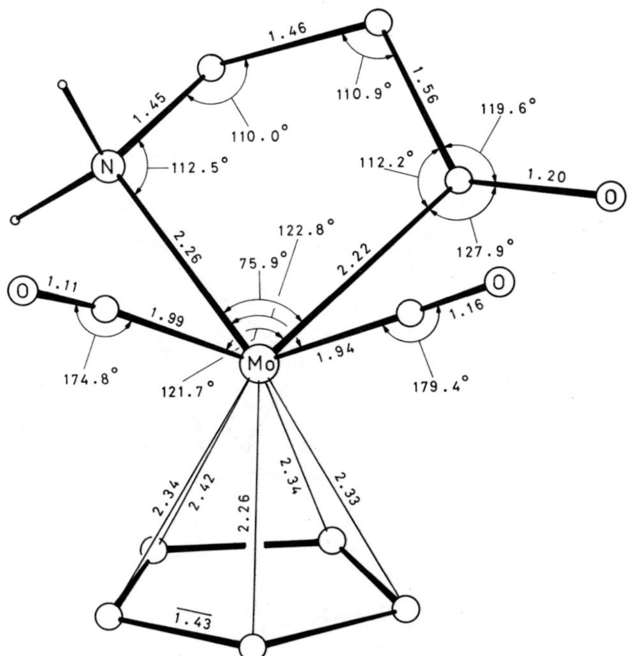

Fig. 30. Molecular structure of $C_5H_5Mo(CO)_2C(O)CH_2CH_2NH_2$ [34].

References on pp. 173/6

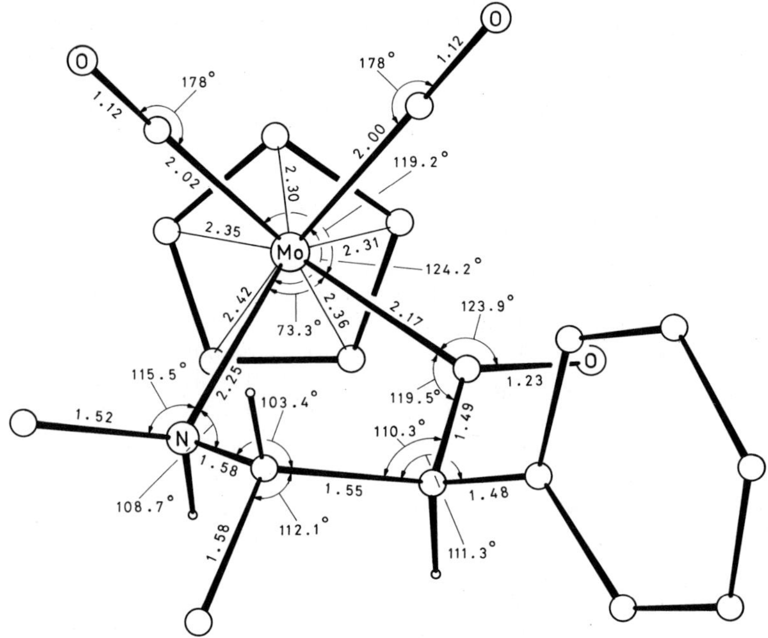

Fig. 31. Molecular structure of $C_5H_5Mo(CO)_2C(O)C(C_6H_5)HC(CH_3)HNHCH_3$ [51].

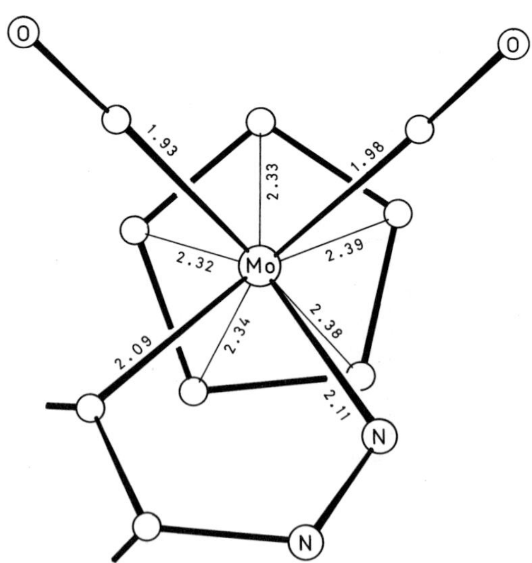

Fig. 32. Molecular structure of $C_5H_5Mo(CO)_2C(OH)C(C(O)OC_2H_5)NHN$ [13].

$C_5H_5Mo(CO)_2C(O)C(C_6H_5)HC(CH_3)HNHCH_3$ (Table **7**, No. **78**) crystallizes in the tetragonal space group $P4_32_12-D_4^8$ (No. 96) with the unit cell parameters a = b = 16.625 (1), c = 14.95 (1) Å; Z = 8 molecules per unit cell, D_{meas} = 1.3, and D_{calc} = 1.26 g/cm³. The molecular structure with the main bond distances and angles is shown in **Fig. 31**. The molecules are connected in the crystal by hydrogen bonding between N and C=O (N···H···O 2.91 Å) [51]; mentioned in [41].

$C_5H_5Mo(CO)_2C(OH)C(C(O)OC_2H_5)NHN$ (Table **7**, No. **86**) crystallizes in the monoclinic space group $Cc-C_s^4$ (No. 9) or $C2/c-C_{2h}^6$ (No. 15) with the unit cell parameters a = 21.36 (1), b = 8.26 (1), c = 16.11 (1) Å, β = 97.8 (3)°; Z = 8 molecules per unit cell, D_{meas} = 1.696 (by flotation), and D_{calc} = 1.699 g/cm³. The molecular structure with the main bond distances and angles is shown in **Fig. 32**; an electron density map for the chelate ring, distances and angles related by inversion, and a projection down another axis are also given. Strong hydrogen bonding was observed with formation of a C-OH···O=CC ring and formation of zigzag chains of molecules along the z axis formed via strong NH···N hydrogen bonds [13].

$[C_5H_5Mo(CO)_2C(OH)C(C(O)OC_2H_5)NHNCH_3]PF_6$ (Table **7**, No. **90**) crystallizes in the triclinic space group $P1-C_1^1$ (No. 1) or $P\bar{1}-C_i^1$ (No. 2) with the unit cell parameters a = 12.02 (1), b = 7.53 (1), c = 10.83 (1) Å, α = 95.5 (3)°, β = 92.8 (3)°, γ = 70.1 (3)°; Z = 2 molecules per unit cell, D_{meas} = 1.877, and D_{calc} = 1.878 g/cm³. The molecular structure with the main bond distances and angles is shown in **Fig. 33**. Hydrogen bonding between C=O and O-H forms a 6-membered ring. Large thermal ellipsoids for the PF_6 anion indicate either disorder or high thermal motion [19].

$C_5H_5Mo(CO)_2C(C_6H_5)_2N(CH_3)C(C_6H_5)=NCH_3$ (Table **7**, No. **107**) crystallizes in the monoclinic space group $P2_1/n-C_{2h}^5$ (No. 14) with the unit cell parameters a = 11.803 (5), b = 12.114 (6), c = 17.198 (6) Å, β = 96.61 (3)°; Z = 4 molecules per unit cell, D_{calc} = 1.45 g/cm³. The molecular

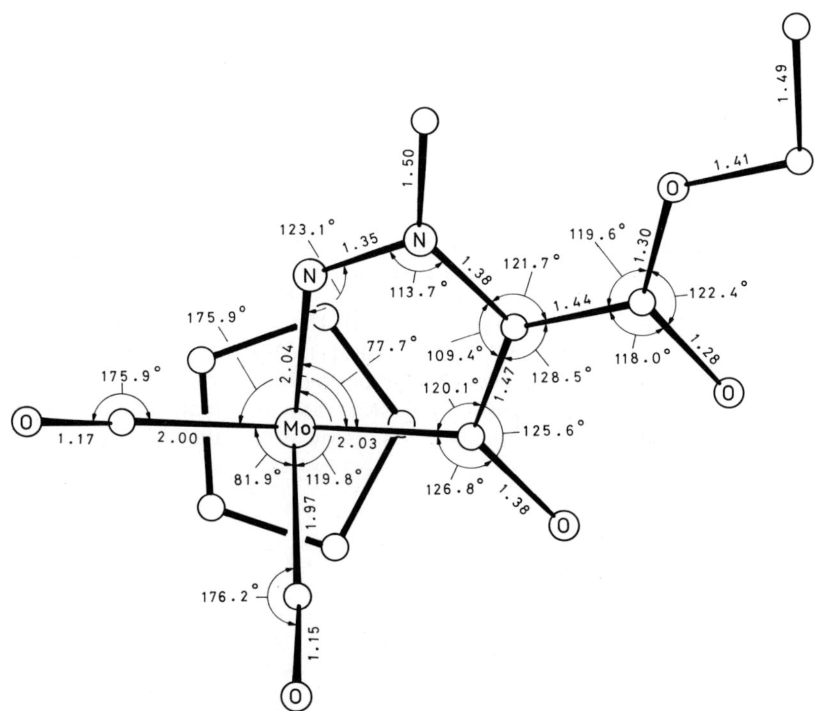

Fig. 33. Molecular structure of $[C_5H_5Mo(CO)_2C(OH)C(C(O)OC_2H_5)NHNCH_3]PF_6$ [19].

References on pp. 173/6

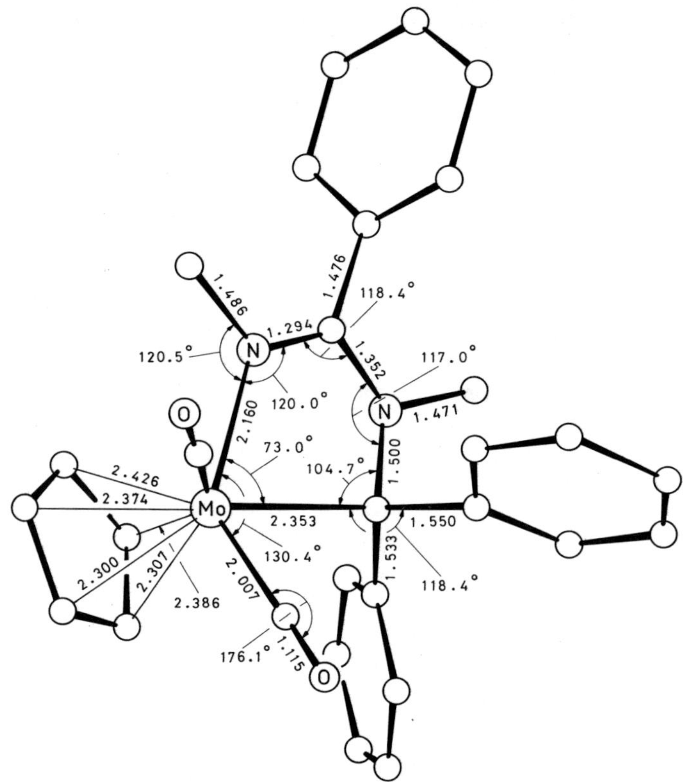

Fig. 34. Molecular structure of $C_5H_5Mo(CO)_2C(C_6H_5)_2N(CH_3)C(C_6H_5)$=$NCH_3$ [80, 82].

structure with the main bond distances and angles is shown in **Fig. 34** [82]. Stereoview [82], packing diagram, and crystal lattice are also given in [80].

$C_5H_5Mo(CO)_2C(CH_3)$=$CHC(OCH_3)O$ (Table **7**, No. **121**). The reaction of $C_5H_5Mo(CO)_3CH_2C{\equiv}CH$ with methanol was also carried out in the presence of $BrCH_2C{\equiv}CR$. The yields of No. 121 and the allyl complex (Formula VIII on p. 165) depend on the nature of R and the dependence of the yields from the $BrCH_2C{\equiv}CR/C_5H_5Mo(CO)_2CH_2C{\equiv}CH$ ratio (mol/mol) is given in the following table [62]. Treatment with 4 N HCl in ether affords methyl crotonate [20].

R	ratio	yield of No. 121	yield of VIII
	0	59%	0%
H	0.5	23%	44%
H	1	0%	85%
D	1	0%	85%
C_6H_5	1	0%	80%

$C_5H_5Mo(CO)_2C(CH_3)$=$C(CH_3)C(C(CH_3)$=$CHCH_3)O$ (Table **7**, No. **125**) crystallizes in the monoclinic space group $P2_1/n-C^5_{2h}$ (No. 14) with the unit cell parameters a = 6.495 (3), b = 16.902 (9), c = 14.002 (7) Å, β = 97.62 (4)°; Z = 4 molecules per unit cell, and D_{calc} = 1.544 g/cm³. The molecular structure with the main bond distances and angles is shown in **Fig. 35** [79].

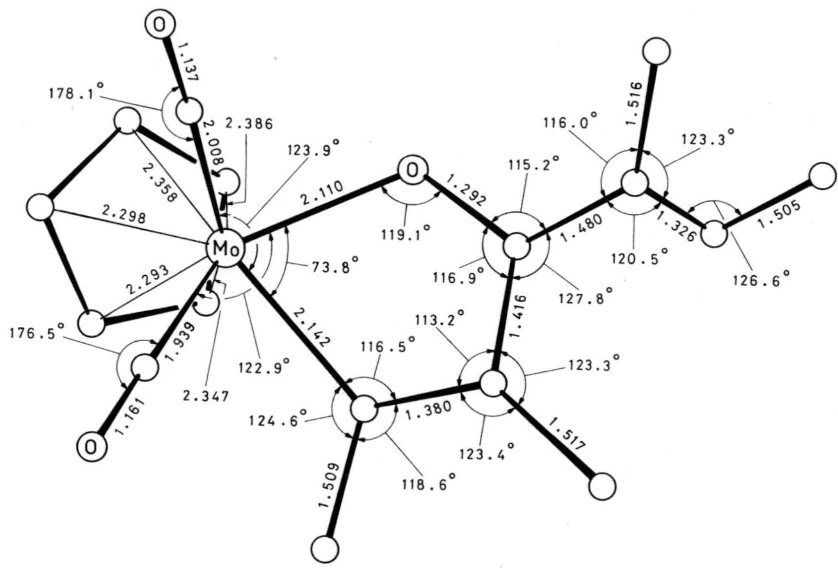

Fig. 35. Molecular structure of $C_5H_5Mo(CO)_2C(CH_3)$=$C(CH_3)C(C(CH_3)$=$CHCH_3)O$ [79].

$C_5H_5Mo(CO)_2C(CH_3)$=$C(CH_3)C(CF_3)O$ (Table **7**, No. **132**). Other products of the preparation were $C_5H_5Mo(CO)_2C(CF_3)C(CH_3)C(CH_3)C(O)O$ (Formula IX; D = CO) and $C_5H_5Mo(CO)OC$-(CF_3)=$C(CH_3)C(CH_3)$=$C(CH_3)C(CH_3)$=CO (Formula X) [64].

Reaction with RN≡C (R = t-C_4H_9 or cyclohexane) affords IX (D = CNR) [46, 64].

IX

X

References on pp. 173/6

$C_9H_7Mo(CO)_2CH{=}C(C_4H_9\text{-}t)C(CH_3)O$ (Table **7**, No. **133**). An other product of the preparation was 13% of $C_9H_7Mo(CO)_2CHC(C_4H_9\text{-}t)C(CH_3)OC(C_4H_9\text{-}t){=}CH$ (Formula XI) [65].

XI XII XIII

The reaction under 80 atm CO pressure in hexane leads to $C_9H_7Mo(CO)_2C(CH_3)C(C_4H_9\text{-}t)CHC(O)O$ (Formula XII) in high yields [65].

$C_5H_5Mo(CO)_2C(CH_3){=}C((CH_2)_4)CO$ (Table **7**, No. **136**). Hydrogenation (1 to 2 atm H_2) in C_6D_6 at 90 °C for four days affords $(C_5H_5Mo(CO)_3)_2$, $(C_5H_5Mo(CO)_2)_2$, and 87% of 2-ethylcyclohexanone. Protonation with CF_3COOH under CO in C_6H_6 affords $C_5H_5Mo(CO)_3OC(O)CF_3$ and 2-ethylidenecyclohexanone. In the absence of CO, the endo and exo isomers of $[C_5H_5Mo(CO)_2(\eta^4\text{-}C(O)(CH_2)_4C{=}CHCH_3)]^+$ (Formula XIII) are formed [69].

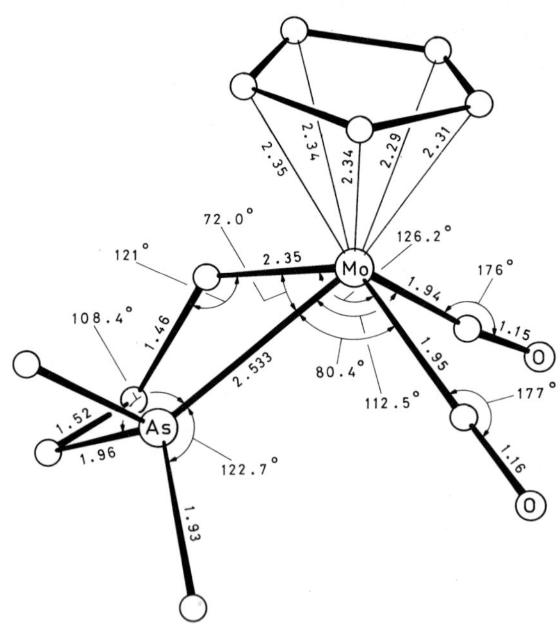

Fig. 36. Molecular structure of $C_5H_5Mo(CO)_2(CH_2)_3As(CH_3)_2$ [42].

References on pp. 173/6

$C_5H_5Mo(CO)_2(CH_2)_3As(CH_3)_2$ (Table **7**, No. **142**) crystallizes in the orthorhombic space group Pcab$-D_{2h}^{15}$ (No. 61) with the unit cell parameters a = 15.903 (5), b = 13.297 (2), and c = 12.925 (2) Å; Z = 8 molecules per unit cell, D_{meas} = 1.77 (1), and D_{calc} = 1.771 g/cm^3. The molecular structure with the main bond distances and angles is shown in **Fig. 36** [42].

References:

[1] King, R. B.; Bisnette, M. B. (J. Am. Chem. Soc. **86** [1964] 1267/8).
[2] King, R. B.; Bisnette, M. B. (Inorg. Chem. **4** [1965] 475/81).
[3] King, R. B.; Bisnette, M. B. (Inorg. Chem. **4** [1965] 486/93).
[4] King, R. B. (J. Am. Chem. Soc. **90** [1968] 1429/37).
[5] King, R. B. (Inorg. Chem. **5** [1966] 2242/3).
[6] King, R. B.; Bisnette, M. B. (Inorg. Chem. **5** [1966] 293/300).
[7] Green, M. L. H.; Sanders, J. R. (J. Chem. Soc. Chem. Commun. **1967** 956/7).
[8] King, R. B. (AD-6642968 [1967] 1/36; C.A. **69** [1968] No. 56601).
[9] de Gil, E. R. (Diss. Univ. of Wisconsin 1968; Diss. Abstr. Intern. B **30** [1969] 160).
[10] de Gil, E. R.; Dahl, L. F. (J. Am. Chem. Soc. **91** [1969] 3751/6).

[11] King, R. B. (Can. J. Chem. **47** [1969] 559/68).
[12] King, R. B. (Org. Mass Spectrom. **2** [1969] 387/99).
[13] Knox, J. R.; Prout, C. K. (Acta Crystallogr. B **25** [1969] 1952/7).
[14] Bruce, M. I.; Iqbal, M. Z.; Stone, F. G. A. (J. Chem. Soc. A **1970** 3204/9).
[15] Faller, J. W.; Anderson, A. S. (J. Am. Chem. Soc. **92** [1970] 5852/60).
[16] Green, M. L. H.; Sanders, J. R. (J. Chem. Soc. A **1971** 1947/51).
[17] Lindner, E.; Stängle, M.; Hiller, W.; Fawzi, R. (Chem. Ber. **122** [1989] 823/7).
[18] Stone, F. G. A. (AD-71462371 [1971] 1/21; C.A. **75** [1971] No. 20455).
[19] Prout, C. K.; Cameron, T. S.; Gent, A. R. (Acta Crystallogr. B **28** [1972] 32/6).
[20] Roustan, J. L.; Charrier, C.; Merour, J. Y.; Benaim, J.; Gianinotti, C. (J. Organometal. Chem. **3** [1972] C37/C40).

[21] Treichel, P. M.; Dean, W. K. (J. Chem. Soc. Chem. Commun. **1972** 804/5).
[22] Wainwright, K. P.; Wild, S. B. (J. Chem. Soc. Chem. Commun. **1972** 571/2).
[23] Beck, W.; Danzer, W.; Höfer, R. (Angew. Chem. **85** [1973] 87/8).
[24] Bruce, M. I.; Goodall, B. L.; Stone, F. G. A. (J. Organometal. Chem. **60** [1973] 343/49).
[25] Beck, W.; Danzer, W. (J. Organometal. Chem. **73** [1974] C56/C58).
[26] Dean, W. K.; Treichel, P. M. (J. Organometal. Chem. **66** [1974] 87/93).
[27] Doedens, R. J. (from King, R. B.; J. Organometal. **100** [1975] 111/25).
[28] Hodges, K. C. (Diss. Univ. Georgia 1974; Diss. Abstr. Intern. B **55** [1975] 4822).
[29] Carrano, C. J.; Cowley, A. H.; Nunn, C. M.; Pakulski, M.; Quashi, S. (J. Chem. Soc. Chem. Commun. **1988** 170/1).
[30] King, R. B.; Douglas, W. M. (Inorg. Chem. **13** [1974] 1339/42).

[31] King, R. B.; Hodges, K. C. (J. Am. Chem. Soc. **96** [1974] 1263/4).
[32] Fong, C. W.; Wilkinson, G. (J. Chem. Soc. Dalton Trans. **1975** 1100/4).
[33] Garber, A. R.; Garrou, P. E.; Hartwell, G. E.; Smas, M. J.; Wilkinson, J. R.; Todd, L. J. (J. Organometal. Chem. **86** [1975] 219/27).
[34] Jones, G. A.; Guggenberger, L. J. (Acta Crystallogr. B **31** [1975] 900/2).
[35] King, R. B.; Hodges, K. C. (J. Am. Chem. Soc. **97** [1975] 2702/12).
[36] King, R. B.; Saran, M. S. (Inorg. Chem. **14** [1975] 1018/26).
[37] Knoth, W. H. (Inorg. Chem. **14** [1975] 1566/72).
[38] Adams, R. D.; Chodosh, D. F. (J. Am. Chem. Soc. **98** [1976] 5391/3).

[39] Adams, R. D.; Chodosh, D. F. (J. Organometal. Chem. **122** [1976] C11/C14).
[40] Alt, H. G. (Angew. Chem. **88** [1976] 800/1).

[41] Beck, W.; Danzer, W.; Liu, A. T.; Huttner, G. (Angew. Chem. **88** [1976] 511/2).
[42] Brotherton, P. D.; Raston, C. L.; White, A. H.; Wild, B. (J. Chem. Soc. Dalton Trans. **1976** 1193/5).
[43] Brunner, H.; Hermann, W. A.; Wachter, J. (J. Organometal. Chem. **107** [1976] C11/C14).
[44] Brunner, H.; Wachter, J. (J. Organometal. Chem. **107** [1976] 307/14).
[45] Danzer, W.; Beck, W.; Keubler, M. (Z. Naturforsch. **31 b** [1976] 1360/6).
[46] Davidson, J. L.; Green, M.; Nyathi, J. Z.; Scott, C.; Stone, F. G. A.; Welch, A. J.; Woodward, P. (J. Chem. Soc. Chem. Commun. **1976** 714/5).
[47] Mickiewiecz, M. M.; Wainwrigth, K. P.; Wild, S. B. (J. Chem. Soc. Dalton Trans. **1976** 262/9).
[48] Adams, R. D.; Chodosh, D. F. (J. Am. Chem. Soc. **99** [1977] 6544/50).
[49] Alt, H. G. (Chem. Ber. **110** [1977] 2862/6).
[50] Alt, H. G. (Z. Naturforsch. **32 b** [1977] 1139/44).

[51] Liu, A. T.; Beck, W.; Huttner, G.; Lorenz, H. (J. Organometal. Chem. **129** [1977] 91/6).
[52] Brunner, H.; Wachter, J. (J. Organometal. Chem. **142** [1977] 133/7).
[53] Chodosh, D. F. (Diss. Univ. N.Y. 1977; Diss. Abstr. Intern. B **38** [1978] 5359).
[54] De Roode, W. H.; Beekfs, M. L.; Oskam, A.; Vrieze, K. (J. Organometal. Chem. **142** [1977] 337/49).
[55] Gaylani, B.; Kilner, M. (J. Less-Common Metals **54** [1977] 175/89).
[56] King, R. B.; Chen, K. N. (Inorg. Chem. **16** [1977] 2648/50).
[57] Adams, R. D.; Chodosh, D. F. (Inorg. Chem. **17** [1978] 41/8).
[58] Alt, H. G.; Schwärzle, J. A.; Kreissl, F. R. (J. Organometal. Chem. **152** [1978] C 57/C 59).
[59] Brunner, H.; Schwägerl, H.; Wachter, J.; Reisner, G. M.; Bernal, I. (Angew. Chem. **90** [1978] 478/9).
[60] Brunner, H.; Wachter, J. (J. Chem. Res. M **1978** 1801/9; J. Chem. Res. S **1978** 136/7).

[61] Brunner, H.; Wachter, J. (J. Organometal. Chem. **155** [1978] C 29/C 33).
[62] Charrier, C.; Collin, J.; Merour, J. Y.; Roustan, J. L. (J. Organometal. Chem. **162** [1978] 57/66).
[63] De Roode, W. H.; Vrieze, K. (J. Organometal. Chem. **153** [1978] 345/57).
[64] Green, M.; Nyathi, J. Z.; Scott, C.; Stone, F. G. A.; Welch, A. J.; Woodward, P. (J. Chem. Soc. Dalton Trans. **1978** 1067/80).
[65] Bottrill, M.; Green, M. (J. Chem. Soc. Dalton Trans. **1979** 820/5).
[66] Brunner, H.; Schwägerl, H.; Wachter, J. (Chem. Ber. **112** [1979] 2079/86).
[67] Brunner, H.; Wachter, J. (J. Organometal. Chem. **175** [1979] 285/92).
[68] Butts, S. B.; Holt, E. M.; Strauß, S. H.; Stimson, R. E.; Alcock, N. W.; Shriver, D. F. (J. Am. Chem. Soc. **101** [1979] 5864/6).
[69] Watson, P. L.; Bergman, R. G. (J. Am. Chem. Soc. **101** [1979] 2055/62).
[70] Brunner, H.; Wachter, J. (J. Chem. Res. S **1980** 328/9).

[71] Brunner, H.; Wachter, J. (J. Organometal. Chem. **201** [1980] 453/7).
[72] Butts, S. B.; Strauß, S. B.; Holt, E. M.; Stimson, R. E.; Alcock, N. W.; Shriver, D. F. (J. Am. Chem. Soc. **102** [1980] 5093/100).
[73] Petillon, F. Y.; Le Floch-Perennou, F.; Guerchais, J. E.; Sharp, D. W. A.; Manojlovic-Muir, L.; Muir, K. W. (J. Organometal. Chem. **202** [1980] 23/37).
[74] Le Gall, J. Y.; Kubicki, M. M.; Petillon, F. Y. (J. Organometal. Chem. **221** [1981] 287/90).
[75] Reisner, G. M.; Bernal, I. (Inorg. Chim. Acta **51** [1981] 201/11).
[76] Sünkel, K.; Ernst, H.; Beck, W. (Z. Naturforsch. **36 b** [1981] 474/81).
[77] Brix, H.; Beck, W. (J. Organometal. Chem. **234** [1982] 151/74).

[78] Adams, H.; Bailey, N. A.; Cahill, P.; Rogers, D.; Winter, M. J. (J. Chem. Soc. Chem. Commun. **1983** 831/3).
[79] Allen, S. R.; Green, M.; Norman, N. C.; Paddick, K. E.; Orpen, A. G. (J. Chem. Soc. Dalton Trans. **1983** 1625/33).
[80] Bernal, I.; Reisner, G. M. (Inorg. Chim. Acta **71** [1983] 65/71).
[81] Brunner, H.; Meyer, W.; Wachter, J. (J. Organometal. Chem. **243** [1983] 437/41).
[82] Brunner, H.; Wachter, H.; Bernal, I.; Reisner, G. M.; Benn, R. (J. Organometal. Chem. **243** [1983] 179/89).
[83] Herberhold, M.; Ehrenreich, W.; Guldner, K.; Jellen, W.; Thewalt, U.; Klein, H. P. (Z. Naturforsch. **38 b** [1983] 1383/7).
[84] Mahe, C.; Patin, H.; Le Marouille, J. Y.; Benoit, A. (Organometallics **2** [1983] 1051/3).
[85] Adams, H.; Bailey, N. A.; Gauntlett, J. T.; Winter, M. J. (J. Chem. Soc. Chem. Commun. **1984** 1360/1).
[86] Cotton, J. D.; Dunstan, P. R. (Inorg. Chim. Acta **88** [1984] 223/7).
[87] Danzer, W.; Höfer, R.; Menzel, H.; Olgemöller, B.; Beck, W. (Z. Naturforsch. **39 b** [1984] 167/79).
[88] Gauntlett, J. T.; Taylor, B. F.; Winter, M. J. (J. Chem. Soc. Chem. Commun. **1984** 420/1).
[89] Groß, E.; Jörg, K.; Fiederling, K.; Göttlein, A.; Malisch, W.; Boese, R. (Angew. Chem. **96** [1984] 705/6).
[90] Lindner, E.; Brösamle, A. (Z. Naturforsch. **39 b** [1984] 535/7).
[91] Lindner, E.; Küster, E. U.; Hiller, W.; Fawzi, R. (Chem. Ber. **117** [1984] 127/41).
[92] Alt, H. G.; Engelhardt, H. E.; Thewalt, U.; Riede, J. (J. Organometal. Chem. **288** [1985] 165/77).
[93] Alt, H. G.; Hayen, H. I. (Angew. Chem. **97** [1985] 506/7).
[94] Brunner, H.; Beier, P.; Frauendorfer, E.; Muschiol, M.; Rastogi, D. K.; Wachter, J.; Minelli, M.; Enemark, J. (Inorg. Chim. Acta **96** [1985] L 5/L 9).
[95] Gauntlett, J. T.; Taylor, B. F.; Winter, M. J. (J. Chem. Soc. Dalton Trans. **1985** 1815/20).
[96] Gauntlett, J. T.; Winter, M. J. (Chem. Uses Molybdenum Proc. 5th Intern. Climax Conf., Newcastle Upon Tyne, Engl., 1985, Abstr. pp. 95/6).
[97] Giulieri, F.; Benaim, J. (Nov. J. Chim. **9** [1985] 335/40).
[98] Jörg, K.; Boese, R.; Malisch, W. (20th Hauptversamml. GdCh, Heidelberg 1985, Abstr. P 2.24).
[99] Jörg, K.; Reich, W.; Boese, R.; Malisch, W.; Baumann, R. (12th Intern. Conf. Organometal. Chem., Vienna 1985, p. 216).
[100] Adams, H.; Bailey, N. A.; Cahill, P.; Rogers, D.; Winter, M. J. (J. Chem. Soc. Dalton Trans. **1986** 2119/26).
[101] Adams, H.; Bailey, N. A.; Osborn, V. A.; Winter, M. J. (J. Chem. Soc. Dalton Trans. **1986** 2127/35).
[102] Alt, H. G.; Hayen, H. I.; Freytag, U. (Chemiedozententag., Würzburg, FRG, 1986, Abstr. B 50).
[103] Amann, S.; Jörg, K.; Malisch, W. (10th Intern. Conf. Phosphorus Chem., Bonn, FRG, 1986, Abstr. B 64).
[104] Brunner, H.; Wachter, J.; Schimdbauer, J.; Sheldrick, G. M.; Jones, P. G. (Angew. Chem. **98** [1986] 339/41).
[105] Brunner, H.; Wachter, J.; Schmidbauer, J.; Sheldrick, G. M.; Jones, P. G. (Organometallics **5** [1986] 2212/9).
[106] Gauntlett, J. T.; Winter, M. J. (Polyhedron **5** [1986] 451/59).
[107] Gudat, D.; Niecke, E.; Krebs, B.; Dartmann, M. (Organometallics **5** [1986] 2376/7).
[108] Malisch, W.; Jörg, K.; Gross, E.; Schmeusser, M.; Meyer, A. (Phosphorus Sulfur **26** [1986] 25/9).

[109] Osborn, V. A.; Parker, C. A.; Winter, M. J. (J. Chem. Soc. Chem. Commun. **1986** 1185/6).
[110] Osborn, V. A.; Winter, M. J. (Polyhedron **5** [1986] 435/7).
[111] Alt, H. G.; Freytag, U.; Herberhold, M.; Hayen, H. I. (J. Organometal. Chem. **336** [1987] 361/70).
[112] Alt, H. G.; Herrmann, G. S.; Engelhardt, H. E.; Rogers, R. D. (J. Organometal. Chem. **331** [1987] 329/39).

1.5.1.3.2.1.11 Compounds with Two ^{1}L Ligands

This section covers compounds with two additional ^{1}L ligands bonded to molybdenum. Anions are obtained when both ^{1}L ligands are alkyl or acyl; if one ^{1}L ligand is alkyl or acyl and the other is isocyanide, carbene, or a η^2-bonded $S{=}CR_2$ ligand, neutral species are formed and two isocyanide ligands give cations. The cis/trans isomer ratios are given only for some compounds; see Table 12. The thioketone derivatives (Nos. 9 to 13) have a cis stereochemistry.

One compound, $C_5H_5Mo(CO)_2(C_5H_5)_3$, is also mentioned in which a structure was proposed with three of the four C_5H_5 groups bonded in a η^1-manner (general type $C_5H_5Mo(CO)_2(^1L)_3$) [8]. However, a structure determination of the corresponding tungsten derivative revealed the presence of a C_5H_5 ligand which functions as η^3-allyl ligand. The complex is therefore placed in Section 1.5.1.3.2.3.3 (Table 12, No. 19) which covers $^5LMo(CO)_2{}^3L$ compounds.

Most of the compounds can be prepared by the following methods:

Method I: $Li[(C_2H_5)_3BH]$ was added to a solution of $C_5H_5Mo(CO)_3R$ (R = CH_3, CD_3, or $(CH_2)_3Br$) in THF cooled to −78 °C and the mixture was warmed to −66 °C [18, 21 to 24]. The reaction of $C_5H_5Mo(CO)_3C(O)C(O)OCH_3$ was carried out at −50 °C [10, 16].

Method II: $C_5H_5Mo(CO)_3{}^1L$ (^1L = CH_3, C_2H_5) and CS_2 were refluxed in C_6H_6 in the dark [12].

Method III: The dinuclear compound (Formula I) and five equivalents of a phosphite were heated in C_6H_6 at 60 °C for 15 h. The product was separated by column chromatography on alumina with hexane. The dinuclear compound (Formula II) also formed can be converted to the title compound by heating or by chromatography on alumina [15].

Method IV: The complexes $^5LMo(CO)_2(CNR)^1L$ and $^5LMo(CO)_2(CNR)C(O)^1L$ were obtained by treating the corresponding $^5LMo(CO)_3{}^1L$ compound with RNC in solution. The complexes were purified in many cases by column chromatography. Further information is given in the table.

I

II

References on p. 184

Table 8
Compounds with Two [1]L Ligands.
An asterisk indicates further information at the end of the table.
For explanations, abbreviations, and units see p. X.

No.	compound	method of preparation (yield) properties and remarks
anionic compounds of the type $[C_5H_5Mo(CO)_2(^1L)_2]^-$		
*1	trans-Li$[C_5H_5Mo(CO)_2(CH_3)C(O)H]$	I (not isolated); small amounts of the cis isomer were also formed [18, 19, 21, 22] ^{1}H NMR (THF, −60°C): 5.00 (s, C_5H_5), 14.3 (br s, C(O)H) [18, 19, 21] ^{13}C NMR (THF, −90°C): −19.6 (CH_3), 94.4 (C_5H_5), 229.8 (CO), 287.7 (br, C(O)H) [18, 21] IR (THF, −50°C): 1833, 1922 (ν(CO)) [18, 21]; a figure is given in [21]
*2	trans-Li$[C_5H_5Mo(CO)_2(CH_3)C(O)D]$	I (not isolated) [21] ^{2}H NMR (THF, −40°C): 13.2 (C(O)D) [21]
*3	trans-Li$[C_5H_5Mo(CO)_2(CD_3)C(O)H]$	I (not isolated) [21] ^{2}H NMR (THF, −30°C): −0.3 (CD_3) [21]
*4	Li$[C_5H_5Mo(CO)_2(C_3H_6Br)C(O)H]$	I (not isolated) [23] ^{1}H NMR (THF, −78°C): 14.1 (C(O)H) [23]
5	Li$[C_5H_5Mo(CO)_2(C(O)C(O)OCH_3)C(O)H]$	I (17% by ^{1}H NMR), not isolated [10, 16] ^{1}H NMR (THF-d_8, −50°C): 12.90 (C(O)H) [10, 16] unstable at room temperature [10]
6	$[P(CH_3)_4][C_5H_5Mo(CO)_2(CH_3)C(O)CH{=}P(CH_3)_3]$	$C_5H_5Mo(CO)_3CH_3$ was allowed to react with 2 equivalents of $CH_2{=}P(CH_3)_3$ [14] alkylation with CH_3SO_3F affords No. 31 [14]
compounds with a three-membered Mo-C-S ring		
7	cis-$C_5H_5Mo(CO)_2(\eta^2\text{-}CS_2)C(O)CH_3$	II (5%) [12] purple solid [12] ^{1}H NMR ($CDCl_3$): 2.48 (s, CH_3), 5.52 (s, C_5H_5) [12] IR (CS_2): 1223 (ν(CS)), 1667 (ν(C=O)), 1926, 1983 (ν(CO)) [12] mass spectrum: ions from loss of $COCH_3$ and consecutive loss of CO [12] light-sensitive [12] no alkylation observed with CH_3I, either neat or in refluxing C_6H_6; with $P(C_6H_5)_3$ in C_6H_6, $C_5H_5Mo(CO)(P(C_6H_5)_3)(\eta^2\text{-}CS_2)C(O)CH_3$ is obtained [12]
8	cis-$C_5H_5Mo(CO)_2(\eta^2\text{-}CS_2)C(O)C_2H_5$	II (< 10%); the reaction is reversible [12] purple solid [12]

References on p. 184

Table 8 (continued)

No.	compound	method of preparation (yield) properties and remarks
8 (continued)		^{1}H NMR ($CDCl_3$): 1.21 (t, CH_3; J(H, H) = 12.5), 2.89 (q, CH_2), 5.49 (s, C_5H_5) [12] IR (CS_2): 1165 (ν(CS)), 1664 (ν(C=O)), 1924, 1980 (ν(CO)) [12] mass spectrum: $[M]^+$ [12]
9	$C_5H_5Mo(CO)_2(C_{10}H_{16}S)C_2H_5$ ($C_{10}H_{16}S$ = [structure: bicyclic thioketone with H_3C, CH_3, CH_3, =S])	III, with $P(OC_2H_5)_3$, $R_2C = C_{10}H_{16}$ (46%) [15] ^{1}H NMR ($CDCl_3$): 0.8 to 2.6 (m, C_2H_5 and $C_{10}H_{16}$), 5.32 (s, C_5H_5) [15] IR ($CHCl_3$): 1828, 1928 (ν(CO)) [15] mass spectrum: $[M-n\ CO]^+$ (n = 0 to 2) [15]
10	$C_5H_5Mo(CO)_2(\eta^2\text{-}S{=}C(C_6H_5)_2)C_2H_5$	III, with $P(OC_2H_5)_3$, R = C_6H_5 (82%) [15] orange solid, m.p. 114 to 116°C [15] ^{1}H NMR ($CDCl_3$): 0.87 (t, CH_3), 1.79 and 2.19 (m, AB part of ABX system, CH_2), 4.89 (s, C_5H_5), 7.10 to 7.30 (m, C_6H_5) [15] ^{13}C NMR ($CDCl_3$): 13.07 (CH_3), 34.09 (CH_2), 64.97 (MoC), 94.24 (C_5H_5), 124.64, 126.37, 127.03, 127.40, 128.06, 128.28, 142.56, and 153.05 (C_6H_5), 244.92 and 246.02 (CO) [15] IR (CH_2Cl_2): 1840, 1942 (ν(CO)) [15] mass spectrum: $[M-n\ CO]^+$ (n = 0 to 2) [15]
11	$C_5H_5Mo(CO)_2(\eta^2\text{-}S{=}C(C_6H_4CH_3\text{-}4)_2)C_2H_5$	III, with $P(OC_2H_5)_3$, R = $C_6H_4CH_3$-4 (55%) [15] orange solid, m.p. 127 to 128°C [15] ^{1}H NMR ($CDCl_3$): 0.92 (t, CH_3), 1.86 and 2.26 (m, AB part of ABX_3 system, CH_2), 2.67 and 2.69 (s, CH_3-4), 4.93 (s, C_5H_5), 6.89 to 7.33 (m, C_6H_4) [15] ^{13}C NMR ($CDCl_3$): 13.19 (CH_3), 20.90 (CH_3-4), 34.04 (CH_2), 65.07 (MoC), 94.28 (C_5H_5), 127.00, 128.15, 128.67, 132.45, 135.98, and 136.47 (C_6H_4), 245.00 and 246.25 (CO) [15] IR ($CHCl_3$): 1833, 1938 (ν(CO)) [15] mass spectrum: $[M-n\ CO]^+$ (n = 0 to 2) [15]
12	$C_5H_5Mo(CO)_2(\eta^2\text{-}S{=}C(C_6H_4OCH_3\text{-}4)_2)C_2H_5$	III, R = $C_6H_4OCH_3$-4, with $P(OC_2H_5)_3$ (92%), with $P(OC_2H_5)_2C_6H_5$ (44%) [15] orange solid, m.p. 81 to 83°C [15] ^{1}H NMR ($CDCl_3$): 0.91 (t, CH_3), 1.78 and 2.18 (m, AB part of ABX_3 system, CH_2), 3.74 and 3.77 (s, OCH_3), 4.94 (s, C_5H_5), 6.71, 6.74, 7.05, and 7.27 (d, C_6H_4; J(H, H) = 8); also given as a diagram in [15]

References on p. 184

Table 8 (continued)

No.	compound	method of preparation (yield) properties and remarks
		^{13}C NMR ($CDCl_3$): 13.26 (CH_3), 34.07 (CH_2), 55.16 and 55.25 (OCH_3), 65.11 (MoC), 94.17 (C_5H_5), 112.72, 113.41, 128.08, 133.56, 135.12, 145.87, 156.81, and 158.07 (C_6H_4), 245.10 and 246.35 (CO) [15] IR (CCl_4): 1844, 1946 (ν(CO)) [15] mass spectrum: $[M-n\ CO]^+$ (n = 0 to 2) [15]
13	$C_5H_5Mo(CO)_2(\eta^2\text{-}S{=}C(C_6H_4OCH_3\text{-}4)_2)C_4H_9\text{-}n$	III, with $P(OC_4H_9\text{-}n)_3$, R = $C_6H_4OCH_3$-4 (57%) [15] crystalline orange solid, m.p. 144 to 146°C [15] ^{1}H NMR ($CDCl_3$): 0.9 to 2.4 (m, C_4H_9-n), 3.78 and 3.81 (s, OCH_3), 4.96 (s, C_5H_5), 6.71, 6.74, 7.05, and 7.29 (d, C_6H_4; J(H, H) = 8) [15] ^{13}C NMR ($CDCl_3$): 13.49 (CH_3), 21.72, 30.82, and 39.24 (CH_2), 55.11 and 55.23 (OCH_3), 66.32 (MoC), 94.21 (C_5H_5), 112.73, 113.41, 128.15, 133.62, 135.24, 146.01, 156.85, and 158.12 (C_6H_4), 245.28 and 246.43 (CO) [15] IR ($CHCl_3$): 1835, 1940 (ν(CO)) [15] mass spectrum: $[M-n\ CO]^+$ (n = 0 to 2), $[M-2\ CO-C_4H_8]^+$ [15]
	compounds of the type $^5LMo(CO)_2(CNR)^1L$	
14	$C_5H_5Mo(CO)_2(CNCH_3)CH_2CN$	$Na[C_5H_5Mo(CO)_2CNCH_3]$ and $ClCH_2CN$ were stirred in THF at −78°C for 15 min; the mixture was then allowed to warm (49% by chromatography on alumina with C_6H_6/THF) [7, 9] cis to trans ratio 0.6 [7, 9] yellow solid, m.p. 78 to 80°C (from hexane or hexane/toluene) [7, 9] ^{1}H NMR ($CDCl_3$): cis isomer: 1.15 (CH_2), 3.57 (CH_3), 5.30 (C_5H_5); trans isomer: 1.23 (CH_2), 3.60 (CH_3), 5.13 (C_5H_5) [9] IR (cyclohexane): 1908, 1913, 1970, 1980 (ν(CO)), 2163, 2205 (ν(NC)) [7, 9]
15	$C_5H_5Mo(CO)_2(CNCH_3)C(O)CH_3$	$K[C_5H_5Mo(CO)_2(C(O)CH_3)CN]\cdot THF$ was allowed to react with 8 equivalents of CH_3I in CH_3CN for 4 h (55%) [5] yellow solid [5] IR (KBr): 1590 (ν(C=O)), 1860, 1955 (ν(CO)), 2110 (ν(NC)) [5] decomposes in solution above 30°C [5]

References on p. 184

Table 8 (continued)

No.	compound	method of preparation (yield) properties and remarks
15 (continued)		insoluble in water, soluble in common organic solvents [5]
16	trans-$C_5H_5Mo(CO)_2(CNC_4H_9\text{-t})CH_2C_6H_5$	complex No. 20 was irradiated for 30 h in THF (25% by chromatography on alumina with C_6H_6) [2] yellow crystals, m.p. 68 to 72°C (from hexane/C_6H_6) [2] ^1H NMR (CS_2): 1.49 (s, t-C_4H_9), 2.84 (s, CH_2), 5.09 (s, C_5H_5), 7.00 (s, C_6H_5) [2]
17	$C_5H_5Mo(CO)_2(CNC_4H_9\text{-t})C(CN){=}C(CN)_2$	IV, with 5.5 equivalents of t-C_4H_9NC in refluxing C_6H_6 for 1 h (50%) [13] red-brown solid, m.p. 115 to 116°C (from CH_2Cl_2/hexane) [13] cis to trans ratio 7:3 [13] ^1H NMR ($CDCl_3$): cis isomer: 1.56 (s, t-C_4H_9), 5.53 (s, C_5H_5); trans isomer: 1.55 (s, t-C_4H_9), 5.41 (s, C_5H_5) [13] ^{13}C NMR ($CDCl_3$): cis isomer: 30.2 (CH_3), 60.2 (**C**($CH_3)_3$), 95.8 (C_5H_5), 103.2 (C=C), 112.0, 115.3, and 120.8 (CN), 195.0 (C=C), 235.7 and 238.7 (CO); trans isomer: 30.2 (CH_3), 61.0 (**C**($CH_3)_3$), 95.2 (C_5H_5), 103.5 (C=C), 112.3, 115.5, and 120.6 (CN), 193.7 (C=C), 227.9 (CO) [13] IR (CH_2Cl_2): 1937, 1993 (ν(CO)), 2169 (ν(NC)), 2226 (ν(CN)) [13]
18	$C_5H_5Mo(CO)_2(CNC_4H_9\text{-t})C(Cl){=}C(CN)_2$	IV, with 2.5 equivalents of t-C_4H_9NC in refluxing C_6H_6 for 75 min (31%); the yield decreased by reaction times >90 min [13] yellow solid (from CH_2Cl_2/hexane), decomposes >105°C [13] ^1H NMR ($CDCl_3$): cis isomer: 1.56 (s, t-C_4H_9), 5.55 (s, C_5H_5); trans isomer: 1.56 (s, t-C_4H_9), 5.42 (s, C_5H_5) [13] ^{13}C NMR ($CDCl_3$): cis isomer: 30.2 (CH_3), 60.0 (**C**($CH_3)_3$), 95.5 (C_5H_5), 96.9 (C=C), 112.4 and 116.1 (CN), ca. 155.7 (NC), 237.3 and 239.2 (CO); trans isomer: 30.5 (CH_3), 60.5 (**C**($CH_3)_3$), 94.7 (C_5H_5), 112.7 and 116.4 (CN), ca. 155.7 (NC), 229.5 and 230.4 (CO) [13] IR (CH_2Cl_2): 1926, 1986 (ν(CO)), 2172 (ν(NC)), 2226 (ν(CN)) [13]
19	$C_5H_5Mo(CO)_2(CNC_4H_9\text{-t})C(O)CH_3$	IV, with t-C_4H_9NC in C_6H_6 for 9 h (56%) [2] or in CH_3CN with an excess of the isonitrile for 19 h

References on p. 184

Table 8 (continued)

No.	compound	method of preparation (yield) properties and remarks
		[6]; $C_5H_5Mo(CO)_2(P(C_6H_5)_3)C(O)CH_3$ was treated with an excess of t-C_4H_9CN in C_6H_6 for 9 h (48%) [2] yellow solid [2, 6], m.p. 101°C (from CH_2Cl_2/hexane) [6], 104 to 105°C (from THF/hexane) [2] ^{1}H NMR ($CDCl_3$): cis isomer: 1.50 (t-C_4H_9), 2.46 (CH_3), 5.34 (C_5H_5); trans isomer: 1.52 (t-C_4H_9), 2.57 (CH_3), 5.19 (C_5H_5) [2, 6]; (CS_2): cis isomer: 1.47 (t-C_4H_9), 2.27 (CH_3), 5.15 (C_5H_5); trans isomer: 1.50 (t-C_4H_9), 2.35 (CH_3), 5.01 (C_5H_5) [2] the cis to trans ratio in $CDCl_3$ is temperature-dependent: 36:64 at +23°C, 33:67 at −3°C, and 32:68 at −10°C [2, 6]; in CS_2 31:69 at +23°C [2] IR (KBr): 1615 (ν(C=O)), 1880, 1962 (ν(CO)), 2140 (ν(CN)) [2]; ($CHCl_3$): 1605 (ν(C=O)), 1891, 1954 (ν(CO)), 2136 (ν(CN)) [6]
*20	$C_5H_5Mo(CO)_2(CNC_4H_9\text{-t})C(O)CH_2C_6H_5$	IV, in C_6H_6 for 9 h (58%) [2] or in CH_3CN with an excess of the isonitrile for 22 h [6]; kinetics of formation in [17, 20] yellow solid [2, 6], m.p. 115°C (from CH_2Cl_2/hexane) [6], 115.5 to 116.5°C (from THF/hexane) [2] ^{1}H NMR: see "Further information" IR (KBr): 1615 (ν(C=O)), 1880, 1958 (ν(CO)), 2140 (ν(CN)) [2]; (CH_2Cl_2): 1612 (ν(C=O)), 1889, 1954 (ν(CO)), 2137 (ν(CN)) [6] irradiation in THF gives No. 16 [2]
21	$C_5H_5Mo(CO)_2(CNC_4H_9\text{-t})C(O)CH_2C_6H_4CH_3\text{-4}$	IV, with ca. 10 equivalents of the isonitrile in CH_3CN at 30°C (ca. 40%, not isolated); kinetics of formation were studied [17, 20]
22	$C_5H_5Mo(CO)_2(CNC_4H_9\text{-t})C(O)CH_2C_6H_4CF_3\text{-3}$	see No. 21
23	$C_5H_5Mo(CO)_2(CNC_4H_9\text{-t})C(O)CH_2C_6H_4CF_3\text{-4}$	see No. 21
24	$C_5H_5Mo(CO)_2(CNC_4H_9\text{-t})C(O)CH_2C_6H_4OCH_3\text{-4}$	IV, with ca. 2 equivalents of the isonitrile in CH_3CN (ca. 40%) [17]; kinetics of formation in [20] ^{1}H NMR ($CDCl_3$): trans isomer: 5.17 (C_5H_5); cis isomer: 5.31 (C_5H_5) [17]

References on p. 184

Table 8 (continued)

No.	compound	method of preparation (yield) properties and remarks
24 (continued)		IR (cyclohexane): 1623 (ν(C=O)), 1889, 1954 (ν(CO)) [17]
25	$CH_3C_5H_4Mo(CO)_2(CNC_4H_9\text{-t})C(O)CH_2C_6H_5$	see No. 21
26	$(CH_3)_5C_5Mo(CO)_2(CNC_4H_9\text{-t})C(O)CH_2C_6H_5$	see No. 21
27	$C_5H_5Mo(CO)_2(CNC_6H_{11}\text{-cyclo})C(O)CH_3$	IV, with 1.3 equivalents of the isonitrile in C_6H_6 for 6 h (48%) [3] cis to trans ratio 39:71 [3] yellow solid (from C_6H_6/hexane), m.p. 80 to 82°C (dec.) [3] ^1H NMR (CS_2): trans isomer: 1.4 to 2.4 (br m, cyclo-C_6H_{11}), 2.63 (s, CH_3), 4.05 to 4.45 (br, cyclo-C_6H_{11}, 1H), 5.15 (s, C_5H_5); cis isomer: 2.55 (s, CH_3), 5.28 (s, C_5H_5) [3] ^{13}C NMR ($CDCl_3$): trans isomer: 50.14 (CH_3), 94.5 (C_5H_5), 233.9 (CO), 267.6 (C=O); cis isomer: 48.03 (CH_3), 94.9 (C_5H_5), 238.7 (CO cis to acyl), 246.0 (CO trans to acyl), 270.4 (C=O) [11] IR (KBr): 1604 (ν(C=O)), 1856, 1950 (ν(CO)), 2150 (ν(CN)) [3]
28	$C_5H_5Mo(CO)_2(CNC_6H_{11}\text{-cyclo})C(O)CH_2C_6H_5$	IV, with 1.2 equivalents of the isonitrile in C_6H_6 for 8 h (4%) [3]; kinetics of formation in CH_3CN in [20] cis to trans ratio 36:64 [3] yellow solid, m.p. 93 to 95°C (dec.) [3] ^1H NMR (CS_2): trans isomer: 1.0 to 2.1 (br m, cyclo-C_6H_{11}), 2.6 to 3.0 (br, cyclo-C_6H_{11}, 1H), 4.04 (s, CH_2), 5.05 (s, C_5H_5), 6.7 to 7.2 (br m, C_6H_5); cis isomer: 3.94 (s, CH_2), 5.20 (s, C_5H_5) [3] IR (KBr): 1614 (ν(C=O)), 1880, 1952 (ν(CO)), 2139 (ν(CN))
29	$(CH_3)_5C_5Mo(CO)_2(CNC_6H_{11}\text{-cyclo})C(O)CH_2C_6H_5$	IV, with 10 equivalents of the isonitrile in CH_3CN at 30°C the kinetics of formation studied [20]
30	$C_5H_5Mo(CO)_2(CNC_6H_3(CH_3)_2\text{-2,6})C(O)CH_3$	IV, in C_6H_6 for 9 h (42%) [2] cis to trans ratio 33:67 [2] yellow solid, m.p. 91.5 to 92.5°C (from THF/hexane) [2]

References on p. 184

Table 8 (continued)

No.	compound	method of preparation (yield) properties and remarks
		^1H NMR (CS_2): trans isomer: 1.34 (s, CH_3-2,6), 2.47 (s, CH_3), 5.17 (s, C_5H_5), 6.15 to 7.0 (m, C_6H_3); cis isomer: 2.42 (s, CH_3), 5.28 (s, C_5H_5) [2] IR (KBr): 1623 (ν(C=O)), 1878, 1948 (ν(CO)), 2119 (ν(CN)) [2]
compound of the type $C_5H_5Mo(CO)_2(=C(OR)R')^1L$		
31	$C_5H_5Mo(CO)_2(=C(OCH_3)CH=P(CH_3)_3)CH_3$	alkylation of No. 6 with CH_3SO_3F; no properties reported [14]
cationic compounds of the type $[C_5H_5Mo(CO)_2(CNR)_2]^+$		
32	$[C_5H_5Mo(CO)_2(CNCH_3)_2]I$	$K[C_5H_5Mo(CO)_2(CN)_2]$ and CH_3I were refluxed in CH_3CN overnight [1] mustard yellow microcrystalline powder, m.p. 180 to 182°C [1] IR: 1945, 2000 (ν(CO)), 2185, 2200 (ν(NC)) air-stable as a solid but rapidly decomposed in solution by bases [1] treatment with aqueous NH_4PF_6 gives No. 33 [1]
33	$[C_5H_5Mo(CO)_2(CNCH_3)_2]PF_6$	obtained by anion exchange of No. 32 with aqueous NH_4PF_6 [1] reaction with $NaBH_4$ in THF at 0°C affords $C_5H_5Mo(CO)_2(CH=NCH_3)_2BH_2$ (Formula III) [4]

III

*Further information:

$Li[C_5H_5Mo(CO)_2(^1L)C(O)H]$ (Table **8**, Nos. **1** to **4**). In solution at temperatures higher than −40°C the complexes rearrange to the hydrido-acyl $Li[C_5H_5Mo(CO)_2(C(O)R)H]$ and ultimately (if R = CH_3, CD_3) to the product $Li[C_5H_5Mo(CO)_2(\eta^2\text{-}O{=}CHR)]$ [18, 21 to 23].

$C_5H_5Mo(CO)_2(CNC_4H_9\text{-}t)C(O)CH_2C_6H_5$ (Table **8**, No. **20**). The solvent-dependent ^1H NMR spectra (δ values in ppm at 23°C) are given in the following table [2, 6]. The CH_2 resonance of the cis isomer splits into two sharp resonances at 3.86 and 3.94 ppm at −50°C in CS_2; the

 References on p. 184

coalescence temperature is −9°C. In $CDCl_3$ the coalescence temperature is found between 61 and 70°C. Figures of the spectra in CS_2 and $CDCl_3$ in the coalescence region are given [2]. The temperature-dependent cis to trans ratios in $CDCl_3$ are 32:68 at −10°C, 33:67 at −3°C, and 36:64 at +23°C [2].

solvent		t-C_4H_9	CH_2	C_5H_5	C_6H_5	cis/trans
C_6H_6	cis	0.9	4.30, 4.37	4.98	6.95 to 7.45	36:64
	trans	0.9	4.57	4.90		
$CHCl_3$	cis	1.48	4.08, 4.15	5.25	6.8 to 7.4	34:66
	trans	1.48	4.24	5.12		
CS_2	cis	1.38	3.91	5.13	6.8 to 7.3	27:73
	trans	1.46	4.07	4.99		

References:

[1] Coffey, C. E. (J. Inorg. Nucl. Chem. **25** [1963] 179/85).
[2] Yamamoto, Y.; Yamazaki, H. (Bull. Chem. Soc. Japan **43** [1970] 143/7).
[3] Yamamoto, Y.; Yamazaki, H. (J. Organometal. Chem. **24** [1970] 717/24).
[4] Treichel, P. M.; Stenson, J. P.; Benedict, J. J. (Inorg. Chem. **10** [1971] 1183/7).
[5] Kruck, T.; Höfler, M.; Liebig, L. (Chem. Ber. **105** [1972] 1174/83).
[6] King, R. B.; Saran, M. S. (Inorg. Chem. **13** [1974] 364/7).
[7] Adams, R. D. (J. Organometal. Chem. **88** [1975] C 38/C 40).
[8] Brunner, H.; Lukas, R. (J. Organometal. Chem. **90** [1975] C 25/C 27).
[9] Adams, R. D. (Inorg. Chem. **15** [1976] 169/74).
[10] Gladysz, J. A.; Selover, J. C. (Tetrahedron Letters **1978** 319/22).

[11] Todd, L. J.; Wilkinson, J. R.; Hickey, J. P.; Beach, D. L.; Barnett, K. W. (J. Organometal. Chem. **154** [1978] 151/7).
[12] Coville, N. J.; Albers, M. O. (J. Organometal. Chem. **172** [1979] C 1/C 3).
[13] King, R. B.; Saran, M. S.; McDonald, D. P.; Diefenbach, S. P. (J. Am. Chem. Soc. **101** [1979] 1138/42).
[14] Malisch, W.; Blau, H. (Chemiedozententag., Darmstadt, FRG, 1979, Abstr. B 47).
[15] Hartgerink, J.; Silavwe, N. O.; Alper, H. (Inorg. Chem. **19** [1980] 2593/7).
[16] Selover, J. C.; Mavri, M.; Parker, D. W.; Gladysz, J. A. (J. Organometal. Chem. **206** [1981] 317/29).
[17] Cotton, J. D.; Dunstan, P. R. (Inorg. Chim. Acta **88** [1984] 223/7).
[18] Gauntlett, J. T.; Taylor, B. F.; Winter, M. J. (J. Chem. Soc. Chem. Commun. **1984** 420/1).
[19] Gauntlett, J. T.; Winter, M. J. (Polyhedron **5** [1986] 451/9).
[20] Cotton, J. D.; Kimlin, H. A. (J. Organometal. Chem. **294** [1985] 213/7).

[21] Gauntlett, J. T.; Taylor, B. F.; Winter, M. J. (J. Chem. Soc. Dalton Trans. **1985** 1815/20).
[22] Gauntlett, J. T.; Winter, M. J. (Chem. Uses Molybdenum Proc. 5th Intern. Climax Conf., Newcastle Upon Tyne, Engl., 1985, Abstr. pp. 95/6).
[23] Adams, H.; Bailey, N. A.; Cahill, P.; Rogers, D.; Winter, M. J. (J. Chem. Soc. Dalton Trans. **1986** 2119/26).

1.5.1.3.2.2 Compounds with Additional ^2L Ligands

The compounds summarized in this chapter contain one ^2L ligand. As ^2L ligands serve alkyne (No. 1), various alkenes (Nos. 2 to 10), η^2-CC-bonded keteneimmonium heteroallenes (Nos. 11 to 18), and η^3-bonded heteroallyl ligands (Nos. 19 to 34); metallacyclic compounds are described with Nos. 35 to 39. At the end of the table, a few compounds containing one ^2L and one additional ^1L ligand are also described. Many of the compounds listed in Table 9 were obtained by the following methods; special methods are given in the table.

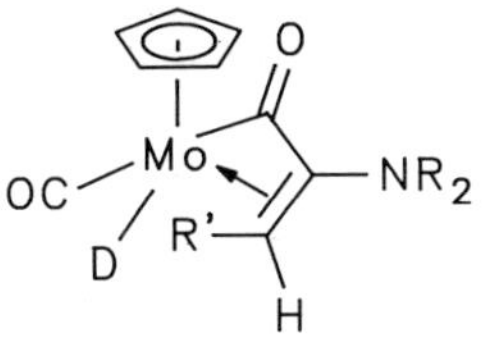

I

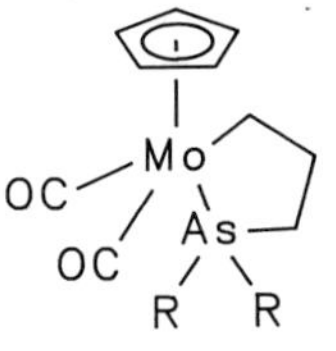

II

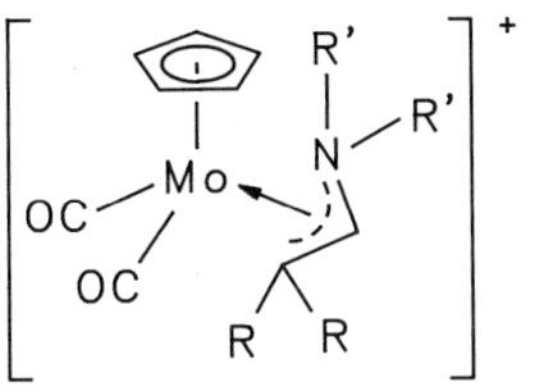

III

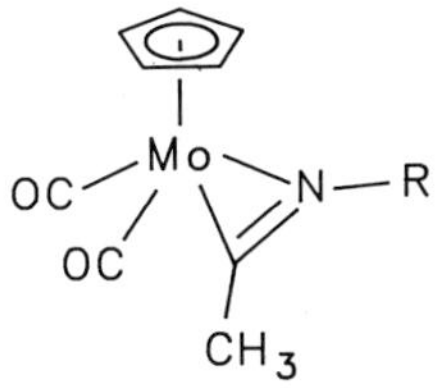

IV

Method I: $C_5H_5Mo(CO)(D)C(R')H{=}C(NR_2)C{=}O$ (Formula I) was heated in cyclohexane at 70 to 80 °C for 30 min (Nos. 11, 12). Alternatively, the complex was stirred in CH_2Cl_2 for several days (No. 14) [23].

Method II: $C_5H_5Mo(CO)_2C_3H_7AsR_2$ (Formula II) was treated with $[C(C_6H_5)_3]X$ (X = PF_6; X = BF_4 in the presence of NH_4PF_6) in CH_2Cl_2 for 5 min. The precipitate was recrystallized from acetone/water [9, 19].

Method III: $[C_5H_5Mo(CO)_2(\eta^3\text{-}CR_2 \cdots CH \cdots NR'_2)]PF_6$ (Formula III, Nos. 32 and 33) was allowed to react with concentrated aqueous NH_3 with supernatant pentane. The organic layer was separated about every 20 min and fresh pentane added. This process was continued for 2 h [11, 13].

Method IV: $C_5H_5Mo(CO)_3C(X){=}C(CN)_2$ (X = Cl or CN) was allowed to react with R_2NR' (R' = H, R = CH_3, R_2 = $(CH_2)_5$; R' = R = C_2H_5) in CH_2Cl_2. The products were isolated by column chromatography on Florisil with CH_2Cl_2/hexane. Hydrolysis to give No. 35 probably proceeds during chromatography [14].

Method V: A solution of two equivalents of $LiN{=}CR_2$ in ether/hexane was added to a solution of $C_5H_5Mo(CO)_3Cl$ in ether cooled at −78 °C. The reaction mixture was allowed to warm and stirred for several hours at ambient temperature [1, 3, 5, 7, 8].

Method VI: The η^2-C=N bonded complex $C_5H_5Mo(CO)_2(\eta^2\text{-}R_2N{=}C{=}C(CH_3)_2)$ was treated with aqueous HPF_6 (ca. 65%) in propionic anhydride for 1.5 to 2 h [11, 13].

Method VII: $(C_5H_5Mo(CO)_3)_2$ and 3 equivalents of 3-phenyl- or 3-(4-tolyl)-2H-azirine were stirred in toluene for 12 h. The products were isolated by chromatography on alumina with ether [33].

Method VIII: $(CO)_5WP(C_6H_5)(CH{=}CH_2)Cl$ and $(C_5H_5Mo(CO)_3)_2$ were heated under reflux in xylene for 1.5 h. The product mixture was chromatographed on silica gel with hexane/CH_2Cl_2 [39].

References on pp. 201/2

Method IX: $C_5H_5Mo(CO)_3CH_2C(O)R$ (R = CH_3, OC_2H_5) was irradiated in C_6D_6 at 6°C for 3 h [32, 36, 37].

Method X: An immediate reaction occurred between $C_5H_5Mo(CO)_2C(CH_3){=}NR$ (Formula IV) and $(NC)_2C{=}C(CN)_2$ in THF [16, 20]. The product was purified by column chromatography on alumina with THF or THF/C_6H_6 [20].

General Remarks. Compounds with η-heteroallyl ligands can formally be considered as coordination of a 4-electron donating heteroallyl anion at the 14 electron fragment $[^5LMo(CO)_2]^+$. In contrast to compounds with allyl ligands (see Section 1.5.3.2.3.2) and 1-heteroatom allyl ligands (Nos. 25 to 34), in which the substituents at the allyl skeleton are approximately in one plane, compounds with the 2-azaallyl ligand of the type $[R_2C{\cdots}N{\cdots}CR_2]^-$ have the groups R in perpendicular planes with a C-N-C angle of 128°; see compounds No. 19 to 24. From NMR studies, for these compounds a different structure in solution from that in the solid state was suggested [8].

Low temperature ^1H NMR studies on compounds with the CC-bonded keteneimmonium ligands $R_2C^1{=}C^2{=}NR'_2$ (Nos. 12 to 14) [23] and NMR-spectroscopically equivalent substituents R at CR_2 and nonequivalent substituents R′ at NR'_2 at these heteroallenes [13, 23] are indicative for a fast rotation of the ligand about the Mo-alkene axis. If the substituents R at CR_2 are different, diastereomers are obtained. C-1 and C-2 of the ligand probably occupy one side of the basal plane of the molecule [23].

Some complexes were not isolated and selected ones were only the subject of theoretical MO calculations. Extended Hückel MO calculations on **$[C_5H_5Mo(CO)_2CH_2{=}C_3H_4\text{-cyclo}]BF_4$** were carried out for different conformations of the η^2-methylenecyclopropane ligand. The barrier of rotation about the axis from the Mo atom to the midpoint of the C=C bond is > 85 kcal/mol [28]. The calculated rotational barrier energy for the d^4 system **$[C_5H_5Mo(CO)_2CH{\equiv}CH]^+$** is 13 kcal/mol; an orbital interaction diagram for the different conformations was given [21].

Table 9
Compounds with ^2L Ligands Bonded to Mo.
An asterisk indicates further information at the end of the table.
For explanations, abbreviations, and units see p. X.

No.	compound	method of preparation (yield) properties and remarks
compounds of the types $[^5LMo(CO)_2{}^2L]^+$, $^5LMo(CO)_2{}^2L$, $^5LMo(CO)_2(^2L)X$, and $^5LMo(CO)_2{}^2L\text{-}X$		
1	$[C_5H_5Mo(CO)_2(\eta^2\text{-}CH_3C{\equiv}CCH_3)]BF_4$	$[C_5H_5Mo(CO)(C_2H_4)CCH_3{\equiv}CCH_3]BF_4$ was allowed to react with CO in solution [38] ^1H NMR (CD_2Cl_2): 3.38 (CH_3), 6.26 (C_5H_5) [38] IR (CH_2Cl_2): 1995, 2050 (ν(CO)) [38]
*2	$C_5H_5Mo(CO)_2C(CN)(P(C_6H_5)_3)CHCN$ 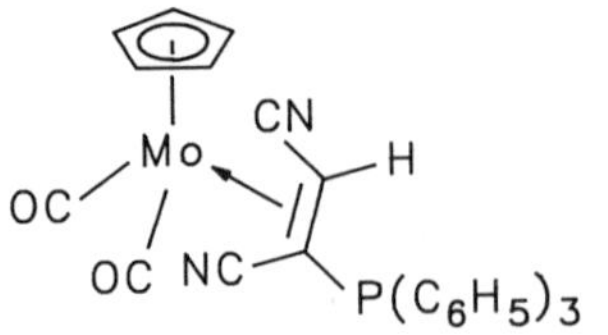	$C_5H_5Mo(CO)_3C(CN){=}CH(CN)$ and $P(C_6H_5)_3$ were irradiated in THF for 1 d (55% by chromatography on Florisil with CH_2Cl_2/THF 10:1) [27] bright yellow microcrystals, m.p. 208°C [27] ^1H NMR ($CDCl_3$): 1.50 (d, CH; J(P, H) = 11.5), 5.33 (s, C_5H_5), 7.50 (m, C_6H_5) [27]

Table 9 (continued)

No.	compound	method of preparation (yield) properties and remarks
		^{13}C NMR ($CDCl_3$): 1.65 (d, MoCP; J(P, C) = 75.2), 10.2 (d, CH; J(P, C) = 3.6), 96.3 (C_5H_5), 123.0 and 125.7 (CN), 232.4 (d, CO; J(P, C) = 4.8), 243.5 (CO) [27] IR (CH_2Cl_2): 1820, 1915 (ν(CO)), 2190, 2210 (ν(CN)) [27]
*3	trans-$C_5H_5Mo(CO)_2(\eta^2\text{-}CH_2{=}CH_2)H$	formed by irradiation of $C_5H_5Mo(CO)_3C_2H_5$ at 12 K in a matrix (paraffin [24], CH_4 with 5% C_2H_4 [26, 30], polyvinyl chloride [29], CH_4 [31], or CO [31]) or in toluene at 77 K for 1 h (not isolated) [24] ^{1}H NMR (toluene-d_8, 223 K): −5.2 (s, MoH), 1.50 (s, C_2H_4), 4.18 (s, C_5H_5) [24] IR (paraffin, 195 K): 1908, 1980 (ν(CO)) [24]; see also "Further information" $C_5H_5Mo(CO)_2(P(OC_6H_5)_3)C_2H_5$ is formed by treatment with $P(OC_6H_5)_3$ for 12 h in toluene at 273 K [24]
*4	$C_5H_5Mo(CO)_2(\eta^2\text{-}CH_2{=}CHCH_3)FBF_3$	$C_5H_5Mo(CO)_2C_3H_5$ (C_3H_5 = allyl) was allowed to react with $HBF_4 \cdot O(C_2H_5)_2$ in CH_2Cl_2 or $HBF_4 \cdot O(CH_3)_2$ in $CDCl_3$ (high yield, not isolated) [35] attempts to isolate as a solid leads to a black insoluble residue [35] dark red solution in CH_2Cl_2 or $CDCl_3$ [35] ^{1}H NMR ($CDCl_3$): 1.98 (d, CH_3; J(H, H) = 5), 2.95 (m, CH_2-Z), 3.70 (m, CH_2-E), 4.95 (m, CH=), 5.65 (s, C_5H_5) [35] IR (CH_2Cl_2): 900 to 1230 (maximum at 1148, $MoFBF_3$), 1973, 2027 (ν(CO)) [35]
5	trans-$C_5H_5Mo(CO)_2(\eta^2\text{-}CH_2{=}CHC_6H_5)Sn(C_6H_5)_3$	trans-$C_5H_5Mo(CO)_2({=}C(C_6H_5)CH_3)Sn(C_6H_5)_3$ rearranges on stirring in THF solution over several hours (good yield); also formed in the reaction of $C_5H_5Mo(CO)_2(C_4H_8O)Sn(C_6H_5)_3$ (C_4H_8O = THF) with $C_6H_5CH{=}CH_2$ [40] ^{1}H NMR ($CDCl_3$): 3.21 (d of d, CH_2=, 1H; J(H, H) = 9, 3), 3.37 (d of d, CH_2, 1H; J(H, H) = 13, 3), 4.54 (d of d, CH=; J(H, H) = 13, 9), 4.90 (s, C_5H_5), 7.23 to 7.34 (m, C_6H_5, H-3,4), 7.50 to 7.58 (m, C_6H_5, H-2) [40] ^{13}C NMR (CD_2Cl_2, 0°C): 37.1 (CH_2=), 59.7 (CH=), 93.0 (C_5H_5), 126.6 to 143.5 (C_6H_5), 231.2 and 232.1 (CO) [40] IR (CH_2Cl_2): 1874, 1935 (ν(CO)) [40] mass spectrum (field desorption): $[M]^+$ [40]

References on pp. 201/2

Table 9 (continued)

No.	compound	method of preparation (yield) properties and remarks
*6	$(CH_3)_5C_5Mo(CO)_2C(C_6H_5)=C(C_6H_5)CHC(O)O$ 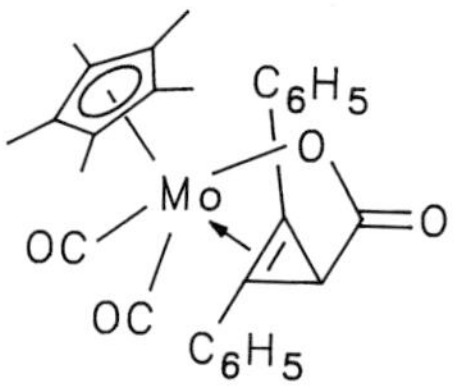	$[(CH_3)_5C_5Mo(CO)_3]BF_4$ was allowed to react with $Na[O_2CCHC(C_6H_5)=C(C_6H_5)]$ in CH_2Cl_2 at −80 °C for 1 h; the mixture was then allowed to warm and stirred overnight (25% by column chromatography on Florisil with acetone) [34] orange crystals (from CH_2Cl_2/hexane), m.p. 124 to 129 °C (dec.) [34] ^1H NMR ($CDCl_3$): 1.67 (s, CH_3), 1.98 (s, CH), 7.39 (m, C_6H_5) [34] ^{13}C NMR ($CDCl_3$): 10.04 (CH_3), 44.22 (CH), 58.64 and 71.91 (**C**C_6H_5), 108.73 ($(CH_3)_5$**C**$_5$), 125.99 to 139.77 (C_6H_5), 183.67 (C(O)O), 222.24 and 228.79 (CO) [34] IR (CH_2Cl_2): 1630 (ν(C=O)), 1961, 2022 (ν(CO)) [34]
	compounds of the types $[^5LMo(CO)_2(^2D)^2L]X$ and $[^5LMo(CO)_2(^2L\text{-}D)]X$	
7	$[C_5H_5Mo(CO)_2(\eta^2\text{-}CH_2=CH_2)P(C_6H_5)_3]X$ (X = AsF_6 or O_3SCF_3)	one decomposition product of $[C_5H_5Mo(CO)_2(P(C_6H_5)_3)=CH_2]O_3SCF_3$ in CD_2Cl_2 at −70 °C; was formed by treating $C_5H_5Mo(CO)_2(P(C_6H_5)_3)C_2H_5$ with $[C(C_6H_5)_3]AsF_6$ (not isolated) [25] ^1H NMR (CD_2Cl_2): 3.53 (d, C_2H_4; J(P, H) = 1.8), 5.28 (d, C_5H_5; J(P, H) = 1.5), 7 to 8 (m, C_6H_5) [25]
8	$[C_5H_5Mo(CO)_2CH_2=CHCH_2As(CH_3)_2]PF_6$ 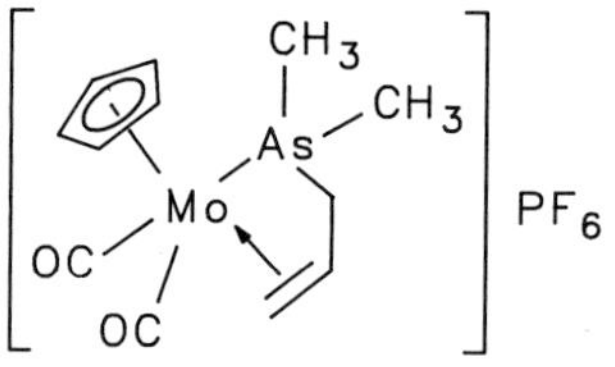	II (32%) [9, 19] fine yellow crystals, m.p. 164 °C (dec.) [19] ^1H NMR ($CDCl_3$): 1.68 and 1.71 (CH_3), 6.12 (C_5H_5) [19] IR (Nujol): 1525 (ν(C=C)), 1930, 2015 (ν(CO)) [9, 19] mass spectrum: $[M]^+$ [9] reaction with KCN in CH_3OH affords cis-$C_5H_5Mo(CO)_2(As(CH_3)_2CH_2CH=CH_2)CN$ and further decomposition [9, 19]; with $NaOCH_3$ in CH_3OH, $(C_5H_5Mo(CO)_2)_2(\mu\text{-}As(CH_3)_2)_2$ is formed [19]
9	$[C_5H_5Mo(CO)_2CH_2=CHCH_2As(C_6H_5)_2]PF_6$ (compare No. 8)	II (39%) [19] fine yellow crystals, m.p. 196 °C (dec.) [19] ^1H NMR ($CDCl_3$): 5.97 (C_5H_5) [19] IR ($CHCl_3$): 1530 (ν(C=C)), 1961, 2020 (ν(CO)) [19]
10	$[C_5H_5Mo(CO)_2CH_2=CHCH_2As(C_6H_4CH_3\text{-}4)_2]PF_6$ (compare No. 8)	II (21%) [19]

References on pp. 201/2

Table 9 (continued)

No.	compound	method of preparation (yield) properties and remarks
		fine yellow crystals, m.p. 182°C (dec.) [19] ^1H NMR ($CDCl_3$): 2.34 and 2.42 (CH_3), 5.48 (C_5H_5) [19]

compounds of the type $^5LMo(CO)_2(\eta^2\text{-}CR_2{=}C{=}NR'_2)$

No.	compound	method of preparation (yield) properties and remarks
11	$C_5H_5Mo(CO)_2(\eta^2\text{-}CH_2{=}C{=}N(C_2H_5)_2)$	I, D = $P(OCH_3)_3$ (high yield) [23] yellow crystals, m.p. 83 to 85°C (from pentane) [23] ^1H NMR ($CDCl_3$): 0.90 (s, CH_2), 1.27 and 1.38 (t, CH_3; J(H, H) = 7.0), 3.82 and 4.13 (q, **CH_2**CH_3), 5.28 (s, C_5H_5); broadening of the CH_2 resonance in the low-temperature spectrum indicates a fast rotation of the ^2L ligand [23] IR (CH_2Cl_2): 1586 (ν(C=N)), 1837, 1926 (ν(CO)) [23] mass spectrum: $[M-n\,CO]^+$ (n = 0 to 2) and organic fragments [23]
12	$C_5H_5Mo(CO)_2(\eta^2\text{-}CH_2{=}C{=}NC_4H_8O)$ (NC_4H_8O = morpholino)	I, D = $P(OCH_3)_3$ (high yield) [23] yellow crystals, m.p. 92 to 94°C (from pentane) [23] ^1H NMR ($CDCl_3$): 0.89 (s, CH_2), 3.6 to 4.2 (m, C_2H_4), 5.23 (s, C_5H_5); broadening of the CH_2 resonance in the low-temperature spectrum indicates a fast rotation of the ^2L ligand [23] IR (CH_2Cl_2): 1582 (ν(C=N)), 1844, 1932 (ν(CO)) [23]
13	$C_5H_5Mo(CO)_2(\eta^2\text{-}CH_3CH{=}C{=}N(C_2H_5)_2)$	$C_5H_5Mo(CO)_2(P(OCH_3)_3)H$ was allowed to react with $CH_3C{\equiv}CN(C_2H_5)_2$ in ether for 1 h [17]; I [23] yellow crystals, m.p. 58 to 62°C, subl. at 70°C in vacuum [17] ^1H NMR ($CDCl_3$): 1.24 and 1.35 (t, **CH_3**CH_2N; J(H, H) = 7), 1.40 (d, **CH_3**CH; J(H, H) = 6), 1.83 (q, CH; J(H, H) = 6), 3.78 and 3.97 (q, CH_2), 5.27 (s, C_5H_5) [17]; the resonance of the CH group broadens at −65°C only slightly [23] ^{13}C NMR ($CDCl_3$): 14.6, 15.0 (**CH_3**CH_2N-Z,E), 23.7 (CH_3), 50.2, 52.0 (CH_2N-Z,E), 90.7 (C_5H_5), 97.5 (=C=), 234.3, 236.3, 249.6 (CO) [23] IR (CH_2Cl_2): 1576 (ν(C=N or C=C)), 1833, 1921 (ν(CO)) [17] mass spectrum: $[M-n\,CO]^+$ (n = 0 to 2) [17]
14	$C_5H_5Mo(CO)_2(\eta^2\text{-}C_2H_5OC(O)CH{=}C{=}N(C_2H_5)_2)$	I, D = CO (100%) [23] yellow solid, m.p. 50 to 53°C (from pentane) [23]

References on pp. 201/2

Table 9 (continued)

No.	compound	method of preparation (yield) properties and remarks
14 (continued)		^{1}H NMR ($CDCl_3$): 1.17 (t, $\mathbf{CH_3}CH_2N$; J(H, H) = 7.0), 1.22 (t, $\mathbf{CH_3}CH_2O$; J(H, H) = 7.0), 1.37 (t, $\mathbf{CH_3}$-CH_2N; J(H, H) = 7.0), 2.28 (s, CH), 3.80 (q, CH_2N), 4.05 (q, CH_2O), 4.18 (q, CH_2N), 5.28 (s, C_5H_5); broadening of the CH resonance in the low-temperature spectrum indicates a fast rotation of the ^{2}L ligand [23] IR (CH_2Cl_2): 1583 (ν(C=N)), 1674 (ν(C(O)OC_2H_5)), 1863, 1951 (ν(CO)) [23]
15	$C_5H_5Mo(CO)_2(\eta^2\text{-}(CH_3)_2C{=}C{=}N(CH_3)_2)$	III, R = R′ = CH_3 (11.6%) [11, 13] yellow solid (from pentane at −78°C) [13] ^{1}H NMR ($CDCl_3$): 1.42 (CCH_3), 2.67 and 2.94 (NCH_3), 5.17 (C_5H_5) [13] IR (hexane): 1830, 1930 (ν(CO)) [13] mass spectrum (70 eV, 60 to 210°C): $[M-n\,CO]^+$ (n = 0 to 2), $[C_5H_5Mo]^+$ [13] decomposes even at lower temperatures [11, 13]
16	$C_5H_5Mo(CO)_2(\eta^2\text{-}(CH_3)_2C{=}C{=}N(CH_2)_5\text{-cyclo})$	III (low yield) [11, 13] yellow, unstable solid (from pentane at −78°C) [13] ^{1}H NMR ($CDCl_3$): 1.43 (CH_3), 1.5 and ca. 3.0 (br m, CH_2), 5.13 (C_5H_5) [13] IR (hexane): 1831, 1926 (ν(CO)) [13] mass spectrum (70 eV, 60 to 210°C): $[M-n\,CO]^+$ (n = 0 to 2), $[C_5H_5Mo]^+$ [13]
17	$C_5H_5Mo(CO)_2(\eta^2\text{-}(NC)_2C{=}C{=}N(CH_3)_2)$	IV, R = CH_3, R′ = H, X = Cl, for 15 min (4%) [14] bright yellow solid, m.p. 169 to 170°C [14] ^{1}H NMR ($CDCl_3$): 3.68 and 3.69 (s, CH_3), 5.43 (s, C_5H_5) [14] ^{13}C NMR (acetone-d_6): 46.0 and 47.6 (CH_3), 95.4 (C_5H_5), 120.0 (CN), 224.4 (C=), 233.3 (CO) [14] IR (CH_2Cl_2): 1590 (ν(C=C)), 1919, 1992 (ν(CO)), 2213 (ν(CN)); (KBr): 2945, 3110, 3116 (ν(CH)) [14]
18	$C_5H_5Mo(CO)_2(\eta^2\text{-}(NC)_2C{=}C{=}N(CH_2)_5\text{-cyclo})$	IV, R = $(CH_2)_5$, R′ = H, X = Cl, for 15 to 22 h (28%) [14] bright yellow solid, m.p. 159 to 161°C [14] ^{1}H NMR ($CDCl_3$): 1.83 (br m, 3 CH_2), 4.04 (br, 2 CH_2), 5.42 (s, C_5H_5) [14] ^{13}C NMR (acetone-d_6): 23.8, 26.9, 56.3, and 57.6 (t, CH_2), 94.4 (d, C_5H_5), 119.6 (s, CN), 219.6 (s, C=), 231.1 (s, CO) [14]

Table 9 (continued)

No.	compound	method of preparation (yield) properties and remarks
		IR (CH_2Cl_2): 1573 (ν(C=C or C=N)), 1916, 1991 (ν(CO)), 2218 (ν(CN)); (KBr): 2865, 2926, 2935, 2964, 3005, 3105 (ν(CH)) [14]
	compounds of the types $^5LMo(CO)_2(\eta^3\text{-}R_2C{\cdots}N{\cdots}CR_2)$ and $^5LMo(CO)_2(\eta^3\text{-}R_2C{\cdots}C{\cdots}E)$ (E = NR′, PR′, and O)	
19	$C_5H_5Mo(CO)_2(\eta^3\text{-}(C_6H_5)_2C{\cdots}N{\cdots}C(C_6H_5)_2)$	V [1, 2, 4, 12], extraction with $CHCl_3$ (83%) [3]; also formed by heating $C_5H_5Mo(CO)_3Cl$ with 2 equivalents of $(CH_3)_3SiN{=}C(C_6H_5)_2$ in $CH_3OC_2H_4OCH_3$ for 6 h at 70°C (5%) [3] golden brown crystals [1], waxy, golden brown crystals, m.p. 203 to 205°C (from $CHCl_3$/hexane at −20°C) [3] ^1H NMR ($CDCl_3$): 4.76 (s, C_5H_5), 7.30 (m, C_6H_5) [1, 3] IR (Nujol): 1845, 1927 (ν(CO)) [3]; ($CHCl_3$): 1842, 1934 (ν(CO)) [1, 3] mass spectrum: $[M-n\,CO]^+$ (n = 0 to 2), $[C_5H_5MoC(C_6H_5)_3]^+$, $[C_5H_5MoC(C_6H_4)_2C_6H_5]^+$, $[C_5H_5MoC(C_6H_4)C_6H_5]^+$, $[C_5H_5MoC_6H_4]^+$, $[C_5H_5Mo(C_6H_5)_4C_2N]^{2+}$ [1, 3] solutions in polar organic solvents are intensively purple-colored even when diluted [1] no reaction is observed with excess $P(C_6H_5)_3$ in refluxing toluene or $CHCl_3$ [1, 3]
*20	$C_5H_5Mo(CO)_2(\eta^3\text{-}4\text{-}CH_3C_6H_4(C_6H_5)C{\cdots}N{\cdots}C(C_6H_5)C_6H_4CH_3\text{-}4)$	V [8] solid, m.p. 183 to 184°C (from hexane or hexane/ether at 0°C) [8] ^1H NMR (CS_2, −40°C): 2.12, 2.16, 2.25, and 2.48 (s, CH_3), 4.42, 4.45, 4.51, and 4.54 (s, C_5H_5), 7.15 (m, C_6H_4 and C_6H_5); (CS_2, +10°C): 2.18 and 2.47 (s, CH_3), 4.46 and 4.57 (s, C_5H_5), 7.15 (m, C_6H_4 and C_6H_5); (CS_2, +70°C): 2.26 (s, CH_3), 4.57 (s, C_5H_5), 7.15 (m, C_6H_4 and C_6H_5) [8] IR (KBr): 1841, 1924 (ν(CO)) [8] mass spectrum (70 eV, 80 to 220°C): $[M-n\,CO]^+$ (n = 0 to 2) [8]
*21	$C_5H_5Mo(CO)_2(\eta^3\text{-}(4\text{-}CH_3C_6H_4)_2C{\cdots}N{\cdots}C(C_6H_4CH_3\text{-}4)_2)$	V [5, 8], starting at room temperature (53%) [7]; also formed by heating $C_5H_5Mo(CO)_2((C_6H_4CH_3\text{-}4)_2CNC(C_6H_4CH_3\text{-}4)_2)\cdot(4\text{-}CH_3C_6H_4)_2CO$ in a vacuum [4, 7]

References on pp. 201/2

Table 9 (continued)

No.	compound	method of preparation (yield) properties and remarks
*21 (continued)		dark purple crystals, m.p. 190 to 191 °C (from ether/hexane at 0 °C) [7] ^{1}H NMR (CS_2, −20 °C): 2.14, 2.18, 2.27, and 2.51 (s, CH_3), 4.46 (s, C_5H_5), 7.08 (m, C_6H_4); (CS_2, +10 °C): 2.20 (s, 3 CH_3), 2.51 (s, 1 CH_3), 4.60 (s, C_5H_5), 7.08 (m, C_6H_4); (CS_2, +70 °C): 2.28 (s, CH_3), 4.68 (s, C_5H_5), 7.08 (m, C_6H_4) [8]; in part in [7] IR (KBr): 1821, 1936 [7], similar data in [8, 10]; (CS_2): 1844, 1938 [7]; (hexane): 1856, 1949 (ν(CO)) [5, 7] ESCA: 228.4 (Mo $3d_{5/2}$), 285.0 (C 1s), 400.0 (N 1s) eV [10] mass spectrum (70 eV, 80 to 220 °C): $[M-n\ CO]^+$ (n = 0 to 2) [5, 8] air-stable in the solid state [7] moderately soluble in hexane and other aliphatic hydrocarbons but very soluble in most organic solvents [7] treatment in Nujol with CO (210 atm) at 185 °C for 45 h affords $Mo(CO)_6$ and traces of $(C_5H_5Mo(CO)_3)_2$ [18]; a substitution reaction with $P(C_6H_5)_3$ failed [7]
*22	$C_5H_5Mo(CO)_2(\eta^3\text{-}4\text{-}CH_3OC_6H_4(4\text{-}CH_3C_6H_4)C{\cdots}N{\cdots}C(C_6H_4CH_3\text{-}4)C_6H_4OCH_3\text{-}4)$	V [8] m.p. 86 to 88 °C (from hexane or hexane/ether at 0 °C) [8] ^{1}H NMR (CS_2, −40 °C): 2.14, 2.21, 2.30, and 2.54 (s, CH_3), 3.55, 3.60, 3.73, and 3.89 (s, OCH_3), 4.49, 4.52, and 4.61 (s, C_5H_5), 7.10 (m, C_6H_4); (CS_2, +10 °C): 2.27 and 2.53 (s, CH_3), 3.63 and 3.88 (s, OCH_3), 4.53 and 4.64 (s, C_5H_5), 7.10 (m, C_6H_4); (CS_2, +70 °C): 2.29 (s, CH_3), 3.67 (s, OCH_3), 4.58 (s, C_5H_5), 7.10 (m, C_6H_4) [8] IR (KBr): 1831, 1930 (ν(CO)) [8] mass spectrum (70 eV, 80 to 220 °C): $[M-n\ CO]^+$ (n = 0 to 2) [8]
*23	$C_5H_5Mo(CO)_2(\eta^3\text{-}(4\text{-}CF_3C_6H_4)_2C{\cdots}N{\cdots}C(C_6H_4CF_3\text{-}4)_2)$	V [8] m.p. 163 to 164 °C (from hexane or hexane/ether at 0 °C) [8] ^{1}H NMR (CS_2): 4.68 (C_5H_5) [8] ^{19}F NMR (toluene, −20 °C): 62.23, 62.79, 62.90, and 63.44 (CF_3); (toluene, +20 °C): 62.67 (2 CF_3), 63.01 and 63.56 (1 CF_3); (toluene, +60 °C): 62.95 (CF_3) [8]

References on pp. 201/2

Table 9 (continued)

No.	compound	method of preparation (yield) properties and remarks
		IR (KBr): 1883, 1962 and 1857, 1954 (ν(CO)); two isomers [8, 10] ESCA (±0.3 eV): 229.3 (Mo $3d_{5/2}$), 286.1 (C 1s), 293.8 (CF_3 C 1s), 401.0 (N 1s), 690.8 (F 1s) eV [10] mass spectrum (70 eV, 80 to 220°C): $[M-n\ CO]^+$ (n = 0 to 2) [8] reaction with CO (210 atm) at 175°C in Nujol for 72 h yields $Mo(CO)_6$ and traces of $(C_5H_5Mo(CO)_3)_2$ [18]
24	$C_5H_5Mo(CO)_2(\eta^3$-(4-$CH_3C_6H_4)_2C$∴N∴$C(C_6H_4CH_3$-4$)_2)$ · (4-$CH_3C_6H_4)_2C$=O	side product of the preparation of No. 21 (<1%) [4, 7] purple crystals, m.p. 120 to 121°C (from ether/hexane at 0°C) [7] ^{1}H NMR (CS_2, −20°C): 2.42 (s, 2 CH_3), 2.14, 2.18, 2.27, and 2.51 (s, 1 CH_3), 4.50 (s, C_5H_5), 7.18 (m, C_6H_4) [7] IR (KBr): 1645 (ν(C=O)), 1832, 1919 (ν(CO)); (CS_2): 1844, 1938 (ν(CO)); (hexane): 1856, 1949 (ν(CO)) [7] seems to be a weak adduct that dissociates in solution or on heating into No. 21 and (4-$CH_3C_6H_4)_2C$=O [4, 7]
25	$C_5H_5Mo(CO)_2(\eta^3$-CH_2∴$C(C_6H_5)$∴NH)	VII (20%, H of N in syn-position) [33] orange crystals (from hexane/ether at −20°C) [33] ^{1}H NMR (CD_2Cl_2): 2.62 (d, anti-CH; J(H, H) = 1.32), 3.62 (d of d, syn-CH; J(H, H) = 1.32, 3.94), 4.34 (br d, NH; J(H, H) = 3.94), 5.49 (s, C_5H_5), 7.2 to 7.5 (m, C_6H_5) [33] ^{13}C NMR (CD_2Cl_2): 30.44 (CH_2), 95.02 (C_5H_5), 101.02 (**C**C_6H_5), 125.12 and 130.11 (C_6H_5), 136.33 (C_6H_5, C-1), 244.46 (CO) [33] IR (CH_2Cl_2): 1863, 1949 (ν(CO)) [33] reacts with $CF_3SO_3CH_3$ to form No. 34; rearranges slowly in solution (C_6H_6 or CH_2Cl_2) to give $C_5H_5Mo(CO)_2$=N=$C(CH_3)C_6H_5$ [33]
26	$C_5H_5Mo(CO)_2(\eta^3$-CH_2∴$C(C_6H_5)$∴NCH_3)	No. 34 was deprotonated with $LiN(C_3H_7$-i$)_2$ in THF at −78°C (70%, CH_3 of N in anti-position) [33] orange crystals (from hexane/ether at −20°C) [33] ^{1}H NMR (CD_2Cl_2): 2.67 (d of q, anti-CH; J(H, H) = 1.65, 0.55), 2.94 (d, syn-CH; J(H, H) = 1.65),

References on pp. 201/2

Table 9 (continued)

No.	compound	method of preparation (yield) properties and remarks
26 (continued)		3.08 (d, NCH_3; J(H, H) = 0.55), 5.53 (s, C_5H_5), 7.32 (s, C_6H_5) [33] ^{13}C NMR (CD_2Cl_2): 30.01 (CH_2), 45.17 (NCH_3), 95.81 (C_5H_5), 102.95 (**C**C_6H_5), 128.47, 128.96, and 129.74 (C_6H_5), 135.21 (C_6H_5, C-1), 248.62 (CO) [33] IR (CH_2Cl_2): 1851, 1937 (ν(CO)) [33]
*27	$C_5H_5Mo(CO)_2(\eta^3\text{-}CH_2\text{∸}C(C_6H_4CH_3\text{-}4)\text{∸}NH)$	VII (20%, H of N in syn-position) [33] orange crystals (from hexane/ether at −20°C) [33] ^1H NMR (CD_2Cl_2): 2.34 (s, CH_3), 2.59 (d, anti-CH; J(H, H) = 1.54), 3.61 (d of d, syn-CH; J(H, H) = 1.54, 3.50), 4.33 (br d, NH; J(H, H) = 3.50), 5.50 (s, C_5H_5), 7.0 to 7.4 (m, C_6H_4) [33] IR (CH_2Cl_2): 1863, 1949 (ν(CO)) [33] rearrangement like No. 25 to give $C_5H_5Mo(CO)_2{=}N{=}C(CH_3)C_6H_4CH_3\text{-}4$ [33]
*28	$C_5H_5Mo(CO)_2CH_2CHP(C_6H_5)W(CO)_5$-anti 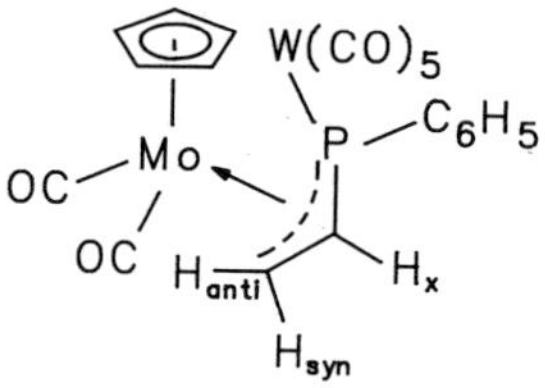	VIII (24%) [39] bright yellow crystals (from hexane/CH_2Cl_2), m.p. 223°C (dec.) [39] ^1H NMR (CD_2Cl_2): 1.54 (m, H_{anti}; J(H_{syn}, H_{anti}) = 2.3, J(H_x, H_{anti}) = 2.1, J(P, H) ≈ 12), 3.24 (m, H_{syn}; ^{3}J(P, H) = 41.4, ^{3}J(H_x, H_{syn}) = 8.9), 4.87 (m, H_x; J(P, H) ≈ 0), 5.10 (s, C_5H_5), 7.3 to 7.7 (m, C_6H_5) [39] ^{13}C NMR (CD_2Cl_2): 41.58 (d, CH_2; J(P, C) = 12.2), 62.38 (d, CH; J(P, C) = 13.7), 94.04 (C_5H_5), 128.76 to 129.71 (C_6H_5), 196.74 (d, WCO-cis; J(P, C) = 9.6) [39] ^{31}P NMR (CD_2Cl_2): −33.24 (s; J(^{183}W, P) = 246.6) [39] IR (CH_2Cl_2): 1945, 1975, 2065 (ν(CO)) [39] mass spectrum (70 eV): $[M-CO]^+$ (31), $[M-7\ CO]^+$ (100) [39]
*29	$C_5H_5Mo(CO)_2CH_2CHP(C_6H_5)W(CO)_5$-syn 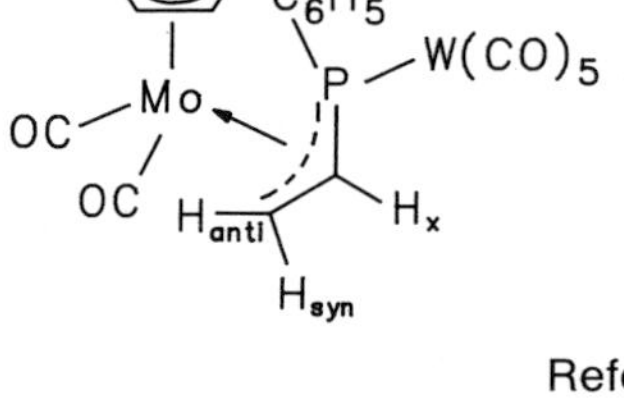	VIII (12%) [39] yellow oil [39] ^1H NMR (C_6D_6): 1.30 (m, H_{anti}; J(H_{syn}, H_{anti}) ≈ 2, J(H_x, H_{anti}) ≈ J(P, H) ≈ 11), 2.46 (m, H_{syn}; J(P, H) = 29.4, J(H_x, H_{syn}) = 8.3), 3.79 (m, H_x; J(P, H) = 30.7), 4.65 (s, C_5H_5), 6.7 to 7.5 (m, C_6H_5) [39]

References on pp. 201/2

Table 9 (continued)

No.	compound	method of preparation (yield) properties and remarks
		^{13}C NMR (C_6D_6): 43.76 (d, CH_2; J(P, C) = 6.3), 59.29 (d, CH; J(P, C) = 9.5), 92.92 (C_5H_5), 128.30 to 132.01 (C_6H_5), 197.44 (d, WCO-cis; J(P, C) = 7.5), 229.52 (MoCO) [39] ^{31}P NMR (C_6D_6): −14.12 (s; J(^{183}W, P) = 239) [39] IR (decalin): 1905, 1930, 1945, 1970, 1980, 2070 (ν(CO)) [39]
30	$C_5H_5Mo(CO)_2(\eta^3\text{-}CH_2\cdots C(CH_3)\cdots O)$	IX [32, 36, 37] ^1H NMR (C_6D_6): 1.91 (s, CH_3), 2.74 (s, CH_2, 1H), 3.29 (s, CH_2, 1H), 4.90 (s, C_5H_5) [36] reacts with $P(C_6H_5)_3$ with formation of $C_5H_5Mo(CO)_2(P(C_6H_5)_3)CH_2C(O)CH_3$ in high yields [32]
31	$C_5H_5Mo(CO)_2(\eta^3\text{-}CH_2\cdots C(OC_2H_5)\cdots O)$	IX [32, 36, 37] ^1H NMR (C_6D_6): 0.87 (t, CH_3; J(H, H) = 7), 2.67 (d, 1H; J(H, H) = 4.5), 3.31 (d, 1H; J(H, H) = 4.5), 3.58 (d of q, 1H; J(H, H) = 7.3), 3.86 (d of q, 1H; J(H, H) = 7.3), 4.89 (s, C_5H_5) [36] reacts with $P(C_6H_5)_3$ with formation of $C_5H_5Mo(CO)_2(P(C_6H_5)_3)CH_2C(O)OC_2H_5$ in high yields [32]
32	$[C_5H_5Mo(CO)_2(\eta^3\text{-}(CH_3)_2C\cdots CH\cdots N(CH_3)_2)]PF_6$	VI (47%) [11, 13] orange crystals (from CH_2Cl_2/pentane), decompose at 64°C [13] ^1H NMR (acetone-d_6): 1.98 and 2.05 (CH_3C), 2.97 and 3.24 (NCH_3), 5.89 (C_5H_5) [13] IR (CH_2Cl_2): 1915, 1998 (ν(CO)) [13] can be deprotonated with concentrated aqueous NH_3 to give No. 15 [11, 13]
33	$[C_5H_5Mo(CO)_2(\eta^3\text{-}(CH_3)_2C(\cdots CH\cdots N(CH_2)_5\text{-cyclo})]PF_6$	VI (22%) [11, 13] orange solid (from CH_2Cl_2/pentane), m.p. 115 to 118°C (dec.) [13] ^1H NMR (acetone-d_6): 1.99 and 2.09 (CH_3C), 1.7 (br, CH_2), 3.3 (br, CH_2), 5.89 (C_5H_5) [13] IR (CH_2Cl_2): 1911, 1990 (ν(CO)) [13] can be deprotonated with concentrated aqueous NH_3 to give No. 16 [11, 13]
34	$[C_5H_5Mo(CO)_2(\eta^3\text{-}CH_2\cdots C(C_6H_5)\cdots NHCH_3)]O_3SCF_3$	$C_5H_5Mo(CO)_2CH_2C(C_6H_5)NH$ (No. 25) was treated with $CF_3SO_3CH_3$ [33]

Table 9 (continued)

No.	compound	method of preparation (yield) properties and remarks
34 (continued)		reaction with $LiN(C_3H_7\text{-}i)_2$ in THF at −78°C affords No. 26 [33]
compounds with Mo in a metallacycle		
35	$C_5H_5Mo(CO)_2C(CN){=}C(CN)CNH_2$ 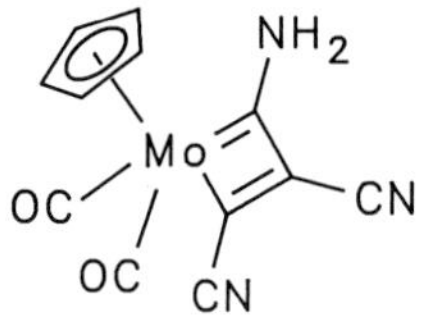	IV, R = R′ = C_2H_5, X = CN, for 17 h in CH_2Cl_2 (25% with water-saturated CH_2Cl_2 only 9%) [14] orange solid, decomposes > 145°C [14] ^{1}H NMR (acetone-d_6): 5.86 (s, C_5H_5), 8.76 and 9.18 (br s, NH_2, $H_{1/2}$ = 20) [14] ^{13}C NMR (acetone-d_6): 65.2 (s, C=), 95.0 (d, C_5H_5), 116.3 and 120.2 (s, CN), 188.7, 213.6 (s, C=), 224.3 (CO) [14] IR (CH_2Cl_2): 1603, 1653 (ν(C=C)), 1966, 2032 (ν(C=C)), 2190, 2209 (ν(CN)); (KBr): 3116 (ν(CH)), 3195, 3215, 3311, 3414 (ν(NH_2)) [14] treatment with D_2O in THF gives the corresponding ND_2 compound (No. 36) [14]
36	$C_5H_5Mo(CO)_2C(CN){=}C(CN)CND_2$	No. 35 was treated four times with D_2O in THF; ca. 85 to 95% deuterium exchange [14] orange solid (from CH_2Cl_2/hexane) [14] IR (CH_2Cl_2): 1635 (ν(C=C)), 1966, 2030 (ν(CO)), 2193, 2209 (ν(CN)); (KBr): 2395, 2434, 2564 (ν(ND_2)), 3110 (ν(CH)) [14]
*37	$C_5H_5Mo(CO)_2{=}C(CF_3)C(CF_3)HCO$ 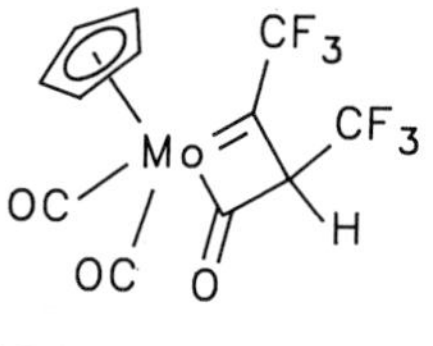	yellow crystals [22] ^{1}H NMR ($CDCl_3$): 3.70 (q, CH; J(F, H) = 7.4), 5.38 (s, C_5H_5) [22] ^{19}F NMR ($CDCl_3$): 56.6 (d, $CHCF_3$; J(F, H) = 7.4), 62.2 (s, CF_3) [22] IR (KBr): 1425, 1440 (ν(C=C), C_5H_5); (CCl_4): 1800 (ν(C=O)), 1995, 2045 (ν(CO)) [22]
38	$Na[C_5H_5Mo(CO)_2C(O)OCH_2CH_2] \cdot CH_3OC_2H_4OCH_3$ 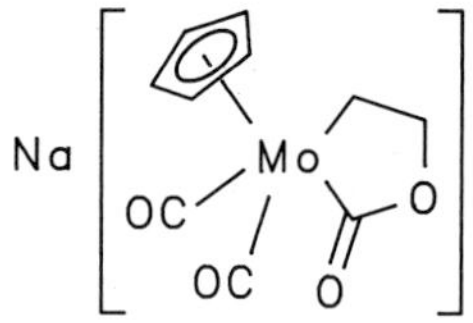	$Na[C_5H_5Mo(CO)_3]$ and ethylene oxide were stirred in $CH_3CN/CH_3OC_2H_4OCH_3$ at 0°C overnight (35%); $C_5H_5Mo(CO)_3CH_2CH_2OH$ was treated with NaH in $CH_3OC_2H_4OCH_3$ (50%) [15] ^{1}H NMR (CD_3CN): 1.0 to 4.70 (m, CH_2), 3.30 and 3.48 ($CH_3OC_2H_4OCH_3$), 5.10 (C_5H_5) [15] IR (Nujol): 1560 (ν(C=O)), 1800, 1900 (ν(CO)) [15] reaction with aqueous HBF_4 in THF affords $[C_5H_5Mo(CO)_3C_2H_4]BF_4$; treatment with $[C_5H_5M(CO)_nC_2H_4]PF_6$ in CH_3CN gives $C_5H_5Mo(CO)_3C_2H_4OC_2H_4M(C_5H_5)(CO)_n$ (M = W, n = 3; M = Fe, n = 2) [15]

References on pp. 201/2

Table 9 (continued)

No.	compound	method of preparation (yield) properties and remarks
39	$C_5H_5Mo(CO)_2(CHNCH_3)_2BH_2$ 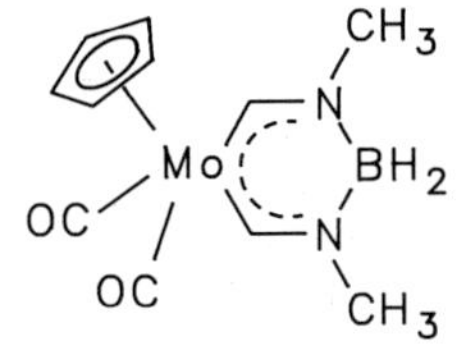	$[C_5H_5Mo(CO)_2(CNCH_3)_2]PF_6$ was allowed to react with 3 equivalents of $NaBH_4$ in THF at 0°C for 24 h (19%) [6] yellow solid, m.p. 132 to 134°C (dec.) [6] ^1H NMR ($CDCl_3$): 3.40 (CH_3), 5.6 (C_5H_5), 11.3 (CH) [6] IR ($CHCl_3$): given from 470 to 3250; 1565 (ν(C=N)), 1865, 1955 (ν(CO)) [6]
40	$C_5H_5Mo(CO)_2(\eta^2\text{-}(NC)_2C{=}C(CN)_2)C({=}NCH_3)CH_3$	X (89%) [16, 20] orange solid (from THF/pentane), m.p. 160 to 170°C (dec.) [20] ^1H NMR (acetone-d_6): 2.68 (CH_3), 3.86 (CH_3), 5.89 (C_5H_5) [20] IR (CH_3CN): 1630 (ν(CN)), 1885, 1970 (ν(CO)), 2215 (ν(CN)) [20]; partly in [16]
41	$C_5H_5Mo(CO)_2(\eta^2\text{-}(NC)_2C{=}C(CN)_2)C({=}NC_6H_5)CH_3$	X (30%) [20] orange solid (from THF/C_6H_6/pentane), m.p. 165 to 170°C (dec.) [20] ^1H NMR (acetone-d_6): 2.45 (CH_3), 5.52 (C_5H_5), 7.0 to 7.6 (m, C_6H_5) [20] IR (acetone): 1615 (ν(C=N)), 1885, 1970 (ν(CO)), 2205 (ν(CN)) [20] undergoes substantial decomposition during chromatography on Al_2O_3 [20]

*Further information:

$C_5H_5Mo(CO)_2C(CN)(P(C_6H_5)_3)CHCN$ (Table **9**, No. **2**) crystallizes in the monoclinic space group $P2_1/n-C^5_{2h}$ (No. 14) with the unit cell parameters a = 13.691 (4), b = 15.892 (4), c = 11.557 (5) Å, β = 100.96 (3)°; Z = 4 molecules per unit cell, and D_{calc} = 1.497 g/cm³. The molecular structure with the main bond distances and angles is shown in **Fig. 37**, p. 198. The C-P bond length indicates the presence of the ylide $=C^--P^+(C_6H_5)_3$ and not the ylene $=C{=}P(C_6H_5)_3$ resonance form [27].

$C_5H_5Mo(CO)_2(C_2H_4)H$ (Table **9**, No. **3**). The ν(CO) absorptions were measured in a matrix at 77 K. The values (in cm^{-1}) obtained are given in the following table. The calculated force and interaction constants for the CO matrix are: k = 1522.0 and k_1 = 57.0 N/m [31].

matrix	ν(CO)
polyvinyl chloride	1885, 1964 [29]
CH_4	1904.7, 1977.4 [31]
CH_4 with 5% C_2H_4	1901.3, 1974.8 [26, 30]
CO	1908.5, 1980 [31]

References on pp. 201/2

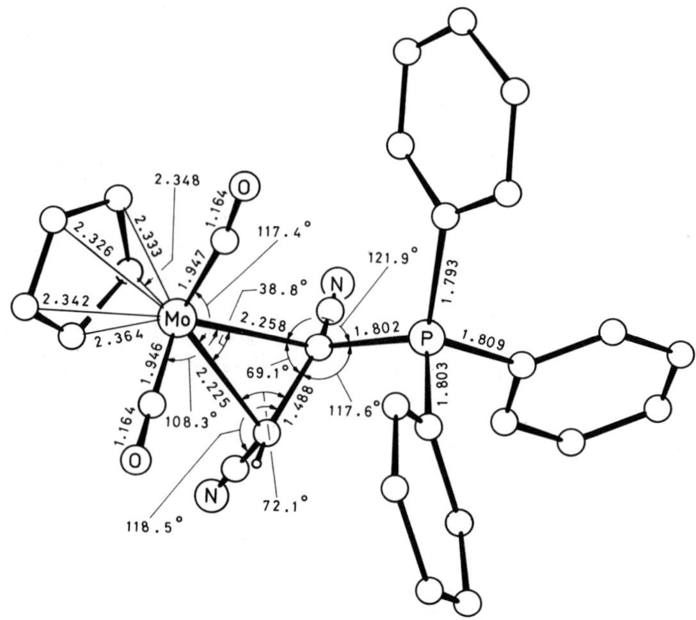

Fig. 37. The molecular structure of $C_5H_5Mo(CO)_2C(CN)(P(C_6H_5)_3)CHCN$ [27].

$C_5H_5Mo(CO)_2(CH_2{=}CHCH_3)FBF_3$ (Table **9**, No. **4**) is an extremely reactive precursor for the organometallic Lewis acid $C_5H_5Mo(CO)_2^+$. The reactions of No. 4, which were carried out in CH_2Cl_2, are compiled in the following table [35].

reactant	product(s)
CO (1 atm)	$[C_5H_5Mo(CO)_3CH_2{=}CHCH_3]BF_4$
CO (6.6 atm)	$[C_5H_5Mo(CO)_3CH_2{=}CHCH_3]BF_4$ and $[C_5H_5Mo(CO)_4]BF_4$ (1:1)
$[N(C_4H_9\text{-}n)_4]I$	$(C_5H_5Mo(CO)_2I)_2$
$P(C_6H_5)_3$	$[C_5H_5Mo(CO)_2(P(C_6H_5)_3)_2]BF_4$ and small amounts of $[C_5H_5Mo(CO)_3P(C_6H_5)_3]BF_4$
$(C_6H_5)_2PC_2H_4P(C_6H_5)_2$	$[C_5H_5Mo(CO)_2P(C_6H_5)_2C_2H_4P(C_6H_5)_2]BF_4$
$C_5H_5Mo(CO)_3CH_3$	$(C_5H_5Mo(CO)_2)_2\mu\text{-}C(O)CH_3$
$C_5H_5Mo(CO)_3H$	traces of $[(C_5H_5Mo(CO)_3)_2H]BF_4$

$(CH_3)_5C_5Mo(CO)_2C(C_6H_5){=}C(C_6H_5)CHC(O)O$ (Table **9**, No. **6**) crystallizes in the monoclinic space group $P2_1/n-C_{2h}^5$ (No. 14) with the unit cell parameters a = 16.638 (5), b = 11.526 (3), c = 15.892 (5) Å, β = 114.54°; Z = 4 molecules per unit cell, and D_{calc} = 1.455 g/cm³. The molecular structure with the main bond distances and angles is shown in **Fig. 38** [34].

$C_5H_5Mo(CO)_2C(R)(R')NC(R)R'$ (Table **9**, Nos. **20, 22, 23**). Variable-temperature ^1H NMR studies showed the molecule to be fluxional. At low temperatures certain orientations of the C(R)R′ groups (R = C_6H_5, R′ = $C_6H_4CH_3$-4; R = $C_6H_4CH_3$-4, R′ = $C_6H_4OCH_3$-4) with respect to the rest of the ligand are preferred (compare Formulas Va to d). Kinetic measurements taken from the ^{19}F NMR spectra of $C_5H_5Mo(CO)_2C(C_6H_4CF_3\text{-}4)_2NC(C_6H_4CF_3\text{-}4)_2$ gave the values 11.6 ± 1.2 kcal/mol for E_a and 10.8 ± 1.0 for log A [8]. Compare also "Further information" on No. 21.

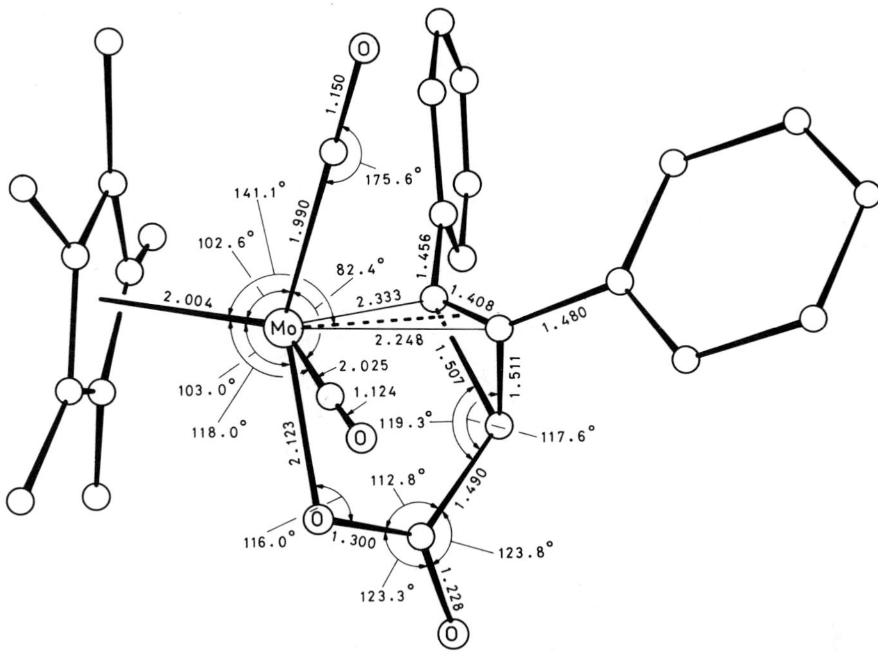

Fig. 38. The molecular structure of $(CH_3)_5C_5Mo(CO)_2C(C_6H_5)$=$C(C_6H_5)CHC(O)O$ [34].

a b

c d

V

$C_5H_5Mo(CO)_2C(C_6H_4CH_3\text{-}4)_2NC(C_6H_4CH_3\text{-}4)_2$ (Table **9**, No. **21**). ^1H NMR studies in the temperature range −20 to +70°C showed the bonding of the aza-allyl to the metal to be of the σ-π type (Formula V; R = R′ = $C_6H_4CH_3$-4). At −20°C, four methyl proton resonances of equal intensity were observed. At +10°C, two signals were observed with an intensity ratio 1:3. At 70°C, these resonances coalesced to a broader time-averaged signal [5]. Epimerization occurs at higher temperatures through a process of rotation and interchange of σ-π bonding [4, 5, 7, 8, 10].

$C_5H_5Mo(CO)_2CH_2C(C_6H_4CH_3\text{-}4)NH$ (Table **9**, No. **27**) crystallizes in the monoclinic space group C2/c − C_{2h}^6 (No. 15) with the unit cell parameters a = 14.890 (6), b = 12.975 (4), c = 15.229 (8)

References on pp. 201/2

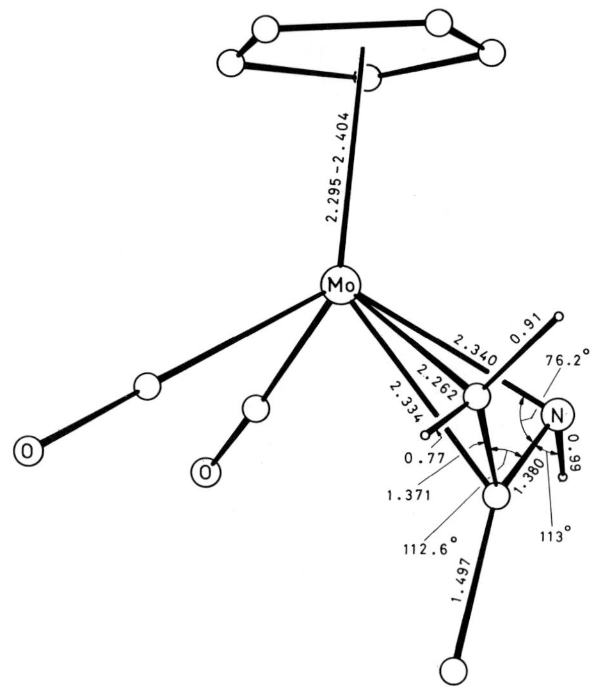

Fig. 39. The molecular structure of $C_5H_5Mo(CO)_2CH_2C(C_6H_4CH_3\text{-}4)NH$; the tolyl group is represented by its first atom [33].

Å, β = 98.73 (4)°; Z = 8 molecules per unit cell, and D_{calc} = 1.60 g/cm^3. The molecular structure with the main bond distances and angles is shown in **Fig. 39** [33].

$C_5H_5Mo(CO)_2CH_2CHP(C_6H_5)W(CO)_5$ (Table **9**, Nos. **28**, **29**). Several products were formed in the preparation (Method VIII). The products in order of elution were (solvent): $(CO)_5WP(C_6H_5)(C_2H_5)H$ (hexane), $C_5H_5Mo(CO)_2(CO\text{-}\mu)(P(C_6H_5)C_2H_5\text{-}\mu)W(CO)_4$ (hexane/CH_2Cl_2 9:1), compound No. 29 (hexane/CH_2Cl_2 8:2), $C_5H_5Mo(CO)_3P(C_6H_5)(CH{=}CH_2)W(CO)_5$ (CH_2Cl_2/hexane 8:2) followed by compound No. 28 (CH_2Cl_2/hexane 1:1) [39].

No. 28 crystallizes in the triclinic space group $P\,\overline{1}-C_i^1$ (No. 2) with the unit cell parameters a = 9.653 (1), b = 9.699 (1), c = 13.417 (2) Å, α = 100.42 (2)°, β = 102.51 (2)°, γ = 110.93 (2)°; Z = 2 molecules per unit cell, and D_{calc} = 1.86 g/cm^3. The P-C_6H_5 bond lies practically in the plane of the allyl unit, and W out of this plane, opposite to Mo. The molecular structure with the main bond distances and angles is shown in **Fig. 40** [39].

$C_5H_5Mo(CO)_2{=}C(CF_3)C(CF_3)HCO$ (Table **9**, No. **37**). $C_5H_5Mo(CO)_3H$ was heated (320 K) or irradiated (2 to 3 d) with a mixture of $CF_3C{\equiv}CCF_3$ and CH_3SSCH_3 in THF. Chromatography of the reaction product on Florisil with CH_2Cl_2 allowed the isolation of the cyclic C,S-bonded compound $C_5H_5Mo(CO)_2C(CF_3){=}C(CF_3)C(O)SCH_3$ (10%), the title complex mixed with the noncyclic isomer $C_5H_5Mo(CO)_3C(CF_3){=}C(CF_3)H$ (15%, the compounds could not be separated), and $C_5H_5Mo(CO)_3C(CF_3){=}C(CF_3)SCH_3$ (10%). The same reaction in the absence of CH_3SSCH_3 gave

References on pp. 201/2

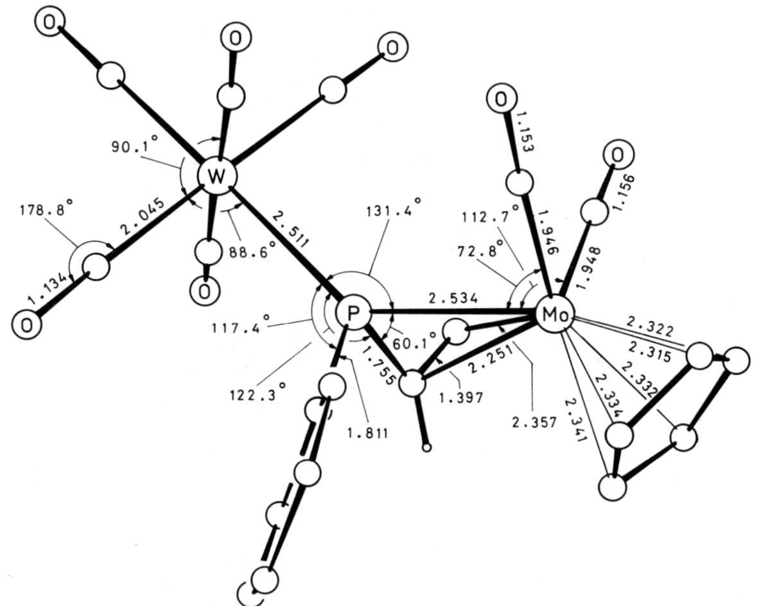

Fig. 40. The molecular structure of $C_5H_5Mo(CO)_2CH_2CHP(C_6H_5)W(CO)_5$-anti [39].

mainly $C_5H_5Mo(CO)_2(\mu\text{-}CF_3C{\equiv}CCF_3)(CO)_2MoC_5H_5$ and No. 37 (<10%) after chromatography [22].

No. 37 and its isomer form an equilibrium between 293 and 333 K; the title complex predominates at low temperature [22].

References:

[1] Famery, K.; Kilner, M. (J. Organometal. Chem. **16** [1969] P 51/P 52).
[2] Famery, K.; Kilner, M.; Midcalf, C.; Payling, C. A. (4th Intern. Conf. Organometal. Chem., Bristol, Engl., 1969, p. 55).
[3] Famery, K.; Kilner, M.; Midcalf, C. (J. Chem. Soc. A **1970** 2279/85).
[4] Keable, H. R.; Kilner, M. (5th Intern. Conf. Organometal. Chem., Moscow 1971, pp. 545/6, Abstr. 208).
[5] Keable, H. R.; Kilner, M. (J. Chem. Soc. Chem. Commun. **1971** 349/50).
[6] Treichel, P. M.; Stenson, J. P.; Benedict, J. J. (Inorg. Chem. **10** [1971] 1183/7).
[7] Keable, H. R.; Kilner, M. (J. Chem. Soc. Dalton Trans. **1972** 153/8).
[8] Keable, H. R.; Kilner, M. (J. Chem. Soc. Dalton Trans. **1972** 1535/40).
[9] Wainwright, K. P.; Wild, S. B. (J. Chem. Soc. Chem. Commun. **1972** 571/2).
[10] Briggs, D.; Clark, D. T.; Keable, H. R.; Kilner, M. (J. Chem. Soc. Dalton Trans. **1973** 2143/7).

[11] Hodges, K. C. (Diss. Univ. Georgia 1974; Diss. Abstr. Intern. B **35** [1975] 4822).
[12] Keable, H. R.; Kilner, M.; Robertson, E. E. (J. Chem. Soc. Dalton Trans. **1974** 639/44).
[13] King, R. B.; Hodges, K. C. (J. Am. Chem. Soc. **97** [1975] 2702/12).

[14] King, R. B.; Saran, M. S. (Inorg. Chem. **14** [1975] 1018/26).
[15] Knoth, W. H. (Inorg. Chem. **14** [1975] 1566/72).
[16] Adams, R. D.; Chodosh, D. F. (J. Am. Chem. Soc. **98** [1976] 5391/3).
[17] Beck, W.; Brix, H.; Köhler, F. H. (J. Organometal. Chem. **121** [1976] 211/23).
[18] Inglis, T.; Kilner, M. (J. Chem. Soc. Dalton Trans. **1976** 562/4).
[19] Mickiewicz, M.; Wainwright, K. P.; Wild, S. B. (J. Chem. Soc. Dalton Trans. **1976** 262/9).
[20] Adams, R. D.; Chodosh, D. F. (J. Am. Chem. Soc. **99** [1977] 6544/50).

[21] Schilling, B. E. R.; Hoffmann, R.; Lichtenberger, D. L. (J. Am. Chem. Soc. **101** [1979] 585/91).
[22] Petillon, F. Y.; Le Floc-Perennou, F.; Guerchais, J. E.; Sharp, D. W. A.; Manojlovic-Muir, L.; Muir, K. W. (J. Organometal. Chem. **202** [1980] 23/37).
[23] Brix, H.; Beck, W. (J. Organometal. Chem. **234** [1982] 151/74).
[24] Kazkaukas, R. J.; Wrighton, M. S. (J. Am. Chem. Soc. **104** [1982] 6005/15).
[25] Kegley, S. E.; Brookhart, M.; Husk, G. R. (Organometallics **1** [1982] 760/2).
[26] Alt, H.; Mahmoud, K. A.; Rest, A. J. (J. Organometal. Chem. **243** [1983] C 5/C 9).
[27] Scordia, H.; Kergoat, R.; Kubicki, M. M.; Guerchais, J. E.; Haridon, P. L. (Organometallics **2** [1983] 1681/7).
[28] Allen, S. R.; Barnes, S. G.; Green, M.; Moran, G.; Trollope, L.; Murall, N. W.; Welch, A. J.; Sharaiha, D. M. (J. Chem. Soc. Dalton Trans. **1984** 1157/69).
[29] Hooker, R. H.; Rest, A. J. (J. Chem. Soc. Dalton Trans. **1984** 761/70).
[30] Mahmoud, K. A.; Rest, A. J.; Alt, H. G. (J. Chem. Soc. Dalton Trans. **1984** 187/97).

[31] Mahmoud, K. A.; Rest, A. J.; Alt, H. G.; Eichner, M. E.; Jansen, B. M. (J. Chem. Soc. Dalton Trans. **1984** 175/86).
[32] Doney, J.-J.; Bergman, R. G.; Heathcock, C. H. (J. Am. Chem. Soc. **107** [1985] 3724/6).
[33] Green, M.; Mercer, R. J.; Morton, C. E.; Orpen, A. G. (Angew. Chem. **97** [1985] 422/3).
[34] Hughes, R. P.; Reisch, J. W.; Rheingold, A. L. (Organometallics **4** [1985] 241/4).
[35] Markham, J.; Menard, K.; Cutler, A. (Inorg. Chem. **24** [1985] 1581/7).
[36] Burkhardt, E. R.; Doney, J. J.; Bergman, R. G.; Heathcock, C. H. (J. Am. Chem. Soc. **109** [1987] 2022/39).
[37] Burckhardt, E. R.; Doney, J. J.; Stack, J. G.; Heathcock, C. H.; Bergman, R. G. (J. Mol. Catal. **41** [1987] 41/57).
[38] Green, M.; Nagle, K. R.; Woolhouse, M.; Williams, D. J. (J. Chem. Soc. Chem. Commun. **1987** 1793/5).
[39] Hugel-Le Goff, C.; Mercier, F.; Ricard, L.; Mathey, F. (J. Organometal. Chem. **363** [1989] 325/33).
[40] Winter, M. J.; Woodward, S. (J. Chem. Soc. Chem. Commun. **1989** 457/8).

1.5.1.3.2.3 Compounds with Additional ^3L Ligands

1.5.1.3.2.3.1 Compounds of the Type $^5LMo(CO)_2{}^1L\text{-}{}^2L$

This section deals with compounds of the general type $^5LMo(CO)_2{}^3L$ in which the ^3L ligand is bonded σ-C(O)-CR=CR$'_2$-π (Formula I). With the exception of No. 7, all of the compounds were prepared as follows. $C_5H_5Mo(CO)_3H$ was allowed to react with 1 to 2 equivalents of an alkyne in THF or ether at −78°C. The reaction mixture was allowed to warm up. The complexes were purified by either concentration, addition of pentane, and crystallization at −78°C or washing of the resulting residue with pentane. The yields were nearly quantitative [1, 2].

References on p. 205

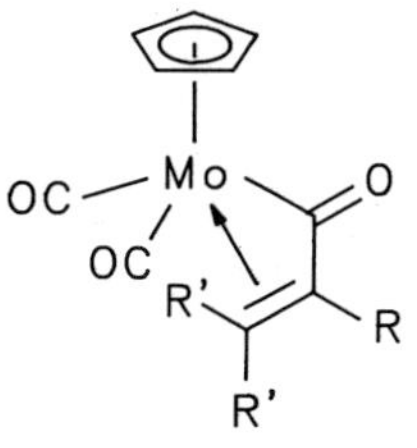

I

Table 10
Compounds of the Type $^5LMo(CO)_2{}^1L\text{-}^2L$.
For explanations, abbreviations, and units see p. X.

No.	compound	remarks on preparation properties and remarks
1	$C_5H_5Mo(CO)_2C(O)C(N(CH_3)_2)$=$CH_2$	with HC≡$CN(CH_3)_2$ in ether [1] red crystals, m.p. 86 to 88°C (from THF/pentane) [1] IR (ether): 1495 (ν(CC or NC)), 1695 (ν(C=O)), 1893, 1965 (ν(CO)) [1] mass spectrum: $[M-n\ CO]^+$ (n = 0 to 2) [1]
2	$C_5H_5Mo(CO)_2C(O)C(N(C_2H_5)_2)$=$CH_2$	with 2 equivalents of HC≡$CN(C_2H_5)$ in THF [1] orange-red crystals, m.p. 81 to 82°C (from pentane at low temperatures) [1] ^{1}H NMR ($CDCl_3$): 1.13 (t, $\mathbf{CH_3}CH_2$; J(H, H) = 7), 3.07 (q, NCH_2), 5.25 (s, C_5H_5); (C_6D_6): 0.84 (t, $\mathbf{CH_3}CH_2$; J(H, H) = 7), 2.77 (q, $\mathbf{CH_2}CH_3$), 4.73 (s, C_5H_5); the CH_2= signal is covered in both spectra by the NCH_2 resonance [1] IR (CH_2Cl_2): 1500 (ν(CN or CC)), 1672 (ν(C=O)), 1882, 1960 (ν(CO)) [1] mass spectrum: $[M-n\ CO]^+$ (n = 0 to 3) [1]
3	$C_5H_5Mo(CO)_2C(O)C(N(C_2H_4)_2O)$=$CH_2$	with HC≡$CN(C_2H_4)_2O$ in THF [2] orange-yellow solid, m.p. 133 to 135°C [2] ^{1}H NMR ($CDCl_3$): 2.83 (t, CH_2; J(H, H) = 5), 2.85 (CH=), 5.27 (s, C_5H_5) [2] IR (CH_2Cl_2): 1676 (ν(C=O)), 1898, 1976 (ν(CO)) [2] mass spectrum: $[M]^+$ [2] protonation with $HBF_4 \cdot O(C_2H_5)_2$ in CH_2Cl_2 at −78°C affords $[C_5H_5Mo(CO)_3$=$C(CH_3)N$-$CH_2CH_2O]BF_4$ [2]
4	$C_5H_5Mo(CO)_2C(O)C(N(C_2H_5)_2)$=$CHCH_3$	with CH_3C≡$CN(C_2H_5)_2$ [1] red crystals, m.p. 35°C (dec.) [1] ^{1}H NMR (C_6D_6): 0.95 (t, $\mathbf{CH_3}CH_2$; J(H, H) = 7), 1.79 (d, CH_3C=; J(H, H) = 6), 2.80 (q, $\mathbf{CH_2}CH_3$), 3.25 (q, HC=; J(H, H) = 6), 4.72 (s, C_5H_5) [1]

References on p. 205

Table 10 (continued)

No.	compound	remarks on preparation properties and remarks
4 (continued)		IR (KBr): 1501 (ν(CN or CC)); (CH_2Cl_2): 1663 (ν(C=O)), 1875, 1957 (ν(CO)) [1] mass spectrum: $[M-n\,CO]^+$ [1] protonation with $HBF_4 \cdot O(C_2H_5)_2$ in THF affords $[C_5H_5Mo(CO)_3C(N(C_2H_5)_2H){=}CHCH_3]BF_4$ [2]
5	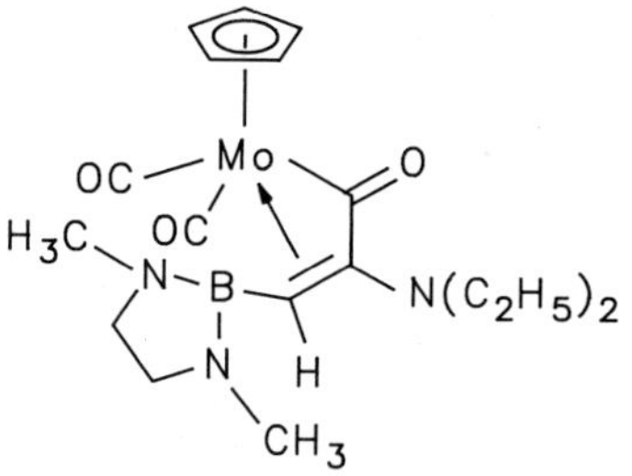 [structure: $C_5H_5Mo(CO)_2$ chelate with C(O), C=C(H) bearing $N(C_2H_5)_2$ and B-bonded 1,3-dimethyl-1,3,2-diazaborolidine; labels: OC, Mo, O, H_3C, OC, N, B, N, H, $N(C_2H_5)_2$, CH_3]	with $(C_2H_5)_2NC{\equiv}CB(N(CH_3)CH_2)_2$ in THF [2] red solid, m.p. 68 to 70°C [2] ^{1}H NMR (C_6D_6): 0.91 (t, **CH_3**CH_2; J(H, H) = 7), 2.5 to 3.1 (m, C_2H_4 and NCH_3), 4.95 (s, C_5H_5) [2] IR (KBr): 1501 (ν(CC)), 1687 (ν(C=O)); (ether): 1889, 1960 (ν(CO)) [2] mass spectrum: $[C_5H_5Mo(CO)_3H]^+$ [2]
6	$C_5H_5Mo(CO)_2C(O)C(N(C_2H_5)_2){=}CHC(O)OC_2H_5$	with $(C_2H_5)_2C{\equiv}CC(O)OC_2H_5$ in ether [1, 2] red crystals, m.p. 66°C (at low temperatures from ether) [1] ^{1}H NMR ($CDCl_3$): 1.14 (t, **CH_3**CH_2N; J(H, H) = 7), 1.25 (t, **CH_3**CH_2O; J(H, H) = 7), 3.08 (q, CH_2N), 4.03 (q, CH_2O), 4.40 (s, =CH), 5.12 (s, C_5H_5) [1] IR (CH_2Cl_2): 1495 (ν(CN or CC)), 1680 (ν(C=O)), 1695 (ν(CO_2)), 1900, 1974 (ν(CO)) [1] mass spectrum: $[M-n\,CO]^+$ (n = 0 to 2) [1] protonation with $HBF_4 \cdot O(C_2H_5)_2$ in THF affords $[C_5H_5Mo(CO)_3{=}C(N(C_2H_5)_2)CH_2C(O)O$-$C_2H_5]BF_4$; decarbonylates at room temperature after some days in CH_2Cl_2 with quantitative formation of $C_5H_5Mo(CO)_2(\eta^2$-$RCH{=}C{=}N(C_2H_5)_2)$ (R = $C(O)OC_2H_5$; see Table 9, No. 14) [2]
7	$C_5H_5Mo(CO)_2C(O)OCH_2CH{=}CH_2$	$Na[C_5H_5Mo(CO)_3]$ was allowed to react with 5 equivalents of epibromohydrin in THF for 12 h (74% by chromatography on Al_2O_3 with hexane/CH_2Cl_2 2:3) [3] orange solid, m.p. 95°C (darkens at 112 to 113°C) [3] ^{1}H NMR (C_6D_6): 2.14 (d, OCH_2, 1H; J(H, H) = 3.8), 2.36 (d, OCH_2, 1H; J(H, H) = 6.5), 3.89 (m, $HC{=}CH_2$, 2H), 4.41 (d, CH_2=, 1H; J(H, H) = 5.9), 4.84 (s, C_5H_5) [3]

References on p. 205

Table 10 (continued)

No.	compound	remarks on preparation properties and remarks
		^{13}C NMR (C_6D_6): 31.8 (CH_2), 65.7 (CH_2=), 79.3 (CH=), 93.1 (C_5H_5), 215.3 (CO_2), 228.4, 235.2 (CO) [3] IR (CH_2Cl_2): 1646 ($\nu(CO_2)$), 1949, 2012 (ν(CO)) [3] reaction with $P(C_6H_5)_3$ under thermal or photochemical conditions affords $(C_5H_5)_2Mo_2(CO)_5P(C_6H_5)_3$ [3]

References:

[1] Beck, W.; Brix, H.; Köhler, F. H. (J. Organometal. Chem. **121** [1976] 211/23).
[2] Brix, H.; Beck, W. (J. Organometal. Chem. **234** [1982] 151/74).
[3] Pannell, K. H.; Cea-Olivares, R.; Toscano, R. A.; Kapoor, R. N. (Organometallics **6** [1987] 1821/2).

1.5.1.3.2.3.2 Compounds with One Open Allylic ^3L Ligand

All of the compounds summarized in this chapter contain a ^3L ligand bonded in the allylic fashion, the three carbon atoms are not all part of the cyclic system (for cyclic ^3L complexes see Chapter 1.5.3.2.3.3). Also included in this chapter, at the end of the table, is one complex (No. 103) with an oxabutadiene system. The complexes have been arranged in order of increasing complexity of the ^3L ligand. Complexes with ^5L ligands others than cyclopentadienyl but having the same ^3L group are described in order of increasing substitution on the ^5L ligand; C_9H_7 means indenyl. Most of the compounds were prepared by one or more of the following methods. Further information on preparation is given in the table.

Method I: Rearrangement of the σ-coordinated allyl group in a $^5LMo(CO)_3{}^1L$ complex to a η^3-coordinated allyl group in $^5LMo(CO)_2{}^3L$, with the loss of one CO. The $^5LMo(CO)_3{}^1L$ complex was obtained in many cases by the reaction of $Na[^5LMo(CO)_3]$ with the appropriate organic halide.
a. $^5LMo(CO)_3{}^1L$ was irradiated.
b. $^5LMo(CO)_3{}^1L$ was heated.
c. $^5LMo(CO)_3{}^1L$ was treated with $(CH_3)_3NO$ in THF for 4 h.

Method II: $^3LMo(CO)_2(NCCH_3)_2X$ was allowed to react with a ^5L source.
a. With MC_5H_5 or $MC_5(CH_3)_5$ (M = Li, Na) in THF for several hours.
b. With $(CH_3)_3SnC_5H_5$ in boiling THF for several hours.

Method III: $C_5H_5Mo(CO)_3CH_2C{\equiv}CR$ was allowed to react with R′EH (E = O, R′ = CH_3, $CH_2C_6H_5$, C_6H_5, t-C_4H_9, or $CH_2C{\equiv}CH$; E = S, R′ = CH_3, $CH_2C_6H_5$, or C_6H_5) in THF to yield $C_5H_5Mo(CO)_2CH_2C(C(O)XR')CHR$ [30, 34]. The presence of free $BrCH_2$-$C{\equiv}CR$ was necessary to suppress for the formation of the carbene complex (Formula I) [30].

References on pp. 248/50

Method IV: $Na[C_5H_5Mo(CO)_3]$ was allowed to react with $BrCH_2CH_2C(R){=}C{=}C(R')R''$ in THF at 65°C for 3 to 10 h. The complexes were purified by column chromatography on alumina with C_6H_6/pentane and recrystallization from CH_2Cl_2/hexane [15, 19, 38].

Method V: $[C_5H_5Mo(CO)_2{}^4L]BF_4$ (^4L = buta-1,3-diene derivative) was allowed to react with $Na[M(CO)_5]$ (M = Mo, Re) in THF at −78°C. After removal of the solvent the residue was extracted with CH_2Cl_2 [67, 70].

Method VI: $[(CH_3)_5C_5Mo(CO)_2(NCCH_3)_2]BF_4$ and cis/trans-$(CH_3)_3SiOCH{=}CHCH{=}CH_2$ were stirred in CH_2Cl_2 for 2 d. The resulting mixture of exo and endo isomers of $(CH_3)_5C_5Mo(CO)_2C_3H_4C(O)H$-syn, after column chromatography on alumina, was allowed to react with $CH_2{=}P(C_6H_5)_3$ in THF at −78°C to afford a mixture of exo and endo η^3-pentadienyl complexes [71].

Method VII: $[{}^5LMo(CO)_2C(CH_2)_n(CR_2)_{3-n}]BF_4$ (see Formula II) was allowed to react with nucleophiles. The products were isolated by column chromatography on alumina with hexane, pentane, or ether [48].
a. With an excess $Na[BH_4]$ in THF for 3 h [48].
b. Two equivalents of $LiCu(CH_3)_2$ in ether were added to a solution of the complex in ether cooled at −78°C and the mixture was stirred for 1 h at ambient temperature [48].
c. The complex was placed on Al_2O_3 and eluted with CH_2Cl_2 [48].
d. $NaSC_6H_5$ in THF was added dropwise to a suspension of the complex in THF cooled at −30°C and the mixture was allowed to warm up [48].

Method VIII: $[{}^5LMo(CO)_2{}^4L]BF_4$ (^4L = buta-1,3-diene derivative) was allowed to react with excess base in CH_2Cl_2 or $Na[BH_4]$ in THF [27, 28]. The products were purified by column chromatography on alumina with ether [27].

OC Mo O OR' OC R — I

OC Mo OC — II

H C H C OC Mo C OC CO C R C R' BF4 — III

Method IX: Reactions of $[C_5H_5Mo(CO)_3CH_2{=}C{=}C{=}C(R)R']BF_4$ (Formula III) with nucleophiles.
a. With amines in ether [54].
b. With water [54].
c. With $NaOCH_3$ in methanol for 15 min [54].

Method X: $Li[C_5H_5Mo(CO)_2C(O)HCH_3]$ and $BrCH_2C_6H_4R$ were stirred in THF for 4 h [49, 53, 60, 66]. The product was separated by column chromatography on alumina with petroleum ether/CH_2Cl_2 (7:1) [53].

Method XI: $C_5H_5Mo(CO)_3CH_2C{\equiv}C(CH_2)_nC(R)(R')OH$ (n = 0, 1) was isomerized in CH_2Cl_2 at 50°C or by column chromatography on alumina with pentane [32, 51].

Not included in Table 11 are the complexes obtained by the irradiation of $C_5H_5Mo(CO)_3H$ in the presence of dienes. The observed changes in the IR spectra (no data given) have been interpreted to be the result of the formation of $C_5H_5Mo(CO)_2{}^3L$ compounds (^3L derived from trans-penta-1,3-diene, penta-1,4-diene, trans,trans-hexa-2,4-diene, or 1,2,3,4,5-pentamethylcyclopentadiene) [37].

For complexes No. 87 to 90, an incorrect structure with one σ,π-coordinated $C(O)(CH_2)_2C(R)=C=C(R')R''$ ligand was proposed in [15] and corrected, together with a discussion of the possible mechanism of formation, see [19].

The allyl fragment can be coordinated either endo or exo (Formula IV). The isomer ratio depends on the solvent, temperature, and the substituents on the allyl fragment. The complexes obtained by Method III show only averaged ^1H NMR spectra for the endo and exo isomers at ambient temperature but four ν(CO) absorptions in the IR spectra, indicating the presence of the two isomers [30, 34]. If not otherwise stated, all compounds in Table 11 are mixtures of both isomers.

If a single group R is in 1-position (Nos. 10 to 31), four possible isomers can be present, due to exo and endo arrangement and syn or anti position of the group (Formula IV); for the isomer distribution see the individual compounds.

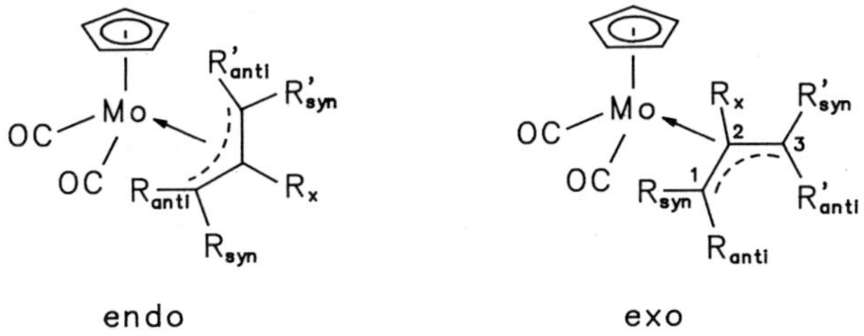

endo exo

IV

The complexes were described as slightly soluble in petroleum ether but well soluble in ether, benzene, CS_2, and the common polar organic solvents. The compounds may be stored for a long time under N_2 at lower temperatures [23].

Many complexes of the type $^5LMo(CO)_2{}^3L$ react with [NO]X (X = PF_6, BF_4) in solution with formation of $[^5LMo(CO)(NO)^3L]X$ [25, 47, 51, 68]. Hydride abstraction with $[C(C_6H_5)_3]BF_4$ leads to the corresponding $[^5LMo(CO)_2{}^4L]BF_4$ (^4L = buta-1,3-diene derivative) complexes [28].

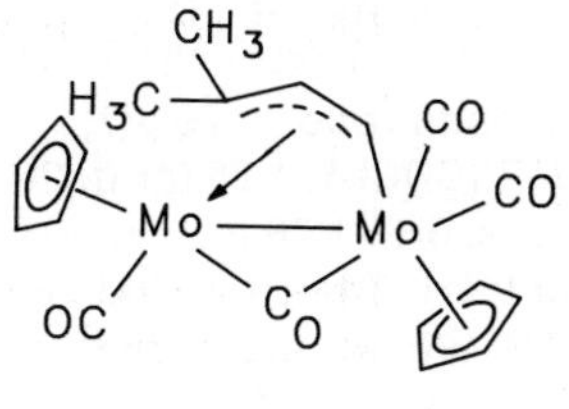

V 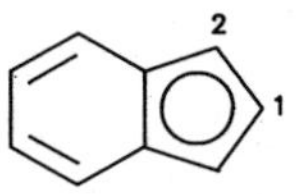 VI

Explanation for Table 11. C_9H_7 means indenyl; the carbon atoms are numbered as shown in Formula VI.

References on pp. 248/50

Table 11
Compounds of the Type $^5LMo(CO)_2{}^3L$ with One Open Allylic ^3L Ligand.
An asterisk indicates further information at the end of the table.
For explanations, abbreviations, and units see p. X.

No.	compound	method of preparation (yield) properties and remarks
$^3L = C_3H_5$		
*1	$C_5H_5Mo(CO)_2C_3H_5$	Ia, b, c; II a, b; see "Further information" lemon yellow crystals [1], yellow solid [17, 23]; m.p. 134°C (dec.) [1], 165 to 168°C [23], 175 to 176°C [22]; subl. at 35°C in a vacuum [1], at 60°C/10^{-3} Torr [10] conductivity (acetone, 1 to 10×10^{-4} M): $\Lambda = 0.5\ cm^2 \cdot \Omega^{-1} \cdot mol^{-1}$ [17] ^1H NMR: see "Further information" ^{13}C NMR (CS_2/C_6F_6 85:15, +7°C): exo isomer: 39.6 (CH), 67.1 (CH_2), 91.2 (C_5H_5); endo isomer: 36.2, 85.6, 90.0 [23]; the values from [23] were initially given relative to CS_2 (192.4 ppm); ($CDCl_3$, isomer not specified): 38.8, 66.2, 90.2, 236.2 (CO) [56] ^{95}Mo NMR (acetone-d_6, vs. 2 M Na_2MoO_4 in D_2O): exo isomer: −1832; endo isomer: −1658; ratio 4:1 [74] IR (CS_2 or CCl_4): given from 741 to 3900; 1871, 1886, 1961 (ν(CO)) [1]; (cyclohexane): 1882, 1898, 1960 [17]; (cyclohexane): exo isomer: 1889, 1963; endo isomer: 1903, 1970 (ν(CO)) [23]; similar data in [4, 56, 63] polarographic measurement (2×10^{-3} M, vs. 10^{-3} M $AgClO_4$|Ag): $E_{1/2} = -2.9$ V, n = 1.4 electrons, $E_{p,c} = -3.1$ V, $E_{p,a} = -1.0$ V [3]
*2	$CH_3C(O)C_5H_4Mo(CO)_2C_3H_5$	IIa, X = Cl, in refluxing THF for 2 h and then 16 h at 25°C (55%); formed in the reaction of $C_3H_5Mo(CO)_2(NCCH_3)Cl$ with LiC_5H_5 (6%) [63] yellow, air-sensitive crystals (from hexane/ CH_2Cl_2), m.p. 98 to 101°C (dec.) exo to endo ratio > 20 [63] ^1H NMR ($CDCl_3$): 1.25 (br d, 2H; J(H, H) = 10.5), 2.30 (s, 3H), 2.76 (d, 2H; J(H, H) = 6.8), 3.82 (t of t, 1H; J(H, H) = 10.5, 6.8), 5.36 and 5.70 (t, 2H; J(H, H) = 2.3) [63] IR (CH_2Cl_2): given from 809 to 3080; 1670 (ν(C=O)), 1867, 1949 (ν(CO)) exo isomer [63] mass spectrum: $[M-n\,CO]^+$ (n = 0, 1) and other ions; base peak is at m/e = 28 [63] reaction with $[NO]PF_6$ in CH_3CN yields $[CH_3C(O)C_5H_4Mo(CO)(NO)C_3H_5]PF_6$ [63]

Table 11 (continued)

No.	compound	method of preparation (yield) properties and remarks
3	$CD_3C(O)C_5H_4Mo(CO)_2C_3H_5$	IIa, with cis-$Mo(CO)_2(C_3H_5)(CD_3CN)_2Cl$ and LiC_5H_5; the mechanism of formation was discussed [63] ^1H NMR ($CDCl_3$): identical to No. 2 except the acyl singlet at 2.30 was missing [63] ^2H NMR ($CDCl_3$): 2.30 (br s, CD_3) [63] IR (CH_2Cl_2): given from 1670 to 3020; 1670 (ν(C=O)), 1867, 1950 (ν(CO)) [63]
4	$C_{10}H_{19}C_5H_4Mo(CO)_2C_3H_5$ $C_{10}H_{19}C_5H_4$ 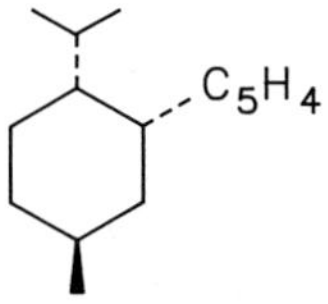 =	IIa, X = Cl, 4 h (65%) [42, 45] yellow needles (from hexane) [42, 45], m.p. 59 to 60°C [42], 66 to 68°C [45] optical rotation (hexane): $[\alpha]_D^{25}$ = +44.5° (0.0113 M) [42], +44.9° (0.247 M) [45] ^1H NMR ($CDCl_3$, −40°C): exo isomer: 0.676, 0.883, and 0.884 (d, CH_3; J(H, H) ≈ 6), 0.88 (d, H_{anti}; J(H, H) = 10.5), 2.79 (d, H_{syn}; J(H, H) = 6.6), 4.05 (m, H_x), 5.04, 5.13, 5.22, and 5.39 (m, C_5H_4); endo isomer: 0.74 (d, CH_3; J(H, H) = 6), 2.72 (d, H_{syn}; J(H, H) = 6.2), 3.62 (m, H_x), 5.35, 6.33, 6.50, and 6.51 (m, C_5H_4), H_{anti} and two CH_3 obscured; K_{eq} (exo/endo) = 4 [45] IR (cyclohexane): exo isomer: 1878.3, 1953.9 (ν(CO)); endo isomer: 1894.3, 1962.7 (ν(CO)) [45] reaction with $[NO]PF_6$ in CH_3CN at 0°C affords $[C_{10}H_{19}C_5H_4Mo(CO)(NO)C_3H_5]PF_6$ [42, 45]
*5	$C_9H_7Mo(CO)_2C_3H_5$	I, reaction of $Na[C_9H_7Mo(CO)_3]$ with allyl chloride in THF for ca. 16 h (21%) [2, 6]; IIa for 18 h [23]; IIb for 1.5 h (80%) [22] yellow crystals (from CH_2Cl_2/hexane at −78°C) [2], m.p. 119 to 121°C [2, 23], 129°C [22], subl. 70 to 100°C/ 0.1 Torr [2] exo to endo ratio 3.27 in $CDCl_3$ at 0°C [23] ^1H NMR (CS_2): 0.85 (d, CH_2; J(H, H) = 11), 2.24 (d, CH_2; J(H, H) = 7), 3.30 (m, CH), 5.56 (C_9H_7, H-1; J(H, H) = 2), 5.87 (C_9H_7, H-2), 6.99 (C_9H_7, six-membered ring) [2]; see "Further information" ^{13}C NMR (CS_2/C_6F_6 85:15, 7°C): endo isomer: 54.4 (CH_2); exo isomer: 48.3 (CH_2) [23]; the values were initially given relative to CS_2 (192.4 ppm) IR (halocarbon mull): given from 743 to 1937; 1855, 1867, 1937 (ν(CO)) [2]; (cyclohexane): exo isomer: 1887, 1963 (ν(CO)); endo isomer: 1902, 1971 (ν(CO)) [23]

References on pp. 248/50

Table 11 (continued)

No.	compound	method of preparation (yield) properties and remarks
*5 (continued)		UV (cyclohexane): λ_{max} (ε) = 221 (37900), 287 (7520) [2] mass spectrum: $[M-n\ CO]^+$ (n = 0 to 2); figure given [12] slightly soluble in petroleum ether but soluble in ether, C_6H_6, CS_2, and polar organic solvents [23] stable for a month under N_2 at low temperatures [23] reaction with $[NO]PF_6$ in CH_3CN at 0°C affords unstable $[C_9H_7Mo(CO)(NO)C_3H_5]PF_6$ [25]
^{3}**L = C_3H_4X** (X = halogen)		
6	$C_5H_5Mo(CO)_2C_3H_4Cl$-2	IIa, X = Cl, 18 h [23] yellow solid, m.p. 108 to 109°C (from ether/hexane) [23] exo to endo ratio 0.14 in C_6H_6 at 5°C [23] ^{1}H NMR ($CDCl_3$): exo isomer: 1.37 (H_{anti}), 3.26 (H_{syn}), 5.29 (C_5H_5); endo isomer: 2.31 (H_{anti}), 3.16 (H_{syn}), 5.18 (C_5H_5) [16, 23] ^{95}Mo NMR (acetone-d_6, vs. 2 M Na_2MoO_4 in D_2O): exo isomer: −1709; endo isomer: −1521; ratio 1:15 [74] IR (cyclohexane): exo isomer: 1897, 1970 (ν(CO)); endo isomer: 1920, 1981 (ν(CO)) [23]
7	$C_9H_7Mo(CO)_2C_3H_4Cl$-2	IIa, X = Cl, 18 h [23] yellow solid, m.p. 95 to 98°C (from ether/hexane) [23] exo to endo ratio 0.01 in C_6H_6 at 5°C [23] ^{1}H NMR ($CDCl_3$): endo isomer: −0.39 (H_{anti}), 4.02 (H_{syn}), 5.71 and 6.19 (indenyl) [16, 23] IR (cyclohexane): 1914, 1979 (ν(CO)) [23]
*8	$C_5H_5Mo(CO)_2C_3H_4Br$-2	IIa, X = Cl, 18 h [23] yellow solid, m.p. 101 to 102°C (from ether/hexane) [23] exo to endo ratio 0.17 in $CDCl_3$ at 0°C [23] ^{1}H NMR ($CDCl_3$): exo isomer: 1.44 (H_{anti}), 3.19 (H_{syn}), 5.31 (C_5H_5); endo isomer: 2.35 (H_{anti}), 3.12 (H_{syn}), 5.14 (C_5H_5) [16, 23] ^{95}Mo NMR (acetone-d_6, vs. 2 M Na_2MoO_4 in D_2O): exo isomer: −1744; endo isomer: −1540; ratio 1:12.5 [74] IR (cyclohexane): exo isomer: 1900, 1971 (ν(CO)); endo isomer: 1921, 1983 (ν(CO)) [23]

References on pp. 248/50

Table 11 (continued)

No.	compound	method of preparation (yield) properties and remarks
9	$C_9H_7Mo(CO)_2C_3H_4Br$-2	IIa, X = Cl, 18 h [23] yellow solid, m.p. 136 to 140°C (from ether/hexane) [23] exo to endo ratio 0.02 in C_6H_6 at 5°C [23] ^{1}H NMR ($CDCl_3$): endo isomer: −0.25 (H_{anti}), 3.79 (H_{syn}), 5.45 and 5.91 (C_9H_7) [16, 23] IR (cyclohexane): exo isomer: 1898, 1967; endo isomer: 1916, 1980 (ν(CO)) [16, 23]
^{3}**L = C_3H_4R** (R in 1-position)		
*10	$C_5H_5Mo(CO)_2C_3H_4CH_3$-1	Ib, in THF at 50°C for 4 h (37%) [18]; Ic (82%) [56]; IIa, X = Cl, 18 h [23]; see "Further information" yellow crystals (from hexane, CH_2Cl_2/pentane, or hexane/ether) [18, 23, 46, 59], m.p. 70°C [18], 73 to 76°C [46] exo to endo ratio 7.0 in CS_2 at −10°C [23]; complete isomerization into the more stable syn isomer after warming to 100°C [18]; see "Further information" ^{1}H NMR (cyclohexane): CH_3 in anti position: 0.95 (d, CH_3), 1.32 (m, H'_{anti}), 2.80 (9 lines, H'_{syn}), 3.83 (13 lines, H_{syn}), 4.0 (8 lines, H_x), 5.20 (s, C_5H_5), with $J(H'_{syn}, H_x)$ = 7.0, $J(H'_{anti}, H_x)$ = 10.2, $J(H_{syn}, H_x)$ = 7.9, $J(H'_{anti}, H'_{syn})$ = 6.0; CH_3 in syn position: 1.26 (m, H'_{anti}), 1.75 (m, CH_3 and H_{anti}), 2.54 (4 lines, H'_{syn}), 3.90 (m, H_x), 5.22 (s, C_5H_5), with $J(H'_{syn}, H_x)$ = 6.8, $J(H'_{anti}, H'_{syn})$ = 2.2, $J(H'_{syn}, H_{anti})$ = 0.2, $J(H'_{anti}, H_x)$ = 10.2, $J(H_{anti}, H_x)$ = 7.9 [18]; see "Further information" ^{13}C NMR ($CDCl_3$): 14.6 (CH_3), 35.0 (C_3H_4), 57.6 (C_3H_4), 66.2 (C_3H_4, central CH), 91.5 (C_5H_5) [46]; (CD_2Cl_2, −70°C): 21.1 (CH_3), 34.1 (CH_2), 64.4 (**C**HCH_3), 69.3 (central CH), 92.3 (C_5H_5), 205.1 and 208.3 (CO) [59] ^{95}Mo NMR (acetone-d_6, vs. 2 M Na_2MoO_4 in D_2O): exo isomer: −1789; endo isomer: −1600; ratio 7:1 [74] IR (KBr): 1490 (ν(C=C)) [18]; (cyclohexane): exo isomer: 1879, 1953; endo isomer: 1894, 1960 (ν(CO)) [23, 63], similar data in [18, 46, 56]; (hexane): 1880, 1952 (ν(CO)) [46] mass spectrum: $[M-n\ CO]^+$ (n = 0 to 2) [46]

Table 11 (continued)

No.	compound	method of preparation (yield) properties and remarks
*10 (continued)		treatment with $[C(C_6H_5)_3]BF_4$ in CH_2Cl_2 affords $[C_5H_5Mo(CO)_2C_4H_6]BF_4$ (C_4H_6 = buta-1,3-diene) [28]
11	$C_5H_5Mo(CO)_2C_3H_4CHD_2$-1	Ib, with $BrCD_2CH_2CH{=}CH_2$ in THF at 50°C for 4 h (30%) [18] ^1H NMR (C_6H_{12}): major isomer: 0.95 (CHD_2-anti), 1.35 (4 lines, H'_{anti}), 2.82 (m, H'_{syn}), 3.9 (13 lines, H_{syn}), 4.00 (H_x), 5.24 (C_5H_5), with $J(H_x, H'_{syn})$ = 7.0, $J(H_x, H'_{anti})$ = 10.5, $J(H_x, H_{syn})$ = 8.0, $J(H'_{syn}, H'_{anti})$ = 2.6, $J(H_{syn}, H'_{syn})$ = 1.8 [18]; for the isomers, see "Further information" for No. 10
12	$CH_3C_5H_4Mo(CO)_2C_3H_4CH_3$-1	$CH_3C_5H_4Mo(CO)_2(=C(CH_2)_3O$-cyclo)I was allowed to react with $Li[(C_2H_5)_3BH]$ in THF at −78°C (ca. 10%) [59] yellow solid, m.p. 31 to 32°C (from pentane) [59] ^1H NMR ($CDCl_3$): major isomer: 0.76 (d of d, H_{anti}; J(H, H) = 10.3), 1.76 (m, CH_3-syn), 1.96 (m, CH_3), 2.30 (d of d, H_{syn}; J(H, H) = 7.3), 3.80 (t of d, H_x; J(H, H) = 10.7), 5.11 and 5.19 (m, C_5H_4, 2H) [59] IR (hexane): 1879, 1893, 1954, 1957 (ν(CO)) [59] mass spectrum: $[M]^+$ [59]
13	$CH_3C(O)C_5H_4Mo(CO)_2C_3H_4CH_3$-1	IIa, X = Cl, 2 h under reflux followed by 16 h at 25°C (high yield); formed in the reaction of 1-$CH_3C_3H_4Mo(CO)_2(NCCH_3)_2Cl$ with LiC_5H_5 in THF [63] yellow solid (from CH_2Cl_2/hexane), m.p. 95 to 97 (with slight decomposition) [63] exo to endo ratio > 20 [63] ^1H NMR ($CDCl_3$): 0.96 (m, 1H), 1.59 (br s, 1H), 1.79 (br d, 3H; J(H, H) ≈ 5), 2.33 (s, 3H), 2.45 (m, 1H), 3.84 (m, 1H), 5.38 (br t, 2H; J(H, H) = 2.2), 5.72 (br d of d, 2H; J(H, H) ≈ 2, 4) [63] IR (CH_2Cl_2): given from 815 to 3050; 1673 (ν(C=O)), 1869, 1948 (ν(CO)) [63] reaction with $[NO]PF_6$ in CH_3CN at 0°C affords $[CH_3C_5H_4Mo(CO)(NO)C_3H_4CH_3\text{-}1]PF_6$ [63]
*14	$C_9H_7Mo(CO)_2C_3H_4CH_3$-1	IIa, X = Cl, 18 h [23]; VIII (40%) [27] yellow solid [23, 27], m.p. 109 to 111°C (from ether/hexane) [23] exo to endo ratio 3.6 in CS_2 at 0°C [23]; rapid equilibration at room temperature, k (exo, endo) = 20 s^{-1} [52]

Table 11 (continued)

No.	compound	method of preparation (yield) properties and remarks
		^{1}H NMR (CS_2): exo isomer: 0.52 (H_x), 0.82 (H'_{anti}), 1.55 (H_{anti}), 1.63 (CH_3-syn), 1.88 (H'_{syn}), 5.59, 5.73, and 5.95 (C_9H_7), with J(H'_{anti}, H'_{syn}) = 2.4, J(H'_{anti}, H_x) = 10.0, J(H'_{syn}, H_x) = 7.1, J(H_{anti}, H_x) = 9, J(H_{anti}, CH_3) = 5.5; endo isomer: −0.67 (H'_{anti}), −0.16 (H_{anti}), 1.94 (CH_3-syn), 3.16 (H_x), 3.27 (H'_{syn}), 5.47, 5.73, and 5.95 (C_9H_7), with J(H'_{anti}, H'_{syn}) < 0.6, J(H_x, H'_{anti}) = 9, J(H_x, H'_{syn}) = 7, J(H_x, H_{anti}) = 10, J(H_{anti}, CH_3) = 5.5 [23]; ($CDCl_3$, 32 °C): −0.3 (m, H_x), 0.8 (d, CH_3-anti; J(H'_{syn}, CH_3) = 7), 1.4 (d of d, H'_{anti}; J(H_x, H'_{anti}) = 12.0), 2.20 (d of t, H'_{syn}; J(H'_{anti}, H'_{syn}) = 8), 3.19 (q of d, H_{syn}; J(H_x, H_{syn}) = 7, J(H_{syn}, CH_3) = 7), 6.6 (t, C_9H_7, C-1; J(H, H) = 3), 5.9 (m, C_9H_7, H-2), 7.0 (m, C_9H_7, six-membered ring) [27] IR (cyclohexane): exo isomer: 1878, 1953 (ν(CO)); endo isomer: 1890, 1960 (ν(CO)) [23]; (hexane): 1875, 1951 (ν(CO)) [27] mass spectrum: $[M-n\,CO]^+$ (n = 0 to 2) [27]
15	$(CH_3)_5C_5Mo(CO)_2C_3H_4CH_3$-1	$(CH_3)_5C_5Mo(CO)_3CH_3$ was irradiated in hexane at 263 K for 20 min in the presence of excess buta-1,3-diene (3% by chromatography on alumina with hexane/ether) [65] yellow crystals (from hexane/ether) [65] ^{1}H NMR ($CDCl_3$): 1.09 (CH_3-anti), 1.50 (CH_3), 1.56 (H'_{anti}), 1.99 (H'_{syn}), 2.48 (H_x), 2.99 (H_{syn}), with J(H'_{anti}, H'_{syn}) = 2.5, J(H_x, H'_{syn}) = 7.3, J(H_x, H_{syn}) = 7.5, J(H_{syn}, CH_3) = 6.5, J(H'_{anti}, H_x) = 11.0, J(H_x, CH_3) = 0.8 [65] IR (hexane): 1867, 1944 (ν(CO)) [65]
16	$C_5H_5Mo(CO)_2C_3H_4C_2H_5$-1	VIII, ^{4}L = $C_4H_5CH_3$-1, with Na[BH_4] at −78 °C (30%) [28] exo allyl with C_2H_5 in anti position [28] ^{1}H NMR: 0.96 and 1.30 (ABX_3 multiplet of diastereotopic CH_2), 2.83 (d of t, H_{syn}; J(H, H) = 2.0, 7.0) [28]
17	$C_5H_5Mo(CO)_2C_3H_4CH_2C(CH_3)_2C(O)H$-1	$[C_5H_5Mo(CO)_2C_4H_6]BF_4$ (C_4H_6 = buta-1,3-diene) was allowed to react with i-butyraldehyde in CH_3CN at 0 °C (76%) [28] ^{1}H NMR ($CDCl_3$): 0.26 (d of d, diastereotopic CH_2, 1H; J(H, H) = 14.5, 12.5), 1.04 and 1.07 (s, diastereotopic CH_3), 1.25 (d of d, H_{anti}; J(H,

References on pp. 248/50

Table 11 (continued)

No.	compound	method of preparation (yield) properties and remarks
17 (continued)		H) = 11.0, 1.5), 2.06 (d of d, 1H, diastereotopic CH_2; J(H, H) = 14.5, 3.0), 2.87 (d of d of d, H_{syn}; J(H, H) = 7.5, 2.5, 1.5), 3.59 (complex t, H_{syn}), 4.07 (d of t, H_x; J(H, H) = 11.0, 7.5), 5.26 (s, C_5H_5), 9.40 (s, C(O)H) [28] IR (cyclohexane): 1733 (ν(C=O) 1884, 1957 (ν(CO)) [28]
*18	$C_5H_5Mo(CO)_2C_3H_4CH{=}CH_2$-1 ^3L = (structure with H atoms labelled 1H–7H)	Ia, ^1L = $CH_2CH{=}CHCH{=}CH_2$, in ether at −20°C for 6 h (ca. 50% by chromatography) [62] yellow crystals (from pentane) exo to endo ratio 1:9 in toluene at −60°C; vinyl group in syn position [62] ^1H NMR (toluene-d_8, −60°C): exo isomer: 0.77 (d of d, H-1), 1.88 (t, H-4), 2.20 (d of d, H-2), 3.35 (t of d, H-3), 4.42 (s, C_5H_5), 4.86 (d of d, H-7), 5.31 (d of t, H-5), 5.35 (d of d, H-6), with J(H-1, 3) = J(H-3, 4) = 10.0, J(H-2, 3) = 7.0, J(H-1, 2) = 2; J(H-4, 5) = J(H-5, 7) = 10.0, J(H-5, 6) = 16.2, J(H-6, 7) = 1.5; endo isomer: 1.33 (d, H-1), 2.40 (d, H-2), 2.90 (t, H-4), 3.45 (t, H-3, overlapping with H-3 of exo isomer, irradiation turns into a d), 4.42 (s, C_5H_5), 4.98 (d, H-7), 5.06 (d, H-6), 6.25 (d of t, H-5), with J(H-1, 3) = J(H-3, 4) = J(H-4, 5) = J(H-5, 7) = 10.0, J(H-2, 3) = 7.0, J(H-5, 6) = 16.0; (toluene-d_8, 100°C): 0.98 (br d, H-1), 2.36 (br m, H-4), 2.40 (d of d, H-2), 3.75 (t of d, H-3), 4.70 (d, H-6), 4.80 (s, C_5H_5), 5.20 (d, H-7), 5.75 (br m, H-5), with J(H-2, 3) = 7.0, J(H-1, 2) = 1.8, J(H-1, 3) = J(H-3, 4) = 10, J(H-5, 6) = 16; coalescence of the exo and endo resonances at 100°C, $\Delta G^{\neq}$ = 18.2 ± 0.3 kcal/mol [62] ^{13}C NMR (C_6D_6): exo isomer: 34.5 (CH-1, 2), 66.5 (CH-4), 69.2 (CH-3), 91.7 (C_5H_5), 111.3 (CH-6, 7), 139.2 (CH-5), 237 and 240 (CO) [62] IR (pentane): both: 1625 (ν(C=C)); exo isomer: 1890, 1960 (ν(CO)); endo isomer: 1905, 1975 (ν(CO)) [62] mass spectrum (12 eV): $[M-n\,CO]^+$ (n = 0, 1) [62] irradiation for 6 h in ether at −20°C affords the $^5LMo(CO)^5L'$ complex $C_5H_5Mo(CO)CH_2$┄CH┄CH┄CH┄CH_2 [62]
*19	$C_9H_7Mo(CO)_2C_3H_4CH{=}CH_2$-1	Ib, during chromatography on Al_2O_3 with ether (2%); IIa, X = Cl, 12 h at 0°C (30 to 45%) [73]

Table 11 (continued)

No.	compound	method of preparation (yield) properties and remarks
		yellow crystals (from ether), exo to endo ratio 6:1 [73] ^{1}H NMR ($CDCl_3$): exo isomer: 0.57 (t of d, H-3), 1.12 (d of d, H-1), 2.03 (d of d, H-2), 2.28 (t, H-4), 4.56 (d of d, H-7), 5.32 (d of d, H-6), 5.68 (t of d, H-5), 5.75 (t, 1H), 5.93 (d, 1H), 6.13 (t, 1H), 7.21 to 7.39 (m, 4H), with J(H-1, 2) = 2.1, J(H-1, 3) = 10.5, J(H-2, 3) = 6.8, J(H-3, 4) = 10.4, J(H-4, 3) = 10.5, J(H-5, 7) = 10.2, J(H-5, 6) = 16.5, J(H-6, 7) = 1.2; endo isomer: −0.42 (d, H-1), 0.65 (t, H-4), 3.41 (d, H-2), 3.69 (t of d, H-3), 4.90 (d of d, H-7), 5.23 (d of d, 1H), 6.33 (t of d, H-5), 5.93 (d, 1H), 6.13 (t, 1H), 7.21 to 7.39 (m, 4H), with J(H-1, 3) = J(H-3, 4) = 9.2, J(H-2, 3) = 6.8 [73]; for numbering, see No. 18 IR (CH_2Cl_2): 1610 (ν(C=C)), 1860, 1940 (ν(CO)) [73] mass spectrum (12 eV): $[M-n\ CO]^+$ (n = 0 to 2) [73] UV irradiation in ether between −20 and +20°C in the presence of 1.5 to 3 equivalents of a ^{2}D ligand affords $C_9H_7Mo(CO)(^2D)C_3H_4CH{=}CH_2$ ($^2D = P(CH_3)_3$ or $P(CH_3)_2C_6H_5$); no reaction was observed under thermal conditions in toluene [73]
20	$C_{13}H_9Mo(CO)_2C_3H_4CH{=}CH_2$-1 ($C_{13}H_9$ = fluorenyl)	Ib, during chromatography on Al_2O_3 with ether (0.5 to 2%); IIa, X = Cl, with $NaC_{13}H_9$ for 16 h (30 to 45%) [73] yellow crystals (from ether at 0°C) [73] only exo isomer; vinyl group in syn position [73] ^{1}H NMR ($CDCl_3$): exo isomer: 0.57 (d of d, H-1), 0.85 (d of d, H-2), 1.51 (t, H-4), 3.17 (t of d, H-3), 4.69 (d of d, H-6), 5.05 (t of d, H-5), 7.05 (m, 2H), 7.20 (m, 2H), 7.57, 7.617, 7.62, and 7.65 (d, 1H), with J(H-1, 2) = 2.4, J(H-1, 3) = 10.2, J(H-2, 3) = 7.1, J(H-3, 4) = J(H-4, 5) = 10.4, J(H-5, 6) = 16.4, J(H-5, 7) = 10.0, J(H-6, 7) = 1.3; it was suggested that the allyl moiety lies between the two phenyl rings as indicated by the absence of shielding effects of the aromatic six-membered rings [73]; for numbering, see No. 18 IR (CH_2Cl_2): 1614 (ν(C=C)), 1864, 1943 (ν(CO)) [73] mass spectrum (12 eV): $[M-n\ CO]^+$ (n = 0 to 2) [73]

References on pp. 248/50

Table 11 (continued)

No.	compound	method of preparation (yield) properties and remarks
20 (continued)		no stable $C_{13}H_9Mo(CO)(^2D)C_3H_4CH{=}CH_2$ complex could be prepared (^2D = $P(CH_3)_3$ or $P(CH_3)_2C_6H_5$) [73]
*21	$(CH_3)_5C_5Mo(CO)_2C_3H_4CH{=}CH_2$-1	VI [71] endo to exo ratio 7:3; vinyl group in syn position ^1H NMR (CD_2Cl_2): exo isomer: 1.04 (d of d, H-1), 1.76 (s, CH_3), 2.16 (d of d, H-4), 3.20 (d of d of d, H-3), 4.81 (d of d, H-6), 5.13 (d of d, H-7), 5.75 (d of d of d, H-5), with J(H-1, 3) = 10.1, J(H-2, 3) = 7.1, J(H-3, 4) = 10.3, J(H-4, 5) = 10.5; endo isomer: 0.55 (d, H-1), 1.68 (s, CH_3), 2.80 (d, H-2), 3.93 (d of d of d, H-3), 4.72 (d, H-6), 4.99 (d of d, H-7), 6.13 (d of d of d, H-5), with J(H-1, 3) = 10.0, J(H-2, 3) = 6.3, J(H-3, 4) = 9.9, J(H-4, 5) = 10.2 [71]; for numbering, see No. 18 ^{13}C NMR (CD_2Cl_2): exo isomer: 11.1 (CH_3), 39.1 (CH-1, 2), 67.3 (CH-4), 77.2 (CH-3), 104.2 ($\mathbf{C}_5(CH_3)_5$), 110.8 (CH-6, 7), 137.8 (CH-5), 240.1 and 242.7 (CO) [71]
*22	$(CH_3)_5C_5Mo(CO)_2C_3H_4CH{=}CH_2$-1	VI [71] only the exo isomer, vinyl group in anti position [71]
23	exo/endo-$C_5H_5Mo(CO)_2C_3H_4CH{=}CHCH_3$-1 ^3L = ^3H ^5H ^2H CH_3 ^1H ^4H ^6H	Ia, in ether at −20°C for 4 h (low yield); Ic, in CH_2Cl_2 for 6 h (60%) [69] exo to endo ratio 8.5 (−60°C); R in syn position [69] yellow blocks (from pentane at −25°C) [69] ^1H NMR (toluene-d_8): exo isomer: 0.74 (d of d, H-1), 1.58 (d, CH_3), 2.16 (t, H-4), 2.27 (d of d, H-2), 3.55 (t of d, H-3), 4.55 (s, C_5H_5), 5.21 (d of d, H-5), 5.72 (m, H-6), with J(H-1, 2) = 1.8, J(H-1, 3) = 10.8, J(H-2, 3) = 6.4, J(H-3, 4) = J(H-4, 5) = 10.4, J(H-5, 6) = 15.8, J(H-6, CH_3) = 6.4; endo isomer: 0.82 (d, H-1), 1.83 (d, CH_3), 2.35 (d, H-2), 3.01 (t, H-4), 3.52 (t of d, H-3), 4.55 (s, C_5H_5), 5.45 (m, H-6), 5.83 (d of d, H-5), with J(H-1, 3) = 10.8, J(H-2, 3) = 6.6, J(H-3, 4) = J(H-4, 5) = 10.5, J(H-5, 6) = 16.2 [69] ^{13}C NMR (C_6D_6): exo isomer: 18.2 (CH_3), 33.7 (CH-1, 2), 67.8 and 68.4 (CH-3 and CH-4), 91.6 (C_5H_5), 123.9 (CH-6), 132.9 (CH-5) [69] IR (pentane): both: 1623 (ν(C=C)); exo isomer: 1890, 1962 (ν(CO)); endo isomer: 1905, 1970 (ν(CO)) [69]

Table 11 (continued)

No.	compound	method of preparation (yield) properties and remarks
		mass spectrum (12 eV): $[M-n\ CO]^+$ (n = 0 to 2) [69] irradiation in ether for 18 h at −20°C affords mainly No. 72 along with small amounts of the $^5LMo(CO)_2{}^5L'$ complexes $C_5H_5Mo(CO)_2CH(CH_3)\cdots CH\cdots CH\cdots CH\cdots CH_2$ (two isomers) [69]
24	$(CH_3)_5C_5Mo(CO)_2C_3H_4CH_2CH{=}CHC_5H_{11}$-1	$[^5LMo(CO)_2C_4H_6]BF_4$ (C_4H_6 = buta-1,3-diene) was treated with $LiCu(CH{=}CHC_5H_{11}\text{-}E)_2$; syn isomer [55] hydride abstraction with $[C(C_6H_5)_3]BF_4$ in CH_2Cl_2 affords $[^5LMo(CO)_2CH_2{=}CH\text{-}CH{=}CH\text{-}CH{=}CH\text{-}C_5H_{11}]BF_4$ [55]
25	$C_9H_7Mo(CO)_2C_3H_4(CH_2CH{=}CHC_5H_{11})$-1	see No. 24 [55]
*26	$(CH_3)_5C_5Mo(CO)_2C_3H_4C(O)H$-1 ^3L 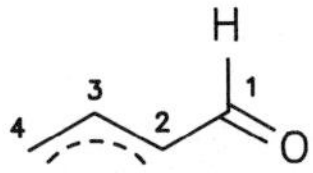 =	VI (50%) [71] exo to endo ratio 7:3; R in syn position [71] ^1H NMR (CD_2Cl_2): exo isomer: 1.64 (d of d, H'_{anti}; $J(H_x, H'_{anti})$ = 11.5), 1.76 (d of d, H_{anti}; $J(C(O)H, H_{anti})$ = 8.4, $J(H_x, H_{anti})$ = 9.4), 1.91 (s, CH_3), 2.09 (d of d, H'_{syn}; $J(H_x, H'_{syn})$ = 7.2), 3.71 (d of d of d, H_x), 8.83 (d, C(O)H); endo isomer: 1.18 (d, H'_{anti}; $J(H_x, H'_{anti})$ = 10.6), 1.60 (d of d, H_{anti}; $J(H_x, H_{anti})$ = 8.3, $J(C(O)H, H_{anti})$ = 7.9), 1.87 (s, CH_3), 3.04 (d, H'_{syn}; $J(H_x, H'_{syn})$ = 6.4), 4.34 (d of d of d, H_x), 8.91 (d, C(O)H) [71] ^{13}C NMR (CD_2Cl_2): exo isomer: 10.8 (CH_3), 47.6 (C-4), 64.9 (C-2), 91.3 (C-3), 104.1 ($\mathbf{C}_5(CH_3)_5$), 195.3 (C-1), 240.8 and 241.7 (CO); endo isomer: 11.4 (CH_3), 44.0 (C-4), 62.1 (C-2), 80.9 (C-3), 105.3 ($\mathbf{C}_5(CH_3)_5$), 193.4 (C-1), 238.9 and 239.3 (CO) [71] IR (CH_2Cl_2): 1666 (ν(C=O)), 1878, 1975 (ν(CO)) [71]
*27	$(CH_3)_5C_5Mo(CO)_2C_3H_4C(O)H$-1 ^3L 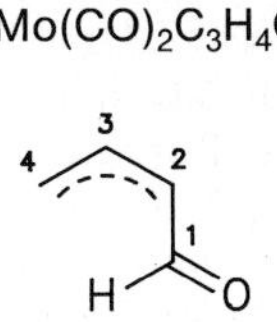 =	VI (4%), major product is No. 26 [71] exo product; R in anti position [71] ^1H NMR (CD_2Cl_2): 1. 5 (d of d, H'_{anti}; $J(H_x, H'_{anti})$ = 11.6), 1.88 (s, CH_3), 2.28 (d of d of d, H'_{syn}; $J(H_x, H'_{syn})$ = 8.4), 3.42 (m, H_{anti}), 3.51 (d of d of d, H_x; $J(H_{anti}, H_x)$ = 7.1), 7.00 (d, C(O)H; $J(H_{anti}, C(O)H)$ = 7.8) [71] ^{13}C NMR (CD_2Cl_2): 10.5 (CH_3), 42.5 (C-4), 65.1 (C-2), 80.2 (C-3), 105.0 ($\mathbf{C}_5(CH_3)_5$), 184.8 (C-1), 237.4 and 237.7 (CO) [71]

References on pp. 248/50

Table 11 (continued)

No.	compound	method of preparation (yield) properties and remarks
*28	$C_9H_7Mo(CO)_2C_3H_4CH_2C(O)CH_3$-1	$C_9H_7Mo(CO)_3CH_3$ was allowed to react with excess buta-1,3-diene for 6 d in a sealed tube (36%) [33] exo isomer; R in anti position [3] yellow crystals, subl. 80°C/0.1 Torr [33] ^{1}H NMR ($CDCl_3$): −0.2 (m, H_x; J(H'_{anti}, H_x) = 12.0, J(H'_{syn}, H_x) = 8.0, J(H_{syn}, H_x) = 9.0), 1.05 (d of d, CH_2, 1H; J(H_{syn}, H) = 11.0, J(H, H) = 17.0), 1.3 (d of d, H'_{anti}; J(H'_{syn}, H'_{anti}) = 2.0, J(H_x, H'_{anti}) = 12.0), 2.0 (s, CH_3), 2.3 (d of t, H'_{syn}; J(H_{syn}, H'_{syn}) = 2.0), 2.7 (d of d, CH_2, 1H; J(H_{syn}, H) = 4.0, J(H, H) = 17), 3.2 (m, H_{syn}), 5.6 (t, H-1; J(H, H) = 3.0), 6.0 (d, H-2), 7.05 (m, C_9H_7, six-membered ring); two CH_2 resonances indicate that rotation about the C-C bond is hindered by the acyl group [33] IR (hexane): given from 768 to 1954; 1723 (ν(C=O)), 1879, 1954 (ν(CO)) [33]
29	$C_5H_5Mo(CO)_2C_3H_4Si(CH_3)_3$-1, syn	Ia, in hexane for 2.5 h (50%) [26] waxy solid (from hexane) [26] ^{1}H NMR ($CDCl_3$): 0.14 (s, CH_3), 0.37 (d of q, H_{anti}), 0.94 (d of q, H'_{anti}), 2.58 (d of d, H'_{syn}), 3.60 (oct, H_x), 4.73 (s, C_5H_5), with J(H_{anti}, H'_{anti}) = 0.3, J(H_{anti}, H_x) = 12.4, J(H'_{anti}, H'_{syn}) = 1.4, J(H'_{anti}, H_x) = 10.0, J(H'_{syn}, H_x) = 6.0 [26] IR: 1882, 1964 (ν(CO)) [26]
^{3}**L** = $\mathbf{C_3H_4R}$ (R in 2-position)		
*30	$C_5H_5Mo(CO)_2C_3H_4CH_3$-2	Ic (68%) [56]; IIa, 18 h [23] yellow solid, m.p. 79 to 81°C (from ether/hexane) [23] exo to endo ratio 0.11 in $CDCl_3$ [23] and 0.38 in CH_3CN at 0°C [25] ^{1}H NMR (toluene, ca. −8°C): exo isomer: 1.06 (H_{anti}), 1.49 (CH_3), 2.66 (H_{syn}), ca. 4.63 (C_5H_5); endo isomer: 1.69 (H_{anti}), 1.86 (CH_3), 2.81 (H_{syn}), 4.63 (C_5H_5); (toluene, +80°C): 1.71 (H_{anti}), 1.81 (CH_3), 2.80 (H_{syn}), 4.80 (C_5H_5) [9]; see "Further information" ^{95}Mo NMR (acetone-d_6, vs. 2 M Na_2MoO_4 in D_2O): exo isomer: −1752; endo isomer: −1573; ratio 1:11 [74] IR (pentane): exo isomer: 1886 (only shoulder); endo isomer: 1895, 1962 (ν(CO)) [23, 56, 63];

Table 11 (continued)

No.	compound	method of preparation (yield) properties and remarks
		similar data in cyclohexane [16] and in pentane [9] reaction with $[NO]PF_6$ in CH_3CN at 0°C affords $[C_5H_5Mo(CO)(NO)C_3H_4CH_3\text{-}2]PF_6$ [25, 58]
31	$CH_3C(O)C_5H_4Mo(CO)_2C_3H_4CH_3\text{-}2$	IIa, X = Cl, for 2 h under reflux followed by 16 h at 25°C (high yield); formed in the reaction of $1\text{-}CH_3C_3H_4Mo(CO)_2(CNCH_3)_2Cl$ with LiC_5H_5 in CH_3CN [63] yellow solid, m.p. 101 to 103°C (with slight decomposition) [63] exo to endo ratio < 0.05 [63] ^1H NMR ($CDCl_3$): 1.73 (s, 3H), 1.78 (br s, 2H); 2.29 (s, 3H), 2.91 (2H), 5.32 and 5.71 (t, 2H; J(H, H) = 2.4) [63] ^{13}C NMR ($CDCl_3$): 23.4 (q), 26.2 (q), 41.6 (t), 90.0 (d), 94.7 (d), 102.2 (s), 105.6 (br s), 194.7 (s), 237.6 (CO) [63] IR (CH_2Cl_2): given from 812 to 3050; 1676 (ν(C=O)), 1872, 1961 (ν(CO)) [63]
32	$C_{10}H_{19}C_5H_4Mo(CO)_2C_3H_4CH_3\text{-}2$ $C_{10}H_{19}C_5H_4$ 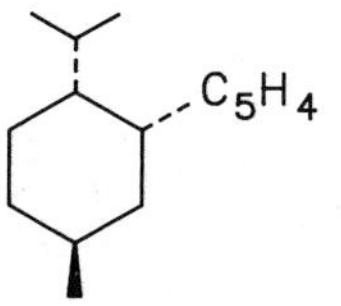 =	IIa, X = Cl (60%) [45]
33	$C_9H_7Mo(CO)_2C_3H_4CH_3\text{-}2$	IIa, X = Cl, 18 h [23] yellow solid, m.p. 117 to 118°C (from ether/hexane) [23] exo to endo ratio < 0.01 in C_6H_6 at 5°C [23] ^1H NMR ($CDCl_3$): −0.72 (H_{anti}), 1.49 (CH_3), 3.46 (H_{syn}), 5.43 and 5.87 (C_9H_7) [23] IR (cyclohexane): 1894, 1965 (ν(CO)) [23]
34	endo-$(CH_3)_5C_5Mo(CO)_2C_3H_4CH_3\text{-}2$	VIIa (75%) [48] yellow crystals (from hexane at −78°C) [48] ^1H NMR (C_6D_6): 0.6 (s, H_{anti}), 1.6 (s, $C_5(CH_3)_5$), 1.8 (s, CH_3), 3.0 (s, H_{syn}) [48] ^{13}C NMR (C_6D_6): 10.9 (($\mathbf{C}H_3)_5C$), 47.7 (CH_2), 102.1 ($\mathbf{C}_5(CH_3)_5$), 106.5 ($\mathbf{C}HCH_3$) [48] IR (hexane): 1877, 2047 (1947?) (ν(CO)) [48]
35	endo-$(CH_3)_5C_5Mo(CO)_2C_3H_4C_2H_5\text{-}2$	VIIb (88%) [48] yellow crystals (from pentane at −78°C) [48]

References on pp. 248/50

Table 11 (continued)

No.	compound	method of preparation (yield) properties and remarks
35 (continued)		^{1}H NMR (C_6D_6): 0.58 (s, H_{anti}), 1.10 (t, **CH_3**CH_2; J(H, H) = 5.0), 1.61 (t, $(CH_3)_5C_5$), 1.83 (q, **CH_2**CH_3), 3.0 (s, H_{syn}) [48] IR (pentane): 1880, 1950 (ν(CO)) [48]
36	$C_5H_5Mo(CO)_2C_3H_4C_3H_7$-i-2	VIIa, R = CH_3 (65%) [48] ^{1}H NMR (C_6D_6): 1.05 (d, CH_3; J(H, H) = 6.8), 1.56 (br s, H_{anti}), 2.05 (sept, **CH**$(CH_3)_2$), 2.75 (br s, H_{syn}), 4.61 (s, C_5H_5) [48] ^{13}C NMR (C_6D_6): 24.1 (CH_3), 34.3 (CH_2), 35.7 (**CH**$(CH_3)_2$), 90.6 (C_5H_5), 92.6 (**C**C_3H_7-i), 241.8 (CO) [48] IR (hexane): 1887, 1959 (ν(CO)) [48]
37	$(CH_3)_5C_5Mo(CO)_2C_3H_4CH_2OH$-2	VIIc (75%) [48] yellow crystals (from CH_2Cl_2) [48] exo to endo ratio 0.5 [48] ^{1}H NMR (C_6D_6): exo isomer: 0.67 (s, H_{anti}), 1.51 (s, CH_3), 3.20 (s, H_{syn}), 3.75 (s, CH_2OH); endo isomer: 0.54 (s, H_{anti}), 1.51 (s, CH_3), 3.06 (s, H_{syn}), 3.67 (s, CH_2OH) [48] ^{13}C NMR (C_6D_6): 10.0 (CH_3), 47.0 (CH_2), 75.0 (CH_2OH), 101.0 (**C_5**$(CH_3)_3$), 240.0 (CO) [48] IR (CH_2Cl_2): 1863, 1939 (ν(CO)) [48]
38	endo-$(CH_3)_5C_5Mo(CO)_2C_3H_4CH_2SC_6H_5$-2	VIId (78%) [48] yellow crystals (from hexane at −78°C) [48] ^{1}H NMR ($CDCl_3$): 0.64 (s, H_{anti}), 1.82 (s, CH_3), 2.87 (s, CH_2S), 3.08 (s, H_{syn}), 7.2 to 7.5 (m, C_6H_5) [48] ^{13}C NMR (C_6D_6): 10.6 (CH_3), 45.4 (CH_2S), 47.4 (CH_2), 102.3 (**C_5**$(CH_3)_5$), 105.6 (**C**CH_2S), 127 to 137 (C_6H_5), 242.0 (CO) [48] IR (hexane): 1880, 1950 (ν(CO)) [48]
39	$C_5H_5Mo(CO)_2C_3H_4C(O)OCH_3$-2	III, for 24 h (85%) [30, 34] m.p. 116°C [30] exo to endo ratio 22:78 in $CDCl_3$ at −46°C [34] ^{1}H NMR ($CDCl_3$): 1.60 (m, H_{anti}), 3.36 (m, H_{syn}), 3.68 (s, CH_3), 5.22 (s, C_5H_5) [30]; ($CDCl_3$, −46°C): exo isomer: 1.07 (H_{anti}), 3.50 (H_{syn}), 3.83 (CH_3), 5.25 (C_5H_5); endo isomer: 1.84 (H_{anti}), 3.30 (H_{syn}), 3.68 (CH_3), 5.32 (C_5H_5) [34] IR (THF): 1720 (ν(C=O)), 1880 to 1910, 1960 to 1980 (ν(CO)) [30]
40	$C_5H_5Mo(CO)_2C_3H_4C(O)OCH_2C{\equiv}CH$-2	III, for 72 h (52%) [30] m.p. 88°C [30]

References on pp. 248/50

Table 11 (continued)

No.	compound	method of preparation (yield) properties and remarks
		^{1}H NMR ($CDCl_3$): 1.64 (m, H_{anti}), 2.47 (t, ≡CH), 3.38 (m, H_{syn}), 4.72 (d, OCH_2), 5.27 (s, C_5H_5) [30] IR (CH_2Cl_2): 1710 (ν(C=O)), 1880 to 1910, 1960 to 1980 (ν(CO)) [30]
41	$C_5H_5Mo(CO)_2C_3H_4C(O)OCH_2C_6H_5$-2	III, for 60 h [30] m.p. 78 to 80°C [30] ^{1}H NMR ($CDCl_3$): 1.60 (m, H_{anti}), 3.39 (m, H_{syn}), 5.13 (s, OCH_2), 5.17 (s, C_5H_5), 7.34 (s, C_6H_5) [30] IR (THF): 1715 (ν(C=O)), 1880 to 1910, 1975 to 1975 (ν(CO)) [30]
42	$C_5H_5Mo(CO)_2C_3H_4C(O)OC_4H_9$-t-2	III, for 40 h (40%) [30] m.p. 111°C [30] ^{1}H NMR ($CDCl_3$): 1.46 (s, t-C_4H_9), 1.60 (m, H_{anti}), 3.34 (m, H_{syn}), 5.23 (s, C_5H_5) [30] IR (CH_2Cl_2): 1705 (ν(C=O)), 1880 to 1905, 1955 to 1975 (ν(CO)) [30]
43	$C_5H_5Mo(CO)_2C_3H_4C(O)OC_6H_5$-2	III, for 72 h (45%) [30] oil [30] ^{1}H NMR ($CDCl_3$): 1.81 (m, H_{anti}), 3.40 (m, H_{syn}), 5.22 (s, C_5H_5), 7.22 (m, C_6H_5) [30] IR (CH_2Cl_2): 1710 (ν(C=O)), 1880 to 1910, 1960 to 1980 (ν(CO)) [30]
44	$C_5H_5Mo(CO)_2C_3H_4C(O)SCH_3$-2	III, for 24 h (24%) [30] m.p. 185°C [30] ^{1}H NMR (C_6D_6): 1.25 (m, H_{anti}), 2.05 (SCH_3), 3.42 (m, H_{syn}), 4.63 (s, C_5H_5) [30] IR (THF): 1665 (ν(C=O)), 1890 to 1920, 1965 to 1980 (ν(CO)) [30]
45	$C_5H_5Mo(CO)_2C_3H_4C(O)SCH_2C_6H_5$-2	III, for 24 h (24%) [30] m.p. 87°C [30] exo to endo ratio 34:76 at −33°C in $CDCl_3$ [34] ^{1}H NMR ($CDCl_3$): 1.52 (m, H_{anti}), 3.34 (m, H_{syn}), 4.15 (s, SCH_2), 5.15 (s, C_5H_5), 7.28 (m, C_6H_5) [30]; ($CDCl_3$, −33°C): exo isomer: 1.05 (H_{anti}), 3.43 (H_{syn}), 4.17 (CH_2), 5.0 (C_5H_5); endo isomer: 1.78 (H_{anti}), 3.27 (H_{syn}), 4.14 (CH_2), 5.25 (C_5H_5) [34] IR (THF): 1660 (ν(C=O)), 1890 to 1920, 1965 to 1980 (ν(CO)) [30]
46	$C_5H_5Mo(CO)_2C_3H_4C(O)SC_6H_5$-2	III, for 36 h (62%) [30] m.p. 165°C [30] ^{1}H NMR ($CDCl_3$): 1.59 (m, H_{anti}), 3.40 (m, H_{syn}), 5.26 (s, C_5H_5), 7.40 (s, C_6H_5) [30]

References on pp. 248/50

Table 11 (continued)

No.	compound	method of preparation (yield) properties and remarks
46 (continued)		IR ($CHCl_3$): 1660 (ν(C=O)), 1880 to 1920, 1960 to 1980 (ν(CO)) [30]
47	$C_5H_5Mo(CO)_2C_3H_4Si(CH_3)_3$-2	I, in THF for 1 h (15% by trap to trap distillation) [26] ^{1}H NMR ($CDCl_3$): 0.15 (s, CH_3), 1.73 (s, H_{anti}), 2.69 (br s, H_{syn}), 5.23 (s, C_5H_5) [26] IR: 1897, 1938 (ν(CO)) [26]
^{3}L = 1,1- or 1,2-disubstituted allyl system		
48	exo-$C_5H_5Mo(CO)_2C_3H_3(CH_3)_2$-1,1	IIa, X = Cl, for 18 h [23] yellow solid, m.p. 65 to 69°C (from ether/hexane) [23] exo to endo ratio > 100 in $CDCl_3$ at 0°C [23] ^{1}H NMR ($CDCl_3$): 0.97 (CH_3-anti), 1.12 (H'_{anti}), 1.78 (CH_3-syn), 2.54 (H'_{syn}), 3.99 (H_x), 5.14 (C_5H_5), with J($H'_{anti'}$, H'_{syn}) = −3.1, J(H_x, H'_{anti}) = 10.6, J(H_x, H'_{syn}) = 7.2, J(H_x, CH_3-anti) = 0.4 [23] ^{95}Mo NMR (acetone-d_6, vs. 2M Na_2MoO_4 in D_2O): −1688 [74] IR (cyclohexane): 1874, 1949 (ν(CO)) [23] reaction with $[C(C_6H_5)_3]BF_4$ in CH_2Cl_2 affords exo/endo-$[C_5H_5Mo(CO)_2C_4H_5CH_3\text{-}2]BF_4$ ($C_4H_5CH_3$-2 = 2-methylbuta-1,3-diene) [28]
49	exo-$C_9H_7Mo(CO)_2C_3H_3(CH_3)_2$-1,1	IIa, X = Cl, for 18 h [23]; VIII, with Na[BH_4], ^{4}L = 2-methylbuta-1,3-diene, for 1 h (77%) [27] yellow solid [23, 27], m.p. 117 to 118°C (from ether/hexane) [23], subl. 80°C/0.1 Torr [27] exo to endo ratio > 100 in $CDCl_3$ at 0°C [23] ^{1}H NMR ($CDCl_3$): −0.05 (H_x), 0.69 (CH_3-anti), 1.17 (H'_{anti}), 1.50 (CH_3-syn), 1.72 (H'_{syn}), 5.53, 5.69, and 5.91 (C_9H_7), with J(H'_{anti}, H'_{syn}) = −2.7, J(H_x, H'_{anti}) = 10.7, J(H_x, H'_{syn}) = 7.4, J(H_x, CH_3-anti) = 0.3; spectrum given as a diagram in [23]; ($CDCl_3$, +32°C): 0.3 (m, H_x), 0.8 (s, CH_3-anti), 1.3 (d of d, H_{anti}), 1.50 (s, CH_3-syn), 1.80 (d of d, H_{syn}), 5.60, 5.69, and 5.90 (m, H-1, 2), 6.8 to 7.4 (m, C_9H_7, 6-membered ring), with J(H'_{anti}, H'_{syn}) = 3.0, J(H_x, H_{anti}) = 10.0, J(H_x, H'_{syn}) = 8.0 [27] IR (cyclohexane): 1874, 1949 (ν(CO)) [23]; similar data in [27] mass spectrum: $[M-n\,CO]^+$ (n = 0 to 2) [27]

Table 11 (continued)

No.	compound	method of preparation (yield) properties and remarks
50	exo-$C_9H_7Mo(CO)_2C_3H_3(CH_3)CH_2CHC(O)(CH_2)_3$ ^3L 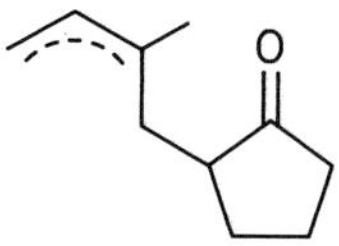 =	VIII, with 1-morpholinocyclopent-1-ene, ^4L = penta-1,3-diene (43%) [27] yellow crystals [27] ^1H NMR ($CDCl_3$, +32°C): 0.3 (m, H_x), 1.3 (s, CH_3), 0.9 to 2.1 (br m, H_{anti}, H_{syn}, and CH_2), 3.50 (m, CH), 5.4 to 6.0 (m, H-1, 2), 6.9 to 7.4 (m, C_9H_7, 6-membered ring) [27] IR (hexane): 1739 (ν(C=O)), 1881, 1953 (ν(CO)) [27] mass spectrum: $[M-n\ CO]^+$ (n = 0 to 2) [27]
51	$(CH_3)_5C_5Mo(CO)_2C_3H_3(CH_3\text{-syn},1)C_2H_5\text{-}2$	VIIa, ^4L = $C(CHCH_3)_2CH_2$ (35%) [48] ^1H NMR (C_6D_6): 0.22 (s, H'_{anti}), 1.04 (t, $\mathbf{CH_3}CH_2$; J(H, H) = 6.0), 1.21 (q, H_{anti}; J(CH_3-syn, H_{anti}) = 6.1), 1.60 (s, $(CH_3)_5C_5$), 1.69 (m, $\mathbf{CH_2}CH_3$, 1H), 1.82 (d, CH_3-syn), 2.23 (m, $\mathbf{CH_2}CH_3$, 1H), 2.88 (s, H'_{syn}) [48] ^{13}C NMR (C_6D_6): 10.66 (($\mathbf{C}H_3)_5C_5$), 15.3 ($\mathbf{C}H_3CH_2$), 16.6 (CH_3-syn), 28.3 ($\mathbf{C}H_2CH_3$), 44.1 (CH_2), 62.3 (CH), 102.0 ($\mathbf{C}_5(CH_3)_5$), 118.0 ($\mathbf{C}C_2H_5$), 244.0 (CO) [48] IR (hexane): 1850, 1940 (ν(CO)) [48]
52	$(CH_3)_5C_5Mo(CO)_2C_3H_3(CH_3\text{-syn},1)C(CH_3)(OH)H\text{-}2$	VIIc, ^4L = $C(CHCH_3)_2CH_2$ (70%) [48] ^1H NMR (C_6D_6): 1.15 (q, H_{anti}; J(CH_3-syn, H_{anti}) = 6.2), 1.35 (d, C($\mathbf{CH_3}$)(OH)H; J(CH, CH_3) = 6.6), 1.55 (s, $(CH_3)_5C_5$), 2.0 (d, CH_3-syn), 2.88 (s, H'_{syn}), 4.16 (q, $C(CH_3)(OH)\mathbf{H}$) [48] ^{13}C NMR (C_6D_6): 10.4 (($\mathbf{C}H_3)_5C_5$), 15.4 (C($\mathbf{C}H_3$)(OH)H), 21.1 (CH_3-syn), 40.3 (CH_2), 62.3 (CH), 69.5 ($\mathbf{C}(CH_3)(OH)H$), 101.9 ($\mathbf{C}_5(CH_3)_5$), 112.5 ($\mathbf{C}CH$), 243.3 and 243.9 (CO) [48] IR (hexane): 1872, 1944 (ν(CO)) [48]
53	$C_5H_5Mo(CO)_2C_3H_3(CH_3\text{-syn},1)C(O)OH\text{-}2$	III, for 170 h (51%) [30] m.p. 220°C [30] exo to endo ratio 1:1 in $CDCl_3$ at −30°C [20, 34] ^1H NMR ($CDCl_3$): 0.90 (m, H'_{anti}), 2.09 (m, CH_3-anti), 2.09 (m, H_{anti}), 3.22 (d, H'_{syn}), 5.23 (s, C_5H_5), 9.83 (s, OH) [30]; ($CDCl_3$, −30°C): exo isomer: 0.58 (H'_{anti}; J(H'_{syn}, H'_{anti}) = 0.5), 1.53 (H_{anti}), 2.04 (CH_3), 3.19 (H'_{syn}), 5.30 (C_5H_5); endo

References on pp. 248/50

Table 11 (continued)

No.	compound	method of preparation (yield) properties and remarks
53 (continued)		isomer: 1.33 (H'_{anti}; $J(H'_{syn}, H'_{anti})$ = 3.1), 2.18 (CH_3), 2.66 (H_{anti}), 3.03 (H'_{syn}), 5.30 (C_5H_5) [20, 34] IR ($CHCl_3$): 1670 (ν(C=O)), 1880 to 1910, 1960 to 1980 (ν(CO)) [30]
54	$C_5H_5Mo(CO)_2C_3H_3(CH_3$-syn,1)C(O)$OCH_3$-2	III, for 170 h (63%) [30] m.p. 75°C [30] ^{1}H NMR ($CDCl_3$): 1.00 (m, H'_{anti}; $J(H'_{syn}, H'_{anti})$ = 2.0), 2.10 (m, H_{anti} and CH_3-syn), 3.10 (d, H'_{syn}), 5.20 (s, C_5H_5) [20, 30] IR ($CHCl_3$): 1720 (ν(C=O)), 1875 to 1900, 1955 to 1970 (ν(CO)) [30]
55	exo-$C_9H_7Mo(CO)_2C_3H_3(CH_3$-anti,1)C(O)$CH_3$-2	$C_9H_7Mo(CO)_3CH_3$ was allowed to react in liquid methylene cyclopropane for 14 d (17% by chromatography on Al_2O_3 with hexane/ether); the formation pathway was discussed [33] yellow crystals, subl. 80°C/0.1 Torr [33] ^{1}H NMR ($CDCl_3$): 0.9 (d, CH_3-anti; $J(H_{syn}, CH_3)$ = 7.0), 1.25 (d, H'_{anti}; $J(H'_{syn}, H'_{anti})$ = 2.0), 2.25 (s, CH_3CO), 2.9 (t, H'_{syn}; $J(H'_{anti}, H'_{syn})$ = $J(H_{syn}, H'_{syn})$ = 2.0), 3.95 (d of q, H_{syn}), 5.4 and 5.5 (m, H-2), 5.7 (t, H-1), 7.1 to 7.5 (m, C_9H_7, 6-membered ring) [33] IR (hexane): given from 742 to 1897; 1660 (ν(C=O)), 1897, 1964 (ν(CO)) [33] mass spectrum: $[M-n\ CO]^+$ (n = 0 to 2) [33]
56	$C_5H_5Mo(CO)_2C_3H_3(CH_3$-syn,1)C(O)$OC_2H_5$-2	III, for 170 h (48%) [30] m.p. 71 to 79°C [30] exo to endo ratio 51:49 in $CDCl_3$ at −42°C [34] ^{1}H NMR ($CDCl_3$): 0.93 (m, H'_{anti}; $J(H'_{syn}, H'_{anti})$ = 2.0), 1.30 (t, **CH_3**CH_2), 2.12 (m, CH_3-syn and H_{anti}), 3.32 (d, H'_{syn}), 4.21 (q, **CH_2**CH_3), 5.21 (s, C_5H_5) [20, 30]; ($CDCl_3$, −42°C): exo isomer: 0.53 (H'_{anti}), 1.51 (H_{anti}), 2.05 (CH_3-syn), 3.22 (H'_{syn}), 5.27 (C_5H_5); endo isomer: 1.33 (H'_{anti}), 2.20 (CH_3-syn), 2.60 (H_{anti}), 3.07 (H'_{syn}), 5.29 (C_5H_5) [34] IR ($CHCl_3$): 1720 (ν(C=O)), 1900 to 1920, 1960 to 1980 (ν(CO)) [30]; in part in [20]
57	$C_5H_5Mo(CO)_2C_3H_3(C{\equiv}CCH_3$-syn,1)C(O)$OCH_3$-2	III, for 24 h at 50°C (60% by chromatography on Al_2O_3 with ether/pentane 75:25) [50]

Table 11 (continued)

No.	compound	method of preparation (yield) properties and remarks
		yellow crystals, m.p. 105 °C [50] ^{1}H NMR ($CDCl_3$): 1.26 (H'_{anti}), 1.93 (CH_3-syn), 2.3 (H_{anti}), 3.06 (H'_{syn}), 3.73 (OCH_3), 5.26 (C_5H_5) [50] IR ($CHCl_3$): 1915, 1975 (ν(CO)) [50] mass spectrum: $[M-n\ CO]^+$ (n = 0 to 2), $[M-2\ CO-CH_3OH]^+$ [50] reaction with HBF_4 in ether at −40 °C affords $C_5H_5Mo(CO)_2CH_2{=}C(C(O)OCH_3)CH{=}C{=}CHCH_3$; treatment with $BrCH_2C{\equiv}CH$ in methanol gives $C_5H_5Mo(CO)_2C(C(O)OCH_3)$-$(CH_2OCH_3)$⋯CH⋯C=CHCH$_3$ (No. 78) [50]
58	$C_5H_5Mo(CO)_2C_3H_3(C_6H_5\text{-syn},1)C(O)OCH_3\text{-}2$	III, for 170 h in the presence of CH_3OH and only 24 h in the presence of $BrCH_2C{\equiv}CH$ (50%) [30] m.p. 95 °C [30] ^{1}H NMR ($CDCl_3$): 1.17 (m, H'_{anti}; J(H'_{anti}, H'_{syn}) = 2.0), 3.11 (d, H_{anti}), 3.44 (m, H'_{syn}), 3.65 (s, OCH_3), 5.29 (s, C_5H_5), 7.24 (m, C_6H_5-syn) [20, 30]

^{3}L = O, C, R'', R', R **and** $^3CF_3$, ^{2}F, $^4CF_3$, ^{1}F, $^5CF_3$

No.	compound	method of preparation (yield) properties and remarks
59	$C_5H_5Mo(CO)_2C_3H_2({=}C(C_6H_5)CH_3\text{-}1)C(O)NHC_2H_5\text{-}2$	IXa, amine = $C_2H_5NH_2$ [54] two isomers possible; see No. 62 ^{1}H NMR ($CDCl_3$): 1 to 1.4 (**CH_3**CH_2), 2.4 ($CH_3C{=}$), 3.2 to 3.5 (CH_2), 5.2 (C_5H_5); H'_{anti} and H'_{syn} are covered [54] IR (CH_2Cl_2): 1650 (ν(C=O)), 1920, 1990 (ν(CO)) [54] mass spectrum: $[M-n\ CO]^+$ (n = 0, 2), $[M-3\ CO-C_2H_5NH_2]^+$ [54]
60	$C_5H_5Mo(CO)_2C_3H_2({=}C(C_6H_5)CH_3\text{-}1)C(O)N(C_2H_5)_2\text{-}2$	IXa, amine = $(C_2H_5)_2NH$ [54] two isomers possible; see No. 62 ^{1}H NMR ($CDCl_3$): 1 to 1.3 (**CH_3**CH_2), 2.42 (CH_3), 3.2 to 3.6 (CH_2), 4.86 (C_5H_5), 7.28 (C_6H_5); H'_{anti} and H'_{syn} are covered [54] IR (CH_2Cl_2): 1600 (ν(C=O)), 1920, 1990 (ν(CO)) [54] mass spectrum: $[M-n\ CO]^+$ (n = 0, 2), $[M-3\ CO-(C_2H_5)_2NH]^+$ [54]

References on pp. 248/50

Table 11 (continued)

No.	compound	method of preparation (yield) properties and remarks
61	$C_5H_5Mo(CO)_2C_3H_2(=C_{13}H_8\text{-}1)C(O)N(C_2H_5)_2\text{-}2$ $C_{13}H_8$ = [structure: fluorenylidene]	IXa, amine = $(C_2H_5)_2NH$ [54] ^1H NMR ($CDCl_3$): 0.9 to 1.3 (CH_3), 2.5 (H_{anti}), 3.2 to 3.5 (CH_2N), 3.7 (H_{syn}), 5.3 (C_5H_5), 7.1 to 7.8 (m, $C_{13}H_8$) [54] IR (CH_2Cl_2): 1600 (ν(C=O)), 1940, 1995 (ν(CO)) [54] mass spectrum: $[M-n\ CO]^+$ (n = 0 to 2), $[M-3\ CO-(C_2H_5)_2NH]^+$ [54]
62	$C_5H_5Mo(CO)_2C_3H_2(=C(C_6H_5)CH_3\text{-}1)C(O)OH\text{-}2$	IXb [54] two isomers with different orientations of the CH_3 and C_6H_5 groups towards Mo [54] ^1H NMR ($CDCl_3$): CH_3 directed to Mo: 2.24 (CH_3), 2.36 (H_{anti}), 3.28 (d, H_{syn}; J(H, H) = 1.08), 4.0 (OH), 5.26 (C_5H_5), 7.1 to 7.5 (C_6H_5); C_6H_5 directed to Mo: 2.14 (H_{anti}), 2.37 (CH_3), 3.23 (d, H_{syn}; J(H, H) = 1.8), 4.3 (OH), 4.81 (C_5H_5), 7.26 (C_6H_6) [54] IR (CH_2Cl_2): 1590 to 1620 (ν(C=O)), 1920, 1985 (ν(CO)) [54]
63	$C_5H_5Mo(CO)_2C_3H_2(=C(C_6H_5)CH_3\text{-}1)C(O)OCH_3\text{-}2$	IXc [54] two isomers with different orientations of the CH_3 and C_6H_5 groups towards Mo [54] ^1H NMR ($CDCl_3$): CH_3 directed to Mo: 2.32 (H_{anti} and CH_3), 3.5 (OCH_3 and H_{syn}), 5.35 (C_5H_5), 7.1 to 7.4 (C_6H_5); C_6H_5 directed to Mo: 2.2 (H_{anti}), 2.46 (CH_3), 3.3 (H_{syn}), 4.92 (C_5H_5), 5.72 (OCH_2), 7.2 to 7.4 (C_6H_5) [54] IR (CH_2Cl_2): 1700 (ν(C=O)), 1930, 1990 (ν(CO)) [54] mass spectrum: $[M-n\ CO]^+$ (n = 0, 2), $[M-3\ CO-CH_3OH]^+$ [54]
64	$C_5H_5Mo(CO)_2C_3H_2(=C_{13}H_8\text{-}1)C(O)OCH_3\text{-}2$ $C_{13}H_8$ = [structure: fluorenylidene]	IXc [54] ^1H NMR ($CDCl_3$): 2.6 (H_{anti}), 3.5 (OCH_3), 3.65 (H_{syn}), 5.4 (C_5H_5), 7.1 to 7.8 ($C_{13}H_8$) [54] IR (CH_2Cl_2): 1710 (ν(C=O)), 1960, 2000 (ν(CO)) [54] mass spectrum: $[M-n\ CO]^+$ (n = 0 to 2), $[M-3\ CO-CH_3OH]^+$ [54]
65	$C_5H_5Mo(CO)_2C_3F_2(=C(CF_3)_2\text{-}1)CF_3\text{-}2$	I, $(CF_3)_2C=C=C(CF_3)_2$ was added to a solution of $Na[C_5H_5Mo(CO)_3]$ in THF cooled to −70°C; the mixture was allowed to warm and stirred overnight (9% by chromatography on alumina with

Table 11 (continued)

No.	compound	method of preparation (yield) properties and remarks
		petroleum ether) [24, 29]; $C_5H_5Mo(CO)_2$-$(P(C_6H_5)_3)C(=C(CF_3)_2)C(CF_3)=CF_2$ was refluxed in n-heptane for 24 h (42%) [31] light yellow [29], lemon yellow crystals [31], m.p. 69 to 70°C [31], 71 to 72°C (from pentane) [29] ^{19}F NMR (C_6F_6 as internal standard): 57.6 and 61.8 (F-4, 5), 58.5 (F-3), 79.3 and 89.5 (F-1, 2), with J(F-4, 5) = J(F-3, 5) = 7, J(F-1, 3) < 0.5, J(F-2, 3) = 10, J(F-1, 2) = 127.6 [29] IR (KBr): 1100 to 1400 (ν(CF)); (cyclohexane): 1690 (ν(C=C)), 1990, 2030, 2068 (ν(CO)) [29]; ($CHCl_3$): 1680 (ν(C=C)), 2025, 2065 (ν(CO)) [31]

^3L = 1,3-disubstituted allyl system

No.	compound	method of preparation (yield) properties and remarks
66	$C_5H_5Mo(CO)_2C_3H_3(CH_3\text{-anti},1)CH_3\text{-syn},3$	Ib, ^1L = cis/trans-pent-3-en-1-yl, in THF at 50°C for 4 h (14% by sublimation) [18]; also obtained in the reaction of $[C_5H_5Mo(CO)(CH_3C{\equiv}C$-$C_2H_5)_2]BF_4$ with $NaBH_4$ in THF under CO (54%) [46] yellow crystals [46], subl. 25°C/10^{-4} Torr [18] ^1H NMR (cyclohexane): 1.00 (d, CH_3-anti), 1.80 (m, H_{anti}), 1.84 (m, CH_3-syn), 3.45 (5 lines, H_{syn}), 4.00 (t, H_x), 5.24 (s, C_5H_5), with J(H_{anti}, CH_3-anti) = 6.5, J(H_x, H_{anti}) = 8.0, J(H_x, H_{syn}) = 9.8 [18]; (C_6D_6): 1.06 (d, CH_3-anti), 1.56 (d, CH_3-syn), 2.10 (m, H_{anti}), 3.14 (m, H_{syn}), 3.48 (d of d, H_x), 4.63 (s, C_5H_5), with J(H_{anti}, CH_3-anti) = 6.2, J(H_x, H_{anti}) = 9.6, J(H_x, H_{syn}) = 7.9 [46] ^{13}C NMR (C_6D_6): 15.5 (CH_3-anti), 20.8 (CH_3-syn), 48.2 (CH_{syn}), 60.0 (CH_{anti}), 69.2 (CH_x), 92.1 (C_5H_5) [46] IR (KBr): 1445 (ν(C=C)), (cyclohexane): 1870, 1945 (ν(CO)) [18]; (hexane): 1877, 1951 (ν(CO)) [46] mass spectrum: $[M-n\,CO]^+$ (n = 1, 2) [46] reaction with $[C(C_6H_5)_3]BF_4$ affords exo/endo-$[C_5H_5Mo(CO)_2C_4H_5CH_3]BF_4$ [28]
67	$C_5H_5Mo(CO)_2C_3H_3(CH_3)_2\text{-syn},1,3$	IIa, X = Cl, for 18 h [23] yellow solid, m.p. 146 to 148°C (from ether/hexane) [23] exo to endo ratio 17.5 in $CDCl_3$ at 0°C [23]

Table 11 (continued)

No.	compound	method of preparation (yield) properties and remarks
67 (continued)		^{1}H NMR ($CDCl_3$): exo isomer: 1.39 (H), 1.69 (CH_3), 3.98 (H_x), 5.23 (C_5H_5), with J(H_x, H) = 9.0, J(H, CH_3) = 6.0; endo isomer: 1.85 (CH_3), 2.60 (H), 3.42 (H_x), 5.14 (C_5H_5), with J(H_x, H) = 9.6, J(H, CH_3) = 5.7 [23] [95]Mo NMR (acetone-d_6, vs. 2 M Na_2MoO_4 in D_2O): exo isomer: −1752; endo isomer: −1709; ratio 4:1 [74] IR (cyclohexane): exo isomer: 1872, 1945 (ν(CO)); endo isomer: 1883, 1950 (ν(CO)) [23] activation parameters and rate constants for isomerization in CS_2 at 25°C: endo to exo: $\Delta F^{\neq}$ = 15.3 ± 0.2 kcal/mol, k = 40 ± 12 s^{-1}; exo to endo: $\Delta F^{\neq}$ = 17.0 ± 0.2 kcal/mol, k = 2 ± 1 s^{-1} [23]
68	exo-$C_{10}H_{19}C_4H_5Mo(CO)_2C_3H_3(CH_3)_2$-syn,1,3 $C_{10}H_{19}C_5H_4$ = C_5H_4 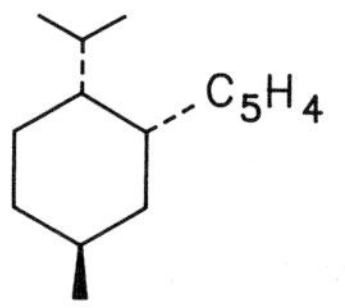	IIa, X = Br, for 10 h at 25°C (75% together with isomer containing CH_3 in anti positions; 85% of No. 68 in the mixture) [47] yellow crystals, m.p. 54 to 56°C (from pentane) optical rotation (pentane, 0.112 M): $[\alpha]_D$ = +49.5° [47] ^{1}H NMR ($CDCl_3$): 0.71 (d, $C_{10}H_{19}$, CH_3; J(H, H) = 6.3), 0.93 (d, $C_{10}H_{19}$, CH_3; J(H, H) = 6.0), 0.94 (d, $C_{10}H_{19}$, CH_3; J(H, H) = 5.9), 1.73 and 1.76 (d, CH_3-syn; J(H, H) = 6.3), 3.98 (t, H_x; J(H, H) = 9.5), 5.01, 5.05, 5.24, and 5.41 (m, C_5H_4); H_{anti} are obscured by neomenthyl resonances [47] IR (cyclohexane): 1872, 1948 (ν(CO)) [47] reaction with $[NO]PF_6$ in CH_3CN at 0°C affords $[C_{10}H_{19}C_5H_4Mo(CO)(NO)C_3H_3(CH_3)_2\text{-1,3}]PF_6$ [47]
69	exo-$C_9H_7Mo(CO)_2C_3H_3(CH_3$-syn,1)$CH_3$-anti,3	III, [4]L = trans-$C_4H_5CH_3$-1, with $Na[BH_4]$ (67%) [27] yellow crystals [27] ^{1}H NMR ($CDCl_3$): 0.1 (d of d, H_x; J(H_{anti}, H_x) = 10.0, J(H_{syn}, H_x) = 7.0), 0.9 (d, CH_3-anti; J(H_{syn}, CH_3-anti) = 7.0), 1.60 (d, CH_3-syn; J(H_{anti}, CH_3-syn) = 7.0), 2.1 (m, H_{anti}), 2.5 (q, H_{syn}), 5.6 (m, H-2), 5.9 (m, H-1), 7.9 to 7.1 (m, C_9H_7, 6-membered ring) [27] IR (hexane): 1872, 1947 (ν(CO)) [27] mass spectrum: $[M-n\,CO]^+$ (n = 0 to 2) [27]

References on pp. 248/50

Table 11 (continued)

No.	compound	method of preparation (yield) properties and remarks
70	exo-$C_9H_7Mo(CO)_2C_3H_3(CH_3$-syn,1)$CH_2CHC(O)(CH_2)_3$-anti,3 ^{3}L = 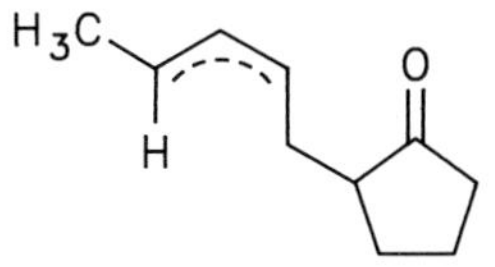	VIII, ^{4}L = cis-$C_4H_5CH_3$-1, with ca. 2 equivalents of 1-morpholinocyclopent-1-ene (66%) [27] yellow crystals (from ether/hexane at −78 °C) [27] ^{1}H NMR ($CDCl_3$, 32 °C): 0.0 (m, H_x), 1.6 (d, CH_3-syn; J(H_{anti}, CH_3-syn) = 7.0), 1.5 to 2.6 (m, H_{anti}, H_{syn}, CH_2, and CH), 5.6, 5.7, and 6.0 (m, H-1,2), 7.1 (m, C_9H_7, 6-membered ring) [27] IR: 1746 (ν(C=O)), 1871, 1944 (ν(CO)) [27] mass spectrum: $[M-CO]^+$ (n = 0 to 2) [27]
71	$C_5H_5Mo(CO)_2C_3H_3(CH_3$-1)$C_3H_7$-i-3	IIa, X = O_2CCH_3 [68] crystal data: orthorhombic space group P $2_12_12_1$ − D_2^4 (No. 19) with the unit cell parameters a = 8.049(1), b = 13.156(2), c = 13.532(4) Å; Z = 4 molecules per unit cell [68] reaction with $[NO]^+$ affords $[C_5H_5Mo(CO)(NO)C_3H_3(CH_3)C_3H_7\text{-i}]^+$ [68]
72	$C_5H_5Mo(CO)_2C_3H_3(CH_3$-syn,1)$CH{=}CH_2$-syn,3 ^{3}L = 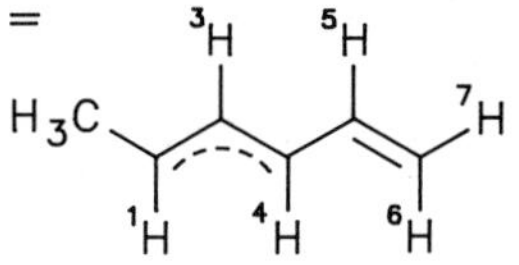	No. 23 was irradiated in ether at −20 °C for 18 h (67% by chromatography on Al_2O_3 at 0 °C with pentane) [69] exo isomer [69] pale yellow crystals (from pentane) [69] ^{1}H NMR (toluene-d_8): 1.42 (d, CH_3), 1.48 (m, H-1), 2.00 (t, H-4), 3.58 (t, H-3), 4.70 (s, C_5H_5), 4.80 (d of d, H-7), 5.23 (d of d, H-6), 5.55 (d of t, H-5), with J(H-1, CH_3) = 6.8, J(H-1, 3) = J(H-3, 4) = 10.4, J(H-4, 5) = 10.5, J(H-5, 6) = 16.8, J(H-5, 7) = 10.4 [69] ^{13}C NMR (C_6D_6): 20.3 (CH_3); 58.0 (CH-1), 61.0 (CH-4), 71.7 (CH-3), 91.5 (C_5H_5), 110.5 (CH-6, 7), 139.6 (CH-5) [69] IR (pentane): 1617 (ν(C=C)), 1885, 1956 (ν(CO)) [69] mass spectrum (12 eV): $[M-n\,CO]^+$ (n = 0, 1) [69]
	^{3}L = trisubstituted allyl system	
73	$C_5H_5Mo(CO)_2C_3H_2(CH_3)_3$-1,1,2	IIa, X = Cl, for 18 h [23] yellow solid, m.p. 140 to 143 °C (from ether/hexane) [23] exo to endo ratio 14.1 in CS_2 at −5 °C [23] ^{1}H NMR (CS_2): exo isomer: 1.10 (CH_3-anti), 1.28 (H_{anti}; J(H_{syn}, H_{anti}) = 2.7), 1.90 (CH_3-x), 2.12

References on pp. 248/50

Table 11 (continued)

No.	compound	method of preparation (yield) properties and remarks
73 (continued)		(CH_3-syn), 2.62 (H_{syn}), 5.12 (C_5H_5); endo isomer: 1.38 (CH_3-anti), 2.04 (CH_3-x), 2.27 (CH_3-syn), 2.42 (H_{anti}; J(H_{syn}, H_{anti}) $\leq$ 0.6), 2.90 (H_{syn}), 5.12 (C_5H_5) [23] ^{95}Mo NMR (acetone-d_6, vs. 2 M Na_2MoO_4 in D_2O): exo isomer: −1657; endo isomer: −1448; ratio 20:1 [74] IR (cyclohexane): exo isomer: 1875, 1950 (ν(CO)); endo isomer: 1885, 1960 (ν(CO)) [23] activation parameters and rate constants for isomerization CS_2 at 25°C: endo to exo: $\Delta F^{\neq}$ = 14.8 ± 0.2 kcal/mol, k = 90 ± 30 s^{-1}; exo to endo: $\Delta F^{\neq}$ = 16.3 ± 0.2 kcal/mol, k = 7 ± 3 s^{-1} [23] reaction with $[C(C_6H_5)_3]BF_4$ in CH_2Cl_2 affords $[C_5H_5Mo(CO)_2C_4H_4(CH_3)_2]BF_4$ ($C_4H_4(CH_3)_2$ = 2,3-dimethylbuta-1,3-diene) [28]
74	exo-$C_9H_7Mo(CO)_2C_3H_2(CH_3)_3$-1,1,2	IIa, X = Cl [23]; VIII, ^{4}L = cis-2,3-dimethylbutadiene, with $Na[BH_4]$ (80%) [27] yellow crystals [27], m.p. 76 to 78°C (from ether/hexane) [23] exo to endo ratio $\leq$ 50 in CS_2 at −5°C [23] ^{1}H NMR (CS_2): 0.65 (CH_3-anti), 0.99 (H_{anti}; J(H_{syn}, H_{anti}) = 2.7), 1.06 (CH_3-x), 2.08 (CH_3-syn), 2.16 (H_{syn}), 5.25, 5.37, and 5.74 (C_9H_7) [23]; ($CDCl_3$, +32°C): 0.7 (s, CH_3-anti), 1.1 (d, H_{anti}; J(H_{syn}, H_{anti}) = 2.0), 1.2 (s, CH_3-syn), 1.29 (d, H_{syn}), 2.10 (s, CH_3-x), 5.39 (m, H-2), 5.8 (t, H-1; J(H, H) = 3.0), 6.9 to 7.5 (m, C_9H_7, 6-membered ring) [27] IR (cyclohexane): 1876, 1950 (ν(CO)) [23]; (hexane): 1873, 1946 (ν(CO)) [27] mass spectrum: $[M-n\,CO]^+$ (n = 0 to 2) [27]
75	$C_5H_5Mo(CO)_2C_3H_2(CH_3)_2C(O)CH{=}CHCH{=}CHCH_3$ ^{3}L = [structure: O, H_3C, CH_3, CH_3, H]	$C_5H_5Mo(CO)_2$(μ-$CHCHC(CH_3)_2$)(μ-CO)-$Mo(CO)C_5H_5$ (Formula V, p. 207) was allowed to react with 50 equivalents of trans-penta-1,3-diene in CH_2Cl_2 for 48 h at 50°C (25% by chromatography on Al_2O_3 with hexane/ether 1:1) [61] orange-yellow crystals (from THF/hexane at −78°C) [61] ^{1}H NMR ($CDCl_3$): 1.19 (s, CH_3), 1.88 (d, CH_3; J(H, H) = 3.9), 1.89 (s, CH_3), 2.49 (d, H_{anti}; J(H_x, H_{anti}) = 9.5), 5.09 (d, H_x), 5.16 (s, C_5H_5), 6.21

Table 11 (continued)

No.	compound	method of preparation (yield) properties and remarks
		(d, 1H; J(H, H) = 4.6), 6.26 (d, 1H; J(H, H) = 4.2), 7.26 (m, 1H) [61] ^{13}C NMR ($CDCl_3$): 18.8, 23.2, and 30.3 (CH_3), 49.9, 71.9 (CH), 83.9 (**C**$(CH_3)_2$), 94.6 (C_5H_5), 128.2, 130.6, 139.7, and 141.2 (CH=), 196.1 (C=O), 238.4 and 242.5 (CO) [61] IR (CH_2Cl_2): 1623, 1651 (ν(C=C)), 1859, 1947 (ν(CO)) [61] mass spectrum: $[M]^+$ [61]
76	$(CH_3)_5C_5Mo(CO)_2C_3H_2(CH_3)_3$-1,2,3 (1,3-syn)	VIIa (65%) [48] ^{1}H NMR (C_6D_6): 0.88 (q, H_{anti}; J(CH_3-syn, H_{anti}) = 6.3), 1.60 (s, $(CH_3)_5C_5$), 1.74 (s, CH_3-x), 1.88 (d, CH_3-syn) [48] ^{13}C NMR (C_6D_6): 10.5 ((**C**$H_3)_5C_5$), 16.0 (CH_3-syn), 28.3 (CH_3-x), 60.6 (**C**CH_3), 101.9 (**C**$_5(CH_3)_5$), 119.4 (**C**HCH_3), 244.6 (CO) [48] IR (hexane): 1865, 1939 (ν(CO)) [48]
77	$(CH_3)_5C_5Mo(CO)_2C_3H_2((CH_3)_2$-syn,1,3)$CH_2OH$-2	VIIc (72%) [48] ^{1}H NMR (C_6D_6): 0.86 (q, H_{anti}; J(CH_3-syn, H_{anti}) = 6.2), 1.55 (s, $(CH_3)_5C_5$), 1.90 (d, CH_3-syn), 3.90 (s, CH_2O) [48] ^{13}C NMR (C_6D_6): 10.5 ((**C**$H_3)_5C_5$), 15.4 (CH_3), 61.7 (CH), 66.4 (CH_2OH), 101.9 (**C**$_5(CH_3)_5$), 128.9 (**C**CH_2OH), 243.0 (CO) [48] IR (hexane): 1850, 1934 (ν(CO)) [48]

^3L = 1,1,3,3-tetrasubstituted allyl system

No.	compound	method of preparation (yield) properties and remarks
78	$C_5H_5Mo(CO)_2C_3H(=CHCH_3$-1)($CH_2OCH_3$-anti,3)C(O)$OCH_3$-syn,3 ^3L = O, CH_3O, 3, CH_3, H^1, H^2, H^4, CH_3O	III, ^1L = $CH_2C{\equiv}CC{\equiv}CCH_3$, in CH_3OH at 50°C in the presence of $BrCH_2C{\equiv}CH$ for 2 h (27% by chromatography on Al_2O_3 with ether/pentane 75:25); also obtained by treating $[C_5H_5Mo(CO)_2CH_2{=}C(C(O)OCH_3)CH{=}C{=}CH$-$CH_3]BF_4$ with $NaOCH_3$ in CH_3OH at −70°C (75%) [50] ^{1}H NMR ($CDCl_3$): 1.92 (m, CH_3), 3.38 (s, CH_3CO), 3.40 (s, OCH_3), 3.42 (H-2; J(H-1, 2) = 10), 4.16 (H-3; J(H-4, 3) = 2.7, J(CH_3, H-3) = 2.2), 4.40 (H-1), 5.3 (C_5H_5), 6.84 (H-4; J(CH_3, H-4) = 6.8) [50] IR ($CHCl_3$): 1980 (?), 1970 (ν(CO)) [50] mass spectrum: $[M-n\ CO]^+$ (n = 1, 2), $[M-2\ CO-CH_3OH]^+$ [50]

References on pp. 248/50

Table 11 (continued)

No.	compound	method of preparation (yield) properties and remarks
^3L 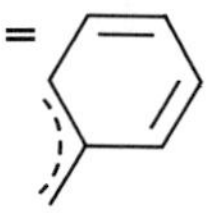 =	**and derivatives**	
79	endo-$C_5H_5Mo(CO)_2CH_2C_6H_5$	Ia, in n-hexane for 5 d (36%); Ib, subl. 100 °C/0.1 Torr (4%) [5, 6]; Ic (51%) [56]; X (63%) [49, 53, 60] dark red crystals, m.p. 83 to 85 °C [5] ^1H NMR (C_6F_5Br, −30 °C): 1.81 and 2.82 (CH_2; J(H, H) = 3), 5.20 (C_6H_5, 1H), 5.31 (C_5H_5), 6.31, 6.88, 7.03, and 7.09 (C_6H_5) [5], similar data in [16]; (C_6F_5Br, +20 °C): 2.0 and 2.7 (CH_2), 5.23 (C_5H_5), 6.18 and 6.93 (C_6H_5); (C_6F_5Br, +64 °C): 2.31 (br s, CH_2), 5.20 (C_5H_5), 5.75 and 6.99 (C_6H_5); $E_a \approx$ 6 kcal/mol (± 30%) for rotation; figure of the spectra also given [5] IR (cyclohexane): 1873, 1965 (ν(CO)) [4, 5]; (halocarbon mull): given from 693 to 1954; 1833, 1935, 1954 (ν(CO)) [5] decomposes as a solid at room temperature; solutions are noticeably oxidized after exposure to air for several minutes, and in CS_2 decomposition is complete within 1 h [5]
80	endo-$C_9H_7Mo(CO)_2CH_2C_6H_5$	exo content < 1% [5] ^1H NMR (C_6D_6, 5 °C): −0.21 (H_{anti}), 2.13 (H'_{anti}), 4.14 (H_{syn}) [16] IR (cyclohexane): 1887, 1960 (ν(CO)) [16]
*81	$C_5H_5Mo(CO)_2CH_2C_6H_4CH_3$-2 ^3L 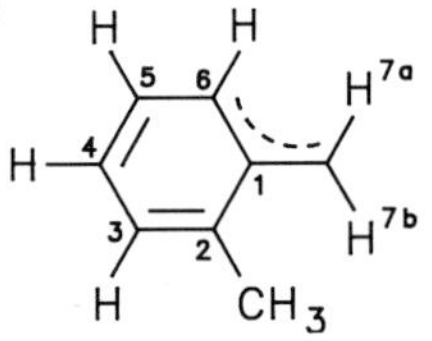 = (H-7a, H-7b at C-7; H at C-3, C-4, C-5, C-6; CH_3 at C-2)	X (moderate yield) [66] ^1H NMR (C_6D_6): 1.22 (H-7b; J(H-7a, b) = 2), 2.07 (CH_3), 3.34 (H-7a), 4.41 (H-6), 4.70 (C_5H_5), 6.70 (H-3; J(H-3, 4) = 6.5), 6.82 (H-5; J(H-5, 6) = 6), 6.96 (H-4; J(H-4, 5) = 8) [66]
*82	$C_5H_5Mo(CO)_2CH_2C_6H_4CH_3$-3 ^3L 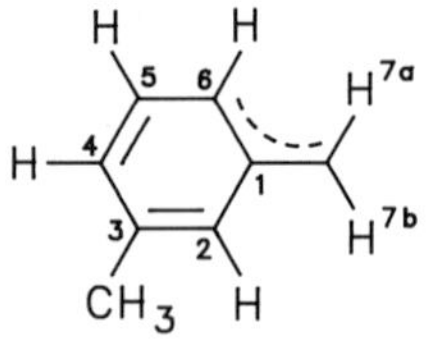 = (H-7a, H-7b at C-7; H at C-2, C-4, C-5, C-6; CH_3 at C-3)	X (moderate yield) [66] two isomers (exo/endo) in a 1:2.2 ratio at −20 °C in toluene-d_8 [66] ^1H NMR (toluene-d_8, −20 °C): major isomer: 1.32 (H-7b), 2.01 (CH_3), 3.15 (H-7a), 4.00 (H-2), 4.56 (C_5H_5), 6.60 (H-6; J(H-5, 6) = 8), ca. 6.70 (H-4 and H-5); minor isomer: 1.81 (H-7a), 2.14 (CH_3), 2.55 (H-7b), 4.41 (C_5H_5), 4.81 (H-6), 5.64 (H-2), 6.70 (H-4), 6.74 (H-5) [66]

References on pp. 248/50

Table 11 (continued)

No.	compound	method of preparation (yield) properties and remarks
*83	$C_5H_5Mo(CO)_2CH_2C_6H_4CH_3$-4	Ib, in vacuum at 105°C for 1.3 h (11% by chromatography on SiO_2 with C_6H_6/petroleum ether 1:3) [11] red-orange solid, m.p. 98 to 100°C, subl. 77°C/0.04 Torr [11] ^1H NMR ($CDCl_3$): ca. 2.3 (CH_3), ca. 5.3 (C_5H_5) [11] IR (cyclohexane): 1875, 1950 (ν(CO)) [11] stable for several months when stored at ca. 5°C under N_2 [11] no reaction of the remaining diene part of the benzyl system is observed with maleic anhydride, $(CN)_2C{=}C(CN)_2$, or $CH_3OC(O)C{\equiv}C$-$C(O)OCH_3$ [11]
*84	$C_5H_5Mo(CO)_2CH_2C_6H_3(C_3H_7\text{-}i)_2$-3,5	Ib, in a vacuum at 110°C for 2.2 h (23% by chromatography on SiO_2 with C_6H_6/petroleum ether 1:3) [11] dark red crystals, m.p. 69 to 71°C, subl. 78°C/0.04 Torr [11] ^1H NMR (toluene-d_8): ca. 1.2 (CH_3), ca. 2.7 (br m, **CH**$(CH_3)_2$; J(H, H) ≈ 6.8), ca. 4.9 (C_5H_5); temperature-dependent spectra are given as a diagram at −24, +25, and +72°C [11] IR (cyclohexane): 1880, 1955 (ν(CO)) [11]
85	$C_5H_5Mo(CO)_2CH(Sn(C_6H_5)_3)C_6H_5$ ^3L =	formed in an equilibrium (ratio 3:1) by dissolving trans-$C_5H_5Mo(CO)_2(Sn(C_6H_5)_3){=}CHC_6H_5$ in THF [72] ^1H NMR ($CDCl_3$): 3.25 (s, H-7; J(Sn, H) = 10), 5.09 (s, C_5H_5), 5.29 (d of d, H-2; J(H, H) = 6, 1.5), 6.87 (m, H-6), 6.97 (m, H-3), 7.14 (m, H-4, 5), 7.32 to 7.42 (m, SnC_6H_5, H-3, 4), 7.50 to 7.58 (m, SnC_6H_5, H-2) [72] ^{13}C NMR (CD_2Cl_2, −50°C): 26.0 (C-7), 75.2 (C-2), 92.7 (C_5H_5), 110.5 (C-1), 124.4 (C-6), 128.4 (SnC_6H_5, C-3, 4; J(Sn, C) = 43), 128.9 (C-3), 133.0 (C-5), 133.2 (C-4), 136.4 (s, SnC_6H_5, C-2; J(Sn, C) = 35), 140.4 (SnC_6H_5, C-1), 243.8 and 244.3 (CO) [72] IR (CH_2Cl_2): 1868, 1945 (ν(CO)) [72]
^3L is a part of a five- or six-membered ring		
86	$C_5H_5Mo(CO)_2CH_2C_5H_7$-cyclo ^3L = 	Ia, in ether [19] mechanism of formation was given, together with a comparison of NMR shifts (no data) [19]

References on pp. 248/50

Table 11 (continued)

No.	compound	method of preparation (yield) properties and remarks
87	$C_5H_5Mo(CO)_2CH_2CC(O)CH_2CH_2CH$-cyclo ^3L 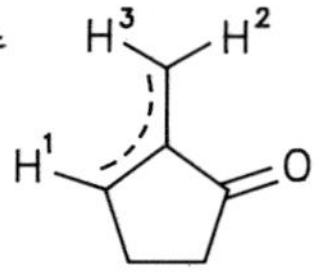 = [structure: H^3, H^2, H^1, O]	IV (55%) [38], 72 h reaction time (40%) [15] m.p. 120°C [38], 124°C [15] exo to endo ratio 3:2 in $CDCl_3$ at −40°C; the ratio is solvent-dependent; in C_2Cl_4 the endo isomer predominates [38] ^{1}H NMR ($CDCl_3$, −40°C): exo isomer: 1.28 (d, H-3; J(H-2, 3) = 3), 3.01 (d, H-2), 3.60 (virtual d, H-1; separated by 4 Hz), 5.10 (s, C_5H_5); endo isomer: 2.24 (br s, H-3), 3.44 (br s, H-2), 4.40 (virtual d, H-1; separated by 3.5 Hz), 5.30 (s, C_5H_5) [35, 38]; spectrum at −40 and +37°C given as a diagram in [38]; (C_2Cl_4, −10°C): exo isomer: 1.23 (d, H-3; J(H-2, 3) = 3), 3.01 (d, H-2), 3.48 (virtual d, H-1; separated by 4 Hz), 5.10 (s, C_5H_5); endo isomer: 5.20 (s, C_5H_5) [38] IR ($CHCl_3$): 1690 (ν(C=O)), 1900, 1965 (ν(CO)); (C_2Cl_4): 1700 (ν(C=O)); exo isomer: 1911, 1982 (ν(CO)); endo isomer: 1884, 1965 (ν(CO)) [15, 38] mass spectrum: $[M]^+$ [38]
88	$C_5H_5Mo(CO)_2CHDCC(O)CH_2CHDCH$-cyclo (compare No. 87)	IV, reaction with $BrCH_2CHDCH{=}C{=}CHD$ [38] ^{1}H NMR: spectrum in $CDCl_3$ at −45°C given as a figure in [38]
89	exo-$C_5H_5Mo(CO)_2C(CH_3)HCC(O)CH_2CH_2CH$-cyclo ^3L 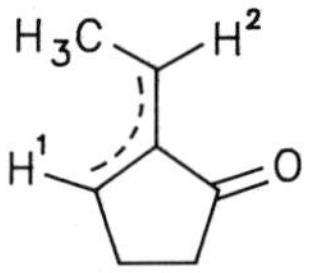 = [structure: H_3C, H^2, H^1, O]	IV (74%) [38], 72 h reaction time (78%) [15] m.p. 148°C [15, 38] ^{1}H NMR ($CDCl_3$, 37°C): 1.15 (d, CH_3; J(H-2, CH_3) = 6.5), 3.84 (q, H-2), 3.84 (pseudo d, H-1; separated by 4.5 Hz), 5.10 (s, C_5H_5) [38]; similar data in [15] IR ($CHCl_3$): 1690 (ν(C=O)), 1880, 1950 (ν(CO)) [38]; similar data in [15] mass spectrum: $[M]^+$ [38]
90	exo-$C_5H_5Mo(CO)_2CH_2CC(O)CH_2CH_2CCH_3$-cyclo ^3L 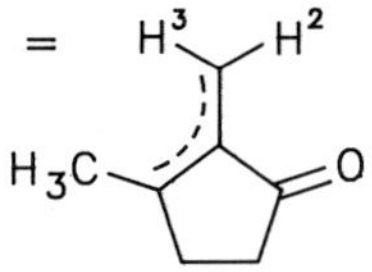 = [structure: H^3, H^2, H_3C, O]	IV (45%) [15, 38] m.p. 154°C [15, 38] ^{1}H NMR ($CDCl_3$, 37°C): 1.40 (d, H-3; J(H-2, 3) = 3), 1.45 (s, CH_3), 3.05 (d, H-2), 5.05 (s, C_5H_5) [38]; similar data in [15] IR ($CHCl_3$): 1690 (ν(C=O)), 1880, 1950 (ν(CO)) [38]; similar data in [15]

Table 11 (continued)

No.	compound	method of preparation (yield) properties and remarks
91	exo-$C_5H_5Mo(CO)_2C(CH_3)HCC(O)CH_2CH_2CCH_3$-cyclo 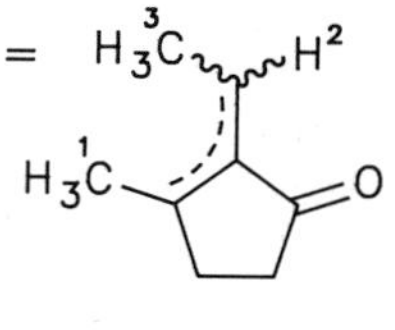	IV (25%); a mixture of isomers with CH_3-3 in syn (47%) and anti (53%) position [38] m.p. 150°C (mixture of the two isomers), 214°C [38] (pure CH_3-syn, 3; obtained after thermal isomerization and recrystallization from CH_2Cl_2/hexane) [19, 38] ^1H NMR (C_2Cl_4, 100°C): CH_3-anti, 3 isomer: 1.55 (d, CH_3-3, J(H-2, CH_3-3) = 7), 1.80 (s, CH_3-1), 4.1 (q, H-2), 5.0 (s, C_5H_5); CH_3-syn, 3 isomer: 1.3 (s, CH_3-1), 2.1 (s, H-2 and CH_3-3), 5.1 (s, C_5H_5) [38] mass spectrum: $[M]^+$ [38]
92	$C_5H_5Mo(CO)_2CH_2CC(O)OCH_2CH$-cyclo 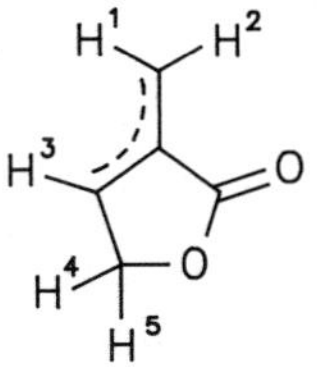	XI (64%) [32, 51] yellow crystals [32, 51] exo to endo ratio 4:1 at −20°C in $CDCl_3$ [32, 51] ^1H NMR ($CDCl_3$, −20°C): exo isomer: 1.45 (H-1), 3.31 (H-2), 3.57 (H-3), 4.59 (H-4), 5.16 (H-5), 5.25 (C_5H_5); endo isomer: 2.32 (H-1), 3.48 (H-2), 4.28 (H-3), 4.78 (H-4), 5.11 (H-5), 5.36 (C_5H_5) [32, 51] IR ($CHCl_3$): 1750 (ν(C=O)), 1860, 1940 (ν(CO)) [32, 51] reaction with $[NO]BF_4$ in ether at low temperatures affords the corresponding complex $[C_5H_5Mo(CO)(NO)^3L]BF_4$; oxidation with $[NH_4]_2[Ce(NO_3)_6]$ in THF at 50°C affords the free dimer of the coordinated ^3L ligand [51]
93	$C_5H_5Mo(CO)_2CH_2CC(O)OC(C_6H_5)HCH$-cyclo 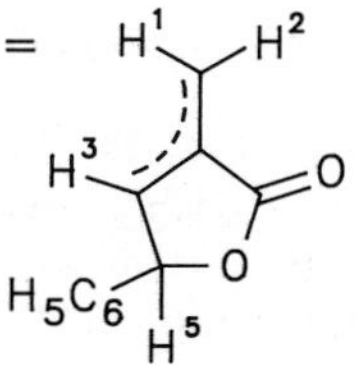	XI (30%) [32] ^1H NMR ($CDCl_3$): 1.50 (H-1), 3.35 (H-2), 3.85 (H-3), 4.90 (C_5H_5), 6.40 (H-5), 7.35 (C_6H_5) [32] IR ($CHCl_3$): 1760 (ν(C=O)), 1890, 1970 (ν(CO)) [32]
94	$C_5H_5Mo(CO)_2CH_2CC(O)OC(CH_3)_2CH$-cyclo 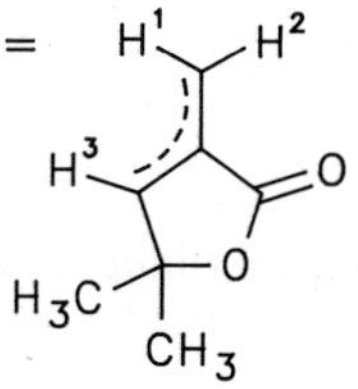	XI (53%) [32] ^1H NMR ($CDCl_3$): 1.43 (H-1), 1.52 and 1.75 (CH_3), 3.15 (H-2), 3.40 (H-3), 5.20 (C_5H_5) [32] IR ($CHCl_3$): 1750 (ν(C=O)), 1860, 1940 (ν(CO)) [32]

References on pp. 248/50

Table 11 (continued)

No.	compound	method of preparation (yield) properties and remarks
95	$C_5H_5Mo(CO)_2CH_2CC(O)OCH_2CH_2CH$-cyclo ^3L 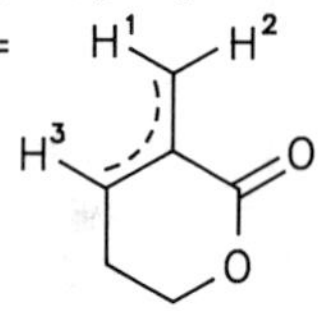 = (structure with H¹, H², H³, O)	XI (40%) [32] ^1H NMR ($CDCl_3$): 0.85 (H-1), 2.10 (H-2), 2.40 (CH_2), 2.85 (H-3), 3.85 (CH_2), 5.20 (C_5H_5) [32] IR ($CHCl_3$): 1710 (ν(C=O)), 1880, 1950 (ν(CO)) [32]
96	$C_5H_5Mo(CO)_2CH_2SC_4H_3$-cyclo ^3L 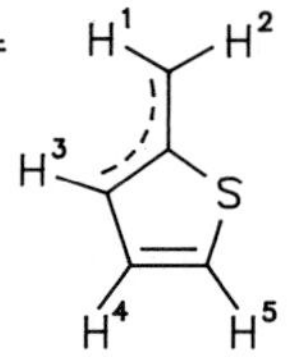 = (structure with H¹, H², H³, S, H⁴, H⁵)	Ia, in hexane for 26 h (53%) [13, 14] red-orange crystals, m.p. 102 to 103 °C (from hexane), subl. 85 °C/0.1 Torr [14] ^1H NMR (toluene-d_8): 1.92 (H-1; J(H, H) = 4), 3.35 (d of d, H-4; J(H, H) = 6, 2), 3.51 (d, H-2; J(H, H) = 4), 4.51 (d, H-3; J(H, H) = 2), 4.64 (s, C_5H_5), 6.67 (d, H-5; J(H, H) = 6); the spectra are temperature-independent between −60 and +105 °C [14] IR (cyclohexane): given from 660 to 1967; 1882, 1895, 1963, 1967 (ν(CO)) [14] mass spectrum: $[M]^+$ (26), $[M-CO]^+$ (7), $[M-2\,CO]^+$ (100), $[C_5H_5MoCH_2SC_2H]^+$ (27), $[C_5H_5MoC_2HS]^+$ (6), $[C_5H_5MoCHS]^+$ (40), $[C_5H_5MoS]^+$ (26), $[C_3H_3MoS]^+$ (11), $[C_5H_5Mo]^+$ (16), $[C_3H_3Mo]^+$ (10) [14]
97	$C_5H_5Mo(CO)_2CH_2C_4H_3S$-cyclo ^3L 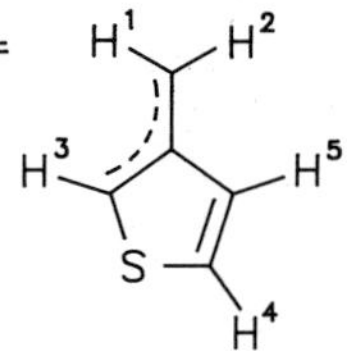 = (structure with H¹, H², H³, H⁵, S, H⁴)	Ia, in hexane for 30 h (35%) [13, 14] red-orange crystals, m.p. 125 to 127 °C, subl. 85 °C/ 0.1 Torr [14] ^1H NMR (toluene-d_8): 1.53 (H-1), 3.18 (d, H-2; J(H, H) = 3), 4.56 (C_5H_5), ca. 4.70 (H-3), 6.03 (d, H-5; J(H, H) = 6), 6.50 (d of d, H-4; J(H, H) = 3, 6); spectrum is temperature-independent between −60 and 105 °C [14] IR (CH_2Cl_2): given from 700 to 3090; 1884, 1897, 1970 (ν(CO)) [14] mass spectrum: $[M]^+$ (28), $[M-CO]^+$ (6), $[M-2\,CO]^+$ (100), $[C_5H_5MoC_2HSCH_2]^+$ (6), $[C_5H_5MoCHS]^+$ (39), $[C_5H_5MoS]^+$ (23), $[C_5H_5Mo]^+$ (15), $[C_3H_3Mo]^+$ (14) [14]

^{3}**L** = (allyl–CH^1H^2–$M(CO)_5$ with R) **and** (allyl–$C(^1H)(^2H)$–$M(CO)_5$) (M = Mn, Re)

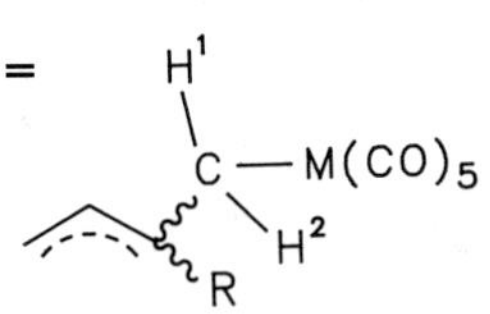

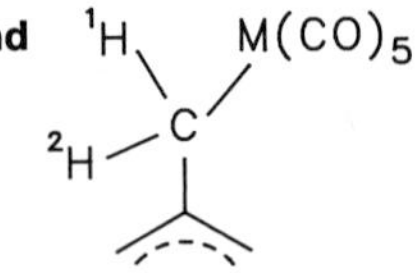

No.	compound	method of preparation (yield) properties and remarks
98	exo-$C_5H_5Mo(CO)_2C_3H_4CH_2Mn(CO)_5$	V, ^4L = cis-C_4H_6, for 15 min [67, 70] orange powder, decomposes > 86 °C

References on pp. 248/50

Table 11 (continued)

No.	compound	method of preparation (yield) properties and remarks
		^{1}H NMR (CD_2Cl_2): 0.40 (m, H-1), 1.60 (m, H-2), 1.80 (m, H_{anti}), 2.83 (m, H_{anti}), 4.08 (m, H_{syn}), 5.05 (m, H_x), 5.23 (s, C_5H_5) [70] IR (CH_2Cl_2): 1855, 1935, 1977, 2007, 2047, 2076, 2104 (ν(CO)) [70] decomposes rapidly in solution [70]
99	exo-$C_5H_5Mo(CO)_2C_3H_4CH_2Re(CO)_5$-1	V, ^{4}L = cis-C_4H_6, for 45 min (62%) [70] orange-yellow powder, decomposes > 128 °C [70] ^{1}H NMR (CD_2Cl_2): 0.20 (br t, H-1), 1.60 (br d of d, H-2), 1.82 (d of d, H_{anti}), 2.70 (m, H_{anti}), 3.91 (m, H_{syn}), 5.20 (m, H_x), 5.21 (C_5H_5), with J(H-1, 2) = −10.9, J(H_{anti}, H_x) = 11.5 and 1.8, J(H_{anti}, H-1) = 3.7 [70] ^{13}C NMR (CD_2Cl_2): 91.4 (C_5H_5), 185.7 ($ReCO_{eq}$) [70] IR (CH_2Cl_2): 1843, 1928, 1980, 2012, 2045, 2125 (ν(CO)) [70] decomposes on standing in solution [70]
100	exo-$C_5H_5Mo(CO)_2C_3H_4CH_2Re(CO)_5$-2	V, ^{4}L = $C(CH_2)_3$, for 60 min (61%) [70] yellow-brown powder, decomposes > 147 °C [70] ^{13}C NMR (CD_2Cl_2): 36.1 (CH-1, 2), 91.2 (C_5H_5), 94.2 (CH_2), 125.5 (C_x), 181.7 ($ReCO_{ax}$), 185.8 ($ReCO_{eq}$), 242.0 (MoCO) [70] ^{1}H NMR and IR spectra are given in "Organo-rhenium Compounds" 2, 1989, p. 179
101	exo-$C_5H_5Mo(CO)_2C_3H_3(CH_3$-anti,1)$CH_2Re(CO)_5$-1	V, ^{4}L = cis-$H_2C{=}CHC(CH_3){=}CH_2$, for 2 h (48%) [70] bright yellow solid, decomposes > 140 °C [70] ^{13}C NMR (CD_2Cl_2): 91.9 (C_5H_5), 186.2 ($ReCO_{eq}$) [70] ^{95}Mo NMR (CD_2Cl_2, vs. 2 M Na_2MoO_4 in D_2O at pH 11): 1497 ($H_{1/2}$ = 250) [70] ^{1}H NMR and IR spectra are given in "Organo-rhenium Compounds" 2, 1989, p. 179
102	exo-$C_5H_5Mo(CO)_2C_3H_3(CH_3$-syn,1)$CH_2Re(CO)_5$-syn,3	V, ^{4}L = cis-$C_4H_5CH_3$-1, for 2 h (66%) [70] orange-yellow crystals, decompose > 115 °C [70] ^{1}H NMR (CD_2Cl_2): 0.19 (br d, H-1), 1.20 (d, H-2), 1.81 (d, CH_3), 1.86 (d of d, H_{anti}), 2.35 (br m, H_{anti}), 3.89 (br d of d, H_x), 5.23 (s, C_5H_5), with J(H-1, 2) = −9.9, J(H_x, H_{anti}) = 10.1 and 7.9,

References on pp. 248/50

Table 11 (continued)

No.	compound	method of preparation (yield) properties and remarks
102 (continued)		$J(H_{anti}, H\text{-}1) = 4.0$ and 3.9, $J(H_{anti}, CH_3) = 5.7$, $J(H\text{-}2, H_x) = 0.6$ [70] ^{13}C NMR (CD_2Cl_2): 91.9 (C_5H_5), 185.9 ($ReCO_{eq}$) [70] IR (CH_2Cl_2): 1839, 1922, 1978, 2010, 2045, 2124 (ν(CO)) [70]
103	exo-$C_5H_5Mo(CO)_2C_3H_3(CH_3$-anti,1)$C(CH_3)HRe(CO)_5$-syn,3	V, ^4L = cis-$CH_3CH{=}CHCH{=}CHCH_3$, for 0.5 h (not obtained pure) [70] dark yellow powder [70] IR (Nujol): 1820, 1900, 1971, 2001, 2090, 2119 (ν(CO)) [70] decomposes rapidly in solution and in the solid state above 10°C [70]
104	exo-$C_5H_5Mo(CO)_2C_3H_2(CH_3$-anti,1)($CH_3$-anti,3)$CH_2Re(CO)_5$-3	V, ^4L = $H_2C{=}C(CH_3)CH{=}CHCH_3$, for 1.5 h (low yield) [70] unstable yellow powder (from ether at 0°C) [70] IR (Nujol): 1820, 1902, 1970, 1992, 2040, 2099, 2118, 2140 (ν(CO)) [70] could not be completely purified due to decomposition in solution [70]
105	exo-$C_5H_5Mo(CO)_2C_3H_2(CH_3$-anti,1)($CH_3$-2)$CH_2Re(CO)_5$-syn,3	V, ^4L = $H_2C{=}CHC(CH_3){=}CHCH_3$, for 30 min (low yield) [70] unstable dark yellow oil [70] IR (Nujol): 1838, 1913, 1972, 2009, 2080, 2120 (ν(CO)) [70] rapid decomposition in solution and in the solid state [70]
106	exo-$C_5H_5Mo(CO)_2C_3H_2(CH_3$-2)($CH_3$-anti,1)$CH_2Re(CO)_5$-1	V, ^4L = $H_2C{=}C(CH_3)C(CH_3){=}CH_2$, for 1.5 h (33%) [70] yellow solid, decomposes > 78°C [70] ^{1}H NMR (CD_2Cl_2): 0.85 (br m, H-1), 1.66 (br m, H-2), 1.80 (H_{anti}), 2.00 (CH_3-anti), 2.11 (CH_3-x), 2.81 (d, H_{syn}), 5.16 (C_5H_5) [70] ^{13}C NMR (CD_2Cl_2): 92.8 (C_5H_5), 186.3 ($ReCO_{eq}$) [70] IR (CH_2Cl_2): 1841, 1925, 1982, 2008, 2046, 2100, 2129, 2148 (ν(CO)) [70]

References on pp. 248/50

Table 11 (continued)

No.	compound	method of preparation (yield) properties and remarks
^{3}L = oxabutadiene		
107	exo/endo-$[C_5H_5Mo(CO)_2C(CH_3)H{=}C(CH_2)_4C{=}O]PF_6$ ^{3}L = (structure: H–C(CH$_3$)= attached to cyclohexane ring bearing O=)	$C_5H_5Mo(CO)_2C(CH_3){=}C(CH_2)_4C{=}O$ was treated with CF_3COOH in the presence of NH_4PF_6 in CH_2Cl_2/CH_3OH (1:2) at 0°C for 1 to 5 h (50%) [40] orange solid, m.p. 183 to 186°C (dec.), darkens above 170°C [40] ^{1}H NMR (CD_2Cl_2): 1.91 (d, CH_3; J(H, H) = 6.5), 2.11 (m, $(CH_2)_2$), 2.71 and 2.88 (m, $CH_2C{=}O$ and $CH_2C{=}$), 3.51 (q, CH=), 5.91 (s, C_5H_5); the product obtained in the reaction with CF_3COOD shows no resonance at 3.51 and a singlet at 1.91 [40] IR (CH_2Cl_2): 2010, 2050 (ν(CO)) [40] the PF_6 salt is quite stable in the absence of $[CF_3COO]^-$ and coordinating solvents [40] the single isomer obtained (not assigned exo or endo) shows slow isomerization and ca. 50% decomposition in CH_2Cl_2 over a period of several weeks at 30°C [40]

*Further information:

$C_5H_5Mo(CO)_2C_3H_5$ (Table **11**, No. **1**). A solution of $C_5H_5Mo(CO)_3Cl$ and C_3H_5Br in CH_2Cl_2 was added dropwise over 8 h to a stirred mixture of $[N(CH_3)_4]OH \cdot 5\,H_2O$ in CH_2Cl_2 at 40 to 45°C. The mixture was heated for 18 h to give a 23% yield, besides $C_5H_5Mo(CO)_3C_3H_5$-σ and unreacted chloride. Slow addition of $C_5H_5Mo(CO)_3Cl$ in C_6H_6 to a stirred C_6H_6/H_2O system at 45°C containing a 10-fold excess of C_3H_5Br, a 5-fold excess of $[(C_2H_5)_3NCH_2C_6H_5]OH$, and a reaction time of 8 h gave a 95% yield. Other conditions or CH_2Cl_2 as solvent gave lower yields [43]. $C_5H_5Mo(CO)_2C_3H_5$ was also prepared by several other methods which are collected in the following table.

reactants or method	conditions (yield)
Method Ia	pure complex for 6 h (ca. 50%) [1]
Method Ib	pure complex at 60°C in a vacuum for 4 h or in refluxing xylene for 2 h (ca. 3%) [1]
Method Ic	(75%) [56]
Method IIa (X = Cl)	in THF for 18 h (72%) [10, 23]
Method IIb (X = Cl, R = CH_3)	in boiling THF for 0.5 h (98%) [22]
$C_5H_5Mo(CO)_3Cl/C_3H_5Sn(CH_3)_3$	in THF for 120 h (60%) [22]
$C_5H_5Mo(CO)_2C(O)(CH_3)H/C_3H_5Br$	in solution [60]
$Na[C_5H_5Mo(CO)_3]/C_3H_5Cl$ (excess)	in THF for 1.5 h (3%); main product was $C_5H_5Mo(CO)_3C_3H_5$-σ [1]
$(C_5H_5Mo(CO)_3)_2/P(OC_3H_5)_n(C_6H_5)_{3-n}$ (n = 1 to 3)	in refluxing C_6H_6 for 2 h (by-product) [17]

References on pp. 248/50

NMR studies suggest an intramolecular rearrangement which interconverts conformers arising from the two orientations of the allyl moiety (Formula IV). The half-life at 0°C is ca. 10^{-1} s [25]. Selected exo/endo conformer ratios, free energies of activation, and rate constants are given in the following table. The thermodynamic parameters for this process are $\Delta H^{+} = -13.1 \pm 0.1$ kcal/mol, $\Delta S^{+} = -2.0 \pm 0.4$ cal · mol^{-1} · K^{-1} in $CDCl_3$ and $\Delta H^{+} = -0.7 \pm 0.1$ kcal/mol, $\Delta S^{+} = 0.7 \pm 0.3$ cal · mol^{-1} · K^{-1} in CS_2 [23]; $E_a = 12.3 \pm 1.2$ kcal/mol and log A = 10.6 ± 1 [9].

solvent	t (°C)	exo/endo	ΔF^{+}; k (exo → endo) (kcal/mol; s^{-1})	ΔF^{+}; k (endo → exo) (kcal/mol; s^{-1})
$CDCl_3$	0	4.27 [23]		
$CDCl_3$	25	3.48	15.2 ± 0.1; 45 ± 5	16.0 ± 0.1; 12 ± 1 [23]
CS_2	0	2.45 [23]		
CS_2	25	2.21	15.4 ± 0.1; 35 ± 4	15.8 ± 0.1; 16 ± 2 [23]
CD_3CN	0	4.7 [25]		
C_6D_{12}	25		15.4 ± 0.1; 31 ± 4	15.9 ± 0.1; 14 ± 1 [23]
none		0.38 [28]		

Figures of the ^{1}H NMR in $CDCl_3$ between −50 to +40°C and in C_6D_6 between +40 to +130°C were given in [8]. At 130°C in C_6D_6, an averaged spectrum of the AA′BB′X type was observed for the allyl ligand [8]; at 86°C in [9, 16, 23]. ^{1}H NMR data with the resonances in δ and J in Hz are given in the following table. Earlier values (C_6H_6 and CS_2) are given in [1].

isomer	solvent (t)	H_{anti}	H_{syn}	H_x	C_5H_5	$J(H_{anti}, H_{syn})$	$J(H_{anti}, H_x)$	$J(H_{syn}, H_x)$
exo	C_6H_6 (+5°C)	0.92	2.60	3.51	4.61	2.2	10.8	7.3 [9, 16, 23]
endo	C_6H_6 (+5°C)	1.54	2.67	3.51	4.65	<0.6	10.5	6.4 [9, 16, 23]
exo	$CDCl_3$	0.88	2.78	3.92	5.10 [23]			
endo	$CDCl_3$	1.76	2.72	3.58	5.10 [23]			
—	$CDCl_3$	1.0	2.90	4.0	5.15 [56]			
—	acetone				5.26 [17]			

The exo isomer crystallizes in the monoclinic space group P 2_1/m$-C^2_{2h}$ (No. 11) with the unit cell parameters a = 6.372 (1), b = 11.439 (3), c = 7.257 (2) Å, β = 95.09 (1)°; Z = 2 molecules per unit cell. The molecule lies on a crystallographic mirror plane which bisects the CO-Mo-CO angle, the ^{3}L ligand, and the ^{5}L ligand. The molecular structure with the main bond distances and angles is shown in **Fig. 41** [41].

Extended Hückel MO calculations were discussed in connection with the related compounds $C_5H_5Mo(CO)(^2D)C_3H_5$ and $[C_5H_5Mo(CO)(NO)C_3H_5]^+$ and the nucleophilic attack on the allyl fragment [39].

The pure complex is fairly stable in air, only slight decomposition being noticeable after a day. It is well soluble in common solvents [1, 23]; solutions kept in air decompose within a few hours [1]. Decomposition occurs in halogenated solvents above 50°C [6].

Reaction with gaseous HCl in petroleum ether for 20 min gives a red precipitate. Extraction of the precipitate with $CHCl_3$ in air affords $C_5H_5MoO_2Cl$ [1]. Protonation with $HBF_4 \cdot O(C_2H_5)_2$ affords the reactive complex $C_5H_5Mo(CO)_2(CH_2{=}CHCH_3)FBF_3$ in quantitative yields [57]. A similar reaction in the presence of dienes in ether affords $[C_5H_5Mo(CO)_2{}^4L]BF_4$ (^{4}L = buta-

References on pp. 248/50

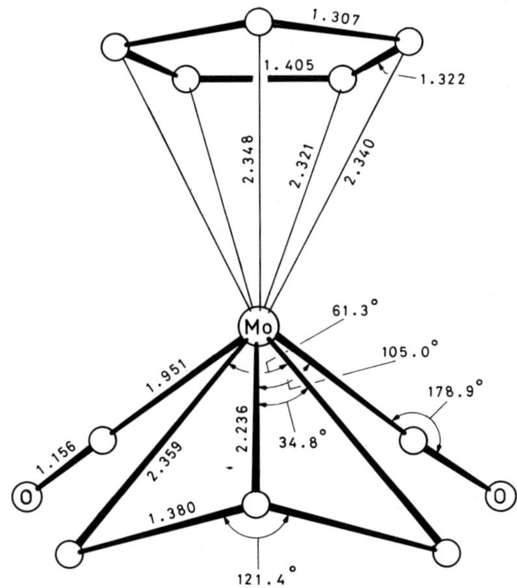

Fig. 41. The molecular structure of exo-$C_5H_5Mo(CO)_2C_3H_5$ [41].

diene, isoprene, cyclohexa-1,3-diene, or cycloocta-1,5-diene) [64]. Reaction with cyclo-$C_6H_{11}NC$ at room temperature affords polycyclohexyl isocyanide [21]. Treatment with $NOPF_6$ in CH_3CN at 0°C affords the thermodynamically less stable isomer, endo-[C_5H_5Mo-(CO)(NO)C_3H_5]PF_6 [25, 58].

exo-$CH_3C(O)C_5H_4Mo(CO)_2C_3H_5$ (Table **11**, No. **2**) crystallizes in the monoclinic space group P 2_1/n$-C_{2h}^5$ (No. 14) with the unit cell parameters (determined at -159°C) a = 10.236(2), b = 8.008(2), c = 13.793(3) Å, β = 92.41(1)°; Z = 4 molecules per unit cell, and D_{calc} = 1.771 g/cm^3. The molecular structure with the main bond distances and angles is shown in **Fig. 42**, p. 242. The allyl ligand is exo-configured and the acyl group on the ^{5}L ligand is coplanar with the ^{5}L ring [63].

$C_9H_7Mo(CO)_2C_3H_5$ (Table **11**, No. **5**). Selected exo to endo ratios, rate constants, and activation parameters are given in the following table. Other thermodynamic parameters for the exo/endo isomerization are: $\Delta H^{\neq} = -1.4 \pm 0.2$ kcal/mol, $\Delta S^{\neq} = -2.7 \pm 0.8$ cal · mol^{-1} · K^{-1} in $CDCl_3$ and $\Delta H^{\neq} = -1.8 \pm 0.3$ kcal/mol, $\Delta S^{\neq} = -6.2 \pm 0.9$ cal · mol^{-1} · K^{-1} in methylcyclohexane [23].

solvent	t (°C)	exo/endo	$\Delta F^{\neq}$; k (exo → endo) (kcal/mol; s^{-1})	$\Delta F^{\neq}$; k (endo → exo) (kcal/mol; s^{-1})
$CDCl_3$	0	3.27		
$CDCl_3$	25	2.65	16.2 ± 0.2; 9 ± 1	16.8 ± 0.2; 4 ± 1
methylcyclo-	25	0.98	16.7 ± 0.1; 4 ± 1	16.7 ± 0.1; 4 ± 1
hexane	0	1.30		

The large differences in the chemical shifts of the allyl protons, compared to those of the cyclopentadienyl complexes, may be ascribed to the ring current effect of the additional six-

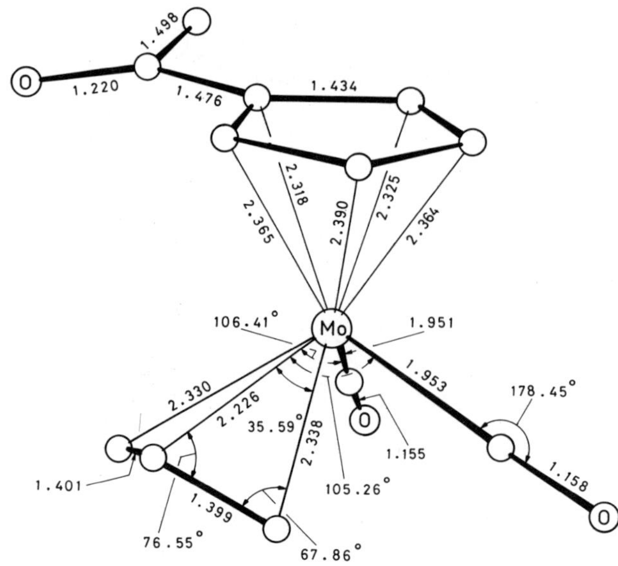

Fig. 42. The molecular structure of $CH_3C(O)C_5H_4Mo(CO)_2C_3H_5$ [63].

membered ring in the indenyl moiety. For a detailed discussion see [23]. ^1H NMR data (δ in ppm and J in Hz) are given in the following table [16, 23].

isomer	solvent	H_x	H_{anti}	H_{syn}	C_9H_7	$J(H_{anti}, H_{syn})$	$J(H_{syn}, H_x)$	$J(H_{anti}, H_x)$
exo	C_6H_6	0.15	1.15	2.39		2.20	7.50	11.4
endo	C_6H_6	3.37	−0.68	3.60		<0.70	6.70	10.70
exo	$CDCl_3$	0.12	0.97	2.32	5.51, 5.84			
endo	$CDCl_3$	3.37	−0.78	3.46	5.46, 5.83			

$C_5H_5Mo(CO)_2C_3H_4Br$-2 (Table **11**, No. **8**). Selected exo to endo ratios, rate constants, and activation parameters are given in the following table. Other thermodynamic parameters for the exo/endo isomerization are: $\Delta H^{\ddagger} = 1.0 \pm 0.2$ kcal/mol, $\Delta S^{\ddagger} = 0.2 \pm 0.6$ cal · mol^{-1} · K^{-1} in $CDCl_3$ and $\Delta H^{\ddagger} = 1.1 \pm 0.1$ kcal/mol, $\Delta S^{\ddagger} = 1.2 \pm 0.6$ cal · mol^{-1} · K^{-1} in CS_2 [23].

solvent	t (°C)	exo/endo	$\Delta F^{\ddagger}$; k (exo → endo) (kcal/mol; s^{-1})	$\Delta F^{\ddagger}$; k (endo → exo) (kcal/mol; s^{-1})
$CDCl_3$	25	0.199	16.5 ± 0.1; 5 ± 1	15.7 ± 0.1; 21 ± 1
$CDCl_3$	0	0.170		
CS_2	25	0.270	16.3 ± 0.1; 7 ± 1	15.5 ± 1; 28 ± 2
CS_2	0	0.227		

$C_5H_5Mo(CO)_2C_3H_4CH_3$-1 (Table **11**, No. **10**) was also obtained by several other reactions. It was prepared by the irradiation of trans-$(C_5H_5Mo(CO)_3)_2C_4H_8$ in toluene [44]. Treating $[C_5H_5Mo(CO)(CCH_3{\equiv}CCH_3)_2]BF_4$ with $NaBH_4$ in THF under CO for 1 h gave a 55% yield [46]. An

11% yield was obtained in the reaction of $C_5H_5Mo(CO)_2(=C(CH_2)_3O\text{-cyclo})I$ with $Li[(C_2H_5)_3BH]$ in THF at −78°C. Traces of the compound were formed in a similar reaction with $C_5H_5Mo(CO)_3(CH_2)_3Br$ [59].

The CH_3 group in 1-position allows four possible isomers arising from exo and endo arrangement and syn or anti position of the CH_3 group (Formula IV). In [23] the spectra were interpreted in terms of exo/endo isomerism with the syn orientation as the thermodynamically more stable (about 95%) isomer; rearrangement to the anti form should not take place because of a high energy (ca. 25 kcal/mol) π-σ-π process. In [18] the NMR spectra were only interpreted as arising from syn-anti isomerism and nothing was reported about the possible exo/endo arrangements. The product described in [46] was not specified but formulated as the endo-anti isomer and that in [59] as the exo-syn species.

^1H NMR spectrum ($CDCl_3$): exo isomer: δ = 1.26 (m, 1H), 1.70 (m, 4H), 2.46 (d of d, 1H), 3.5 to 4.00 (m, 1H), 5.15 (s, C_5H_5) ppm; endo isomer: δ = 0.87 (d, 3H), 1.25 (m, 1H), 2.70 (d of d, 1H), 3.8 to 4.0 (m, 2H), 5.15 (s, C_5H_5) ppm [56]. Other ^1H NMR data (δ in ppm and J in Hz) are given in the following table.

	exo, CH_3-syn [23]	endo, CH_3-syn [23]	CH_3-anti [46]
solvent	CS_2	CS_2	toluene-d_8
H'_{anti}	0.72	1.60	1.30 (d of d)
$H_{anti\ or\ syn}$	1.67	2.76	2.42 (d of d of d)
CH_3	1.84	1.98	1.02 (d)
H'_{syn}	2.53	2.60	3.46 (m)
H_x	3.90	3.50	3.32 (m)
C_5H_5	5.17	5.12	4.38 (s)
$J(H'_{anti}, H'_{syn})$	2.4	<0.6	2.5
$J(H_x, H'_{anti})$	10.1	10.0	10.6
$J(H_x, H'_{syn})$	6.9	6.4	7.1
$J(H_x, H_{anti\ or\ syn})$	9.6	9.5	8.0
$J(H_{anti\ or\ syn}, CH_3)$	5.9	5.9	6.4
$J(H_{syn}, H'_{syn})$			1.7

exo-$C_9H_7Mo(CO)_2C_3H_4CH_3$-1 (Table **11**, No. **14**). Solutions of the complex show the presence of exo (major) and endo (minor) isomers. Crystallization results only in the exo-syn isomer which crystallizes in the monoclinic space group P $2_1/n-C^5_{2h}$ (No. 14) with the unit cell parameters a = 11.978(3), b = 7.992(1), c = 14.109(4) Å, β = 91.85(3)°; Z = 4 molecules per unit cell, and D_{calc} = 1.59 g/cm^3. The molecular structure with the main bond distances and angles is shown in **Fig. 43**, p. 244; another view was given in [52].

exo-$C_5H_5Mo(CO)_2C_3H_4CH{=}CH_2$-1 (Table **11**, No. **18**) crystallizes in the monoclinic space group P $2_1/n-C^5_{2h}$ (No. 14) with the unit cell parameters a = 12.647(5), b = 6.703(2), c = 13.817(3) Å, β = 102.59(3)°; and Z = 4 molecules per unit cell. The molecular structure with the main bond distances and angles is shown in **Fig. 44**, p. 244 [62].

exo-$C_9H_7Mo(CO)_2C_3H_4CH{=}CH_2$-1 (Table **11**, No. **19**) crystallizes in the monoclinic space group P $2_1/n-C^5_{2h}$ (No. 14) with the unit cell parameters a = 7.955(10), b = 12.531(3), c = 13.987(5) Å, β = 100.86(7)°; Z = 4 molecules per unit cell, and D_{calc} = 1.624 g/cm^3. The molecular structure with the main bond distances and angles is shown in **Fig. 45**, p. 245 [73].

$(CH_3)_5C_5Mo(CO)_2C_3H_4CH{=}CH_2$-1,syn/anti (Table **11**, Nos. **21** and **22**) were protonated with CF_3SO_3H in CH_2Cl_2. The syn-CH=CH_2 isomer gave the trans,trans-coordinated penta-1,3-diene

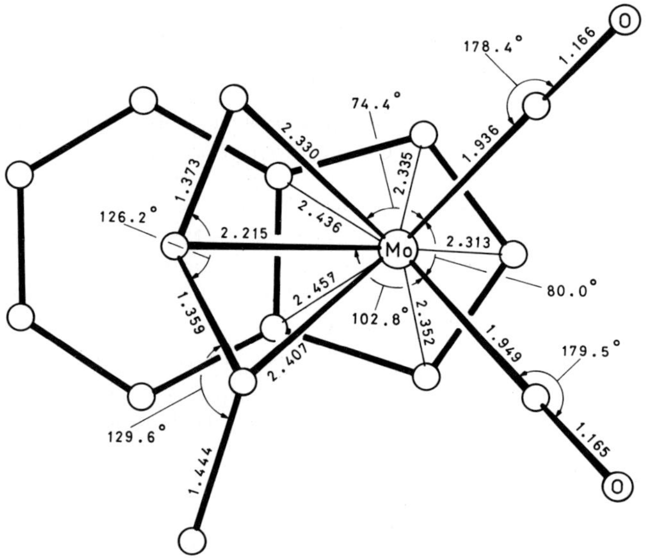

Fig. 43. The molecular structure of $C_9H_7Mo(CO)_2C_3H_4CH_3$-1 [52].

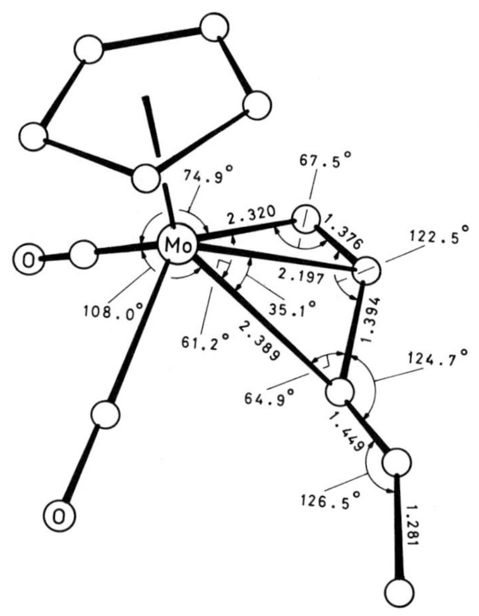

Fig. 44. The molecular structure of $C_5H_5Mo(CO)_2C_3H_4CH{=}CH_2$ [62].

complex at −78 °C and the anti-CH=CH$_2$ isomer gave the cis,trans-coordinated penta-1,3-diene at 25 °C, both with retention of the endo and exo configuration of the precursor [71].

$(CH_3)_5C_5Mo(CO)_2C_3H_4C(O)H$-1,syn/anti (Table **11**, Nos. **26** and **27**). The syn isomers were protonated in CD_2Cl_2 at −78°C with CF_3SO_3H, resulting in a quantitative conversion into

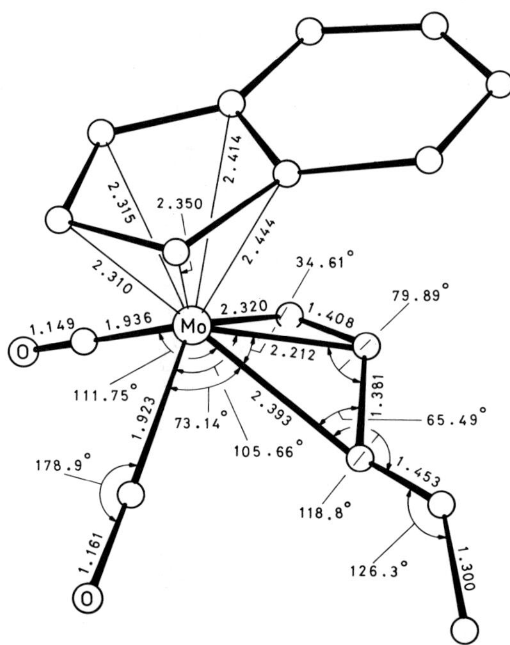

Fig. 45. The molecular structure of $C_9H_7Mo(CO)_2C_3H_4CH{=}CH_2$ [73].

$[(CH_3)_5C_5Mo(CO)_2CH{=}CHCH{=}CHOH\text{-trans}]^+$ with the same exo to endo ratio as the precursor. The deprotonation was reversible. At low temperatures with $N(C_2H_5)_3$ the starting complexes were obtained without isomerization, but at 25°C the anti isomers were obtained. Reaction with the Wittig reagent $(C_6H_5)_3P{=}CH_2$ in THF at −78°C affords Nos. 21 and 22 [71].

$C_5H_5Mo(CO)_2C_3H_4CH_3$-2 (Table **11**, No. **30**). The endo conformer is thermodynamically preferred due to a minimum steric interaction between the cyclopentadienyl and 2-methylallyl ligands [25]. Selected exo to endo ratios, rate constants, and activation parameters are given in the following table [23]. Other thermodynamic parameters for the exo/endo isomerization are: $\Delta H^{\ddagger} = 0.8 \pm 0.2$ kcal/mol, $\Delta S^{\ddagger} = -1.5 \pm 0.9$ cal · mol^{-1} · K^{-1} in $CDCl_3$ and $\Delta H^{\ddagger} = 1.6 \pm 0.2$ kcal/mol, $\Delta S^{\ddagger} = 0.6 \pm 0.7$ cal · mol^{-1} · K^{-1} in CS_2 [23].

solvent	t (°C)	exo/endo	$\Delta F^{\ddagger}$; k (exo → endo) (kcal/mol; s^{-1})	$\Delta F^{\ddagger}$; k (endo → exo) (kcal/mol; s^{-1})
$CHCl_3$	0	0.114		
$CHCl_3$	25	0.128	16.1 ± 0.1; 10 ± 1	15.0 ± 0.1; 60 ± 10
CS_2	0	0.072		
CS_2	25	0.092	16.1 ± 0.1; 10 ± 1	14.8 ± 0.1; 90 ± 10

Activation parameters for exo/endo interconversion were determined by matching the experimental and computed line shapes of CH_3 (between +13.5 and +54.5°C) and C_5H_5

References on pp. 248/50

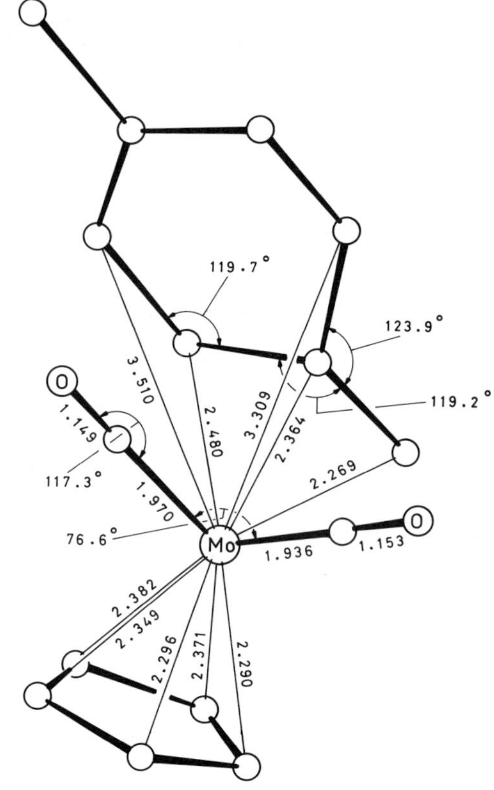

Fig. 46. The molecular structure of $C_5H_5Mo(CO)_2CH_2C_6H_4CH_3$-4 [7].

(between −5.5 to +9.5°C; given in parentheses): E_a = 16.8 ± 0.4 (12.3 ± 1.2) kcal/mol, log A = 14.5 ± 0.3 (10.6 ± 1.0) [9]. ^{1}H NMR data (δ in ppm) are given in the following table.

isomer	solvent (t)	H_{anti}	H_{syn}	CH_3	C_5H_5
—	C_6H_6 (5°C)	1.79	2.91	1.96	[16]
exo	$CDCl_3$	0.94	2.38	1.75	5.14 [23]
endo	$CDCl_3$	1.92	2.76	1.70	5.16 [23]

$C_5H_5Mo(CO)_2CH_2C_6H_4CH_3$-2/3 (Table **11**, Nos. **81**, **82**). It was shown that $C_5H_5Mo(CO)_2$-$CH_2C_6H_4CH_3$-2 exists in solution as one fluxional isomer. High-temperature ^{1}H NMR experiments were carried out in toluene-d_8 at 97°C. Irradiation of H-1 produced a small decrease in the intensity of H-2, and vice versa (Formula VII) corresponding to an exchange rate between 0.2 and 0.02 s^{-1} and $\Delta G^{\ddagger}_{370}$ = 23.6 ± 1.0 kcal/mol. In contrast, $C_5H_5Mo(CO)_2CH_2C_6H_4CH_3$-3 exists as two isomers at −20°C which undergo interconversion at room temperature with $\Delta G^{\ddagger}_{293}$ = 15.7 kcal/mol. The dynamic process is an allyl rotation [66].

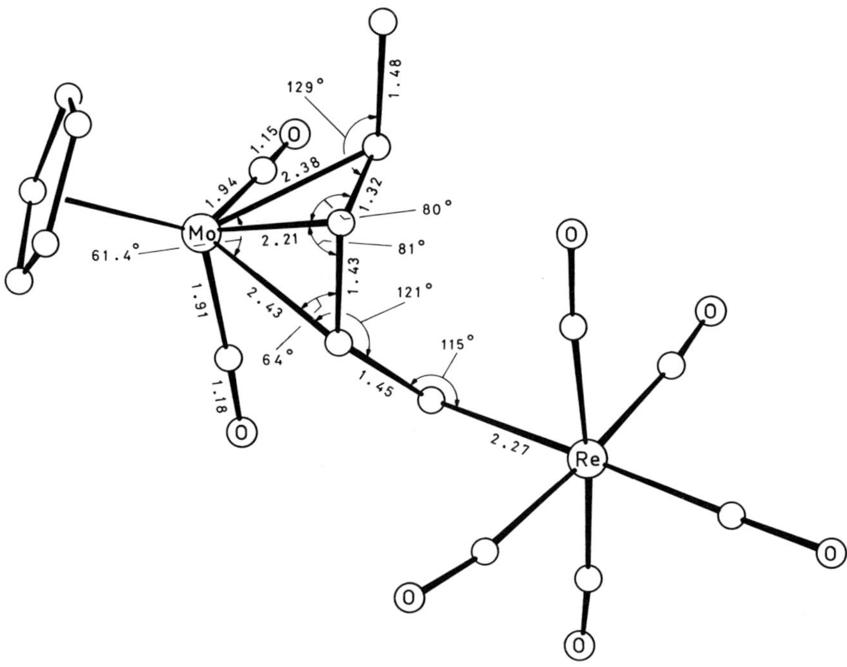

Fig. 47. The molecular structure of $C_5H_5Mo(CO)_2C_3H_3(CH_3\text{-syn},1)CH_2Re(CO)_5\text{-syn},3$ [70].

VII

$C_5H_5Mo(CO)_2CH_2C_6H_4CH_3$-4 and **$C_5H_5Mo(CO)_2CH_2C_6H_3(C_3H_7\text{-i})_2$-3,5** (Table **11**, Nos. **83** and **84**). The fluxional behavior was studied by ^1H NMR. Activation parameters for the site exchanges (Formula VII) are given in the following table [11]. $\Delta G^{\ddagger}_{348} \leq 18.8$ kcal/mol was calculated with this data [66].

No.	solvent	protons	log A	E_a (kcal/mol)	$\Delta H^{\ddagger}$ (kcal/mol)	$\Delta S^{\ddagger}$ ($cal \cdot mol^{-1} \cdot K^{-1}$)
85	$CDCl_3$	1, 2	16.6 ± 0.2	19.2 ± 0.3	18.6 ± 0.3	15.6 ± 1.3
		3, 4	16.2 ± 0.3	18.8 ± 0.4	18.2 ± 0.4	14.0 ± 1.4
86	$CDCl_3$	3, 4	14.9 ± 0.3	18.7 ± 0.4	18.1 ± 0.4	7.5 ± 1.3
	toluene-d_8	3, 4	14.6 ± 0.3	18.3 ± 0.5	17.7 ± 0.5	6.2 ± 1.3
	acetone-d_6	3, 4	14.9 ± 0.3	19.1 ± 0.4	18.4 ± 0.4	7.6 ± 1.3

Deep red monoclinic prisms of No. 83 were obtained by recrystallization from pentane/CH_2Cl_2. They crystallize in the space group $P\,2_1/n - C^5_{2h}$ (No. 14) with the unit cell parameters

References on pp. 248/50

a = 6.334(3), b = 17.179(1), c = 12.861(2) Å, β = 106.98°; Z = 4 molecules per unit cell, and D_{calc} = 1.607 g/cm^3. The molecular structure with the main bond distances and angles is shown in **Fig. 46**, p. 246 [7].

$C_5H_5Mo(CO)_2C_3H_3(CH_3$-syn,1)$CH_2Re(CO)_5$-syn,3 (Table **11**, No. **102**). The molecular structure is shown in **Fig. 47**, p. 247 [70].

References:

[1] Cousins, M.; Green, M. L. H. (J. Chem. Soc. **1963** 889/94).
[2] King, R. B.; Bisnette, M. B. (Inorg. Chem. **4** [1965] 475/81).
[3] Dessy, R. E.; King, R. B.; Waldrop, M. (J. Am. Chem. Soc. **88** [1966] 5112/7).
[4] King, R. B. (Inorg. Chem. **5** [1966] 2242/3).
[5] King, R. B.; Fronzaglia, A. (J. Am. Chem. Soc. **88** [1966] 709/12).
[6] King, R. B. (AD-6642968 [1967] 1/36; C.A. **69** [1968] No. 56601).
[7] Cotton, F. A.; La Prade, M. D. (J. Am. Chem. Soc. **90** [1968] 5418/22).
[8] Davidson, A.; Rode, W. C. (Inorg. Chem. **6** [1968] 2124/5).
[9] Faller, J. W.; Incorvia, M. J. (Inorg. Chem. **7** [1968] 840/2).
[10] Hayter, R. G. (J. Organometal. Chem. **13** [1968] P 1/P 3).

[11] Cotton, F. A.; Marks, T. J. (J. Am. Chem. Soc. **91** [1969] 1339/46).
[12] King, R. B. (Can. J. Chem. **47** [1969] 559/68).
[13] King, R. B.; Kapoor, R. N. (5th Intern. Conf. Organometal. Chem., Bristol, Engl., 1969, Abstr. K3).
[14] King, R. B.; Kapoor, R. N. (Inorg. Chem. **8** [1969] 2534/9).
[15] Benaim, J.; Merour, J. Y.; Roustan, J. L. (Tetrahedron Letters **15** [1971] 983/6).
[16] Faller, J. W.; Jakubowski, A. (J. Organometal. Chem. **31** [1971] C 75/C 78).
[17] Haines, R. J.; Du Preez, A. L.; Marais, I. L. (J. Organometal. Chem. **28** [1971] 97/104).
[18] Merour, J. Y.; Charrier, C.; Benaim, J.; Roustan, J. L.; Commereue, D. (J. Organometal. Chem. **39** [1972] 321/8).
[19] Roustan, J. L.; Benaim, J.; Charrier, C.; Merour, J. Y. (Tetrahedron Letters **19** [1972] 1953/6).
[20] Roustan, J. L.; Charrier, C.; Merour, J. Y.; Benaim, J.; Giannotti, C. (J. Organometal. Chem. **38** [1972] C 37/C 40).

[21] Yamamoto, Y.; Yamazaki, H. (unpublished results; cited in Coord. Chem. Rev. **8** [1972] 225/39).
[22] Abel, E. W.; Moorhouse, S. (J. Chem. Soc. Dalton Trans. **1973** 1706/11).
[23] Faller, J. W.; Chen, C.-C.; Mattina, M. J.; Jakubowski, A. (J. Organometal. Chem. **52** [1973] 361/86).
[24] Nesmeyanov, A. N.; Kolobova, N. E.; Zlozina, I. B.; Solodova, M. Ya.; Anisimov, K. N. (Tezisy Dokl. 12th Vses. Chugaevskoe Soveshch. Khim. Kompleksn. Soedin., Novosibirsk 1975, pp. 474/5; C.A. **89** [1978] No. 5580).
[25] Faller, J. W.; Rosan, A. M. (J. Am. Chem. Soc. **98** [1976] 3388/9).
[26] Pannell, K. H.; Lappert, M. F.; Stanley, K. (J. Organometal. Chem. **112** [1976] 37/48).
[27] Bottrill, M.; Green, M. (J. Chem. Soc. Dalton Trans. **1977** 2365/71).
[28] Faller, J. W.; Rosan, A. M. (J. Am. Chem. Soc. **99** [1977] 4858/9).
[29] Kolobova, N. E.; Zlotina, I. B.; Solodova, M. Ya. (Izv. Akad. Nauk SSSR Ser. Khim. **1977** 2334/6; Bull. Acad. Sci. USSR Div. Chem. Sci. **1977** 2168/70).
[30] Charrier, C.; Collin, J.; Merour, M.; Roustan, J. L. (J. Organometal. Chem. **162** [1978] 57/66).

[31] Kolobova, N. E.; Zlotina, I. B.; Yudin, E. N. (Izv. Akad. Nauk SSSR Ser. Khim. **1978** 681/3; Bull. Acad. Sci. USSR Div. Chem. Sci. **1978** 588/90).

[32] Benaim, J.; Giulieri, F. (J. Organometal. Chem. **165** [1979] C 28/C 32).

[33] Bottrill, M.; Green, M. (J. Chem. Soc. Dalton Trans. **1979** 820/5).

[34] Collin, J.; Charrier, C.; Pouet, M. J.; Cadiot, P.; Roustan, J. L. (J. Organometal. Chem. **168** [1978] 321/36).

[35] Collin, J.; Roustan, J. L.; Cadiot, P. (J. Organometal. Chem. **169** [1979] 53/62).

[36] Gibson, D. H.; Hsynu, W.-L.; Lin, D.-S. (J. Organometal. Chem. **172** [1979] C 7/C 12).

[37] Mills III, W. C.; Wrighton, M. S. (J. Am. Chem. Soc. **101** [1979] 5830/2).

[38] Roustan, J. L.; Merour, J. Y.; Charrier, C.; Benaim, J.; Cadiot, P. (J. Organometal. Chem. **169** [1979] 39/52).

[39] Schilling, B. E. R.; Hoffmann, R.; Faller, J. W. (J. Am. Chem. Soc. **101** [1979] 592/8).

[40] Watson, P. L.; Bergman, R. G. (J. Am. Chem. Soc. **101** [1979] 2055/62).

[41] Faller, J. W.; Chodosh, D. F.; Katahira, D. (J. Organometal. Chem. **187** [1980] 227/31).

[42] Faller, J. W.; Shoo, Y. (J. Am. Chem. Soc. **102** [1980] 5396/8).

[43] Gibson, D. H.; Hsu, W.-L.; Almed, F. U. (J. Organometal. Chem. **215** [1980] 379/401).

[44] Bailey, N. A.; Chell, P. L.; Mukhopadhyay, A.; Tabbron, H. E.; Winter, M. J. (J. Chem. Soc. Chem. Commun. **1982** 215/7).

[45] Faller, J. W.; Shvo, Y.; Chao, K.; Murray, H. H. (J. Organometal. Chem. **226** [1982] 251/75).

[46] Allen, S. R.; Green, M; Norman, N. C.; Paddick, K. E.; Orpen, A. G. (J. Chem. Soc. Dalton Trans. **1983** 1625/33).

[47] Faller, J. W.; Chao, K.-H. (J. Am. Chem. Soc. **105** [1983] 3893/8).

[48] Allen, S. R.; Barnes, S. G.; Green, M.; Moran, G.; Trollope, L.; Murrall, N. W.; Welch, A. J.; Sharaiha, D. M. (J. Chem. Soc. Dalton Trans. **1984** 1157/69).

[49] Gauntlett, J. T.; Taylor, B. F.; Winter, M. J. (J. Chem. Soc. Chem. Commun. **1984** 420/1).

[50] Giulieri, F.; Benaim, J. (J. Organometal. Chem. **276** [1984] 367/76).

[51] Benamou, C.; Benaim, J. (J. Organometal. Chem. **280** [1985] 377/87).

[52] Faller, J. W.; Crabtree, R. H.; Habib, A. (Organometallics **4** [1985] 929/35).

[53] Gauntlett, J. T.; Taylor, B. F.; Winter, M. J. (J. Chem. Soc. Dalton Trans. **1985** 1815/20).

[54] Giulieri, F.; Benaim, J. (Nouv. J. Chim. **9** [1985] 335/40).

[55] Green, M.; Greenfield, S.; Kersting, M. (J. Chem. Soc. Chem. Commun. **1985** 18/20).

[56] Luh, T.-Y.; Wong, C. S. (J. Organometal. Chem. **287** [1985] 231/3).

[57] Markham, J.; Menard, K.; Cutler, A. (Inorg. Chem. **24** [1985] 1581/7).

[58] Vanarsdale, W. E.; Winter, R. E. K.; Kochi, J. K. (J. Organometal. Chem. **296** [1985] 31/54).

[59] Adams, H.; Bailey, N. A.; Cahill, P.; Rogers, D.; Winter, M. J. (J. Chem. Soc. Dalton Trans. **1986** 2119/26).

[60] Gauntlett, J. T.; Winter, M. J. (Polyhedron **5** [1986] 451/9).

[61] Green, M.; Mercer, R. J.; Orpen, A. G.; Schaverien, C. J.; Williams, I. D. (J. Chem. Soc. Dalton Trans. **1986** 1971/82).

[62] Lee, G. H.; Peng, S.-M.; Lee, T.-W.; Liu, R.-S. (Organometallics **5** [1986] 2378/80).

[63] Vanarsdale, W. E.; Kochi, J. K. (J. Organometal. Chem. **317** [1986] 215/32).

[64] Gusev, O. V.; Krivykh, V. V.; Petrowskii, P. V.; Rybinskaya, M. I. (Izv. Akad. Nauk SSSR Ser. Khim. **1987** 1655/7; Bull. Acad. Sci. USSR Div. Chem. Sci. **1987** 1532/4).

[65] Kreiter, C. G.; Wendt, G.; Sheldrick, W. S. (J. Organometal. Chem. **333** [1987] 47/59).

[66] Mann, B. E.; Shaw, S. D. (J. Organometal. Chem. **326** [1987] C 13/C 16).

[67] Müller, H.-J.; Beck, W. (J. Organometal. Chem. **330** [1987] C 13/C 16).

[68] Faller, J. W.; Linebanier, D. (Organometallics **7** [1988] 1670/2).

[69] Lee, T.-W.; Liu, R.-S. (Organometallics **7** [1988] 878/83).
[70] Müller, H.-J. (Diss. Univ. München 1988, pp. 1/150).

[71] Benyunes, S. A.; Green, M.; Grimshire, M. J. (Organometallics **8** [1989] 2268/70).
[72] Winter, M. J.; Woodward, S. (J. Chem. Soc. Chem. Commun. **1989** 457/8).
[73] Lee, G.-H.; Peng, S.-M.; Liu, F.-C.; Liu, R.-S. (J. Organometal. Chem. **377** [1989] 123/32).
[74] Faller, J. W.; Whitmore, B. C. (Organometallics **5** [1986] 752/5).

1.5.1.3.2.3.3 Compounds with One Cyclic ^{3}L Ligand

Method I: $[^3LMo(CO)_2(CNCH_3)_3]PF_6$ and TlC_5H_5 were stirred in THF overnight. The products were purified by column chromatography on Florisil with hexane [35].

Method II: $^3LMo(CO)_2(NCCH_3)_2X$ was allowed to react with LiC_5H_5 (or substituted derivative) in THF [6, 26, 47].

Method III: $^5LMo(CO)_2C_3(C_6H_5)_3CO$ (Formula I) was treated with $(C_5H_5)_2TiCH_2C(C_4H_9\text{-t})H\text{-}CH_2$ (Formula II) in toluene at 65°C for 4 h [30].

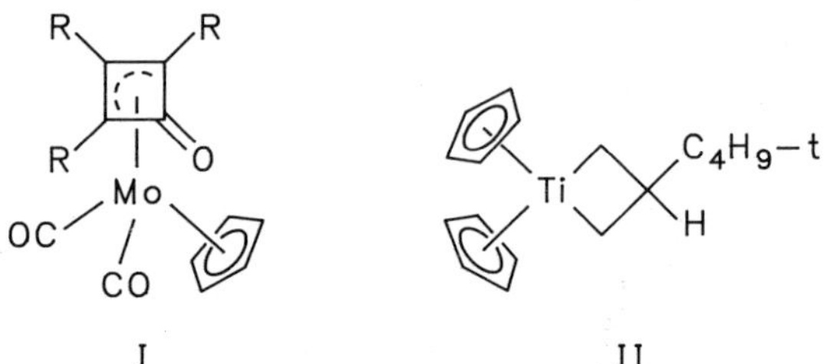

Method IV: Solid $[cyclo\text{-}C_3R_3]X$ (X = ClO_4, BF_4, or PF_6) was added to a solution of $Na[(CH_3)_5C_5Mo(CO)_3]$ in THF cooled at −80°C. The mixture was stirred cold for 1 h, then allowed to warm, and stirred overnight. The product was purified by column chromatography on Florisil with ether followed by recrystallization from CH_2Cl_2/hexane [34].

Method V: $(CH_3C_5H_4Mo(CO)_3)_2$ and the diene were irradiated in pentane at 253 K for 40 to 60 min. The products were separated by chromatography on alumina with pentane [43].

Method VI: $[^5LMo(CO)_2{}^4L]X$ (X = BF_4 or PF_6; ^4L = a cyclohexa-1,3-diene or cyclohepta-1,3-diene derivative) was allowed to react with MR in THF [17, 26, 32, 36, 44, 46, 47]. Further information is given in the table. The products were purified in many cases by thin layer chromatography on silica with CH_2Cl_2 or ethyl acetate in hexane [46, 47].

Method VII: The corresponding methyl ester complex was dissolved in methanol/THF (1:1) and stirred while an excess of an aqueous solution of KOH was added. Stirring was continued for 8 to 10 h. The reaction mixture was poured onto dilute ice/HCl and the product was extracted with ether [36, 47].

Method VIII: The allyl complex, $^5LMo(CO)_2C_6H_7(R\text{-}4)CN\text{-}6$, was decyanated by treatment with 1.2 equivalents of n-, s-, or t-LiC_4H_9 in THF for 0.5 h at −78°C to generate the anion, $[^5LMo(CO)_2C_6H_7R\text{-}4]^-$. Water, benzaldehyde, 2-(methoxycarbonyl)-cyclopent-1-en-3-one, $[CH_3OC_6H_6Fe(CO)_3]PF_6$, $[R_2C_6H_4Mn(CO)_3]PF_6$, or di-t-butyl methylenemalonate was added to the reaction mixture. After 30 to 60 min the product was extracted with ether and purified by thin layer chromatography [46].

Method IX: The cyano complex (Nos. 51, 66, 118, or 126) in THF was added dropwise to a solution of $LiN(C_3H_7\text{-}i)_2$ in THF cooled at −78°C. After stirring for 0.5 h, the resulting carbanion complex was quenched with water, D_2O, di-t-butyl methylenemalonate, $C_6H_5CHO/(C_2H_5)_2O \cdot BF_3$, 2-methoxycarbonylcyclopent-2-en-1-one, $[C_5H_5Fe(CO)_2CH_2{=}C(OCH_3)H]BF_4$, or $[R_2C_6H_4Mn(CO)_3]PF_6$. The mixture was allowed to warm. In most cases the product was purified by thin layer chromatography on silica with ethyl acetate/hexane [31, 40, 46].

Method X: To the corresponding phenylsulfonyl ester complex in CH_3OH/THF (4:1) Na_2HPO_4 was added. After 15 min the mixture was cooled and treated with excess sodium amalgam (6% Na). When the reaction was complete (by thin layer chromatography), the mixture was acidified with 10% HCl and poured into water. The product was extracted with ether [36, 47].

The reaction of $^5LMo(CO)_2{}^3L$ (^3L = a cyclohexa-2-en-1-yl derivative) with $[C(C_6H_5)_3]BF_4$, CF_3COOH, or $(C_2H_5)_2O \cdot HBF_4$ leads to hydride abstraction with formation of $[^5LMo(CO)_2{}^4L]^+$ (^4L = cyclohexa-1,3-diene) [33, 41]. The coordinated cyclohept-2-en-1-yl ligand shows a similar chemical behavior [47]. Oxidation of the coordinated cyclohexadienyl or cycloheptenyl ligand with iodine or $NOPF_6$ gives free cyclohexene and cycloheptene derivatives [32, 36, 46, 47]; these reactions are described under "Further information". In contrast, the coordinated cyclooctadiene ligand did not react with $NOPF_6$ by this or other techniques [47].

General Remarks. The designations endo and exo are used ambigously in this series of compounds. Thus, in No. 22 the endo, exo isomers (no sp^3 carbon atom is present in the five-membered ring of the ligand) are related to the previous section (Formula IV on p. 207); see also the orientation of the C_7H_7 ring in No. 144. In all other compounds, endo and exo is used for the position of the substituents X (e.g. in the allyl ligand C_5H_6X in which X is at an sp^3 carbon atom) either towards (endo) or away (exo) from the Mo atom. For the majority of these compounds for steric reasons no isomerism concerning the orientation of the allyl ligand with respect to the ^5L ligand is known and the orientations of the allyl rings follow those in the **Fig. 48** to **Fig. 53** (pp. 296/302) corresponding to the exo assignment used for the open allyl compounds in the previous section. Theoretically, compounds with asymmetric allyl ligand (e.g. C_5H_6X) should form four different isomers each as pairs of enantiomers (center of chirality at CX). In compounds with six-membered rings chair and boat conformers must also be taken in consideration. The ^5L ligand C_9H_7 means indenyl and the numbering follows Formula VI, p. 207.

Table 12
Compounds with One Cyclic ^3L Ligand.
An asterisk indicates further information at the end of the table.
For explanations, abbreviations, and units see p. X.

No.	compound	method of preparation (yield) properties and remarks
^3L = derivative of cyclopropenyl		
*1	$C_5H_5Mo(CO)_2C_2(C_6H_5)_2CC_4H_9\text{-}t$	I (50%); addition of halide ions had no effect [35] orange-red crystals (from hexane), m.p. 110 to 112°C (dec.) [35]

Table 12 (continued)

No.	compound	method of preparation (yield) properties and remarks
*1	(continued)	^{1}H NMR ($CDCl_3$): 1.10 (s, t-C_4H_9), 4.90 (s, C_5H_5), 7.25 (m, C_6H_5) [35] ^{13}C NMR (CD_2Cl_2, −80°C): 31.80 ((**C**H_3)$_3$C), 33.16 (**C**(CH_3)$_3$), 33.62 (**C**C_4H_9-t), 52.64 (**C**C_6H_5), 90.42 (C_5H_5), 126.14, 128.76, 129.50, and 137.49 (C_6H_5), 231.81 (CO); the data indicate rapid rotation of the C_3 ring even at −80°C [35] IR (hexane): 1915, 1978 (ν(CO)) [35] does not react with $P(CH_3)_2C_6H_5$ in C_6H_6 at 80°C or under irradiation [35]
*2	$C_5H_5Mo(CO)_2C_3(C_6H_5)_3$	I (93%) [35]; II, X = Cl [6] orange-red crystals (from hexane), m.p. 143 to 144°C (dec.) [35] ^{1}H NMR (CS_2): 4.84 (s, C_5H_5), 7.2 (m, C_6H_5) [6]; ($CDCl_3$): 5.00 (s, C_5H_5), 7.30 (m, C_6H_5) [35] ^{13}C NMR (CD_2Cl_2, −80°C): 34.01 (**C**C_6H_5), 90.33 (C_5H_5), 126.24, 128.29, 128.89, and 135.38 (C_6H_5), 229.53 (CO); data indicate rapid rotation of the C_3 ring even at −80°C [35] IR (hexane): 1912, 1971 (ν(CO)) [35] reaction with ^{2}D ligands affords $C_5H_5Mo(CO)(^2D)C_3(C_6H_5)_3CO$ (compare Formula I; ^{2}D = $P(CH_3)_2C_6H_5$, $P(OC_6H_5)_3$) [35]
^{3}L = derivative of cyclobutenyl		
3	$(CH_3)_5C_5Mo(CO)_2C_3(CH_3)_3C{=}CH_2$	III (not isolated) [30] reaction with $(C_2H_5)_2O \cdot HBF_4$ at 0°C affords $[(CH_3)_5C_5Mo(CO)_2C_4(CH_3)_4\text{-cyclo}]BF_4$ [30]
4	$C_5H_5Mo(CO)_2C_3(C_6H_5)_3C{=}CH_2$	III (90%) [30] yellow crystals (from hexane) [30] ^{1}H NMR (CD_3CN): 3.32 (s, CH_2), 5.20 (s, C_5H_5), 7.17 to 7.40 (m, C_6H_5) [30] IR (hexane): 1897, 1929 (ν(CO)) [30] reaction with $(C_2H_5)_2O \cdot HBF_4$ in ethereal solution at 0°C affords $[C_5H_5Mo(CO)_2C_3(C_6H_5)_3CCH_3\text{-cyclo}]BF_4$ [30]
5	$(CH_3)_5C_5Mo(CO)_2C_3(CH_3)_3CO$	IV, with [cyclo-$C_3(CH_3)_3$]BF_4 (70%) [34] yellow crystals, m.p. 184 to 190°C [34] ^{1}H NMR ($CDCl_3$): 1.64 (s, 2 CH_3), 1.90 (s, $(CH_3)_5C_5$), 2.02 (s, 1 CH_3) [34] ^{13}C NMR ($CDCl_3$): 9.25 ((**C**H_3)$_5C_5$), 64.3 and 88.1 (**C**CH_3), 102.4 (**C**$_5$(CH_3)$_5$), 172.0 (C=O), 233.1 (CO) [34]

Table 12 (continued)

No.	compound	method of preparation (yield) properties and remarks
		IR (CH_2Cl_2): 1645 (ν(C=O)), 1907, 1971 (ν(CO)) [34]
6	$(CH_3)_5C_5Mo(CO)_2C_2(C_6H_5)_2CHCO$ 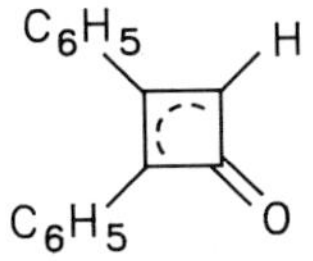	IV, with [cyclo-$C_2(C_6H_5)_2CH$]ClO_4 (67%) [34] yellow crystals, m.p. 188 to 190°C [34] ^{1}H NMR ($CDCl_3$): 1.85 (s, $(CH_3)_5C_5$), 3.65 (s, CH), 7.19 (m, C_6H_5) [34] ^{13}C NMR ($CDCl_3$): 9.3 ((**C**H_3)$_5C_5$), 47.6 (CH), 79.0 and 85.7 (**C**C_6H_5), 103.4 (**C**$_5(CH_3)_5$), 126.4, 127.9, 129.5, 134.3, and 135.2 (C_6H_5), 170.0 (C=O), 231.0 and 232.3 (CO) [34] IR (CH_2Cl_2): 1663 (ν(C=O)), 1924, 1987 (ν(CO)) [34]
7	$C_5(CH_3)_5Mo(CO)_2C_2(C_6H_5)_2C(CH_3)CO$ 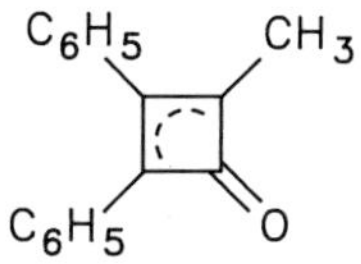	IV, with [cyclo-$C_2(C_6H_5)CCH_3$]BF_4 (31%) [34] yellow crystals, m.p. 163 to 164°C [34] ^{1}H NMR ($CDCl_3$): 1.85 (s, $(CH_3)_5C_5$), 1.90 (s, CH_3), 7.0 to 7.3 (m, C_6H_5) [34] ^{13}C NMR ($CDCl_3$): 9.1 ((**C**H_3)$_5C_5$), 10.1 (CH_3), 63.4 (**C**CH_3), 68.9 and 87.9 (**C**C_6H_5), 102.9 (**C**$_5(CH_3)_5$), 125.7, 127.2, 127.8, 128.0, 128.2, 128.5, 134.7, and 135.0 (C_6H_5), 171.2 (C=O), 232.1 and 233.0 (CO) [34] IR (CH_2Cl_2): 1672 (ν(C=O)), 1920, 1982 (ν(CO)) [34]
8	$(CH_3)_5C_5Mo(CO)_2C_2(C_6H_5)_2C(C_4H_9\text{-}t)CO$ 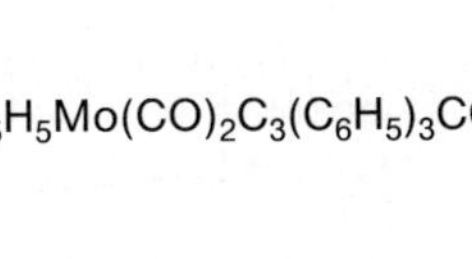	IV, with [cyclo-$C_2(C_6H_5)_2CC_4H_9$-t]PF_6 (71%) [34] yellow crystals, m.p. 198 to 201°C [34] ^{1}H NMR ($CDCl_3$): 1.05 (s, t-C_4H_9), 1.84 (s, $(CH_3)_5C_5$), 7.28 (m, C_6H_5) [34] ^{13}C NMR ($CDCl_3$): 9.1 ((**C**H_3)$_5C_5$), 30.5 ((**C**H_3)$_3$C), 33.5 (**C**$(CH_3)_3$), 68.7 (**C**C_4H_9-t), 83.5 and 90.6 (**C**C_6H_5), 102.0 (**C**$_5(CH_3)_5$), 125.1, 127.1, 127.4, 128.1, 131.9, and 134.2 (C_6H_5), 167.8 (C=O), 232.0 and 235.0 (CO) [34] IR (CH_2Cl_2): 1656 (ν(C=O)), 1910, 1970 (ν(CO)) [34]
9	$C_5H_5Mo(CO)_2C_3(C_6H_5)_3CO$	I; addition of halide ions had no effect (61%) [35] yellow crystals (from CH_2Cl_2/hexane), m.p. 205 to 206°C (dec.) [35] ^{1}H NMR ($CDCl_3$): 5.20 (s, C_5H_5), 7.36 (m, C_6H_5) [35] ^{13}C NMR ($CDCl_3$): 70.12 and 89.36 (**C**C_6H_5), 90.20 (C_5H_5), 125.84, 127.99, 128.52, 131.00, 132.53, and 135.54 (C_6H_5), 168.33 (C=O), 227.78 (CO) [35]

References on pp. 303/4

Table 12 (continued)

No.	compound	method of preparation (yield) properties and remarks
9 (continued)		IR (CH_2Cl_2): 1677 (ν(C=O)), 1945, 2001 (ν(CO)) [35] reaction with $(C_5H_5)_2TiCH_2C(C_4H_9\text{-t})HCH_2$ (Method III) affords No. 4 [30]
10	$(CH_3)_5C_5Mo(CO)_2C_3(C_6H_5)_3CO$	IV, with [cyclo-$C_3(C_6H_5)_3$]PF_6 (86%) [34] yellow crystals, m.p. 197 to 199°C [34] ^{1}H NMR ($CDCl_3$): 1.77 (s, $(CH_3)_5C_5$), 7.0 to 7.4 (m, C_6H_5) [34] ^{13}C NMR ($CDCl_3$): 8.7 (($\mathbf{C}H_3)_5C_5$), 68.4 and 88.3 ($\mathbf{C}C_6H_5$), 103.2 ($\mathbf{C}_5(CH_3)_5$), 125.2, 126.3, 127.5, 131.2, 134.0, and 135.1 (C_6H_5), 167.9 (C=O), 231.5 (CO) [34] IR (CH_2Cl_2): 1664 (ν(C=O)), 1928, 1987 (ν(CO)) [34]
^{3}L = derivative of cyclopentenyl		
*11	$C_5H_5Mo(CO)_2C_5H_7$	obtained by several reactions; see "Further information" orange solid, subl. 40°C/10^{-3} Torr [13] ^{1}H NMR: 1.90 (m, CH_2), 3.50 to 3.63 (m, C_3H_3 of C_5H_7), 4.65 (s, C_5H_5) [13] IR (alkane): 1882, 1958 [13]; 1880, 1950 (ν(CO)), also as a diagram in [22] mass spectrum: $[M]^+$ [22]
12	$CH_3C_5H_4Mo(CO)_2C_5H_7$	V, with cyclopenta-1,3-diene (2%) [43] yellow crystals (from pentane) [43] ^{1}H NMR (toluene-d_8, 213 K): 1.32 (CH_3), 1.87 (AA′BB′ system, H-4′, 5′), 2.02 (AA′BB′ system, H-4, 5), 3.24 (H-2), 3.37 (H-1, 3), 4.41 and 4.56 (AA′BB′ system, C_5H_4) [43] IR: 1880, 1950 (ν(CO)) [43]
13	$C_5H_5Mo(CO)_2C_5H_6Re(CO)_5$-4,exo	VI, X = BF_4, ^{4}L = cyclo-C_5H_6, MR = Na[$Re(CO)_5$], at −78°C for 15 min [42, 44] yellow solid, decomposes > 95°C [44] ^{1}H NMR (C_6D_6): 2.33 (m, H-5, exo), 2.54 (d of d, H-5, endo), 2.59 (d of d, H-1), 3.42 (t of d, H-4), 4.64 (s, C_5H_5), 4.72 (d of d, H-2); with J(H-1, 2) ≈ 1.9, J(H-2, 3) ≈ 3.7, J(H-3, 4) ≈ 3.4, J(H-4, 5) ≈ 3.2, J(H-4, 5 exo) ≈ 3.9, J(H-5, endo, exo) ≈ 4 [44] IR (CH_2Cl_2): 1857, 1952, 2012, 2071, 2101, 2132 (ν(CO)) [44] decomposes in solution [44]

Table 12 (continued)

No.	compound	method of preparation (yield) properties and remarks
14	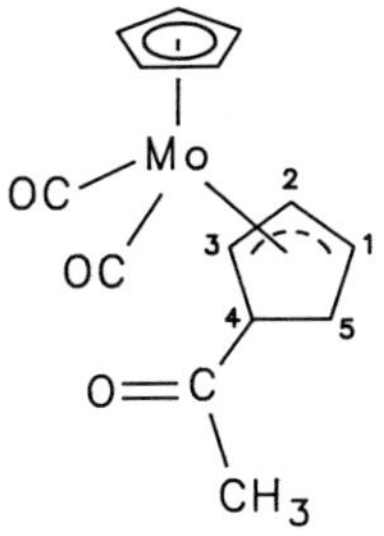	$C_5H_5Mo(CO)(C_5H_6\text{-cyclo})C(O)CH_3$ was refluxed for 100 min in pentane under CO (8%) [39] red crystals (from n-hexane at 243 K) [39] ^1H NMR (C_6D_6): 1.94 (H-5, exo), 2.04 (CH_3), 2.39 (H-5, endo), 2.92 (H-4, exo), 3.24 (H-1), 3.53 (H-3), 3.55 (H-2), 4.58 (C_5H_5); with J(H-1, 2) = 2.9, J(H-2, 3) = 4.0, J(H-1, 3) = 2.7, J(H-3, 4) = 3.6, J(H-5, exo, endo) = 14.6, J(H-1, 5) = 3.6, J(H-4 exo, 5 endo) = 3.7, J(H-4 exo, 5 exo) = 8.6 [39] IR (n-hexane): 1718 (ν(C=O)), 1884, 1893, 1955, 1963 (ν(CO)) [39]
15	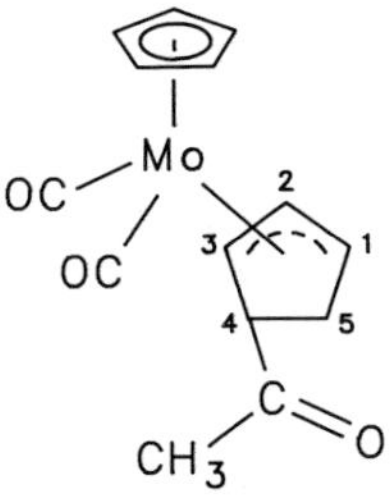	$C_5H_5Mo(CO)(C_5H_6\text{-cyclo})C(O)CH_3$ was refluxed for 90 min in pentane (4% by chromatography) [39] ^1H NMR (C_6D_6): 1.69 (CH_3), 1.76 (H-5, endo), 2.13 (H-5, exo), 2.67 (H-4, endo), 3.38 (H-1), 3.40 (H-3), 3.75 (H-2), 4.62 (C_5H_5); with J(H-1, 2) = 3.9, J(H-2, 3) = 3.8, J(H-1, 3) = 3.0, J(H-4 endo, 5 exo) = 3.3, J(H-5 exo, 5 endo) = 14.0, J(H-4 endo, 5 endo) = 7.9, J(H-1, 5 exo) = 3.4 [39] IR (n-hexane): 1720 (ν(C=O)), 1883, 1893, 1953, 1962 (νCO)) [39]
16	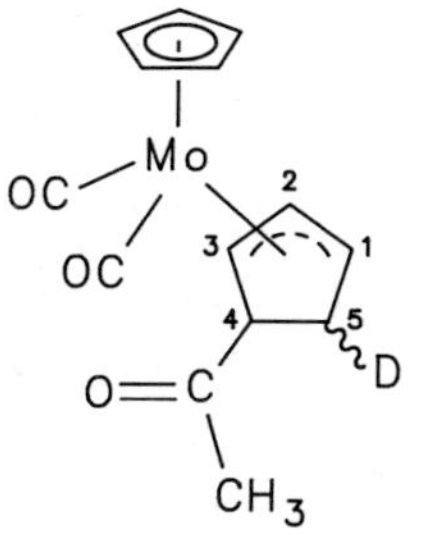	$C_5H_5Mo(CO)(C_5H_5D\text{-cyclo})C(O)CH_3$ (isomer mixture; 40% exo-D, 60% endo-D) was allowed to react in pentane with 1 atm CO [39] deuterium 40% in exo position and 60% in endo position [39]
17	$C_5H_5Mo(CO)_2C_5H_5D_2$	$(C_5H_5)_2MoD_2$ was heated in toluene at 80 °C under 1 atm CO pressure for 24 h [20] deuterium on the methylene carbons; the cis positions are preferred by the deuterium atoms [20]
18	$C_5H_5Mo(CO)_2C_3(CF_3)_3C(O)C{=}CF_2$ 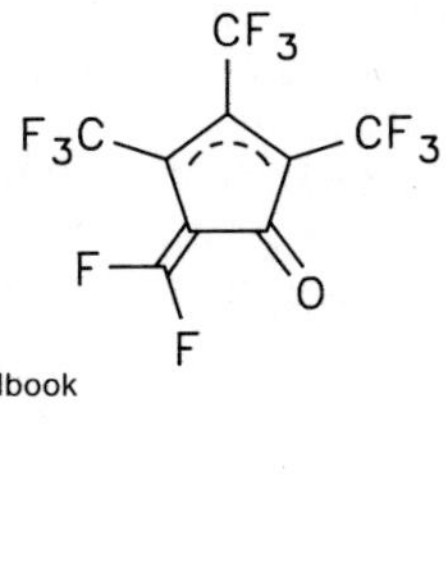	by-product of the reaction of $C_5H_5Mo(CCF_3{\equiv}CCF_3)_2Cl$ with 3 equivalents of $Co_2(CO)_8$ in ether for 48 h, isolated from the insoluble tar by extraction with CH_2Cl_2 (2 to 3%) [25] yellow crystals (from CH_2Cl_2/hexane at −20 °C) [25] ^1H NMR ($CDCl_3$): 5.32 (s, C_5H_5) [25]

References on pp. 303/4

Table 12 (continued)

No.	compound	method of preparation (yield) properties and remarks
18 (continued)		^{19}F NMR ($CDCl_3$): −81.58 (d, CF_2; J(F, F) = 8.5), −75.84 (d of q, CF_2; J(F, F) = 26.6, 8.5), −56.26 (d of q, CF_3; J(F, F) = 26.6, 9.8), −55.71 (q, CF_3; J(F, F) = 8.9), −48.82 (q of q, CF_3; J(F, F) = 8.9, 9.9) [25] IR: 1686 (ν(C=C)), 1748 (ν(C=O)), 2026, 2068 (ν(CO)) [25]
19	$C_5H_5Mo(CO)_2C_3H_3(CHC_5H_5)_2$	a solution of NaC_5H_5 (4 equivalents) in THF was added slowly to a solution of $C_5H_5Mo(CO)_2Cl_3$ in THF cooled at 0°C; the mixture was stirred for 3 h at room temperature followed by addition of $NaBH_4$ and further stirring for 15 h (30% by chromatography on SiO_2 with C_6H_6) [18, 48]; the compound was first described as $C_5H_5Mo(CO)_2(\eta^1\text{-}C_5H_5)_3$ [48]; see Section 1.5.1.3.2.1.11 yellow crystals (from ether/pentane) [18, 48] ^1H NMR (toluene-d_8): 2.4, 3.3, 3.9, and 4.0 (br m, $\eta^1\text{-}C_5H_5$), 4.59 (s, $\eta^5\text{-}C_5H_5$); no dynamic processes were detected at higher or lower temperatures [48] IR (THF): 1870, 1945 (ν(CO)) [48] mass spectrum: $[M-n\,CO]^+$ (n = 0 to 2) [48] volatile, air-stable solid but decomposes in solution under air [48] very soluble in C_6H_6, toluene, and THF but less soluble in pentane and hexane [48]
^{3}L = five-membered lactone		
20	$C_5H_5Mo(CO)_2C(CF_3)(CCH_3)_2C(O)O$ CH3; F3C; CH3; O; O	$C_5H_5Mo(CO)_2C(O)CF_3$ and but-2-yne were heated in hexane at 60°C for 24 h (20% by chromatography on Florisil with ether; 50% under 10 atm CO pressure) [15, 19] yellow crystals (from hexane at −20°C), m.p. 150 to 151°C (dec.) [19] ^1H NMR ($CDCl_3$): 1.90 and 2.56 (s, CH_3), 5.44 (s, C_5H_5) [19] ^{19}F NMR (acetone-d_6): −64.2 (CF_3) [19] IR (CCl_4): 1755 (ν(C=O)), 1945, 2000 (ν(CO)) [19]
21	$C_5H_5Mo(CO)_2C(CH_3)(CSi(CH_3)_3)_2C(O)O$ Si(CH3)3; H3C; Si(CH3)3; O; O	$C_5H_5Mo(CO)_3CH_3$ and $(CH_3)_3SiC{\equiv}CSi(CH_3)_3$ were irradiated in hexane for 5 d (10% by chromatography on Florisil with ether) [19] yellow crystals, m.p. 155 to 156°C (from ether/hexane at −78°C) [19] ^1H NMR ($CDCl_3$): 0.43 and 0.47 (s, $Si(CH_3)_3$), 2.30 (s, CH_3), 5.30 (s, C_5H_5) [19] IR (CCl_4): 1716 (ν(C=O)), 1911, 1972 (ν(CO)) [19]

Table 12 (continued)

No.	compound	method of preparation (yield) properties and remarks
22	$C_9H_7Mo(CO)_2C(CH_3)C(C_4H_9\text{-t})CHC(O)O$-endo 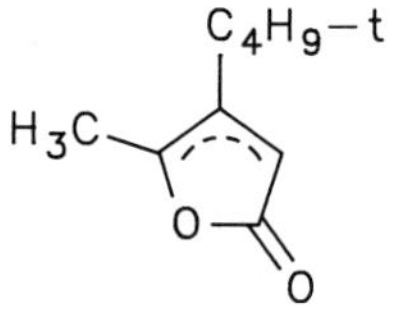	a solution of $C_9H_7Mo(CO)_2CH{=}C(C_4H_9\text{-t})$-$C(CH_3)O$ in hexane was pressured to 80 atm CO for 6 h (81%) [21] yellow crystals (from hexane at −78°C) [21] ^{1}H NMR ($CDCl_3$): 1.22 (s, t-C_4H_9), 1.60 (s, CH_3), 3.08 (s, CH), 5.66 (d of d, H-1; J(H, H) = 3.0), 5.86 (d, H-2), 5.92 (d, H-2′), 7.2 (m, C_9H_7, 4H) [21] IR (Nujol): 1709 (ν(C=O)); (n-hexane): 1819, 1948 (ν(CO)) [21]

^{3}L = cyclohexenyl and monosubstituted derivatives

No.	compound	method of preparation (yield) properties and remarks
*23	$C_5H_5Mo(CO)_2C_6H_9$-exo 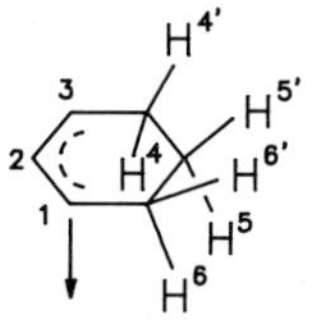	II, X = Br, for 2 h at 25°C (79% by column chromatography on Al_2O_3 with CH_2Cl_2) [26]; VIII, with H_2O (96%) [40, 46] yellow crystals, m.p. 95 to 95.5°C (from pentane) [26] exo to endo ratio > 100 [26] ^{1}H NMR ($CDCl_3$): 0.33 (d of d of d, H-5; J(H, H) = 14.5, 11.5, 6.8), 0.96 (d of d of d of d, H-5′; J(H, H) = 14.5, 6.0, 0.2), 1.67 (d of d of d of d, H-4; J(H, H) = 14.5, 6.9, 1.9, 0.2), 1.91 (d of d of d of d, H-4′; J(H, H) = 14.5, 11.5, 6.0, 1.8), 3.68 (m, H-1, 3), 4.16 (t, H-2; J(H, H) = 7.1), 5.29 (s, C_5H_5); also given as a diagram in [26] ^{95}Mo NMR (acetone-d_6, vs. 2 M Na_2MoO_4 in D_2O): −1824 (exo isomer) [38] IR (cyclohexane): 1875, 1947 (ν(CO)) [26]
24	$CH_3C_5H_4Mo(CO)_2C_6H_9$	V (2%) [43] yellow crystals (from pentane) [43] ^{1}H NMR ($CDCl_3$): 0.22 (H-5′), 0.83 (H-5), 1.58 (AA′BB′XY system, H-4′, 6′), 1.79 (AA′BB′XY system, H-4, 6), 1.74 (CH_3 on ^{5}L), 3.36 (H-1, 3), 3.66 (H-2), 4.96 (AA′BB′ system, C_5H_4, 2H), 5.05 (AA′BB′ system, C_5H_4, 2H) [43] IR: 1881, 1954 (ν(CO)) [43]
25	$C_9H_7Mo(CO)_2C_6H_9$-exo	VI, with excess of Na[BH_4] (42%) [17] yellow crystals [17] ^{1}H NMR ($CDCl_3$, 32°C): −0.6 (t, H-2; J(H-2, 3) = 8.0), 0.2 and 0.8 (m, CH_2, 1H), 1.6 (m, CH_2, 4H), 3.1 (br d, H-1, 3; J(H-2, 3) = 8), 5.6 (t,

References on pp. 303/4

Table 12 (continued)

No.	compound	method of preparation (yield) properties and remarks
25 (continued)		C_9H_7, H-1; J(H, H) = 3.0), 5.9 (d, C_9H_7, H-2), 7.0 (m, C_9H_7, 4H) [17] IR (hexane): 1875, 1948 (ν(CO)) [17]
26	$C_5H_5Mo(CO)_2C_6H_8D$-4,exo	VI, with $Na[D_3BCN]$ at 0°C for 0.5 h (83% by chromatography on alumina with CH_2Cl_2) [26]; VIII, with D_2O (96%) [40, 46] yellow crystals [26] ^1H NMR ($CHCl_3$): major differences with No. 23 are the loss of vicinal coupling at H-5, 5′, the loss of geminal coupling at H-4, and the decreased intensity of H-4′ [26] ^2H NMR ($CHCl_3$): 1.90 (s, H-4, 6); also given as a diagram in [26] IR: (cyclohexane): 1875, 1947 (ν(CO)) [26] oxidation with O_2 in $CDCl_3$ affords free 3-deuteriocyclohexene [26]
27	$C_5H_5Mo(CO)_2C_6H_8CH_3$-4	VI, with CH_3MgBr at 0°C for 15 min (61% by chromatography on silica with CH_2Cl_2) [26]; $C_5H_5Mo(CO)_2C_6H_7(CH_3\text{-}6)CN$-4 was allowed to react with LiR (R = n-, s-, or t-C_4H_9) in THF for 0.5 h followed by hydrolysis (97%) [40, 46] chair conformation with CH_3 in axial position [26] bright yellow solid, m.p. 102 to 104°C [26, 46] ^1H NMR ($CDCl_3$): 0.52 (d of d of d of d, H-5; J(H, H) = 13.7, 11.5, 6.7, 6.3), 0.72 (d of d of d of d, H-5′; J(H, H) = 13.7, 6.1, 1.4, 1.4), 1.15 (d, CH_3; J(H, H) = 7.1), 1.58 (d of d of d of d, H-6; J(H, H) = 15.0, 6.3, 3.4, 1.4), 1.91 (q of d of d of d, H-4; J(H, H) = 7.1, 6.7, 2.9, 1.4), 2.03 (d of d of d of d, H-6′; J(H, H) = 15.0, 11.5, 6.1, 2.6), 3.60 (br d, H-3; J(H, H) = 5.8), 3.71 (br d, H-1; J(H, H) = 6.3), 4.14 (d of d, H-2; J(H, H) = 6.3, 5.8), 5.28 (s, C_5H_5) [26] IR (cyclohexane): 1877, 1951 (ν(CO)) [26] reaction with $[C(C_6H_5)_3]PF_6$ in CH_2Cl_2 at 0°C affords $[C_5H_5Mo(CO)_2C_6H_7CH_3]PF_6$ [26]
28	$C_5H_5Mo(CO)_2C_6H_7$(D-4,exo)CH_3-6,endo	$C_5H_5Mo(CO)_2C_6H_7(CH_3\text{-}6)CN$-4 was allowed to react with LiR (R = n-, s-, or t-C_4H_9) in THF for 0.5 h followed by hydrolysis with D_2O (89%) [40, 46] yellow solid [46] ^1H NMR: similar to No. 27, with the exception of the loss of vicinal coupling of H-5, 5′, the

Table 12 (continued)

No.	compound	method of preparation (yield) properties and remarks
		loss of geminal coupling at the H-4 resonance, and the loss of the H-4′ resonance [46] IR: similar to No. 27 [46]
29	$C_9H_7Mo(CO)_2C_6H_8CH_3$-1	VI, with $K[(s\text{-}C_4H_9)_3BH]$, ^4L = 2-methylcyclohexa-1,3-diene [33] reaction with $[C(C_6H_5)_3]BF_4$ in CH_2Cl_2 affords $[C_9H_7Mo(CO)_2C_6H_7CH_3\text{-}1]BF_4$ [33]
30	$C_9H_7Mo(CO)_2C_6H_6CH_3$-4 (structure: ring positions 1, 2, 3, 5, 6 with CH_3 at C-4)	VI, ^4L = 1-methylcyclohexa-1,3-diene, with $LiN(C_3H_7\text{-}i)_2$ (good yields) or with $LiC(CH_3)HC(O)NC(CH_3)HC(C_6H_5)HOCO$ (11%) [33] yellow-orange colored [33] ^{1}H NMR (C_6D_6): −0.63 (t, H-2; J(H, H) = 6.8), 1.80 (q, CH_3; J(H, H) = 2.0), 2.48 (m, H-6), 3.24 (d, H-3; J(H, H) = 6.4), 3.33 (m, H-1), 4.22 (br s, H-5), 4.99 (t, C_9H_7, H-1; J(H, H) = 2.7), 5.34 and 5.48 (m, C_9H_7, H-2), 6.56 (m, C_9H_7, 4H) [33] IR (CH_2Cl_2): 1850, 1930 (ν(CO)) [33] reaction with $(C_2H_5)_2O \cdot HBF_4$ produces the precursor [33]
*31	$C_5H_5Mo(CO)_2C_6H_8CH_2COOH$-4	VII (97%) [36] yellow crystalline solid, m.p. 173 °C [36] ^{1}H NMR ($CDCl_3$): 0.50 (d of t, H-5, exo; J(H, H) = 14, 6), 0.87 (br d of d, H-5, endo; J(H, H) = 14, 6), 1.6 (br d of m, H-6, exo; J(H, H) = 14), 1.97 (br t of d, H-6, endo; J(H, H) ≈ 14, 5), 2.31 (m, 1H), 2.48 (m, diastereotopic CH_2, 2H), 3.50 (br d, H-1 or H-3, 1H; J(H, H) = 7.1), 3.73 (br d, H-3 or H-1, 1H; J(H, H) = 7.1), 4.17 (br d, H-2; J(H, H) = 7.1), 5.30 (s, C_5H_5) [36] IR (CCl_4): 1735, 1860, 1940, 3450 [36]
32	$C_5H_5Mo(CO)_2C_6H_8CH_2C(O)OCH_3$-4	X, with No. 38 (85%) [36] yellow crystalline solid, m.p. 106 °C [36] ^{1}H NMR ($CDCl_3$): 0.75 (m, H-5, exo), 0.78 (d of d, H-5, endo; J(H, H) = 13, 6.5), 1.54 (m, H-6, exo), 1.95 (m, H-6, endo), 2.28 (m, H-4), 2.44 (m, diastereotopic CH_2, 2H), 3.54 (br d, H-1 or H-3, 1H; J(H, H) = 7.2), 3.66 (s, CH_3), 3.71 (br d, obscured, H-3 or H-1, 1H), 4.15 (t, H-2; J(H, H) = 7.14), 5.28 (s, C_5H_5) [36] IR (CCl_4): 1750, 1860, 1940 [36] saponification affords No. 31 [36]

References on pp. 303/4

Table 12 (continued)

No.	compound	method of preparation (yield) properties and remarks
*33	$C_5H_5Mo(CO)_2C_6H_8CH_2CH(C(O)OC_4H_9\text{-}t)_2$-4,exo	VIII, R = H, addition of $CH_2{=}C(C(O)OC_4H_9\text{-}t)_2$ (86%) [40, 46] yellow solid [46] ^{1}H NMR ($CDCl_3$): 0.5 (m, H-5, exo), 0.9 (m, H-5, endo), 1.45 (s, t-C_4H_9), 2.0 (m, 2H), 2.46 (m, H-4, endo), 3.30 (t, 1H; J(H, H) = 7.6), 3.61 (m, H-1), 3.69 (m, H-3), 4.15 (t, H-2; J(H, H) = 7.2), 5.28 (s, C_5H_5) [46] IR ($CHCl_3$): 1720, 1850, 1935 [46] mass spectrum: $[M]^+$, $[M-CO]^+$ (5), $[M-2\,CO]^+$ (36), base peak at m/e = 358 [46]
*34	$C_5H_5Mo(CO)_2C_6H_8CH(C(O)CH_3)C(O)OCH_3$-4	VI, with $NaCH(C(O)CH_3)C(O)OCH_3$ for 15 min (80%) [32, 36] a 2:1 mixture of diastereomers; 1:1 after thin layer chromatography on silica with ethyl acetate/hexane [36] yellow crystalline solid, m.p. 136°C [36] ^{1}H NMR ($CDCl_3$): major isomer: 0.6 (m, H-5, 5′), 1.55 (br d, H-6, exo; J(H, H) = 15), 1.87 (m, H-4), 2.16 (s, $C(O)CH_3$), 2.57 (m, H-6, endo), 3.5 (m, obscured H-1 or H-3, 1H), 3.44 (d, CH; J(H, H) = 10), 3.72 (m, obscured, H-3 or H-1, 1H), 3.77 (s, $C(O)OCH_3$), 4.16 (t, H-2; J(H, H) = 7), 5.29 (s, C_5H_5); minor isomer: 2.30 (s, $(O)CH_3$), 3.32 (br d, 1H; J(H, H) = 7), 3.47 (d, 1H; J(H, H) = 10), 3.67 (s, 3H), 4.14 (t, 1H; J(H, H) = 7), 5.28 (s, C_5H_5) [36] IR (CCl_4): 1710, 1740, 1860, 1940 [36] mass spectrum: $[M]^+$, $[M-2\,CO]^+$ [36]
35	$C_5H_5Mo(CO)_2C_6H_8CH(C(O)OCH_3)COOH$-4	VII (98%) [36] yellow foam, two diastereomers [36] ^{1}H NMR ($CDCl_3$): 0.55 (m, H-5, exo), 0.87 (m, H-5, endo), 1.56 (m, H-6, exo), 1.64 (m, H-6, endo), 2.52 (m, H-4), 3.43 (d, CH; J(H, H) = 9.9), 3.47 (overlapping d, H-1, 3; J(H, H) = 6.7), 3.72 (s, $C(O)OCH_3$, minor isomer), 3.81 (s, $C(O)OCH_3$, major isomer), 4.24 (t, H-2; J(H, H) = 6.7), 5.34 (s, C_5H_5), 9.0 (br s, OH) [36] IR (CCl_4): 1745, 1862, 1945 [36]

Table 12 (continued)

No.	compound	method of preparation (yield) properties and remarks
*36	$C_5H_5Mo(CO)_2C_6H_8CH(C(O)OCH_3)_2$-4	VI, with 1 equivalent of $NaCH(C(O)OCH_3)_2$ for 15 min (85%) [32, 36] yellow crystals, m.p. 132°C (from ether/pentane) [36] ^1H NMR ($CDCl_3$): 0.61 (m, H-5, exo), 0.8 (d of d, H-5, endo; J(H, H) = 15, 5.7), 1.5 (br d, H-6, exo; J_{gem} = 15), 1.91 (t of d, H-6, endo; J_{gem} = $J_{5\text{-endo}}$ = 15, $J_{5\text{-exo}}$ = 5.2), 2.5 (m, H-4, endo), 3.36 (d, CH; J(H, H) = 10.1), 3.5 and 3.7 (m, H-1 and H-3, each 1H), 3.69 and 3.78 (s, CH_3), 4.18 (t, H-2; J(H, H) = 7.2), 5.29 (s, C_5H_5) [36] IR (CCl_4): 1735, 1750, 1860, 1940 [36] mass spectrum: $[M-n\,CO]^+$ (n = 0 to 2) [36] saponification yields No. 35 [36]
37	$C_5H_5Mo(CO)_2C_6H_8C(C_6H_5)(OH)H$-4,exo	VIII, R = H, addition of C_6H_5CHO (81%) [40, 46] a 6:5 mixture of two diastereomers, which could be separated by thin-layer chromatography with 40% ethyl acetate in hexane [46] ^1H NMR ($CDCl_3$): major diastereomer: 0.4 (m, H-5, exo), 0.7 (d of d, H-5, endo; J(H, H) = 15.0, 6.0), 1.6 (m, H-6, exo), 2.0 (m, H-4, endo, H-6, endo, and OH exchanged with D_2O), 3.76 (m, H-1 and H-3), 4.30 (t, H-2; J(H, H) = 7.2), 4.53 (d, $C(C_6H_5)(OH)\mathbf{H}$; J(H, H) = 7.4), 5.27 (s, C_5H_5); minor diastereomer: 0.4 (m, H-5, exo), 1.3 (d of d, H-5, endo; J(H, H) = 6.0, 15.0), 1.6 (m, H-6, exo), 2.0 (m, H-4, endo, H-6, endo, and OH exchanged with D_2O), 3.05 (m, H-3), 3.74 (m, H-1), 4.17 (t, H-2; J(H, H) = 7.2), 4.43 (d, $C(C_6H_5)(OH)\mathbf{H}$; J(H, H) = 7.4), 5.22 (s, C_5H_5) [46] IR ($CHCl_3$): 1860, 1945, 3500 [46] mass spectrum: $[M-n\,CO]^+$ (n = 0, 1), other peaks given without assignment [46]
38	$C_5H_5Mo(CO)_2C_6H_8CH(S(O)OC_6H_5)C(O)OCH_3$-4	VI, with $NaC(S(O)OC_6H_5)(C(O)OCH_3)H$ for 0.5 h (90%) [32, 36] 8:1 mixture of diastereomers [36] yellow foam solid [36] ^1H NMR ($CDCl_3$): 0.56 (m, H-5, exo), 0.63 (m, H-5, endo), 1.56 (m, H-6, exo), 1.86 (m, H-6, endo), 2.48 (m, H-4, major isomer), 2.66 (m, H-4, minor isomer), 3.41 (s, CH_3, minor isomer), 3.60 (s, CH_3, major isomer), 3.22 and 3.7 (m, H-1 and H-3), 3.95 (d, CH, major

References on pp. 303/4

Table 12 (continued)

No.	compound	method of preparation (yield) properties and remarks
38 (continued)		isomer; J(H, H) = 9.1), 4.07 (d, CH, minor isomer; J(H, H) = 9.1), 4.09 (t, H-2, major isomer; J(H, H) = 6.8), 4.20 (t, H-2, minor isomer; J(H, H) = 6.8), 5.23 (s, C_5H_5, major isomer), 5.28 (s, C_5H_5, minor isomer), 7.51 and 7.85 (m, C_6H_5) [36] IR (CCl_4): 1145, 1330, 1450, 1740, 1860, 1940 [36] reaction with Na_2HPO_4 in CH_3OH/THF followed by treatment with sodium amalgam affords No. 32 (Method X) [36]
39	$C_5H_5Mo(CO)_2C_6H_8C(C(O)OCH_3)(COOH)CH_3$-4	VII, for 8 d (70%) [36] only one diastereomer detected [36] yellow oil, could not be crystallized [36] ^{1}H NMR (dimethyl sulfoxide-d_6): 0.39 (m, H-5, exo), 0.83 (m, H-5, endo), 1.41 (s, CH_3C), 1.47 (m, obscured, H-6, exo), 1.89 (m, H-6, endo), 2.65 (m, H-4), 3.32 (s, OCH_3), 3.66 and 3.69 (t, H-1 and H-3; J(H, H) = 7.3), 4.56 (t, H-2; J(H, H) = 7.3), 5.47 (s, C_5H_5), 13.86 (br s, OH) [36] IR (CCl_4): 1720, 1860, 1940 [36]
*40	$C_5H_5Mo(CO)_2C_6H_8C(C(O)OCH_3)_2CH_3$-4	VI, with $NaC(C(O)OCH_3)_2CH_3$ (high) [32, 36] yellow crystalline solid, m.p. 108°C [36] ^{1}H NMR ($CDCl_3$): 0.61 (m, H-5, exo), 0.81 (m, H-5, endo), 1.60 (m, H-6, exo), 1.92 (m, H-6, endo), 2.72 (m, H-4, endo), 3.65 (s, OCH_3), 3.43 and 3.71 (m, H-1 and H-3), 3.70 (s, OCH_3), 4.3 (t, H-2; J(H, H) = 7.2), 5.26 (s, C_5H_5) [36] IR: 1735 ($\nu(CO_2)$), 1860, 1940 (ν(CO)) [36] treatment with aqueous KOH in CH_3OH/THF yields No. 39 in 70% yield [36]
41	$C_5H_5Mo(CO)_2C_6H_8C(CH_3)_2CH_2OH$-4	No. 43 was allowed to react with $LiBH_4$ in THF for 1 h (89%) [26] yellow solid (from CH_2Cl_2/pentane), m.p. 172°C (dec.) [26] ^{1}H NMR ($CDCl_3$): 0.42 (m, H-4), 0.98 (s, CH_3), 1.05 (t, 1H; J(H, H) = 7.0), 1.56 (s, 1H), 1.59 (m, 1H), 1.77 (d of d of d, 1H; J(H, H) = 1.1, 1.5, 7.7), 1.99 (m, 1H), 3.49 (AB system, CH_2O; $\Delta\delta$ = 0.19, J(H, H) = 1.2), 3.68 and 3.74 (virtual d, H-1 and H-3; J(H, H) = 7.3), 4.36 (t, H-2; J(H, H) = 7.3), 5.31 (s, C_5H_5) [26] IR (cyclohexane): 1874, 1948 (ν(CO)) [26]

References on pp. 303/4

Table 12 (continued)

No.	compound	method of preparation (yield) properties and remarks
		treatment with 1 equivalent of $[C(C_6H_5)_3]X$ (X = BF_4 or PF_6) in CH_3CN in the presence of $N(C_2H_5)_3$ affords No. 42 (2.5%) and No. 149 (18%) [26]
42	$C_5H_5Mo(CO)_2C_6H_8C(CH_3)_2CH_2OC(C_6H_5)_3$-4	No. 41 was allowed to react with $[C(C_6H_5)_3]X$ (X = BF_4 or PF_6) in CH_3CN for 1.5 h followed by the addition of $N(C_2H_5)_3$; the mixture was stirred for 1 h (2.5% by column chromatography on Al_2O_3 followed by SiO_2 with $CHCl_3$/hexane 75:25) [26] yellow solid [26] ^1H NMR ($CDCl_3$): 0.28 (m, H-5), 0.94 and 1.08 (s, CH_3), 0.94 (m, H-5), 1.55 (m, H-6), 1.78 (m, H-6), 2.07 (d of d, H-4; J(H, H) = 2.5, 8.0), 2.90 (AB system, CH_2O; $\Delta\delta$ = 0.11, J(H, H) = 9.0), 3.31 (approximately d, H-3; J(H, H) = 7.3), 3.63 (approximately d of d, H-1; J(H, H) = 2, 7), 4.07 (t, H-2; J(H, H) = 7.3), 5.23 (s, C_5H_5), 7.28 (m, C_6H_5, 9H), 7.46 (m, C_6H_5, 6H) [26]
43	$C_5H_5Mo(CO)_2C_6H_8C(CH_3)_2CHO$-4	similar to VI, with (2-methyl-1-propenyl)-1-pyrrolidine in CH_3CN at 0°C for 30 min followed by hydrolysis with water (66% by chromatography on SiO_2 with CH_2Cl_2) [26] yellow solid, m.p. 131 to 132°C (from pentane) [26] ^1H NMR ($CDCl_3$): 0.44 (m, H-5), 0.84 (d of d, H-5; J(H, H) = 6.0, 14.0), 1.13 (d, H-6; J(H, H) = 7.0), 1.57 (m, 1H), 1.87 (m, 1H), 1.98 (approximately d, 1H; J(H, H) = 7.5), 3.36 (approximately d, H-1 or H-3; J(H, H) = 7.3), 3.69 (m, H-3 or H-1), 4.34 (t, H-2; J(H, H) = 7.2), 5.31 (s, C_5H_5), 9.53 (s, CHO) [26] IR (cyclohexane): 1733 (ν(C=O)), 1877, 1949 (ν(CO)) [26] treating with $LiBH_4$ in THF affords No. 41 [26]
44	$C_5H_5Mo(CO)_2C_6H_8CHC(C(O)OCH_3)HC(O)CH_2CH_2$-4,exo 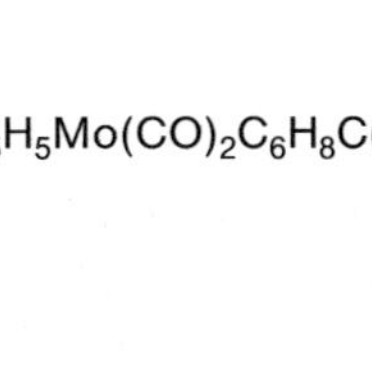 CH₃O O	VIII, R = H, with 2-(methoxycarbonyl)-cyclopent-1-en-3-one (81%) [46] a 9:5 mixture of two diastereomers [46] ^1H NMR ($CDCl_3$): 0.6 (m, 1H), 1.05 (m, 1H), 1.6 (m, 2H), 1.95 (m, 1H), 2.4 (m, 4H), 2.7 (m, **C**HCHC(O)OCH$_3$, major isomer), 2.94 (d, 1H, **C**HC(O)OCH$_3$, minor isomer; J(H, H) = 11.2),

References on pp. 303/4

Table 12 (continued)

No.	compound	method of preparation (yield) properties and remarks
44 (continued)		3.11 (d, C**H**C(O)OCH_3, major isomer; J(H, H) = 11.2), 3.44 (m, H-1), 3.70 (m, H-3), 3.73 (s, CH_3, minor isomer), 3.79 (s, CH_3, major isomer), 4.10 (t, H-2, major isomer; J(H, H) = 7.2), 4.21 (t, H-2, minor isomer; J(H, H) = 7.2), 5.28 (s, C_5H_5, major isomer), 5.31 (s, C_5H_5, minor isomer) [46] IR ($CHCl_3$): 1730, 1755, 1865, 1945 [46] mass spectrum: $[M-n\,CO]^+$ (n = 0, 2), other fragments given without assignment [46]
*45	$C_5H_5Mo(CO)_2C_6H_8C(C(O)OCH_3)C(O)CH_2CH_2CH_2$-4	VI, with $Na[C(C(O)OCH_3)C(O)CH_2CH_2CH_2$-cyclo] (85%) [32, 36] crystals of a single isomer were obtained from $CHCl_3$/ether after addition of pentane [36] ^{1}H NMR ($CDCl_3$): 0.63 (m, H-5), 1.68 (m, H-6, exo), 1.93 (m, H-6, endo), 2.04, 2.17, and 2.31 (m, CH_2), 2.73 (m, H-4), 3.15 (br d, H-1; J(H, H) = 7.13), 3.52 (br d, H-3; J(H, H) = 7.13), 3.75 (s, CH_3), 4.26 (t, H-2; J(H, H) = 7.13), 5.29 (s, C_5H_5) [36] ^{13}C NMR ($CDCl_3$): 24.67, 24.78, 25.79, 35.72, 42.11, 44.23, 57.61, 58.00, 60.01, 63.84, 73.09, 97.42 (C_5H_5), 175.97 (C=O) [36] IR: (CCl_4): 1720, 1750, 1860, 1940 [36] mass spectrum: $[M-n\,CO]^+$ (n = 0 to 2), other fragments given without assignment [36]
46	$C_5H_5Mo(CO)_2C_6H_8(C(C(O)OCH_3)(CH_2)C(O)C_6H_3CH_3)$-4	VI, with the sodium enolate of methyl 4-methylindanone-2-carboxylate [32, 36] two diastereomers in a 2:3 ratio [36] yellow colored foam [36] ^{1}H NMR ($CDCl_3$): major isomer: 0.5 to 0.9 (m, 2H), 1.5 to 2.0 (m, 2H), 2.45 (s, CH_3), 3.15 (AB system, CH_2), 3.3 and 3.7 (d, H-1 and H-3; J(H, H) = 7), 3.67 (s, OCH_3), 4.15 (t, H-2; J(H, H) = 7), 5.23 (s, C_5H_5), 7.2 to 7.6 (m, C_6H_3); minor isomer: 2.42 (s, CH_3), 3.75 (s, OCH_3), 4.39 (t, H-2; J(H, H) = 7), 5.31 (s, C_5H_5), 7.2 to 7.6 (m, C_6H_3), other peaks obscured by the major isomer [36] IR (CCl_4): 1735, 1860, 1940 [36] mass spectrum (field desorption): isotopic pattern of $[M]^+$ [36]
*47	$C_5H_5Mo(CO)_2C_6H_8(C_6H_4OCH_3$-4)-4	VI, with 4-$CH_3C_6H_4$MgBr at 0°C for 0.5 h (95%); traces ($\leq$ 5%) of 4,4′-dimethoxy-

Table 12 (continued)

No.	compound	method of preparation (yield) properties and remarks
		biphenyl could not be removed chromatographically [36] ^{1}H NMR ($CDCl_3$): 0.83 (m, H-5, exo), 1.25 (m, H-5, endo), 1.67 (m, H-6, exo), 2.09 (m, H-6, endo), 3.02 (m, H-4), 3.74 and 3.83 (m, H-1 and H-3; J(H, H) = 7.13), 3.81 (s, CH_3), 4.44 (t, H-2; J(H, H) = 7.13), 5.3 (s, C_5H_5), 6.86 (d of d, C_6H_4, 2H; J(H, H) = 6.72, 2.0), 7.33 (d, C_6H_4, 2H; J(H, H) = 6.72) [36] IR (CCl_4): 1860, 1940 [36]
48	$C_5H_5Mo(CO)_2C_6H_8C_6H_6OCH_3Fe(CO)_3$ 2, 1, 3, 4, 1', 2', 3', 5', Fe(CO)$_3$, OCH$_3$	VIII, with $[CH_3OC_6H_6Fe(CO)_3]PF_6$ (64%) [46] yellow oily complex [46] ^{1}H NMR ($CDCl_3$): 0.5 to 2.0 (m, 6H), 2.65 (m, H-4, endo and H-1′, exo), 3.32 (m, H-5′), 3.46 (s, CH_3), 3.53 (m, H-3), 3.68 (m, H-1), 4.23 (t, H-2; J(H, H) = 7.3), 5.06 (d of d, H-3′; J(H, H) = 6.4, 2.3), 5.28 (s, C_5H_5) [46] IR ($CHCl_3$): 1850, 1955, 2040 [46] mass spectrum: $[M-n\ CO]^+$ (n = 0, 1, 3 to 5), other fragments given without assignment [47]
49	$C_5H_5Mo(CO)_2C_6H_8C_6H_6Mn(CO)_3$ 2, 1, 3, 6, 4, 5, 1', 6', 2', 5', 3', 4', Mn(CO)$_3$	VIII, with $[C_6H_6Mn(CO)_3]PF_6$ (83%) [46] yellow crystalline solid, m.p. 300 °C (from petroleum ether/pentane 1:1) [46] ^{1}H NMR ($CDCl_3$): 0.8 (m, H-5, exo), 1.2 (m, H-5, endo), 1.5 (m, H-6, exo), 1.8 (m, H-6, endo), 2.2 (m, H-4, endo and H-6′), 3.21 (t, H-5′; J(H, H) = 5.4), 3.47 (m, H-3 and H-1′), 3.67 (m, H-1), 4.12 (t, H-2; J(H, H) = 7.2), 4.77 (t, H-4′; J(H, H) = 6.1), 4.88 (t, H-2′; J(H, H) = 6.1), 5.27 (s, C_5H_5), 5.75 (t, H-3′; J(H, H) = 5.4) [46] IR ($CHCl_3$): 1860, 1950, 2020 [46]
50	$C_5H_5Mo(CO)_2C_6H_8C_6H_4(OCH_3)_2Mn(CO)_3$ 2, 1, 3, 6, 4, 5, 1', 6', 5', 3', OCH$_3$, Mn(CO)$_3$, OCH$_3$	VIII, with $[1,3\text{-}(CH_3O)_2C_6H_4Mn(CO)_3]PF_6$ (73%) [46] yellow solid [46] ^{1}H NMR ($CDCl_3$): 0.1 (m, H-5, exo), 0.8 (m, H-5, endo), 1.6 (m, H-6, exo), 1.9 (m, H-6, endo), 2.52 (m, H-4, endo and H-6′), 3.11 (br d, H-5′; J(H, H) = 5.9), 3.27 (br d, H-1′; J(H, H) = 5.9), 3.42 (m, H-3), 3.44 and 3.56 (s, diastereotopic CH_3), 3.66 (m, H-1), 4.15 (t, H-2; J(H, H) = 7.2), 5.27 (s, C_5H_5), 5.69 (s, H-3′) [46] IR ($CHCl_3$): 1855, 1935, 2010 [46] mass spectrum: $[M-n\ CO]^+$ (n = 0 to 2, 4, 5), other fragments given without assignment [46]

References on pp. 303/4

Table 12 (continued)

No.	compound	method of preparation (yield) properties and remarks
*51	$C_5H_5Mo(CO)_2C_6H_8CN$-4,exo	similar to VI, with NaCN in CH_3CN/H_2O (10:1) for 1 h (82%) [46] yellow needle-shaped crystals, m.p. 155 to 156°C (from petroleum ether/pentane 1:1) [46] ^1H NMR ($CDCl_3$): 0.64 (m, H-5, exo), 1.30 (d of d, H-5, endo; J(H, H) = 14.5, 5.8), 1.75 (t of d, H-6, exo; J(H, H) = 12.3, 4.0), 2.17 (t of d, H-6, endo; J(H, H) = 13.5, 4.0), 2.77 (d of d, H-4, endo; J(H, H) = 6.5, 3.0), 3.59 (br d, H-1; J(H, H) = 6.8), 3.86 (m, H-3), 4.36 (t, H-2; J(H, H) = 7.1), 5.25 (s, C_5H_5) [46] IR ($CHCl_3$): 1870, 1950, 2215 [46]
52	$C_5H_5Mo(CO)_2C_6H_8CN$-4,endo	IX, with No. 51 and H_2O (96%) [40, 46] yellow crystalline solid, m.p. 172 to 174°C (from petroleum ether/pentane 1:1) [46] ^1H NMR ($CDCl_3$): 0.72 (m, H-6, exo), 1.45 (m, H-5, endo), 1.85 (m, H-6, endo and H-5, exo), 3.16 (d of d, H-4, exo; J(H, H) = 12.0, 4.0), 3.61 (br d, H-1; J(H, H) = 7.3), 3.75 (m, H-3), 4.20 (t, H-2; J(H, H) = 7.3), 5.34 (s, C_5H_5) [40, 46] IR ($CHCl_3$): 1875, 1955, 2240 [40, 46]
53	$C_5H_5Mo(CO)_2C_6H_7$(D-4,exo)CN-4,endo	IX, with No. 51 and D_2O [40, 46]
54	$C_9H_7Mo(CO)_2C_6H_8OCH_3$-4,endo	VI, with $K[(s\text{-}C_4H_9)_3BH]$ [33, 41]
55	$C_5H_5Mo(CO)_2C_6H_8Re(CO)_5$-4	VI, X = BF_4, MR = $Na[Re(CO)_5]$, at −78°C [42, 44] decomposes at 96°C [44] ^1H NMR (C_6D_6): 1.34 and 1.97 (m, CH_2), 2.70 (m, CHRe), 3.34 (m, H-1), 3.62 (br t, H-3), 4.63 (s, C_5H_5), 4.86 (m, H-2); with J(H-2, 3) ≈ J(H-1, 2) ≈ 7 [44] IR (CH_2Cl_2): 1917, 2044, 2083, 2128 (ν(CO)); (cyclohexane): 1937, 2007, 2046, 2121 (ν(CO)) [44]
	^3L = disubstituted cyclohexenyl	
56	$C_5H_5Mo(CO)_2C_6H_7(CH_3)_2$-4,6	VI, X = PF_6, ^3L = $C_6H_7CH_3$-4, at 0°C for 15 min (27%) [26] yellow solid, m.p. 112 to 113°C [26] ^1H NMR ($CDCl_3$): 0.59 (d of t, H-5′; J(H, H) = 14.3, 3.0), 0.85 (d of t, H-5; J(H, H) = 14.3, 7.1), 1.23 (d, CH_3; J(H, H) = 6.9), 1.92 (d of d of d of d, H-4′ and H-6′; J(H, H) = 7.1, 6.9, 3.0, 2.5), 3.70 (d, H-1 and H-3; J(H, H) = 7.1),

Table 12 (continued)

No.	compound	method of preparation (yield) properties and remarks
		4.16 (t, H-2; J(H, H) = 7.1), 5.28 (s, C_5H_5); for assignment see No. 27; a chair conformation was suggested for the cyclohexenyl ligand based on the observed and calculated vicinal interactions [26] IR (CH_2Cl_2): 1884, 1958 (ν(CO)) [26]
57	$C_9H_7Mo(CO)_2C_6H_7(CH_3)_2$-1,4	VI, ^{4}L = $C_6H_7CH_3$-1, with $LiCu(CH_3)_2$ [33] hydride abstraction with $(C_2H_5)_2O \cdot HBF_4$ in CH_2Cl_2 at −78 °C followed by addition of 2,3-dichloro-5,6-dicyano-1,4-benzoquinone affords $[C_9H_7Mo(CO)_2{}^4L]BF_4$ (^{4}L = 2,5-dimethylcyclohexa-1,3-diene) [33]
58	$C_9H_7Mo(CO)_2C_6H_7(CH_3)_2$-3,4	VI, ^{4}L = $C_6H_7CH_3$-2, with $LiCu(CH_3)_2$ [33] hydride abstraction with $[C(C_6H_5)_3]BF_4$ in CH_2Cl_2 affords $[C_9H_7Mo(CO)_2{}^4L]BF_4$ (^{4}L = 1,6-dimethylcyclohexa-1,3-diene) [33]
59	$C_9H_7Mo(CO)_2C_6H_7(CH_3-1)C_2H_5$-4	VI, ^{4}L = $C_6H_7CH_3$-1, with C_2H_5MgBr [33] hydride abstraction with $(C_2H_5)_2O \cdot HBF_4$ in CH_2Cl_2 at −78 °C followed by addition of 2,3-dichloro-5,6-dicyano-1,4-benzoquinone affords $[C_9H_7Mo(CO)_2{}^4L]BF_4$ (^{4}L = 2-methyl-5-ethylcyclohexa-1,3-diene) [33]
60	$C_5H_5Mo(CO)_2C_6H_7(CH_3-6)CH_2COOH$-4	VII (85%) [36] yellow crystalline solid, m.p. 174 °C [36] ^{1}H NMR (dimethyl sulfoxide-d_6): 0.5 to 0.8 (m, H-5), 1.18 (d, CH_3; J(H, H) = 7.3), 1.79 (m, H-6), 2.1 (m, H-4), 2.3 to 2.45 (m, diastereotopic C**H**$_2$COOH), 2.69 (br d, H-1 and H-3; J(H, H) = 6.9), 4.36 (t, H-2; J(H, H) = 7.1), 5.44 (s, C_5H_5), 12.2 (br, OH); ($CDCl_3$): 0.65 (d of t, H-5, 1H; J(H, H) = 14, 2.5), 0.8 (d of t, H-5, 1H; J(H, H) = 14, 7) [36] IR (CCl_4): 1700, 1860, 1940, 3400 [36]
61	$C_5H_5Mo(CO)_2C_6H_7(CH_3-6)CH_2C(O)OCH_3$-4	X, with No. 64 (70%) [36] could not be crystallized [36] ^{1}H NMR ($CDCl_3$): 0.64 (d of t, H-5, exo; J(H, H) = 14.36, 8), 0.84 (d of t, H-5, endo; J(H, H) = 14.36, 7.16), 1.21 (d, CH_3; J(H, H) = 7.32), 1.92 and 2.30 (m, H-4 and H-6), 2.54 (d, C**H**$_2$C(O)OCH$_3$; J(H, H) = 7.5), 3.68 (s, CH_3), 3.70 (m, obscured H-1 and H-3), 4.15 (t, H-2; J(H, H) = 7.08), 5.28 (s, C_5H_5) [36] IR (CCl_4): 1740, 1860, 1940 [36]

References on pp. 303/4

Table 12 (continued)

No.	compound	method of preparation (yield) properties and remarks
62	$C_5H_5Mo(CO)_2C_6H_7(CH_3\text{-}6)CH(C(O)OCH_3)COOH\text{-}4$	VII (95%) [36] a ca. 1:1 mixture of diastereomers [36] could not be crystallized [36] ^{1}H NMR (dimethyl sulfoxide-d_6): 0.42 (br d, H-5, exo; J(H, H) = 14.0), 0.69 (d of t, H-5, endo; J(H, H) = 14.0, 6.8), 1.13 (d, CH_3; J(H, H) = 7.2), 1.79 (m, H-6), 2.26 (m, H-4), 3.34 (C**H**C(O)OCH_3), 3.57 and 3.66 (s, diastereomeric CH_3), 3.68 (d, H-1; J(H, H) = 7), 3.52 and 3.85 (d, H-3, diastereomers; J(H, H) = 7), 4.40 and 4.42 (t, H-2, diastereomers; J(H, H) = 7), 5.45 (s, C_5H_5), 12.04 (br s, OH) [36] IR (CCl_4): 1730, 1860, 1945, 3400 [36]
63	$C_5H_5Mo(CO)_2C_6H_7(CH_3\text{-}6)CH(C(O)OCH_3)_2\text{-}4$	VI, X = PF_6, ^{4}L = $C_6H_7CH_3$-5, with $NaCH(C(O)OCH_3)_2$ at −78°C for 0.5 h (95%) [36] m.p. 163 to 164°C [36] ^{1}H NMR ($CDCl_3$): 0.59 (d of t, H-5, exo; J(H, H) = 14.84, 2.63), 0.87 (d of t, H-5, endo; J(H, H) = 14.84, 7.09), 1.19 (d, CH_3; J(H, H) = 7.22), 1.89 (m, H-6), 2.56 (m, H-4), 3.48 (d, C**H**$(C(O)OCH_3)_2$; J(H, H) = 11), 3.62 (br d, 1H; J(H, H) = 7.05), 3.69 (s, CH_3), 3.75 (br d, 1H; J(H, H) = 7.05), 3.78 (s, CH_3), 4.17 (t, H-2; J(H, H) = 7.05), 5.28 (s, C_5H_5) [36] IR (CCl_4): 1730, 1750, 1860, 1940 [36] hydrolysis (Method VII) affords No. 62 [36]
64	$C_5H_5Mo(CO)_2C_6H_7(CH_3\text{-}6)C(C(O)OCH_3)(S(O)OC_6H_5)H\text{-}4$	VI, X = PF_6, ^{4}L = $C_6H_7CH_3$-5, with $NaC(S(O)OC_6H_5)(C(O)OCH_3)H$ in THF at −78°C for 0.5 h (95%); contains small amounts of methyl phenylsulfonylacetate which could not be removed [36] a mixture of two diastereomers in a ca. 8:1 ratio [36] ^{1}H NMR ($CDCl_3$): major isomer: 0.85 (m, H-5, exo), 1.23 (d, CH_3; J(H, H) = 7.16), 1.62 (m, H-5, endo), 1.90 (m, H-6), 2.54 (br t, H-4; J(H, H) ≈ 6), 3.28 (d, H-1 or H-3: J(H, H) = 7), 3.66 (s, CH_3), 3.76 (d, H-1 or H-3; J(H, H) = 7), 4.16 (t, H-2; J(H, H) = 7), 4.28 (d, C**H**C(O)OCH_3; J(H, H) = 5.15), 5.26 (s, C_5H_5), 7.6 and 7.9 (m, C_6H_5); minor isomer: 1.15 (d, CH_3; J(H, H) =

References on pp. 303/4

Table 12 (continued)

No.	compound	method of preparation (yield) properties and remarks
		7.15), 2.66 (m, H-4), 3.71 (s, CH_3), 3.98 (d, C**H**C(O)OCH_3; J(H, H) = 6), 4.16 (t, 1H; J(H, H) = 7), 5.34 (s, C_5H_5), other signals obscured by the major isomer [36] IR (CCl_4): 1144, 1330, 1450, 1740, 1860, 1940 [36] reduction affords No. 61 [36]
65	$C_5H_5Mo(CO)_2C_6H_7(CH_3\text{-}6)C(C(O)OCH_3)C(O)(CH_2)_3\text{-}4$ 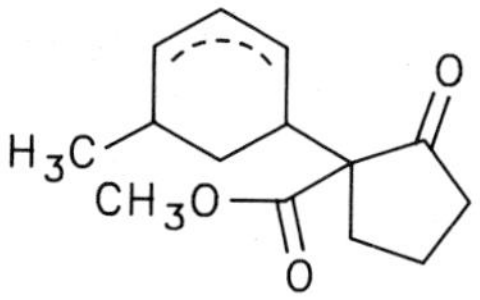	VI, X = PF_6, ^4L = $C_6H_7CH_3$-5, with sodium enolate of methyl 2-oxocyclopentanecarboxylate in THF at −78 °C (80%) [36] a mixture of 2 diastereomers in ca. 2:1 ratio [36] could not be crystallized [36] ^{1}H NMR ($CDCl_3$): 0.60 (m, H-5, exo), 1.00 (d, CH_3C; J(H, H) = 6.84), 1.09 (m, obscured H-5, endo), 1.73, 1.91, and 2.06 (m, CH_2), 2.29 (m, H-6), 2.49 (m, H-4), 3.66 (s, OCH_3, minor isomer), 3.70 (s, OCH_3, major isomer), 3.77 (m, H-1 and H-3), 4.40 (t, H-2; J(H, H) = 6.65), 5.21 (s, C_5H_5), 5.27 (s, C_5H_5) [36] IR (CCl_4): 1735, 1755, 1860, 1940 [36] mass spectrum (field desorption): isotopic pattern of $[M]^+$ [36]
66	$C_5H_5Mo(CO)_2C_6H_7(CH_3\text{-}6,exo)CN\text{-}4,exo$	similar to VI, X = PF_6, ^4L = $C_6H_7CH_3$-5, exo, with NaCN in CH_3CN/H_2O (10:1) for 1 h (95%) [46] yellow crystalline solid, m.p. 145 to 147 °C (from petroleum ether/pentane 1:1) [46] ^{1}H NMR ($CDCl_3$): 0.8 (m, H-5, exo), 1.2 (m, H-5, endo), 1.40 (d, CH_3; J(H, H) = 7.1), 2.10 (m, H-6, endo), 2.75 (d of d, H-4, endo; J(H, H) = 6.5, 3.0), 3.62 (m, H-1), 3.82 (m, H-3), 4.31 (t, H-2; J(H, H) = 7.1), 5.34 (s, C_5H_5) [46] IR ($CHCl_3$): 1870, 1950, 2240 [46] for reactions see Method IX [46]
67	$C_5H_5Mo(CO)_2C_6H_7(CH_3\text{-}6,exo)CN\text{-}4,endo$	IX, with H_2O (95%) [46] yellow crystalline complex, m.p. 139 to 141 °C (from petroleum ether/pentane 1:1) [46] ^{1}H NMR ($CDCl_3$): 0.89 (m, H-5, exo), 1.1 (m, H-5, endo), 1.14 (d, CH_3; J(H, H) = 6.8), 2.1 (m, H-6, endo), 3.27 (d of d, H-4, exo; J(H, H) = 12.0, 4.0), 3.67 (m, H-1 and H-3), 4.19 (t, H-2; J(H, H) = 7.1), 5.34 (s, C_5H_5) [46] IR ($CHCl_3$): 1870, 1950, 2230 [46]

References on pp. 303/4

Table 12 (continued)

No.	compound	method of preparation (yield) properties and remarks
68	$C_5H_5Mo(CO)_2C_6H_6(CH_3$-6,exo)(D-4,exo)CN-4,endo	IX, with D_2O (90%) [46] no further data given [46]
69	$C_5H_5Mo(CO)_2C_6H_7$(CN-4,endo)$CH_2CH(C(O)OC_4H_9$-t$)_2$-4,exo	IX, with di-t-butyl methylenemalonate (88%) [46] yellow crystalline solid, m.p. 185°C (from petroleum ether/pentane 1:1) [46] ^1H NMR ($CDCl_3$): 0.9 (m, H-5, exo), 1.24 (m, H-5, endo), 1.46 and 1.49 (s, t-C_4H_9), 1.74 (m, H-6, exo), 2.0 (m, H-6, endo), 2.30 and 2.45 (AB system, C**H**$_2$CH; J(A, B) = 14.4, J(A, X) = 6.05, J(B, X) = 6.35), 3.53 (m, H-1 and C**H**CH_2), 3.78 (m, H-3), 4.17 (t, H-2; J(H, H) = 7.2), 5.32 (s, C_5H_5) [46] IR ($CHCl_3$): 1725, 1870, 1955, 2230 [46]
*70	$C_5H_5Mo(CO)_2C_6H_7(CH_3$-6)$CH(C(O)OCH_3)_2$-4	VI, X = PF_6, ^4L = $C_6H_7CH_3$-6, with $NaCH(C(O)OCH_3)_2$ (> 80%) [32]
71	$C_5H_5Mo(CO)_2C_6H_7$(CN-4,endo)$C(C_6H_5)$(OH)H-4,exo	IX, with C_6H_5CHO in the presence of $(C_2H_5)_2O \cdot BF_3$ (78%) [40, 46] a mixture of two diastereomers in a 4:1 ratio [46] yellow solid [46] ^1H NMR ($CDCl_3$): major isomer: 0.9 (m, H-5), 1.85 (m, H-6, exo), 2.0 (m, H-6, endo), 2.35 (d, OH; J(H, H) = 4.0; disappeared after shaking with D_2O), 2.82 (br d, H-3; J(H, H) = 7.2), 3.82 (m, H-1), 4.22 (t, H-2; J(H, H) = 7.2), 4.38 (d, C**H**OH; J(H, H) = 4.0; a singlet after shaking with D_2O), 5.25 (s, C_5H_5); minor isomer: 0.8 (m, H-5, exo), 1.0 (d of d, H-5, endo; J(H, H) = 15.0, 6.0), 1.8 (m, H-6, exo), 2.0 (m, H-6, endo), 2.34 (d, OH; J(H, H) = 4.0; disappeared after shaking with D_2O), 3.91 (m, H-1), 4.10 (br d, H-3; J(H, H) = 7.2), 4.37 (t, H-2; J(H, H) = 7.2), 4.61 (d, C**H**OH; J(H, H) = 4.0; a singlet after shaking with D_2O), 5.35 (s, C_5H_5) [40, 46] IR ($CHCl_3$): 1850, 1945, 2240; 3400 and 3600 (ν(OH)) [40, 46]

References on pp. 303/4

Table 12 (continued)

No.	compound	method of preparation (yield) properties and remarks
		mass spectrum: $[M-n\,CO]^+$ (n = 0, 1), other fragments given without assignment [46]
72	$C_5H_5Mo(CO)_2C_6H_7(CN\text{-}4,endo)CHC(C(O)OCH_3)HC(O)CH_2CH_2\text{-}4,exo$ 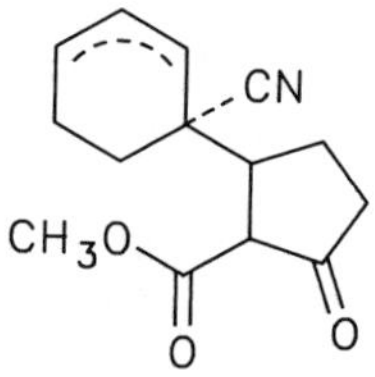	IX, with 2-carbomethoxycyclopent-2-en-1-one (No. 51) (70%) [46] three diastereomers in an 8:9:10 ratio; the major isomer was separated by thin-layer chromatography on silica with 40% ethyl acetate in hexane [46] ^{1}H NMR ($CDCl_3$): major isomer: 0.95 (m, CH_2, 2H), 1.75 (m, 1H), 2.05 (m, CH_2, 2H), 2.3 to 2.7 (m, 3H), 2.93 (d of t, $CH_2C\mathbf{H}C(C(O)OCH_3)H$; J(H, H) = 12.0, 10.4), 3.26 (d, $C\mathbf{H}C(O)OCH_3$; J(H, H) = 10.4), 3.73 (m, H-1, H-3, and CH_3), 4.25 (t, H-2; J(H, H) = 7.2), 5.36 (s, C_5H_5); mixture of the two other isomers: 0.6 (m, 1H), 1.05 (m, 1H), 1.5 (m, CH_2), 1.95 (m, 1H), 2.4 (m, 3H), 2.7 (m, $CH_2C\mathbf{H}C(C(O)OCH_3)H$), 2.93 (d, $C\mathbf{H}C(O)OCH_3$, minor isomer; J(H, H) = 11.2), 3.10 (d, $C\mathbf{H}C(O)OCH_3$, major isomer; J(H, H) = 11.2), 3.43 (m, H-1), 3.70 (m, H-3), 3.74 (s, CH_3, minor isomer), 3.79 (s, CH_3 major isomer), 4.09 (t, H-2, major isomer; J(H, H) = 7.2), 4.21 (t, H-2 minor isomer; J(H, H) = 7.2), 5.28 (s, C_5H_5, major isomer), 5.32 (s, C_5H_5, minor isomer) [46] IR ($CHCl_3$): 1725, 1755, 1855, 1940, 2200 [46] mass spectrum: $[M-CN]^+$, other fragments given without assignment [46]
73	$C_5H_5Mo(CO)_2C_6H_7(CN\text{-}4,endo)C(CH_2Fe(CO)_2C_5H_5)(OCH_3)H\text{-}4,exo$	IX, with $[C_5H_5Fe(CO)_2CH_2{=}C(OCH_3)H]BF_4$ or PF_6 (100%) [46] dark red solid; proved to be unstable [46] ^{1}H NMR ($CDCl_3$): 0.5 to 2.0 (m, 6H), 3.1 (m, $C\mathbf{H}OCH_3$), 3.5 (s, CH_3), 3.65 (d, H-3; J(H, H) = 7.1), 3.85 (m, H-1), 4.3 (t, H-2; J(H, H) = 7.1), 4.85 (s, C_5H_5Fe), 5.30 (s, C_5H_5Mo) [46] IR ($CHCl_3$): 1870, 1950, 2010, 2240 [46] reaction with $(C_2H_5)_2O \cdot HBF_4$ in CH_2Cl_2 at −78°C affords No. 75 [46]
74	$C_5H_5Mo(CO)_2C_6H_7(CN\text{-}4,endo)CH{=}CH_2\text{-}4,exo$	No. 75 was allowed to react with NaI in acetone for 1 h (product could not be separated from $C_5H_5Fe(CO)_2I$); No. 75 was refluxed in CH_3CN

References on pp. 303/4

Table 12 (continued)

No.	compound	method of preparation (yield) properties and remarks
74 (continued)		for 15 min followed by extraction with ether (92%) [46] ^{1}H NMR ($CDCl_3$): 0.99 (m, H-5, exo), 1.31 (d of d, H-5, endo; J(H, H) = 13.7, 5.9), 1.79 (m, H-6, exo), 1.98 (m, H-6, endo), 3.49 (d, H-3; J(H, H) = 7.2), 3.82 (m, H-1), 4.28 (t, H-2; J(H, H) = 7.2), 5.19 (t, 1H; J(H, H) = 9.9), 5.34 (s, C_5H_5), 5.56 (d, 1H; J(H, H) = 16.9), 5.86 (d of d, 1H; J(H, H) = 16.9, 9.9) [46] IR ($CHCl_3$): 1875, 1955, 2230 [46] mass spectrum: $[M-n\,CO]^+$ (n = 0 to 2), other fragments given without assignment [46]
75	$[C_5H_5Mo(CO)_2C_6H_7(CN\text{-}4,endo)CH{=}CH_2Fe(CO)_2C_5H_5\text{-}4,exo]BF_4$	No. 73 was treated with $(C_2H_5)_2O \cdot HBF_4$ in CH_2Cl_2 at −78 °C for 1 h followed by warming up (85%) [46] yellow crystals [46] ^{1}H NMR (CD_3CN): 0.8 to 2.5 (m, 4H), 4.0 (m, H-1 and H-3), 4.5 (t, H-2; J(H, H) = 7.1), 5.45 (s, C_5H_5Fe), 5.65 (s, C_5H_5Mo), 5.2 to 5.8 (m, $CH{=}CH_2$) [46] IR (CH_3CN): 1870, 1950, 2040, 2080 [46] treatment with NaI in acetone gives No. 74 and $C_5H_5Fe(CO)_2I$; refluxing in CH_3CN gives No. 74 and $[C_5H_5Fe(CO)_2NCCH_3]BF_4$ [46]
76	$C_5H_5Mo(CO)_2C_6H_7(CN\text{-}4,endo)C_6H_6OCH_3Fe(CO)_3\text{-}4,exo$ 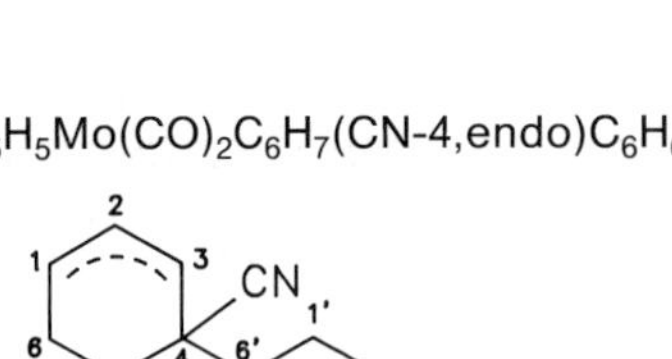 (structure: 1, 2, 3, CN, 1', 2', 3', 5', Fe(CO)$_3$, OCH$_3$)	IX, with $[CH_3OC_6H_6Fe(CO)_3]PF_6$ (34%) [40, 46] yellow, oily solid [46] ^{1}H NMR ($CDCl_3$): 0.8 to 2.2 (m, H-1′ and 3 CH_2), 2.54 (d of d, H-4′ (?); J(H, H) = 6.4, 3.0), 3.26 (m, H-5′), 3.48 (m, H-3), 3.66 (s, CH_3), 3.76 (m, H-1), 4.16 (t, H-2; J(H, H) = 7.3), 5.15 (d of d, H-3′; J(H, H) = 6.4, 2.3), 5.31 (s, C_5H_5) [46] IR ($CHCl_3$): 1870, 1960, 2040, 2240 [46] mass spectrum: $[M-n\,CO]^+$ (n = 0, 4, 5), other fragments given without assignment [46]
77	$C_5H_5Mo(CO)_2C_6H_7(CN\text{-}4,endo)C_6H_6Mn(CO)_3\text{-}exo$ 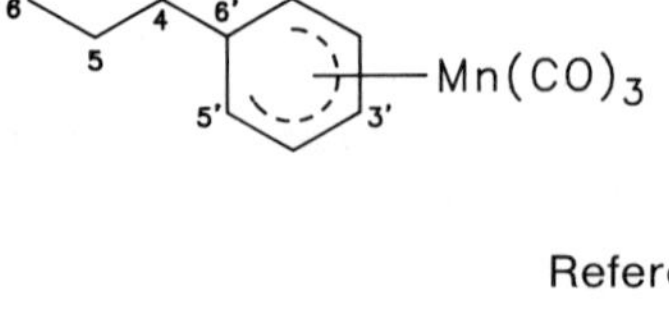 (structure: 1, 2, 3, 4, 5, 6, CN, 1', 3', 5', 6', Mn(CO)$_3$)	IX, with $[C_6H_6Mn(CO)_3]PF_6$ (74%) [46] yellow solid [46] ^{1}H NMR ($CDCl_3$): 0.55 (m, H-5, exo), 1.24 (d of d, H-5, endo; J(H, H) = 14.5, 5.8), 1.73 (m, H-6, exo), 1.84 (m, H-6, endo), 2.70 (t, H-6′; J(H, H) = 5.6), 3.26 (br t, H-5′; J(H, H) = 5.4), 3.47 (d, H-3; J(H, H) = 7.2), 3.58 (br t, H-1′; J(H,

References on pp. 303/4

Table 12 (continued)

No.	compound	method of preparation (yield) properties and remarks
		H) = 5.4), 3.76 (m, H-1), 4.22 (t, H-2; J(H, H) = 7.2), 5.01 (t, H-4′; J(H, H) = 6.1), 5.09 (t, H-2′; J(H, H) = 6.1), 5.31 (s, C_5H_5), 5.85 (t, H-3′; J(H, H) = 5.4) [46] IR ($CHCl_3$): 1870, 1950, 2020, 2230 [46] mass spectrum: $[M-n\ CO]^+$ (n = 0, 4), other fragments given without assignment [46]
78	$C_5H_5Mo(CO)_2C_6H_7(CN\text{-}4,endo)C_6H_4(OCH_3)_2Mn(CO)_3\text{-}4,exo$ 2, 1, 3, CN, 1′, OCH_3, 6, 4, 6′, 5, $Mn(CO)_3$, 5′, 3′, OCH_3	IX, with $[(CH_3O)_2C_6H_4Mn(CO)_3]PF_6$ (76%) [46] yellow solid [46] ^1H NMR ($CDCl_3$): 0.6 (m, H-5, exo), 1.1 (m, H-5, endo), 1.7 (m, H-6, exo), 1.9 (m, H-6, endo), 2.85 (t, H-6′; J(H, H) = 5.9), 3.1 (br d, H-5′; J(H, H) = 5.9), 3.3 (br d, H-1′; J(H, H) = 5.9), 3.47 (m, H-3), 3.55 (s, CH_3-4′), 3.64 (s, CH_3-2′), 3.74 (m, H-1), 4.26 (t, H-2; J(H, H) = 7.2), 5.32 (s, C_5H_5), 5.72 (s, H-3′) [46] IR ($CHCl_3$): 1870, 1940, 2020, 2230 [46] mass spectrum: $[M-n\ CO]^+$ (n = 0 to 2, 4, 5), other fragments given without assignment [46]
79	$C_9H_7Mo(CO)_2C_6H_7(C_6H_5\text{-}4,exo)OCH_3\text{-}4,endo$	VI, X = BF_4, ^4L = $C_6H_7OCH_3$-1, with C_6H_5MgBr (23%); main product is No. 83 [41]
80	$C_9H_7Mo(CO)_2C_6H_7(CH_3\text{-}1)C(CH_3)HC(O)NC(CH_3)HC(C_6H_5)HOCO\text{-}4,exo$ H_3C, CH_3, O, N, O, O, CH_3, C_6H_5	VI, X = BF_4, ^4L = $C_6H_7CH_3$-1, with $LiC(CH_3)HC(O)NC(CH_3)HC(C_6H_5)HOCO$ [33] side product in low yield is No. 30 [33]
81	$C_9H_7Mo(CO)_2C_6H_7(C(O)OCH_3\text{-}1)CH_3\text{-}4,exo$	VI, X = BF_4, ^4L = $C_6H_7C(O)OCH_3$-1, with $LiCu(CH_3)_2$ [33]
82	$C_9H_7Mo(CO)_2C_6H_7(CH_3\text{-}4,exo)OCH_3\text{-}4,endo$	VI, X = BF_4, ^4L = $C_6H_7OCH_3$-1, with CH_3MgBr at 0°C [33, 41] ^1H NMR (C_6D_6): −0.78 (t, H-2; J(H, H) = 7.4), 0.42 to 0.96 (m, CH_2), 1.14 (d, CH_3; J(H, H) =

References on pp. 303/4

Table 12 (continued)

No.	compound	method of preparation (yield) properties and remarks
82 (continued)		0.8), 1.44 to 2.10 (m, CH_2), 3.06 (m, H-1), 3.17 (s, OCH_3), 3.31 (d, H-3; J(H, H) = 7.9), 5.10 (t, H-1′; J(H, H) = 3.1), 5.47 (m, H-2′), 6.58 (m, H-3′) [33]
^{4}L = cyclohexenyl-4-one		
*83	$C_9H_7Mo(CO)_2C_6H_7O$ 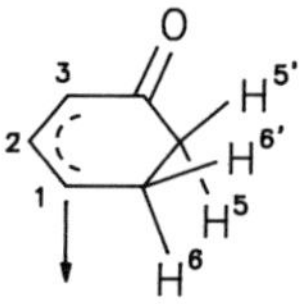	VI, ^{4}L = C_6H_7OR-1, R = CH_3, X = BF_4, with C_6H_5MgBr (43%); R = $Si(CH_3)_3$, X = CF_3SO_3, with $NaOCH_3$ (86%); also observed in the reaction of $[C_9H_7Mo(CO)_2(NCCH_3)_2]BF_4$ with cyclo-$C_6H_7OSi(CH_3)_3$-1 [41] ^{1}H NMR (CD_2Cl_2): 0.21 (d of d, H-2; J(H-3, 2) = 6.3, J(H-1, 2) = 7.3), 1.31 (d of d of d, H-5; J(H-6, 5) = 9.6, J(H-1, 5) = 7.8, J(H-5′, 5) = 18.9), 1.55 (d of d of d, H-5′; J(H-6, 5′) = 1.9, J(H-6′, 5′) = 8.7, J(H-5, 5′) = 18.7), 1.94 (d of d of d of d, H-6; J(H-1, 6) = 2.9, J(H-6, 6′) = 15.6, J(H-5, 6) = 9.6, J(H-5′, 6) = 1.9), 2.16 (d of d of d of d, H-6′; J(H-1, 6′) = 2.9, J(H-6, 6′) = 15.6, J(H-5, 6′) = 7.8, J(H-5′, 6′) = 8.7), 3.20 (d of d, H-3; J(H-2, 3) = 6.3, J(H-1, 3) = 1.5), 3.63 (d of d of d of d, H-1; J(H-3, 1) = 1.5, J(H-2, 1) = 7.3, J(H-6, 1) = J(H-6′, 1) = 2.9) [41] IR (CH_2Cl_2): 1635 (ν(C=O)), 1884, 1960 (ν(CO)) [41]
84	$C_9H_7Mo(CO)_2C_6H_6DO$	formed by treating No. 83 with CD_3OD in the presence of anhydrous Na_2CO_3; exchanged in the 5′ position [41]
85	$C_9H_7Mo(CO)_2C_6H_6(O)CH_3$ 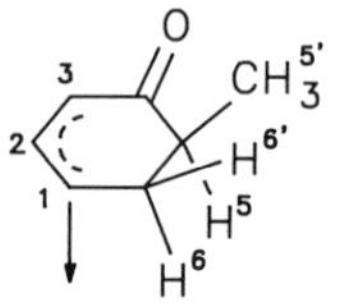	similar to IX, with No. 83 and $LiNR_2$ (R = $Si(CH_3)_3$ or i-C_3H_7) followed by addition of CH_3I (87%) [41] ^{1}H NMR (CD_2Cl_2): 0.23 (d of d, H-2; J(H-3, 2) = 6.3, J(H-1, 2) = 7.5), 0.3 (d, CH_3; J(H5, 5′) = 6.8), 1.46 (d of d of q, H-5; J(H-6, 5) = 7.9, J(H-1, 5) = 8.6, J(H-5′, 5) = 6.8), 1.82 (d of d of d, H-6; J(H-1, 6) = 2.8, J(H-6′, 6) = 14.8, J(H-5, 6) = 7.9), 2.26 (d of d of d, H-6′; J(H-1, 6′) = 2.8, J(H-6, 6′) = 14.8, J(H-5, 6′) = 8.6), 3.22 (d of d, H-3; J(H-2, 3) = 6.3, J(H-1, 3) = 1.5), 3.56 (d of d of d of d, H-1; J(H-2, 1) = 7.5, J(H-6, 1) = J(H-6′, 1) = 2.8, J(H-3, 1) = 1.5) [41]

References on pp. 303/4

Table 12 (continued)

No.	compound	method of preparation (yield) properties and remarks
86	$C_9H_7Mo(CO)_2C_6H_6(O)C_2H_5$	similar to IX, with No. 83 and $LiNR_2$ (R = $Si(CH_3)_3$ or i-C_3H_7) followed by addition of C_2H_5I (87%) [41] see No. 85 with C_2H_5 in CH_3 position [41]
87	$C_9H_7Mo(CO)_2C_6H_6(O)C(C_6H_5)(OH)H$ 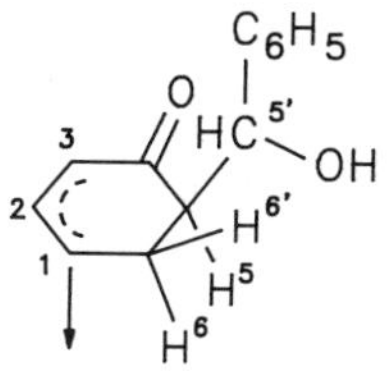	similar to IX, with No. 83 and $LiN(Si(CH_3)_3)_2$ followed by addition of C_6H_5CHO (good yield) [41] yellow crystalline solid [41] ^1H NMR (CD_2Cl_2): 0.16 (d of d of d; H-2, J(H-3, 2) = 5.9, J(H-6′, 2) = 7.3, J(H-1, 2) = 0.8), 1.57 (d of d of d, H-6; J(H-1, 6) = 2.3, J(H-6′, 6) = 14.2, J(H-5, 6) = 8.5), 1.70 (d of d of d, H-5; J(H-6, 5) = 8.5, J(H-6′, 5) = 7.3, J(H-5′, 5) = 3.1), 2.08 (d of d of d of d, H-6′; J(H-2, 6′) = 0.8, J(H-1, 6′) = 2.4, J(H-6, 6′) = 14.2, J(H-5, 6′) = 7.3), 2.93 (br, OH; J(H-5′, OH) = 5.1), 3.27 (d of d, H-3; J(H-2, 3) = 5.9, J(H-1, 3) = 1.5), 3.52 (d of d of d of d, H-1; J(H-3, 1) = 1.5, J(H-2, 1) = 7.3, J(H-6, 1) = 2.3, J(H-6′, 1) = 2.4), 5.02 (br d of d, H-5′; J(H-5, 5′) = 3.1, J(OH, 5′) = 5.1) [41]
	^3L = trisubstituted cyclohexenyl derivatives	
88	$C_9H_7Mo(CO)_2C_6H_6(CH_3)_3$-1,4,6	VI, X = BF_4, ^4L = $C_6H_6(CH_3)_2$-1,6, with $LiCu(CH_3)_2$ [33]
89	$C_9H_7Mo(CO)_2C_6H_6(CH_3)_3$-1,4,4	VI, X = BF_4, ^4L = $C_6H_6(CH_3)_2$-1,4, with $LiCu(CH_3)_2$ at 0°C for 0.5 h (35%) or 10 equivalents of CH_3MgI in CH_2Cl_2/ether (7:1) at 0°C for 15 min (92%) [45, 46] yellow solid [46] ^1H NMR ($CDCl_3$): −0.37 (d, H-2; J(H, H) = 7.4), 0.3 (m, H-5, exo), 0.7 (d of d, H-5, endo; J(H, H) = 14.0, 6.0), 0.89 (s, CH_3-4), 0.99 (s, CH_3-4), 1.51 (s, CH_3-1), 1.6 (m, H-6, endo and H-6, exo), 2.57 (d of d, H-3; J(H, H) = 7.4, 1.7), 5.55 (t, 1H; J(H, H) = 2.9), 5.73 (m, 1H), 6.0 (m, 1H), 7.1 (m, 4H) [46] IR ($CHCl_3$): 1845, 1930 [46] mass spectrum: $[M-n\,CO]^+$ (n = 0 to 2), other fragments given without assignment [46]
90	$C_9H_7Mo(CO)_2C_6H_6((CH_3)_2$-1,4,endo)$CH_2CH{=}CH_2$-4,exo	similar to VI, X = BF_4, ^4L = $C_6H_6(CH_3)_2$-1,4, with 2 equivalents of ethereal $CH_2{=}CHCH_2$-

Table 12 (continued)

No.	compound	method of preparation (yield) properties and remarks
90 (continued)		MgBr in CH_2Cl_2 at −78 °C for 1 h (90%) [45, 46] yellow solid [46] ^{1}H NMR ($CDCl_3$): −0.36 (d, H-2; J(H, H) = 7.4), 0.2 (m, H-5, exo), 0.8 (d of d, H-5, endo; J(H, H) = 14.0, 6.0), 0.94 (s, CH_3-4), 1.50 (s, CH_3-1), 1.6 (m, H-6, endo and H-6, exo), 1.82 and 1.97 (AB system q d, C**H**$_2$CH=; J(A, B) = 14.0, J(A, X) = 7.3, J(B, X) = 7.8), 2.56 (d of d, H-3; J(H, H) = 7.4, 1.7), 4.92 (m, CH=CH_2, 2H), 5.55 (t, 1H; J(H, H) = 2.9), 5.6 to 5.9 (m, 1H CH=CH_2), 5.74 (m, 1H), 6.0 (m, 1H), 7.1 (m, 4H) [46] IR ($CHCl_3$): 1845, 1930 [46] mass spectrum: $[M-n\ CO]^+$ (n = 0 to 2), other fragments given without assignment [46]
91	$C_5H_5Mo(CO)_2C_6H_6(CH_3$-6,exo)(CN-4,endo)$CH_2CH(C(O)OC_4H_9$-t$)_2$-4,exo	IX, with CH_2=$C(C(O)OC_4H_9$-t$)_2$ (85%) [46] yellow solid, m.p. 140 to 142 °C (from petroleum ether/pentane 1:1) [46] ^{1}H NMR ($CDCl_3$): 0.9 (m, H-5, exo), 1.21 (m, H-5, endo), 1.30 (d, CH_3; J(H, H) = 7.4), 1.47 (s, t-C_4H_9), 1.50 (s, t-C_4H_9), 2.14 (m, H-6, endo), 3.56 (d of d, C**H**$(C(O)OC_4H_9$-t$)_2$; J(H, H) = 7.9, 4.0), 3.62 (d of d, H-1; J(H, H) = 7.1, 1.7), 3.81 (t of d, H-3; J(H, H) = 7.1, 2.1), 4.15 (t, H-2; J(H, H) = 7.1), 5.30 (s, C_5H_5) [46] IR ($CHCl_3$): 1725, 1875, 1955, 2230 [46]
*92	$C_9H_7Mo(CO)_2C_6H_6((CH_3)_2$-1,4,endo)$CH_2COOH$-4,exo	VI, X = BF_4, ^{4}L = $C_6H_6(CH_3)_2$-1,4, with $LiCH_2C(O)OSi(CH_3)_3$ followed by extraction with ether and washing with 10% aqueous HCl (85%); VII for 2 d (99%) [46] yellow solid [46] ^{1}H NMR ($CDCl_3$): −0.34 (d, H-2; J(H, H) = 7.5), 0.23 (m, H-5, exo), 0.95 (d of d, H-5, endo; J(H, H) = 14.0, 6.0), 1.14 (s, CH_3-4), 1.51 (s, CH_3-1), 1.50 to 1.64 (m, H-6, endo and H-6, exo), 2.12 and 2.24 (AB system q, CH_2COO; J(A, B) = 12.9), 2.62 (d of d, H-3; J(H, H) = 7.5, 1.7), 5.56 (t, 1H; J(H, H) = 1.9), 5.75 (m, 1H), 6.01 (m, 1H), 7.1 (m, 4H) [46] IR ($CHCl_3$): 1700, 1855, 1935, 2940 [46] mass spectrum: $[M-n\ CO]^+$ (n = 0 to 2), other fragments given without assignment [46]

References on pp. 303/4

Table 12 (continued)

No.	compound	method of preparation (yield) properties and remarks
93	$C_9H_7Mo(CO)_2C_6H_6((CH_3)_2$-1,4,endo)$CH_2C(O)OCH_3$-4,exo	VI, X = BF_4, ^4L = $C_6H_6(CH_3)_2$-1,4, with CH_2=C(OLi)OCH_3 at −78°C for 1 h (81%) [46] yellow solid [46] ^{1}H NMR ($CDCl_3$): −0.36 (d, H-2; J(H, H) = 7.4), 0.24 (m, H-5, exo), 0.94 (d of d, H-5, endo; J(H, H) = 14.0, 6.0), 1.09 (s, CH_3-4), 1.51 (s, CH_3-1), 1.50 to 1.65 (m, H-6, endo and H-6, exo), 2.08 and 2.20 (AB system q, CH_2COO; J(A, B) = 12.9), 2.59 (d of d, H-3; J(H, H) = 7.4, 1.7), 3.61 (s, OCH_3), 5.55 (t, 1H; J(H, H) = 2.9), 5.75 (m, 1H), 6.0 (m, 1H), 7.1 (m, 4H) [46] IR ($CHCl_3$): 1720, 1850, 1930 [46] mass spectrum: $[M-n\,CO]^+$ (n = 0 to 2), other fragments given without assignment [46] saponification affords No. 92 [46]
*94	$C_9H_7Mo(CO)_2C_6H_6((CH_3)_2$-1,4,endo)$CH(CH_3)COOH$-4,exo	VI, X = BF_4, ^4L = $C_6H_6(CH_3)_2$-1,4, with CH_3CH=C(OLi)$OSi(CH_3)_3$ at −78°C for 0.5 h followed by extraction with ether and washing with 10% aqueous HCl (93%) [46] a mixture of two diastereomers in a 2:1 ratio [46] ^{1}H NMR ($CDCl_3$): −0.48 (d, H-2 major isomer; J(H, H) = 7.5), −0.26 (d, H-2 minor isomer; J(H, H) = 7.5), 0.99 (d, C**H**$_3$CH minor isomer; J(H, H) = 7.1), 1.02 (s, CH_3-4, major isomer), 1.05 (s, CH_3-4, minor isomer), 1.20 (d, C**H**$_3$CH major isomer; J(H, H) = 7.1), 1.49 (s, CH_3-1, minor isomer), 1.52 (s, CH_3-1, major isomer), 2.11 (q, C**H**CH_3, minor isomer), 2.19 (q, C**H**CH_3, major isomer; J(H, H) = 7.1), 2.55 (d, of d, H-3, minor isomer; J(H, H) = 7.5, 1.8), 2.62 (d, H-3, major isomer; J(H, H) = 7.5, 1.8), 5.55 (m, 1H), 5.74 (m, 1H), 6.02 (m, 1H), 7.1 (m, 4H), other resonances obscured by the three methyl signals of each isomer [46] IR ($CHCl_3$): 1700, 1844, 1935 [46] mass spectrum: $[M-n\,CO]^+$ (n = 0 to 2), other fragments given without assignment [46]
95	$C_9H_7Mo(CO)_2C_6H_6((CH_3)_2$-1,4,endo)$CH(CH_3)C(O)OCH_3$-4,exo	VI, X = BF_4, ^4L = $C_6H_6(CH_3)_2$-1,4, with CH_3CH=C(OLi)OCH_3 for 0.5 h (89%) [46]

References on pp. 303/4

Table 12 (continued)

No.	compound	method of preparation (yield) properties and remarks
95 (continued)		a mixture of two diastereomers in a 4:3 ratio [46] yellow solid ^{1}H NMR ($CDCl_3$): −0.48 (d, H-2 minor isomer; J(H, H) = 7.5), −0.39 (d, H-2 major isomer; J(H, H) = 7.5), 0.96 (d, C**H**$_3$CH major isomer; J(H, H) = 7.1), 0.99 (s, CH_3-1, minor isomer), 1.03 (s, CH_3-4, major isomer), 1.16 (d, C**H**$_3$CH, minor isomer; J(H, H) = 7.1), 1.52 (s, CH_3-1, major isomer), 2.19 (q, C**H**CH_3, major isomer; J(H, H) = 7.1), 2.36 (d of d, H-3, major isomer; J(H, H) = 1.8, 7.5), 2,62 (d of d, H-3, minor isomer; J(H, H) = 7.5, 1.8), 3.55 (s, OCH_3, minor isomer), 3.66 (s, OCH_3, major isomer), 5.55 (m, 1H), 5.74 (m, 1H), 6.01 (m, 1H), 7.1 (m, 4H), other resonances obscured by the three methyl signals of each isomer [46] IR ($CHCl_3$): 1715, 1850, 1940 [46] mass spectrum: $[M-n\,CO]^+$ (n = 0 to 2), other fragments given without assignment [46]
96	$C_9H_7Mo(CO)_2C_6H_6((CH_3)_2$-1,4,endo)CH=$CH_2$-4,exo	VI, X = BF_4, ^{4}L = $C_6H_6(CH_3)_2$-1,4, with 10 equivalents of CH_2=CHMgBr in THF in CH_2Cl_2 at 0°C for 1 h followed by quenching with H_2O and extraction with ether (37%) [46] ^{1}H NMR ($CDCl_3$): −0.45 (d, H-2; J(H, H) = 7.4), 0.3 (m, H-5, exo), 0.9 (m, H-5, endo), 1.1 (s, CH_3-4), 1.55 (s, CH_3-1), 1.6 (m, H-6, endo and H-6, exo), 2.65 (d of d, H-3; J(H, H) = 7.4, 1.7), 4.78 (d, CH=CH_2, 1H; J(H, H) = 9.9), 4.95 (d, CH=CH_2, 1H; J(H, H) = 16.9), 5.55 (m, 1H), 5.73 (m, 1H), 6.0 (m, 1H), 5.7 (d of d, CH=CH_2, 1H; J(H, H) = 16.9, 9.9), 7.1 (m, 4H) [46] mass spectrum: $[M-n\,CO]^+$ (n = 0 to 2), other fragments given without assignment [46]
97	$C_9H_7Mo(CO)_2C_6H_6((CH_3)_2$-1,4,endo)$CH(C(O)OCH_3)_2$-4,exo	VI, X = BF_4, ^{4}L = $C_6H_6(CH_3)_2$-1,4, with $NaCH(C(O)OCH_3)_2$ for 0.5 h (91%) [45, 46] yellow oily compound [46] ^{1}H NMR ($CDCl_3$): −0.48 (d, H-2; J(H, H) = 7.5), 0.24 (m, H-5, exo), 1.1 (d of d, H-5, endo; J(H, H) = 14.0, 6.0), 1.25 (s, CH_3-4), 1.52 (s, CH_3-1), 1.4 to 1.8 (m, H-6, endo and H-6, exo), 2.63 (d of d, H-3; J(H, H) = 7.5, 1.8), 3.14 (s,

References on pp. 303/4

Table 12 (continued)

No.	compound	method of preparation (yield) properties and remarks
		C**H**$(C(O)OCH_3)_2$), 3.62 (s, OCH_3), 3.72 (s, OCH_3), 5.55 (t, 1H; J(H, H) = 2.9), 5.75 (m, 1H), 6.0 (m, 1H), 7.1 (m, 4H) [46] IR ($CHCl_3$): 1730, 1760, 1860, 1945 [46] mass spectrum: $[M - CH(C(O)OCH_3)_2]^+$, other fragments given without assignment [46]
98	$C_9H_7Mo(CO)_2C_6H_6((CH_3)_2$-1,4,endo)$CH(S(O)OC_6H_5)C(O)OCH_3$-4,exo	VI, X = BF_4, ^4L = $C_6H_6(CH_3)_2$-1,4, with $NaCH(S(O)OC_6H_5)(C(O)OCH_3)$ for 0.5 h (97%) [45, 46] a mixture of two diastereomers in a 3:2 ratio, which could be separated by thin-layer chromatography on silica with 20% ethyl acetate in hexane [46] yellow oily complex [46] ^{1}H NMR ($CDCl_3$): major diastereomer: −0.73 (d, H-2; J(H, H) = 7.4), 0.3 (m, H-5, exo), 1.44 (s, CH_3-4), 1.52 (s, CH_3-1), 1.65 (m, H-6, endo and H-6, exo), 1.9 (m, H-5, endo), 2.08 (d of d, H-3; J(H, H) = 7.4, 1.7), 3.46 (s, OCH_3), 3.65 (s, C**H**S), 5.51 (t, 1H; J(H, H) = 2.9), 5.73 (m, 1H), 6.00 (m, 1H), 7.04 (m, 4H), 7.5 to 7.9 (m, C_6H_5); minor isomer: −0.03 (d, H-2; J(H, H) = 7.4), 0.3 (m, H-5, exo), 0.75 (m, H-5, endo), 1.45 (s, CH_3-4), 1.51 (s, CH_3-1), 1.6 (m, H-6, endo and H-6, exo), 3.26 (d of d, H-3; J(H, H) = 7.4, 1.7), 3.34 (s, OCH_3), 3.89 (s, C**H**S), 5.57 (t, 1H; J(H, H) = 2.9), 5.74 (m, 1H), 6.03 (m, 1H), 7.19 (m, 4H), 7.5 to 7.9 (m, C_6H_5) [46] IR ($CHCl_3$): 1745, 1860, 1945 [46] mass spectrum: $[M - CH(S(O)OC_6H_5)C(O)OCH_3]^+$, other fragments given without assignment [46]
99	$C_9H_7Mo(CO)_2C_6H_6((CH_3)_2$-1,4,endo)$C(CH_3)(S(O)OC_6H_5)C(O)OCH_3$-4,exo	VI, X = BF_4, ^4L = $C_6H_6(CH_3)_2$-1,4, with $NaC(CH_3)(S(O)OC_6H_5)C(O)OCH_3$ for 0.5 h (81%) [45, 46] two diastereomers in a 3:2 ratio, could be separated by thin-layer chromatography with 20% ethyl acetate in hexane [46] yellow oily complex [46] ^{1}H NMR ($CDCl_3$): major isomer: −0.73 (d, H-2; J(H, H) = 7.4), 0.3 (m, H-5, exo), 1.44 (s, CH_3-4), 1.52 (s, CH_3-1), 1.62 (s, C(C**H**$_3$)S), 1.65 (m,

References on pp. 303/4

Table 12 (continued)

No.	compound	method of preparation (yield) properties and remarks
99 (continued)		H-6, endo and H-6, exo), 1.9 (m, H-5, endo), 2.08 (d of d, H-3; J(H, H) = 7.4, 1.7), 3.46 (s, OCH_3), 5.51 (t, 1H; J(H, H) = 2.9), 5.73 (m, 1H), 6.00 (m, 1H), 7.04 (m, 4H), 7.5 to 7.9 (m, C_6H_5); minor isomer: −0.03 (d, H-2; J(H, H) = 7.4), 0.3 (m, H-5, exo), 0.75 (m, H-5, endo), 1.45 (s, CH_3-4), 1.51 (s, CH_3-1), 1.6 (m, H-6, endo and H-6, exo), 1.63 (s, $C(C\mathbf{H}_3)S$), 3.26 (d of d, H-3; J(H, H) = 7.4, 1.7), 3.34 (s, OCH_3), 5.57 (t, 1H; J(H, H) = 2.9), 5.74 (m, 1H), 6.03 (m, 1H), 7.19 (m, 4H), 7.5 to 7.9 (m, C_6H_5) [46] IR ($CHCl_3$): 1745, 1860, 1945 [46] mass spectrum: $[M-C(CH_3)(S(O)OC_6H_5)C(O)$-$OCH_3]^+$, other fragments given without assignment [46]
100	$C_9H_7Mo(CO)_2C_6H_6((CH_3)_2$-1,4,endo)$C(CH_3)C(O)(CH_2)_3$-4,exo 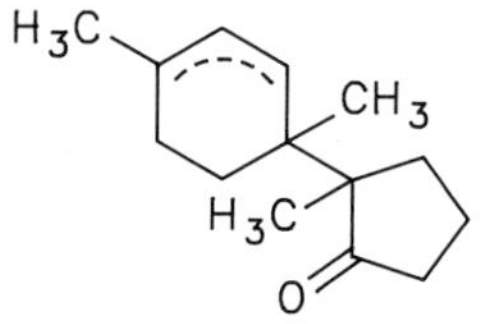	VI, X = BF_4, ^4L = $C_6H_6(CH_3)_2$-1,4, with lithium 1-methyl-2-oxo-cyclopentyl enolate for 0.5 h (83%) [46] a mixture of two diastereomers in a 2:1 ratio [46] yellow complex [46] ^1H NMR ($CDCl_3$): −0.31 (d, H-2, major isomer: J(H, H) = 7.4), −0.25 (d, H-2, minor isomer; J(H, H) = 7.4), 0.5 to 2.2 (m, 10H), 0.95 (s, $C(CH_3)CO$, major isomer), 1.03 (s, CH_3-4, minor isomer), 1.10 (s, CH_3-4, major isomer), 1.25 (s, $C(CH_3)CO$, minor isomer), 1.44 (s, CH_3-1, minor isomer), 1.47 (s, CH_3-1, major isomer), 2.5 (d, H-3, major isomer; J(H, H) = 7.4), 3.1 (d, H-3, minor isomer; J(H, H) = 7.4), 5.57 (m, 1H), 5.74 (m, 1H), 6.0 (m, 1H), 7.1 (m, 4H) [46] IR ($CHCl_3$): 1725, 1845, 1930 [46] mass spectrum: $[M-n\ CO]^+$ (0 to 2), other fragments given without assignment [46]
101	$C_9H_7Mo(CO)_2C_6H_6((CH_3)_2$-1,4,endo)$C_6H_5$-4,exo	VI, X = BF_4, ^4L = $C_6H_6(CH_3)_2$-1,4, in CH_2Cl_2 at 0°C with 10 equivalents of C_6H_5MgBr for 1 h followed by quenching with excess H_2O and extraction with ether (30%) [46] yellow solid [46] ^1H NMR ($CDCl_3$): −0.35 (d, H-2; J(H, H) = 7.4), 0.5 (m, H-5, exo), 0.9 (m, H-5, endo), 1.25 (s, CH_3-4), 1.55 (s, CH_3-1), 1.6 (m, H-6, exo and

References on pp. 303/4

Table 12 (continued)

No.	compound	method of preparation (yield) properties and remarks
		H-6, endo), 2.65 (d of d, H-3; J(H, H) = 7.4, 1.7), 5.55 (t, 1H; J(H, H) = 2.9), 5.74 (m, 1H), 6.0 (m, 1H), 7.1 (m, C_9H_7 and C_6H_5, 9H) [46] IR ($CHCl_3$): 1845, 1930 [46] mass spectrum: $[M-n\,CO]^+$ (n = 0 to 2), other fragments given without assignment [46]
102	$C_9H_7Mo(CO)_2C_6H_6((CH_3)_2$-1,4,endo)CN-4,exo	VI, X = BF_4, ${}^4L$ = $C_6H_6(CH_3)_2$-1,4, with NaCN in CH_3CN/H_2O (10:1) for 1 h followed by extraction with ether (85%) [46] yellow oily solid [46] ^1H NMR ($CDCl_3$): −0.27 (d, H-2; J(H, H) = 7.3), 0.38 (m, H-5, exo), 1.25 (d of d, H-5, endo; J(H, H) = 15.0, 6.0), 1.36 (s, CH_3-4), 1.58 (s, CH_3-1), 1.7 (m, H-6, exo), 1.9 (m, H-6, endo), 2.51 (d of d, H-3; J(H, H) = 7.3, 1.6), 5.53 (t, 1H; J(H, H) = 2.9), 5.78 (m, 1H), 6.01 (m, 1H), 7.14 (m, 4H) [46] IR ($CHCl_3$): 1865, 1945, 2220 [46] mass spectrum: $[M-n\,CO]^+$ (n = 0 to 2), other fragments given without assignment [46]
103	$C_9H_7Mo(CO)_2CHC(C_4H_9$-t)$C(CH_3)OC(C_4H_9$-t)CH 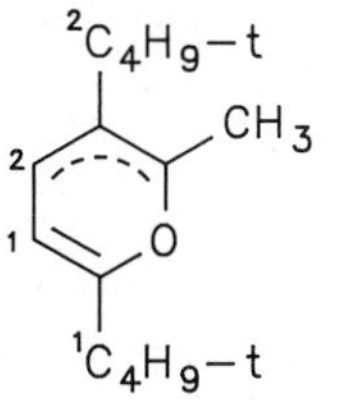	$C_9H_7Mo(CO)_3CH_3$ was stirred in t-$C_4H_9C{\equiv}CH$ for 4 d (13% by chromatography on Al_2O_3 with hexane); also formed by treating $C_9H_7Mo(CO)_2CH{=}C(C_4H_9$-t)$C(CH_3)O$ (C,O bonded) with t-$C_4H_9C{\equiv}CH$ for 7 d (20%) [21] orange crystals (from hexane at −78°C) [21] ^1H NMR (C_6D_6): 1.05 (s, C_4H_9-t-2), 1.2 (s, C_4H_9-t-1), 1.4 (d, H-2; J(H-1, 2) = 2.0), 1.6 (s, CH_3), 5.2 (m, H-2′; J(H-1′, 2′) = 3.0, J(H-2′, 2′) = 2.0), 5.3 (m, H-2′; J(H-1′, 2′) = 3.0, J(H-2′, 2′) = 2.0), 5.5 (t, H-1′), 5.6 (d, H-1; J(H-1, 2) = 2.0), 6.7 (m, H-3′, 3H), 7.1 (m, H-3′, 1H) [21] IR (Nujol): 1635 (ν(C=C)); (hexane): 1875, 1945 (ν(CO)) [21] mass spectrum: $[M-n\,CO]^+$ (n = 0 to 2) [21]
	${}^3L$ = cycloheptenyl derivatives	
104	$C_5H_5Mo(CO)_2C_7H_{11}$ 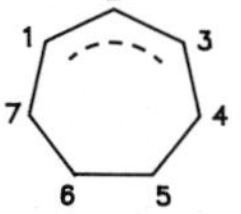	II, for 2 h (88% by chromatography on Al_2O_3 with CH_2Cl_2) [47]; similar to IX, with LiC_4H_9-t and H_2O (86%) [46] yellow crystalline solid, m.p. 101.5 to 102°C [47] ^1H NMR ($CDCl_3$): 0.81 (m, H-5, 6, exo), 1.08 (m, H-5, 6, endo), 2.17 (m, H-4, 7, exo), 2.36 (m,

References on pp. 303/4

Table 12 (continued)

No.	compound	method of preparation (yield) properties and remarks
104 (continued)		H-4, 7, endo), 3.72 (t, H-2; J(H, H) = 8.2), 4.06 (t of d, H-1, 3; J(H, H) = 8.2, 2.1), 5.24 (s, C_5H_5): irradiating the resonance at 4.06 causes collapse at 3.72 to singlet and at 2.17 to d of d with J(H, H) = 7.8 and 4.0, irradiating the resonance at 3.72 causes partial collapse at 4.06 [47] ^{95}Mo NMR (acetone-d_6, vs. 2 M Na_2MoO_4 in D_2O): −1718 (exo isomer) [38] IR (CCl_4): 1865, 1945 (ν(CO)) [47] hydride abstraction with $[C(C_6H_5)_3]PF_6$ in CH_2Cl_2 yields $[C_5H_5Mo(CO)_2C_7H_{10}]PF_6$ [47]
105	$C_5H_5Mo(CO)_2C_7H_{10}D$-4,exo	similar to IX, with LiC_4H_9-t and D_2O (81%) [46] the ^1H NMR and IR spectra compared well with the nondeuterated complex (No. 104) [46]
* 106	$C_5H_5Mo(CO)_2C_7H_{10}CH_3$-4	VI, X = PF_6, ^4L = C_7H_{10}, with 10 equivalents of CH_3MgBr at 0 °C for 20 min followed by addition of water and extraction with ether (85%) [47] yellow crystalline solid, m.p. 70 to 71 °C [47] ^1H NMR ($CDCl_3$): 0.40 (m, H-6), 0.71 (q, H-5, exo; J(H-5, 6) = J(H-4, 5) = 10.1), 1.16 (d, CH_3; J(H, H) = 6.8), 1.24 (m, H-6 and H-7, exo), 2.21 (m, H-4, endo, H-5, endo, and H-7, endo), 3.66 (t, H-2; J(H, H) = 9.5), 3.73 (d, H-3; J(H, H) = 9.5), 4.08 (m, H-1), 5.24 (s, C_5H_5) [47] IR (CCl_4): 1870, 1950 (ν(CO)) [47] hydride abstraction with $[C(C_6H_5)_3]PF_6$ in CH_2Cl_2 affords $[C_5H_5Mo(CO)_2C_7H_9CH_3\text{-}5]PF_6$ [47]
107	$C_5H_5Mo(CO)_2C_7H_{10}CH_2CH{=}CH_2$-4	VI, X = PF_6, ^4L = C_7H_{10}, with $CH_2{=}CHCH_2MgBr$ for 30 min followed by addition of water and extraction with ether (90%) [47] yellow crystalline solid, m.p. 71 to 71.5 °C [47] ^1H NMR ($CDCl_3$): 0.34 (m, H-5, exo), 0.76 (m, H-6), 1.22 (m, H-6 and H-7, exo), 2.16 (m, H-5, endo, H-4, endo, H-7, endo, and **CH_2**CH=), 3.68 (t, H-2; J(H, H) = 8.79), 3.80 (d, H-3; J(H, H) = 8.8), 4.14 (m, H-1), 5.03 (d, CH=CH_2, 1H; J(H, H) = 10.94), 5.04 (d, CH=CH_2-trans, 1H; J(H, H) = 15.74), 5.24 (s, C_5H_5), 5.83 (m, CH=CH_2, 1H) [47] IR (CCl_4): 1860, 1950 (ν(CO)) [47]

Table 12 (continued)

No.	compound	method of preparation (yield) properties and remarks
108	$C_5H_5Mo(CO)_2C_7H_{10}CH_2CH(C(O)OC_4H_9\text{-t})_2$-4,exo	similar to IX, with LiC_4H_9-t and $CH_2{=}C(C(O)OC_4H_9\text{-t})_2$ (86%) [46] ^1H NMR ($CDCl_3$): 0.4 (m, H-6, exo), 0.8 (m, H-5, exo), 1.2 (m, H-6, endo and H-7, exo), 1.45 (s, t-C_4H_9), 1.46 (s, t-C_4H_9), 2.53 (m, H-4 endo), 2.0 to 2.5 (m, H-5, endo, H-7, endo, and **CH_2**CH), 3.29 (t, C**H**$(C(O)OC_4H_9\text{-t})_2$; J(H, H) = 7.2), 3.7 (m, H-2 and H-3), 4.13 (m, H-1), 5.24 (s, C_5H_5) [46] IR ($CHCl_3$): 1720, 1850, 1935 [46] mass spectrum: $[M-n\,CO]^+$ (n = 0, 2), other fragments given without assignment [46]
*109	$C_5H_5Mo(CO)_2C_7H_{10}CH_2COOH$-4	VII (95%) [47] yellow crystalline solid, m.p. 190°C (dec.) [47] ^1H NMR ($CDCl_3$): 0.46 (m, H-6), 0.84 (m, H-5, endo and H-5, exo), 1.21 (m, H-7, exo), 2.15 (m, H-7, endo and H-6), 2.50 (d, C**H_2**COOH; J(H, H) = 6.35), 2.63 (m, H-4, endo), 3.7 (m, H-2 and H-3), 4.14 (d of m, H-1; J(H, H) = 7.24), 5.24 (s, C_5H_5) [47] IR (CCl_4): 1720, 1860, 1945 [47]
110	$C_5H_5Mo(CO)_2C_7H_{10}CH_2C(O)OCH_3$-4	X, with No. 113 (80%) [47] yellow crystalline solid, m.p. 125 to 126°C [47] ^1H NMR ($CDCl_3$): 0.45 (m, H-6), 0.84 (m, H-5, exo), 1.24 (m, H-6 and H-7, exo), 2.16 (m, H-7, endo and H-5, endo), 2.47 (d, C**H_2**$C(O)OCH_3$; J(H, H) = 7.8), 2.61 (m, H-4, endo), 3.69 (s, OCH_3), 3.71 (m, H-2 and H-3), 4.15 (m, H-1), 5.20 (s, C_5H_5) [47] IR (CCl_4): 1740, 1870, 1950 [47]
*111	$C_5H_5Mo(CO)_2C_7H_{10}CH(C(O)OCH_3)_2$-4	VI, X = PF_6, ^4L = C_7H_{10}, $NaCH(C(O)OCH_3)_2$ for 0.5 h followed by addition of H_2O and extraction with ether (95%) [47] yellow oil [47] ^1H NMR ($CDCl_3$): 0.35 (m, H-6), 0.99 (q, H-5, exo; J_{gem} = J(H-5, 6) = J(H-4, 5) = 11.72), 1.26 (m, H-6 and H-7, exo), 2.17 (m, H-5, endo and H-7, endo), 2.91 (m, H-4, endo), 3.51 (d, C**H**$(C(O)OCH_3)_2$; J(H, H) = 6.35), 3.74 (s, OCH_3), 3.77 (s, OCH_3), 3.75 (m, H-2 and H-3), 4.14 (d of d, H-1; J(H, H) = 6.84, 3.15), 5.25 (s, C_5H_5) [47] IR (CCl_4): 1740, 1870, 1950 [47] mass spectrum (field desorption): pattern of $[M]^+$ [47]

References on pp. 303/4

Table 12 (continued)

No.	compound	method of preparation (yield) properties and remarks
*112	$C_5H_5Mo(CO)_2C_7H_{10}CH(C(O)OCH_3)C(O)CH_3$-4	VI, X = PF_6, ^{4}L = C_7H_{10}, $NaCH(C(O)OCH_3)C(O)CH_3$ for 0.5 h followed by addition of H_2O and extraction with ether (90%) [47] a mixture of two diastereomers in a 2:1 ratio; after thin-layer chromatography a 1:1 mixture [47] yellow oil, which could not be crystallized [47] ^{1}H NMR ($CDCl_3$): 0.34 (two d of q, H-6; J_q = 10.8, J_d = 3.64), 0.89 (two q, H-5, exo; J_{gem} = J(H-5, 6) = J(H-4, 5) = 10.35), 1.23 (br d, H-6 and H-7, exo; J(H, H) = 10.96), 2.13 (m, H-5, endo and H-7, endo), 2.23 (s, $C(O)CH_3$, major isomer), 2.31 (s, $C(O)CH_3$, minor isomer), 2.94 (m, H-4, endo), 3.52 (four overlapping d, H-3 and C**H**$C(O)OCH_3$; J(H-2, 3) = 8.19, J(H-4, CH) = 6.35), 3.70 (s, $C(O)OCH_3$, major isomer), 3.76 (s, $C(O)OCH_3$, minor isomer), 3.75 (m, H-2), 4.17 (m, H-1), 5.24 (s, C_5H_5) [47] IR (CCl_4): 1720, 1870, 1950 [47] mass spectrum (field desorption): pattern of $[M]^+$ [47]
113	$C_5H_5Mo(CO)_2C_7H_{10}CH(C(O)OCH_3)S(O)OC_6H_5$-4	VI, X = PF_6, ^{4}L = C_7H_{10}, $NaCH(C(O)OCH_3)S(O)OC_6H_5$ for 0.5 h followed by addition of H_2O and extraction with ether (98%) [47] a mixture of two diastereomers in a 2:1 ratio, changing to 1:1 after thin-layer chromatography due to epimerization [47] yellow oil, could not be crystallized [47] ^{1}H NMR ($CDCl_3$): 0.30 (m, H-6), 0.96 (q, H-5, exo; J_{gem} = J(H-5, 6) = J(H-4, 5) = 13.37), 1.28 (m, H-6 and H-7, exo), 2.11 (m, H-5, endo and H-7, endo), 2.90 (m, H-4, endo, minor isomer), 3.07 (m, H-4, endo, major isomer), 3.60 (s, CH_3, minor isomer), 3.76 (s, OCH_3, major isomer), 3.68 (t, H-2; J(H, H) = 7.18), 3.90 (d, C**H**$(C(O)OCH_3)$; J(H, H) = 7.32), 4.19 (m, H-1 and H-3), 5.21 (s, C_5H_5, major isomer), 5.28 (s, C_5H_5, minor isomer), 7.64 (m, C_6H_5), 7.96 (m, C_6H_5) [47] IR (CCl_4): 1750, 1870, 1950 [47]

Table 12 (continued)

No.	compound	method of preparation (yield) properties and remarks
		mass spectrum (field desorption): pattern of $[M]^+$ [47] reaction with Na_2HPO_4 followed by treatment with excess sodium amalgam affords No. 110 (Method X) [47]
114	$C_5H_5Mo(CO)_2C_7H_{10}C(C(O)OCH_3)C(O)(CH_2)_3$-4 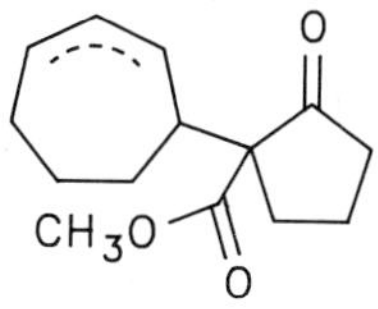	VI, X = PF_6, ^4L = C_7H_{10}, with $NaC(C(O)OCH_3)C(O)(CH_2)_3$ for 0.5 h followed by addition of H_2O and extraction with ether (90%) [47] a mixture of two diastereomers in a 9:1 ratio; the major isomer was obtained pure by recrystallization from $CHCl_3$/ether by addition of pentane [47] yellow crystals, m.p. 135 to 137°C (dec.) [47] ^1H NMR ($CDCl_3$): 0.16 (m, H-6), 0.78 (q, H-5, exo; J_{gem} = J(H-5, 6) = J(H-4, 5) = 11.71), 1.22 (m, CH_2), 1.91 (m, H-7, exo and CH_2), 2.15 (m, H-6 and CH_2), 2.37 (d of t of d, H-5, endo; J_{gem} = 18.54, J_t = 6.54, J_d = 1.41), 2.62 (d of t of d, H-7, endo; J_{gem} = 13.34, J_t = 5.57, J_d = 1.77), 3.00 (d of d, H-4, endo; J(H, H) = 12.11, 2.26), 3.56 (d of d, H-3; J(H, H) = 8.73, 0.98), 3.69 (d of d, H-2; J(H, H) = 8.73, 4.61), 3.75 (s, OCH_3), 4.17 (m, H-1), 5.23 (s, C_5H_5) [47] ^{13}C NMR ($CDCl_3$): 19.49, 23.55, 29.20, 29.69, 30.61, 39.27, 45.56, 52.50, 58.84, 60.37, 64.78, 65.45, 92.60 (C_5H_5), 170.0, 212.0 [47] IR (CCl_4): 1725, 1755, 1875, 1950 [47]
115	$C_5H_5Mo(CO)_2C_7H_{10}(C_6H_4OCH_3$-4)-4	VI, X = PF_6, ^4L = C_7H_{10}, addition of 4-$CH_3C_6H_4MgBr$ in THF to a solution of the diene complex in CH_2Cl_2 for 0.5 h followed by addition of water and extraction with ether (90%); the reaction was also carried out in THF, resulting in contamination with 4,4'-dimethoxybiphenyl [47] yellow crystalline solid, m.p. 112 to 113°C [47] ^1H NMR ($CDCl_3$): 0.77 (m, H-6), 1.04 (m, H-5, exo), 1.25 (d, H-7, exo; J(H, H) = 6.95), 1.38 (m, H-6), 2.36 (m, H-7, endo and H-5, endo), 3.50 (m, H-4, endo), 3.79 (s, OCH_3), 3.94 (m, H-1, H-2, and H-3), 5.26 (s, C_5H_5), 6.88 (d, C_6H_4, 2H; J(H, H) = 8.60), 7.37 (d, C_6H_5, 2H) [47] IR (CCl_4): 1865, 1945 [47]

References on pp. 303/4

Table 12 (continued)

No.	compound	method of preparation (yield) properties and remarks
116	$C_5H_5Mo(CO)_2C_7H_{10}CN$-4,endo	IX, with No. 118 and H_2O (90%) [46] yellow solid, m.p. 156 to 158°C [46] ^1H NMR ($CDCl_3$): 0.8 (m, H-6, exo), 0.9 (m, H-5, exo), 1.55 (m, H-6, endo), 1.7 (m, H-7, exo), 2.2 (m, H-7, endo), 2.6 (m, H-5, endo), 3.53 (t of d, H-4, exo; J(H, H) = 11.2, 3.36), 3.81 (t, H-2; J(H, H) = 8.7), 4.08 (t of d, H-3; J(H, H) = 8.7, 1.5), 4.29 (t, H-1; J(H, H) = 8.5), 5.31 (s, C_5H_5) [46] IR ($CHCl_3$): 1870, 1950, 2240 [46]
117	$C_5H_5Mo(CO)_2C_7H_9$(D-4,exo)CN-4,endo	IX, with No. 118 and D_2O (90%) [46]
*118	$C_5H_5Mo(CO)_2C_7H_{10}CN$-4,exo	similar to VI, X = PF_6, ^4L = C_7H_{10}, with NaCN in CH_3CN/H_2O (10:1) for 17 h (90%) [47] yellow crystalline solid, m.p. 94.5 to 95.5°C [47] ^1H NMR ($CDCl_3$): 0.88 (m, H-6), 1.20 (m, H-5, exo), 1.38 (m, H-7), 2.16 (d of m, H-7, endo and H-6; J_{gem} = 18.5), 2.45 (d of m, H-5, endo, J_{gem} = 16.5), 3.23 (m, H-4, endo), 3.82 (t, H-1; J(H, H) = 7.3), 3.96 (d, H-3; J(H, H) = 7.3), 4.19 (br t, H-1; J(H, H) = 7.3), 5.31 (s, C_5H_5) [47] IR (CCl_4): 1875, 1955, 2220 [47]
119	$C_5H_5Mo(CO)_2C_7H_9$(CN-4,endo)$CH_2CH(C(O)OC_4H_9$-t$)_2$-4,exo	IX, with CH_2=$C(C(O)OC_4H_9$-t$)_2$ (86%) [46] yellow solid, m.p. 170°C (from petroleum ether/pentane 1:1) [46] ^1H NMR ($CDCl_3$): 1.3 (m, H-5, exo and H-6, exo), 1.46 (s, t-C_4H_9), 1.49 (s, t-C_4H_9), 2.15 (m, H-7, endo), 2.55 (m, H-5, endo), 2.61 (d of d, C**H**$_2$CH; J(H, H) = 5.8, 3.0), 3.54 (t, CH_2C**H**; J(H, H) = 5.8), 3.75 (t, H-2; J(H, H) = 8.5), 3.90 (d, H-3; J(H, H) = 8.5), 4.46 (br t, H-1; J(H, H) = 8.1), 5.28 (s, C_5H_5); resonances of H-6, endo and H-7, exo obscured by t-C_4H_9 signals [46] IR ($CHCl_3$): 1725, 1870, 1950, 2240 [46]
120	$C_5H_5Mo(CO)_2C_7H_9$(CN-4,endo)$C_6H_6Mn(CO)_3$-4,exo 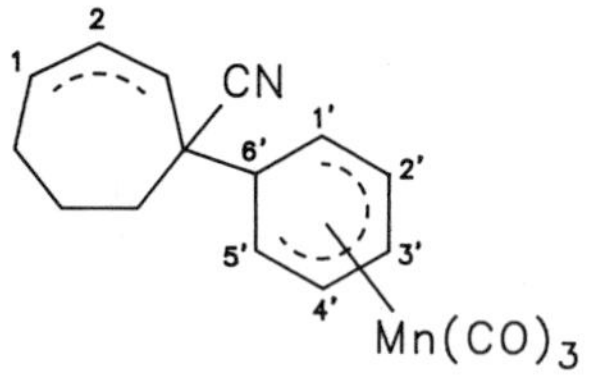 [structure: 1, 2, CN, 1', 2', 3', 4', 5', 6', Mn(CO)$_3$]	IX, with $[C_6H_6Mn(CO)_3]PF_6$ (62%) [46] yellow solid [46] ^1H NMR ($CDCl_3$): 0.9 (m, H-6, exo), 1.2 (m, H-5, exo), 1.4 (m, H-6, endo and H-7, exo), 2.2 (m, H-7, endo), 2.5 (m, H-5, endo), 3.10 (t, H-6′; J(H, H) = 5.6), 3.25 (t, H-5′; J(H, H) = 5.4), 3.65 (t, H-1′; J(H, H) = 5.4), 3.75 (m, H-2 and

References on pp. 303/4

Table 12 (continued)

No.	compound	method of preparation (yield) properties and remarks
		H-3), 4.4 (br t, H-1; J(H, H) = 7.3), 5.05 (t, H-4′; J(H, H) = 6.1), 5.2 (t, H-2′; J(H, H) = 6.1), 5.3 (s, C_5H_5), 5.85 (t, H-3′; J(H, H) = 5.4) [46] IR ($CHCl_3$): 1865, 1950, 2020, 2230 [46] mass spectrum: $[M-CN]^+$, other fragments given without assignment [46]
121	$C_5H_5Mo(CO)_2C_7H_9$(CN-4,endo)$CH(OCH_3)CH_2Fe(CO)_2C_5H_5$-4,exo	IX, with $[C_5H_5Fe(CO)_2CH_2{=}CHOCH_3]BF_4$ (100%); not purified, used directly for the preparation of No. 123 [46] dark red complex [46] ^{1}H NMR ($CDCl_3$): 0.9 to 2.3 (m, 8H), 3.30 (m, **C****H**OCH_3), 3.67 (s, OCH_3), 3.80 (m, H-2 and H-3), 4.39 (m, H-1), 4.85 (s, C_5H_5Fe), 5.29 (s, C_5H_5Mo) [46] IR ($CHCl_3$): 1875, 1960, 2020, 2240 [46] treating with $(C_2H_5)_2O \cdot HBF_4$ ether in CH_2Cl_2 at −78°C affords No. 123 [46]
122	$C_5H_5Mo(CO)_2C_7H_9$(CN-4,endo)$CH{=}CH_2$-4,exo	No. 123 was gently refluxed in CH_3CN for 15 min followed by extraction with ether (93%) [46] yellow crystalline solid, m.p. 148 to 150°C [46] ^{1}H NMR ($CDCl_3$): 1.30 (m, H-5, exo, H-6, endo, and H-6, exo), 1.57 (m, H-7, exo), 2.17 (m, H-7, endo), 2.51 (m, H-5, endo), 3.86 (m, H-2 and H-3), 4.47 (br t, H-1; J(H, H) = 8.0), 5.29 (d, $CH{=}CH_2$,1H; J(H, H) = 10.1), 5.30 (s, C_5H_5), 5.64 (d, $CH{=}CH_2$, 1H; J(H, H) = 16.9), 6.11 (d of d, $CH{=}CH_2$,1H; J(H, H) = 16.9, 10.1) [46] IR ($CHCl_3$): 1870, 1955, 2230 [46]
123	$[C_5H_5Mo(CO)_2C_7H_9$(CN-4,endo)$CH{=}CH_2Fe(CO)_2C_5H_5$-4,exo$]BF_4$	No. 121 was allowed to react with $(C_2H_5)_2O \cdot HBF_4$ in CH_2Cl_2 for 1 h (76%) [46] yellow crystals [46] ^{1}H NMR (CD_3CN): 1.2 to 2.5 (m, 6H), 4.01 (m, H-2 and H-3), 3.53 (m, H-1), 5.41 (s, C_5H_5Fe), 5.78 (s, C_5H_5Mo), 5.2 to 6.3 (m, $CH{=}CH_2$) [46] IR (CH_3CN): 1860, 1950, 2040, 2070 [46] mass spectrum: $[M-C_5H_5Fe(CO)_2BF_4]^+$, other fragments given without assignment [46] heating in CH_3CN gives No. 122 and $[C_5H_5Fe(CO)_2NCCH_3]BF_4$ [46]

References on pp. 303/4

Table 12 (continued)

No.	compound	method of preparation (yield) properties and remarks
124	$C_5H_5Mo(CO)_2C_7H_9(CN\text{-}4,endo)CH_3\text{-}7,exo$	IX, with H_2O (90%) [46] yellow crystalline solid, m.p. 166 to 168°C (from petroleum ether/pentane) [46] ^1H NMR ($CDCl_3$): 0.6 to 0.9 (m, H-5, exo and H-6, exo), 1.14 (d, CH_3-7; J(H, H) = 7), 1.4 to 1.7 (m, H-6, endo and H-7, endo), 2.3 (m, H-5, endo), 3.53 (t of d, H-4, exo; J(H, H) = 11.6, 3.0), 3.76 (t, H-2; J(H, H) = 8.7), 3.98 (d, H-3; J(H, H) = 8.7), 4.05 (br d, H-1; J(H, H) = 8.7), 5.29 (s, C_5H_5) [46] IR ($CHCl_3$): 1870, 1955, 2250 [46]
125	$C_5H_5Mo(CO)_2C_7H_8(D\text{-}4,exo)(CN\text{-}4,endo)CH_3\text{-}7,exo$	IX, with D_2O (80%) [46]
126	$C_5H_5Mo(CO)_2C_7H_9(CN\text{-}4,exo)CH_3\text{-}7,exo$	similar to VI, X = PF_6, ^4L = 5, exo-methylcycloheptadiene, with NaCN in CH_3CN/H_2O (10:1) for 1 h (79%) [46] yellow crystalline solid, m.p. 136 to 138°C (from petroleum ether/pentane 1:1) [46] ^1H NMR ($CDCl_3$): 0.78 (m, H-6, exo), 1.13 (m, H-5, exo), 1.22 (d, CH_3-7; J(H, H) = 6.9), 1.35 (m, H-6, endo), 1.51 (m, H-7, endo), 2.34 (m, H-5, endo), 3.22 (m, H-4, endo), 3.78 (t, H-2; J(H, H) = 8.5), 3.89 (d, H-3; J(H, H) = 7.9), 4.02 (br d, H-1; J(H, H) = 7.9), 5.31 (s, C_5H_5) [46, 47] IR ($CHCl_3$): 1860, 1945, 2240 [46] reaction with $LiN(C_3H_7\text{-}i)_2$ gives a reactive carbanion complex (compare Method IV) [46]
127	$C_5H_5Mo(CO)_2C_7H_9(CN\text{-}4)CH_3\text{-}5$	similar to VI, X = PF_6, ^4L = $C_7H_9CH_3$-5, with NaCN in CH_3CN/H_2O (5:1) for 17 h followed by extraction with ether (82% in a 2:3 mixture with the 7-CH_3 complex) [47] yellow oil [47] ^1H NMR ($CDCl_3$): 0.78 (m, H-6), 1.13 (m, H-5, exo), 1.20 (d, CH_3; J(H, H) = 6.9), 1.34 (m, H-6), 2.08 (m, H-7, endo), 2.34 (m, H-5, endo), 3.22 (m, H-4, endo), 3.9 to 3.7 (m, H-1 and H-2), 4.18 (t of m, H-3; J(H, H) = 6.8), 5.30 (s, C_5H_5) [47] IR (CCl_4): 1875, 1955, 2220 [47] mass spectrum (field desorption): $[M]^+$ [47]

References on pp. 303/4

Table 12 (continued)

No.	compound	method of preparation (yield) properties and remarks
128	$C_5H_5Mo(CO)_2C_7H_8(CN\text{-}4,endo)(CH_2CH(C(O)OC_4H_9\text{-}t)_2\text{-}4,exo)CH_3\text{-}7,exo$	IX, with $CH_2{=}C(C(O)OC_4H_9\text{-}t)_2$ (84%) [46] yellow solid [46] ^1H NMR ($CDCl_3$): 1.12 (d, CH_3-7; J(H, H) = 7.1), 1.46 (s, t-C_4H_9), 1.49 (s, t-C_4H_9), 2.22 (m, H-5, endo), 2.63 (d, C**H**$_2$CH; J(H, H) = 5.9), 3.53 (t, C**H**$(C(O)OC_4H_9\text{-}t)_2$; J(H, H) = 5.9), 3.72 (t, H-2; J(H, H) = 8.8), 3.92 (d, H-3; J(H, H) = 8.8), 4.17 (d, H-1; J(H, H) = 8.8), 5.28 (s, C_5H_5); other resonances obscured by methyl and t-butyl signals [46] IR ($CHCl_3$): 1725, 1870, 1950, 2240 [46]
129	$C_5H_5Mo(CO)_2C_7H_9(CH_3)_2\text{-}4,5$	was obtained in a 1:2 mixture with No. 130 [47] see No. 130
130	$C_5H_5Mo(CO)_2C_7H_9(CH_3)_2\text{-}4,7$	VI, X = PF_6, ^4L = $C_7H_9CH_3$-5, with CH_3MgBr for 20 min at 0°C (85% in a 2:1 mixture with No. 129); the isomers could not be separated [47] yellow solid, m.p. 91 to 93°C [47] ^1H NMR ($CDCl_3$): 0.36 (m, H-6), 0.66 (m, H-5, exo), 0.88 (m, H-6), 0.95 (m, H-5, endo), 1.13 (d, CH_3, No. 129; J(H, H) = 6.80), 1.15 (d, CH_3, No. 130; J(H, H) = 6.80), 2.16 (m, H-4, endo and H-7, endo), 3.54 (t, H-2; J(H, H) = 9.3), 3.66 (d, H-3, regioisomer; J(H, H) = 9.26), 3.80 (d, H-3, regioisomer; J(H, H) = 9.26), 4.07 (m, H-1), 5.20 (s, C_5H_5, No. 130), 5.21 (s, C_5H_5, No. 129) [47] IR (CCl_4): 1860, 1950 [47]
*131	$C_5H_5Mo(CO)_2C_7H_9(CH_3\text{-}7)CH_2COOH\text{-}4$	VII (95%) [47] yellow solid, m.p. 140 to 140.5°C [47] ^1H NMR ($CDCl_3$): 0.96 (m, H-6), 1.06 (m, H-5, exo), 1.19 (d, CH_3; J(H, H) = 6.96), 2.16 (m, H-5, endo and H-6), 2.80 (m, H-7, endo and H-4, endo), 3.62 (t, H-2; J(H, H) = 8.77), 3.73 (t, H-3; J(H, H) = 8.77), 3.85 (m, H-1), 5.26 (s, C_5H_5) [47] IR (CH_2Cl_2): 1710, 1860, 1950, 3500 [47]
132	$C_5H_5Mo(CO)_2C_7H_9(CH_3\text{-}7)CH_2C(O)OCH_3\text{-}4$	X, a mixture with No. 135 (70%) [47] yellow crystalline solid, m.p. 70 to 72°C [47] ^1H NMR ($CDCl_3$): 0.91 (m, H-6), 1.05 (m, H-5, exo), 1.18 (d, CH_3; J(H, H) = 6.94), 1.25 (m, H-6), 2.21 (m, H-5, endo and H-7, endo), 2.52

References on pp. 303/4

Table 12 (continued)

No.	compound	method of preparation (yield) properties and remarks
132 (continued)		(d, $CH_2C(O)OCH_3$; J(H, H) = 9.11), 2.63 (m, H-4, endo), 3.60 (t, H-2; J(H, H) = 8.15), 3.69 (s, OCH_3), 3.75 (m, H-3; J(H, H) = 8.15), 3.86 (d, H-1; J(H, H) = 8.15), 5.25 (s, C_5H_5) [47] IR (CCl_4): 1750, 1860, 1950 [47]
133	$C_5H_5Mo(CO)_2C_7H_9(CH_3\text{-}7)CH(C(O)OCH_3)_2\text{-}4$	VI, X = PF_6, ^4L = $C_7H_9CH_3$-5, with $NaCH(C(O)OCH_3)_2$ for 1.5 h (85% in a 2:1 mixture with No. 134); the isomers were not separated [47] yellow crystalline solid [47] ^1H NMR ($CDCl_3$): 0.39 (m, H-6), 0.88 (m, H-5, exo or H-7, exo), 1.19 (d, CH_3; J(H, H) = 6.95), 1.25 (m, H-6), 2.18 (m, H-5, endo or H-7, endo), 2.29 (m, H-7, endo or H-5, endo), 2.91 (m, H-4, endo), 3.58 (d, $CHC(O)OCH_3$, regioisomer; J(H, H) = 8.30), 3.65 (d, $CHC(O)OCH_3$; J(H, H) = 8.30), 3.77 (s, OCH_3, No. 134), 3.79 (s, OCH_3, No. 133), 3.90 (m, H-2 and H-3), 4.17 (m, H-1), 5.25 (s, C_5H_5, No. 133), 5.26 (s, C_5H_5, No. 134) [47] IR (CCl_4): 1740, 1870, 1950 [47]
134	$C_5H_5Mo(CO)_2C_7H_9(CH_3\text{-}5)CH(C(O)OCH_3)_2\text{-}4$	was obtained in a mixture with No. 133 [47] see No. 133
135	$C_5H_5Mo(CO)_2C_7H_9(CH_3\text{-}7)CH(S(O)OC_6H_5)C(O)OCH_3\text{-}4$	VI, X = PF_6, ^4L = $C_7H_9CH_3$-5, with $NaCH(C(O)OCH_3)S(O)OC_6H_5$ for 1 h (98%) a mixture of diastereomers [47] yellow solid, m.p. 150°C [47] ^1H NMR ($CDCl_3$): 0.31 (m, H-6), 0.92 (m, H-5, exo), 1.15 (d, CH_3, major isomer; J(H, H) = 6.97), 1.17 (d, CH_3, minor isomer; J(H, H) = 6.97), 1.32 (m, H-6), 2.13 (m, H-5, endo), 2.27 (m, H-7, endo), 3.01 (m, H-4, endo), 3.65 (m, CHS), 3.69 (m, H-3), 3.73 (s, CH_3, major isomer), 3.77 (s, OCH_3, minor isomer), 3.92 (m, H-2), 4.23 (m, H-1), 5.21 (s, C_5H_5, major isomer), 5.29 (s, C_5H_5, minor isomer), 7.63 (m, C_6H_5, 3H), 7.97 (m, C_6H_5, 2H) [47] IR (CCl_4): 1750, 1860, 1950 [47] reaction with excess Na_2HPO_4 followed by treatment with sodium amalgam yields No. 132 (Method X) [47]

References on pp. 303/4

Table 12 (continued)

No.	compound	method of preparation (yield) properties and remarks
136	$C_5H_5Mo(CO)_2C_7H_9(CH_3\text{-}7)C(C(O)OCH_3)C(O)(CH_2)_3\text{-}4$ 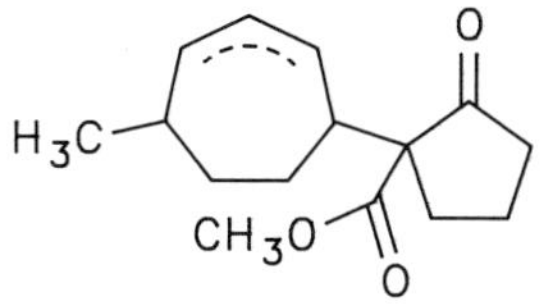	VI, X = PF_6, ^4L = $C_7H_9CH_3$-5, with $NaC(C(O)OCH_3)C(O)(CH_2)_3$ for 0.5 h (95%) [47] a mixture of 2 diastereomers in a 4:1 ratio [47] yellow solid, m.p. 157 to 161°C [47] ^1H NMR ($CDCl_3$): 0.18 (m, H-6), 0.9 (m, H-5, exo), 1.18 (m, CH_2), 1.23 (d, CH_3; J(H, H) = 7.11), 1.95 (m, CH_2), 2.20 (m, H-6 and CH_2), 2.42 (m, H-5, endo), 2.64 (m, H-7, endo), 2.93 (d of d, H-4, endo; J(H, H) = 11.28, 3.58), 3.56 (m, H-2), 3.66 (m, H-3), 3.69 (s, OCH_3, minor isomer), 3.75 (s, OCH_3, major isomer), 3.99 (m, H-1), 4.17 (m, H-1), 5.22 (s, C_5H_5, major isomer), 5.23 (s, C_5H_5, minor isomer) [47] IR (CCl_4): 1725, 1755, 1875, 1950 [47]
*137	$C_5H_5Mo(CO)_2C_7H_9(CH_2CH{=}CH_2\text{-}7)CH_2COOH\text{-}4$	VII (98%) [47] yellow crystalline solid, m.p. 141 to 142°C [47] ^1H NMR ($CDCl_3$): 0.87 (m, H-5, exo and H-6), 1.16 (m, H-6), 2.15 (m, H-7, endo, H-5, endo, and $C\mathbf{H}_2CH{=}$), 2.56 (d of d, diastereotopic $C\mathbf{H}_2COOH$; J(H, H) = 10.8, 6.75), 2.66 (m, H-4, endo), 3.63 (t, H-2; J(H, H) = 8.76), 3.90 (m, H-2 and H-3), 5.03 (d, $CH{=}CH_2$, 1H; J(H, H) = 11.02), 5.05 (d, $CH{=}CH_2$, 1H; J(H, H) = 16.5), 5.26 (s, C_5H_5), 5.77 (m, $CH{=}CH_2$, 1H) [47] IR (CCl_4): 1720, 1860, 1950 [47]
138	$C_5H_5Mo(CO)_2C_7H_9(CH_2CH{=}CH_2\text{-}7)CH_2C(O)OCH_3\text{-}4$	X, with No. 140 (78%) [47] yellow oily product [47] ^1H NMR ($CDCl_3$): 0.6 (m, H-6 and H-5, exo), 2.11 (m, H-7, endo, H-5, endo, and $C\mathbf{H}_2CH{=}$), 2.45 (d of d, diastereotopic $C\mathbf{H}_2C(O)OCH_3$; J(H, H) = 9.88, 7), 2.58 (m, H-4, endo), 3.55 (t, H-2; J(H, H) = 8.35), 3.62 (s, OCH_3), 3.81 (d, H-3; J(H, H) = 8.35), 3.85 (d, H-1; J(H, H) = 8.35), 4.96 (d, $CH{=}CH_2$, 1H; J(H, H) = 11.12), 4.98 (d, $CH{=}CH_2$, 1H; J(H, H) = 16.6), 5.18 (s, C_5H_5), 5.75 (m, $CH{=}CH_2$, 1H) [47] IR (CCl_4): 1740, 1860, 1950 [47] mass spectrum: pattern of $[M]^+$ [47] hydrolysis affords No. 137 (Method VII) [47]

Table 12 (continued)

No.	compound	method of preparation (yield) properties and remarks
139	$C_5H_5Mo(CO)_2C_7H_9(CH_2CH{=}CH_2\text{-}7)CH(C(O)OCH_3)_2\text{-}4$	VI, X = PF_6, ${}^4L$ = $C_7H_9CH_2CH{=}CH_2\text{-}5$, with $NaCH(C(O)OCH_3)_2$ for 10 min (98%) [47] single regioisomer [47] yellow crystalline solid, m.p. 127 to 128°C [47] ^1H NMR ($CDCl_3$): 0.73 (m, H-5, exo and H-6), 1.15 (m H-6 and H-7, exo), 2.19 (m, H-5, endo and C**H**$_2$CH=CH$_2$), 2.90 (m, H-4, endo), 3.56 (d, C**H**(C(O)OCH$_3$)$_2$; J(H, H) = 8.3), 3.65 (d of d, H-2; J(H, H) = 9.06, 8.79), 3.85 (d, H-3; J(H, H) = 9.06), 3.96 (d, H-1; J(H, H) = 8.79), 5.03 (d, CH=CH$_2$, 1H; J(H, H) = 8.24), 5.04 (d, CH=CH$_2$, 1H; J(H, H) = 16.98), 5.24 (s, C_5H_5), 5.78 (m, CH=CH$_2$, 1H) [47] IR (CCl_4): 1740, 1870, 1940 [47]
140	$C_5H_5Mo(CO)_2C_7H_9(CH_2CH{=}CH_2\text{-}7)CH(C(O)OCH_3)S(O)OC_6H_5\text{-}4$	VI, PF_6, ${}^4L$ = $C_7H_9CH_2CH{=}CH_2\text{-}5$, $NaCH(C(O)OCH_3)S(O)OC_6H_5$ for 10 min (98%) [47] a mixture of two diastereomers in a 2:3 ratio [47] yellow colored foam [47] ^1H NMR ($CDCl_3$): 0.90 (m, H-6 and H-5, exo), 1.34 (m, H-6), 2.17 (m, H-5, endo and C**H**$_2$-CH=CH$_2$), 3.02 (m, H-7, endo), 3.62 (m, H-4, endo), 3.70 (s, OCH$_3$, minor isomer), 3.73 (s, OCH$_3$, major isomer), 3.87 (d, C**H**C(O)OCH$_3$; J(H, H) = 8.96), 3.98 (t, H-2; J(H, H) = 8.45), 4.11 (d, H-3; J(H, H) = 8.45), 4.27 (d, diastereomeric H-1; J(H, H) = 8.45), 4.46 (d, diastereomeric H-1; J(H, H) = 8.45), 5.08 (m, CH=CH$_2$, 2H), 5.21 (s, C_5H_5, major isomer), 5.29 (s, C_5H_5, minor isomer), 5.76 (m, CH=CH$_2$, 1H), 7.57 (m, C_6H_5), 7.95 (m, C_6H_5) [47] IR (CCl_4): 1745, 1865, 1945 [47] mass spectrum: pattern of $[M]^+$ [47] reaction with Na_2HPO_4 followed by treatment with sodium amalgam affords No. 138 (Method X) [47]
141	$C_5H_5Mo(CO)_2C_7H_9((C_6H_4OCH_3\text{-}4)\text{-}7)CH_3\text{-}4$	VI, X = PF_6, ${}^4L$ = $C_7H_9(C_6H_4OCH_3\text{-}4)\text{-}5$, with CH_3MgBr for 0.5 h (95%); when THF was used as the solvent, the product was contaminated with products of reductive coupling [47]

References on pp. 303/4

Table 12 (continued)

No.	compound	method of preparation (yield) properties and remarks
		yellow solid [47] ^{1}H NMR ($CDCl_3$): 0.77 (m, H-6), 1.04 (m, H-5, exo), 1.25 (d, CH_3; J(H, H) = 6.95), 1.38 (m, H-6), 2.36 (m, H-4, endo and H-5, endo), 3.60 (m, H-7, endo), 3.81 (s, OCH_3), 3.94 (m, H-1, H-2, and H-3), 5.26 (s, C_5H_5), 6.88 (d, C_6H_4, 2H; J(H, H) = 8.60), 7.37 (d, C_6H_4, 2H) [47] IR (CCl_4): 1860, 1945 [47] mass spectrum: pattern of $[M]^+$ [47]
142	$C_5H_5Mo(CO)_2C_7H_9((C_6H_4OCH_3\text{-}4)\text{-}7)CH(C(O)OCH_3)_2\text{-}4$	VI, X = PF_6, ^{4}L = $C_7H_9(C_6H_4OCH_3\text{-}4)\text{-}5$, with $NaCH(C(O)OCH_3)_2$ for 0.5 h (95%) [47] a single regioisomer [47] yellow crystalline solid, m.p. 58 to 59°C [47] ^{1}H NMR ($CDCl_3$): 0.83 (m, H-6), 1.10 (m, H-5, exo), 1.27 (m, H-5, endo and H-6), 2.97 (m, H-4, endo), 3.45 (m, H-7, endo), 3.60 (d, C**H**C(O); J(H, H) = 7.48), 3.70 (s, $C(O)OCH_3$), 3.80 (s, $C(O)OCH_3$ and OCH_3-4), 3.91 (m, H-1, H-2, and H-3), 5.28 (s, C_5H_5), 6.88 (d, C_6H_4, 2H; J(H, H) = 8.59), 7.37 (d, C_6H_4, 2H) [47] IR (CCl_4): 1738, 1868, 1950 [47]
143	$CH_3C_5H_4Mo(CO)_2C_7H_9$ 	V (3%) [43] light yellow crystals (from pentane) [43] ^{1}H NMR (C_6D_6): 1.30 (CH_3), 1.61 (H-6′), 2.07 (H-6), 2.30 (m, H-7 and H-7′), 3.65 (H-2), 3.26 (H-3), 3.89 (H-1), 4.35, 4.44, and 4.58 (ABCD system, C_5H_4), 5.24 (H-4), 6.23 (H-5) [43] IR: 1882, 1955 (ν(CO)) [43]
*144	$C_5H_5Mo(CO)_2C_7H_7$ 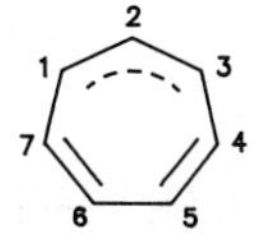	similar to II; cyclo-$C_7H_7Mo(CO)_2I$ was allowed to react with 3 equivalents of NaC_5H_5 in THF for 18 h followed by evaporation and extraction with CH_2Cl_2 (60%; only 10% after sublimation) [1, 2, 5]; obtained in the reaction of $[cyclo\text{-}C_5H_5Mo(CO)_2C(CH_3)HO]^-$ with $[C_7H_7]BF_4$ (13%) or $[cyclo\text{-}C_7H_7Mo(CO)_3]BF_4$ (50%) [27] orange crystals, m.p. 111 to 112°C (from C_6H_6), subl. 90 to 100°C/0.1 Torr [1, 2] ^{1}H NMR (CS_2, −40°C): 5.03 (br s, C_7H_7), 5.32 (s, C_5H_5) [4]; (CS_2, +35°C): 4.79 (s, C_7H_7), 5.07 (s, C_5H_5) [1, 2, 4], similar data in $CDCl_3$ [8]; the spectra in CS_2 were also given as a diagram in [4]; see "Further information"

References on pp. 303/4

Table 12 (continued)

No.	compound	method of preparation (yield) properties and remarks
*144 (continued)		IR (halocarbon mull): given from 704 to 3100; 1893, 1933 (ν(CO)), 3010, 3100 (ν(CH)) [1, 2]; (cyclohexane): 1896, 1911, 1960, 1966 (ν(CO)) [9], similar in [3]; (no solvent stated, with relative intensities in parentheses): 1899 (8), 1914 (4; appears at −60°C), 1961 (10), 1970 (6; appears at −60°C) [7] UV (cyclohexane): λ_{max} (ε) = 306 (11600), 758 (315) [1, 2]
*145		$C_5H_5Mo(CO)_2C_7H_7$ (No. 144) was irradiated for 4 d in ether/$Fe(CO)_5$ (2:1) followed by extraction with CH_2Cl_2 (55% by chromatography on alumina with CH_2Cl_2); formed in the reaction of No. 144 with $Fe_2(CO)_9$ in n-heptane at 80°C for 24 h (5%) [8] red, air-stable crystals (from n-hexane/CH_2Cl_2 at −70°C), m.p. 176°C (dec.) [8] ^{1}H NMR ($CDCl_3$): 4.57 (br s, C_7H_7), 5.13 (s, C_5H_5) [8]; see "Further information" IR (cyclohexane): 1880, 1950, 1975, 1980, 2040 (ν(CO)) [8]

^{3}L = cyclooctenyl

No.	compound	method of preparation (yield) properties and remarks
146	$C_5H_5Mo(CO)_2C_8H_{13}$ 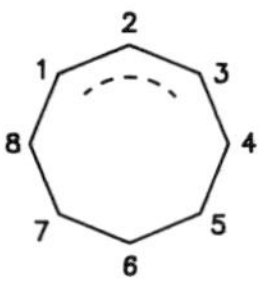	II, X = Br, for 15 h (60 to 80% by chromatography on silica or Soxhlet extraction with petroleum ether) [23, 29]; $(C_5H_5Mo(CO)_3)_2$ and 10 equivalents of cycloocta-1,5-diene were heated in octane for 14 d (10% by column chromatography on Al_2O_3 with hexane) [24] yellow crystals [23, 24, 29], m.p. 115 to 116°C (from hexane/CH_2Cl_2) [24], 117 to 118°C (from petroleum ether) [23, 29] ^{1}H NMR ($CDCl_3$): 1.20, 1.46, 2.18, and 2.30 (m, CH_2), 3.74 (d of d of d, H-1, 3; J(H, H) = 8.1, 8.2, 8.5), 4.15 (t, H-2; J(H, H) = 8.1), 5.23 (s, C_5H_5) [23, 29]; similar in [24] ^{13}C NMR ($CDCl_3$): 23.9, 29.2, and 32.9 (CH_2), 56.7 and 68.1 (CH), 91.7 (C_5H_5), 238.8 (CO) [24] ^{95}Mo NMR (acetone-d_6, vs. 2M Na_2MoO_4 in D_2O): −1646 [38] IR (cyclohexane): 18768, 1954 (ν(CO)) [23, 29]; similar in [24] reaction with NOX (X = BF_4 or PF_6) in CH_3CN at 0°C for a short time affords

Table 12 (continued)

No.	compound	method of preparation (yield) properties and remarks
		$[C_5H_5Mo(CO)(NO)C_8H_{13}]X$ [23, 29, 37]; resistant to hydride abstraction with $[C(C_6H_5)_3]^+$ and a variety of other techniques [47]
147	$CH_3C_5H_4Mo(CO)_2C_8H_{13}$	V (2% by sublimation at 353 K) [43] yellow crystals (from pentane) [43] ^{1}H NMR (C_6D_6): 1.50 (CH_3 and H-5 to H-7), 2.25 (H-4, 8), 2.35 (H-4′, 8′), 3.35 (H-1, H-2, and H-3), 4.57 and 4.78 (AA′BB′ system, C_5H_4) [43] IR: 1878, 1955 (ν(CO)) [43]
148	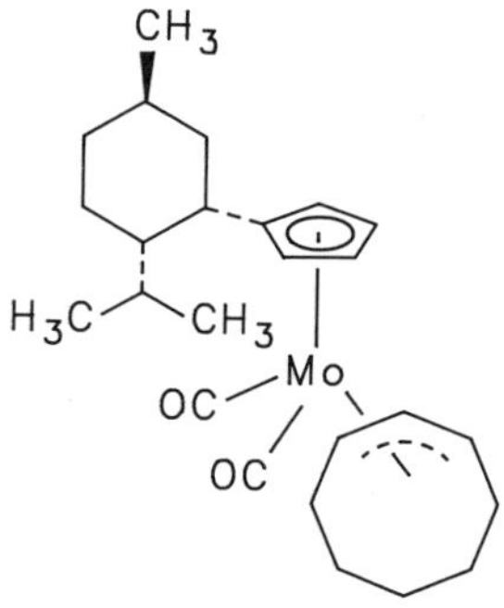	II, X = Br, for 6 h (75% by chromatography on Al_2O_3 with pentane/CH_2Cl_2 1:1) [28] yellow needles, m.p. 89°C (from pentane) [28] optical rotation (pentane, 0.197 M): $[\alpha]_D$ = +44° [28] ^{1}H NMR ($CDCl_3$): 0.705, 0.898, and 0.928 (d, CH_3; J(H, H) = 6), 1.822 (m, 2H), 2.070 (m, 2H), 2.202 (m, C_8H_{13}, 2H), 2.325 (m, C_8H_{13}, 2H), 2.793 (m, 1H), 3.688 (m, C_8H_{13}, 2H), 4.089 (t, C_8H_{13}, 1H; J(H, H) = 8.27), 4.984, 5.080, 5.135, and 5.327 (q, C_5H_4; J(H, H) = 2) [28] IR (cyclohexane): 1886, 1935 (ν(CO)) [28]
^{3}L = cyclononyl derivative		
149	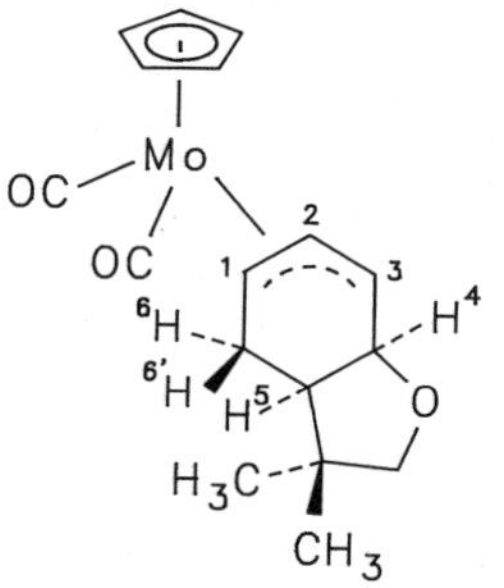	No. 41 was allowed to react with $[C(C_6H_5)_3]BF_4$ in CH_3CN for 1.5 h followed by addition of $N(C_2H_5)_3$ and additional stirring for 1 h (18% by thin-layer chromatography on Al_2O_3 followed by SiO_2 with $CHCl_3$/hexane) [26] yellow crystals (from cyclohexane), m.p. 180°C (dec.) [26] ^{1}H NMR ($CDCl_3$): 0.74 (d of d of d, H-5; J(H-4, 5) = 4.85, J(H-6, 5) = 7.8, J(H-6′, 5) = 9.8), 0.85 (s, CH_3), 1.00 (s, CH_3), 1.79 (approximately d of d of d, H-6; J(H-2, 6) = 1.4, J(H-1, 6) = 3.2, J(H-6, 6′) = 14.5, J(H-5, 6) = 7.8), 1.91 (approximately d of d of d, H-6′; J(H-2, 6′) = 1.4, J(H-1, 6′) = 2.5, J(H-5, 6′) = 9.8), 3.57 (AB system q, OCH_2; Δδ = 0.15 ppm, J(A, B) = 7.5), 3.69 (d of d of d, H-3; J(H-2, 3) = 6.1, J(H-4, 3) = 3.4, J(H-1, 3) = 1.9), 3.89 (approximately d of q, H-1; J(H-2, 1) = 6.1), 4.31 (approximately t, H-2; J(H-3, 2) = 6.9), 4.44 (approximately t, H-4; J(H-5, 4) = 4.9), 5.31 (s, C_5H_5) [26] IR (cyclohexane): 1874, 1948 (ν(CO)) [26]

References on pp. 303/4

Table 12 (continued)

No.	compound	method of preparation (yield) properties and remarks
149 (continued)		mass spectrum (relative intensities): $[M]^+$ (6.4), $[M-CO]^+$ (13.0), $[M-2\ CO]^+$ (37.3) [26] the structure was confirmed by single-crystal X-ray diffraction analysis [26]
supplement		
150	$C_5H_5Mo(CO)_2C_6H_7$	treatment of the cationic dienyl complex $[C_5H_5Mo(CO)_2C_6H_8]^+$ with $N(C_2H_5)_3$ [49] reacts with carbonyl compounds in the presence of Lewis acids to give substituted cationic dienyl compounds; it forms an adduct with $[C(C_6H_5)_3]PF_6$ [49]

*Further information:

$C_5H_5Mo(CO)_2C_2(C_6H_5)_2CC_4H_9$-t (Table **12**, No. **1**) crystallizes in the monoclinic space group $P2_1/n-C^5_{2h}$ (No. 14) with the unit cell parameters a = 8.488(2), b = 16.394(4), c = 16.668(4) Å, β = 95.28(2)°; Z = 4 molecules per unit cell, and D_{calc} = 1.374 g/cm³. The molecular structure with the main bond distances and angles is shown in **Fig. 48** [35].

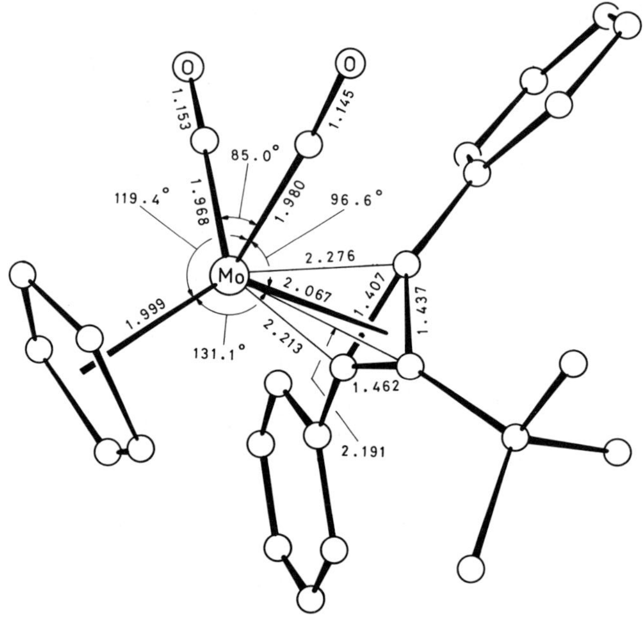

Fig. 48. The molecular structure of $C_5H_5Mo(CO)_2C_2(C_6H_5)_2CC_4H_9$-t [35].

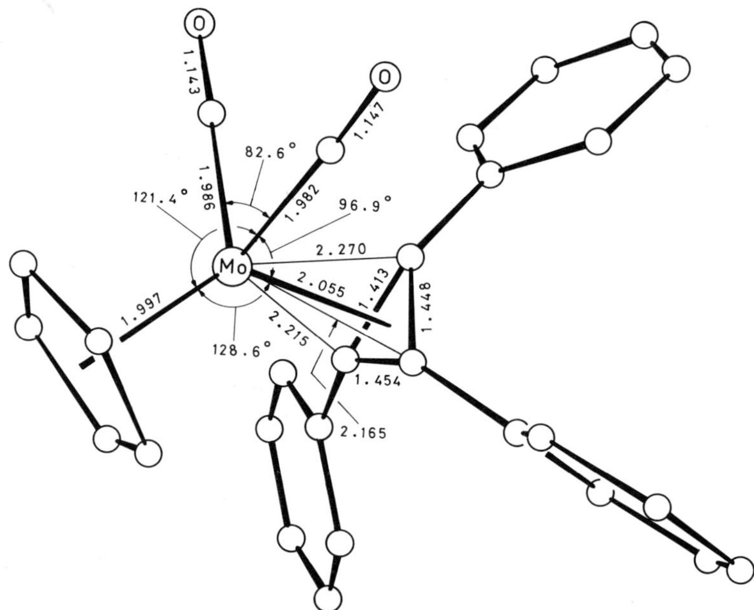

Fig. 49. The molecular structure of $C_5H_5Mo(CO)_2C_3(C_6H_5)_3$ [35].

$C_5H_5Mo(CO)_2C_3(C_6H_5)_3$ (Table **12**, No. **2**) crystallizes in the monoclinic space group $P2_1/n - C^5_{2h}$ (No. 14) with the unit cell parameters a = 9.545(2), b = 16.580(3), c = 14.463(3) Å, β = 98.47(2)°; Z = 4 molecules per unit cell, and D_{calc} = 1.421 g/cm³. The molecular structure with the main bond distances and angles is shown in **Fig. 49** [35].

$C_5H_5Mo(CO)_2C_5H_7$ (Table **12**, No. **11**) was obtained in the reaction of $(LiHMo(C_5H_5)_2)_4$ with CO in 5% yield [12]. $(C_5H_5)_2MoCl_2$ was treated with two equivalents of sodium amalgam in toluene under 150 atm CO and 50 atm H_2 at 65°C for 20 h to give No. 11 in 20% yield [13]. $(C_5H_5)_2HMo\{\mu\text{-}(C_6H_{11}\text{-cyclo})MgBr_2Mg(O(C_2H_5)_2)_2\}_2MoH(C_5H_5)$ was allowed to react in toluene/THF (1:1) with a stream of CO for 48 h. The product was isolated by column chromatography (first fraction) and recrystallized in 40% yield [16]. It was also observed in the reaction of $(C_5H_5)_2MoH_2$ with 1 atm CO at 80°C for 24 h in toluene in a 1:2 ratio with $(C_5H_5)_2MoCO$, besides traces of $[(C_5H_5)_2Mo(CO)H][C_5H_5Mo(CO)_3]$ [20]. A 20% yield of No. 11 was obtained by prolonged irradiation of $Mo(CO)_6$ or $C_5H_5Mo(CO)_3H$ in the presence of cyclopentadiene [22]. Traces were formed in the reaction of $C_5H_5Mo(CO)(C_5H_6)C(O)CH_3$ (C_5H_6 = cyclopentadiene) in THF at 309 K for 1.5 h. A 4% yield was isolated by chromatography on Al_2O_3 with ether/hexane (1:1) [39].

$C_5H_5Mo(CO)_2C_6H_9$ (Table **12**, No. **23**) crystallizes in the monoclinic space group $P2_1/n - C^5_{2h}$ (No. 14) with the unit cell parameters a = 8.586(3), b = 9.063(1), c = 15.353(2) Å, β = 97.38(1)°; Z = 4 molecules per unit cell, and D_{calc} = 1.672 g/cm³. The molecular structure with the main bond distances and angles is shown in **Fig. 50**, p. 298. The ring adopts a chair configuration; the data were discussed with respect to the observed and calculated vicinal coupling constants [26].

Reaction with $[C(C_6H_5)_3]PF_6$ in CH_2Cl_2 affords $[C_5H_5Mo(CO)_2C_6H_8\text{-cyclo}]PF_6$. Treatment with $NOPF_6$ in CH_3CN at 0°C affords $[C_5H_5Mo(CO)(NO)C_6H_9]PF_6$. Oxidation with O_2 in $CDCl_3$ for 7 d or treatment with CF_3COOD in a variety of solvents yields free 3-deuteriocyclohexene [26].

References on pp. 303/4

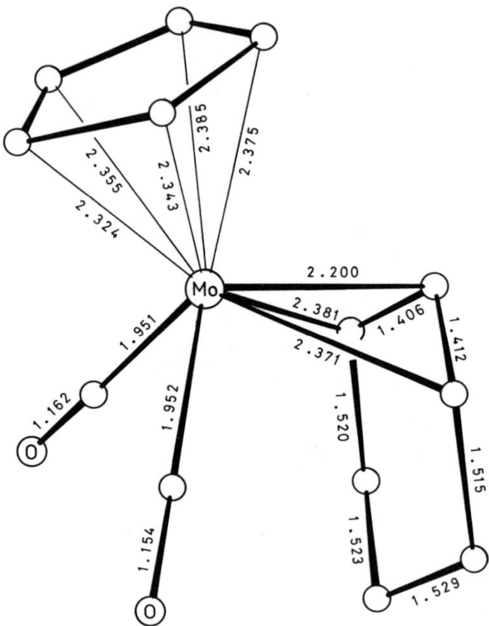

Fig. 50. The molecular structure of $C_5H_5Mo(CO)_2C_6H_9$ [26].

$C_5H_5Mo(CO)_2C_6H_8R$-4 (Table **12**, Nos. **31**, **33**, **34**, **36**, **40**, **47**, **70**) were oxidized by iodine or ceric ammonium nitrate with formation of free cyclohexene derivatives. To a stirred solution of the complex (Nos. 31, 33, 34, 40, 47, 70) in CH_3CN was added an excess of iodine. After 30 min the reaction was complete (by IR). No. 31 gave the free lactone (Formula III: $^1R = {}^2R = {}^3R = H$) [32, 36]; Nos. 33, 40, and 47 gave the cyclohexene derivative shown in Formula IV ($R^1 = I$, $R^2 = CN$, $R^3 = CH_2CH(C(O)OC_4H_9\text{-t})_2$; $R^1 = I$, $R^2 = C(C(O)OCH_3)_2CH_3$, $R^3 = H$; $R_1 = I$, $R^2 = C_6H_4OCH_3$-4, $R^3 = H$) [32, 36, 46]; and No. 34 the derivative shown in Formula V [36]. The oxidation of No. 70 gave the mixture of cyclohexenes shown in Formula VI [32]. Oxidation of No. 36 with ceric ammonium nitrate in acetone yielded the mixture of cyclohexene derivatives

III IV V

VI

References on pp. 303/4

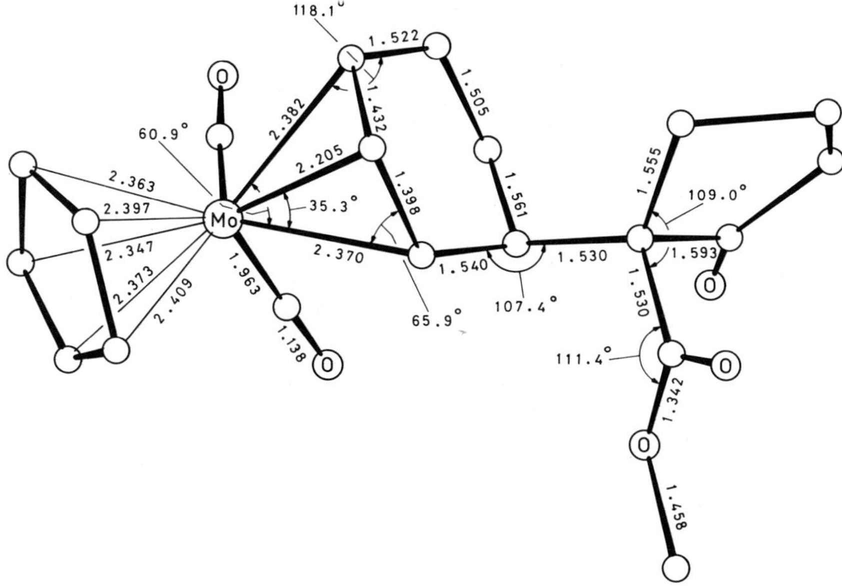

Fig. 51. The molecular structure of $C_5H_5Mo(CO)_2C_6H_8C(C(O)OCH_3)C(O)CH_2CH_2CH_2$-4 [36].

shown in Formulas III (R^1, R^2 = H or $C(O)OCH_3$, R^3 = H) and IV (R^1 = OH, R^2 = $CH(C(O)OCH_3)_2$, R^3 = H) [36].

$C_5H_5Mo(CO)_2C_6H_8C(C(O)OCH_3)C(O)CH_2CH_2CH_2$-4 (Table **12**, No. **45**) crystallizes in the monoclinic space group $P2_1/n-C_{2h}^5$ (No. 14) with the unit cell parameters a = 15.053(5), b = 8.327(3), c = 15.583(3) Å, β = 108.26(2)°. The molecular structure with the main bond distances and angles is shown in **Fig. 51** [36].

The reaction with iodine in CH_3CN yields the cyclohexene derivative shown in Formula IV: R^1 = I, R^2 = $C(C(O)OCH_3)C(O)CH_2CH_2CH_2$-cyclo, R^3 = H [32, 36].

$C_5H_5Mo(CO)_2C_6H_8CN$-4,exo (Table **12**, No. **51**) was deprotonated with $LiN(C_3H_7\text{-}i)_2$ (prepared in situ) in THF at −78°C for 0.5 h followed by quenching with excess H_2O or D_2O to yield Nos. 52 and 53 [40, 46]. The complex is decyanated with n-, s-, or t-LiC_4H_9 in THF at −78°C to yield an anionic intermediate which was allowed to react with various electrophiles (Method VIII). These reactions are given in the following table. Addition occurs at the 4-position of the cyclohexa-1,3-diene ligand [46].

reaction with	product
H_2O	No. 23
D_2O	No. 26
$CH_2{=}C(C(O)OC_4H_9\text{-}t)_2$	No. 33
$C_6H_5C(O)H$	No. 37
2-(methoxycarbonyl)-cyclohex-1-en-3-one	No. 44
$[C_5H_5Mo(CO)_2C_6H_8\text{-cyclo}]PF_6$	$C_5H_5Mo(CO)_2C_6H_8C_6H_8Mo(CO)_2C_5H_5$
$[CH_3OC_6H_6Fe(CO)_3]PF_6$	No. 48
$[R_2C_6H_4Mn(CO)_3]PF_6$	Nos. 49, 50 (R = H, OCH_3)

References on pp. 303/4

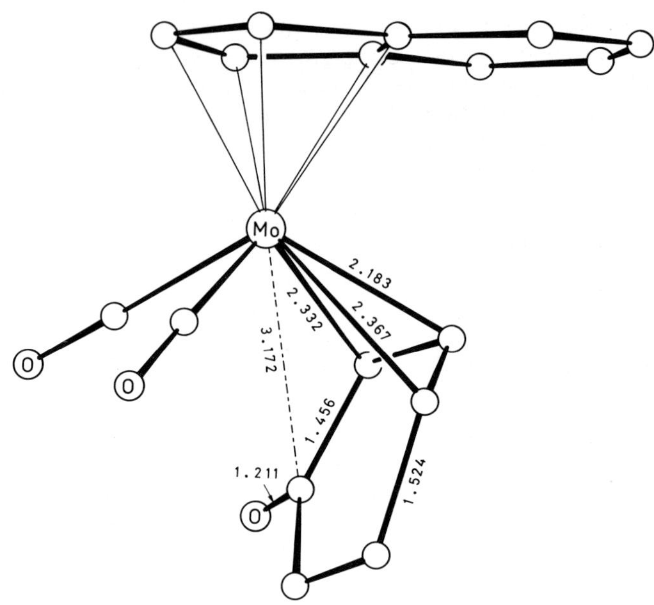

Fig. 52. The molecular structure of $C_9H_7Mo(CO)_2C_6H_7O$ [41].

$C_9H_7Mo(CO)_2C_6H_7O$ (Table **12**, No. **83**) crystallizes in the monoclinic space group $P2_1-C_2^2$ (No. 4) with the unit cell parameters a = 7.177(2), b = 17.063(8), c = 12.123(5) Å, β = 103.83(4)°; Z = 4 molecules per unit cell, D_{calc} = 1.714, and D_{meas} = 1.724 g/cm^3. The molecular structure with main bond distances and angles is shown in **Fig. 52**. The unit cell contains two crystallographically independent molecules of the opposite chirality but otherwise similar geometry [41].

Protonation of No. 83 either with $(C_2H_5)_2O \cdot HBF_4$ or CF_3SO_3H affords the complex $[C_9H_7Mo(CO)_2C_6H_7OH\text{-}1]^+$. Stirring in CD_3OD in the presence of anhydrous Na_2CO_3 results in H-D exchange at the 5′ position (No. 84). Treatment with $LiNR_2$ (R = $Si(CH_3)_3$ or i-C_3H_7) in THF at −78°C followed by addition of R′I (R′ = CH_3 or C_2H_5) affords Nos. 85 and 86. Reaction with C_6H_5CHO affords No. 87 in good yields. Alkylation with $R'OSO_2CF_3$ leads to $[C_9H_7Mo(CO)_2C_6H_7OR'\text{-}1]^+$ (R′ = CH_3, $Si(CH_3)_3$) [41].

$C_9H_7Mo(CO)_2C_6H_6((CH_3)_2$-1,4,endo)$CH_2COOH$-4,exo, $C_9H_7Mo(CO)_2C_6H_6((CH_3)_2$-1,4,endo)$C(CH_3)$-(COOH)H-4,exo, $C_5H_5Mo(CO)_2C_7H_{10}CH_2COOH$-4, $C_5H_5Mo(CO)_2C_7H_9(CH_3$-7)$CH_2COOH$-4, C_5H_5Mo-$(CO)_2C_7H_9(CH_2CH{=}CH_2$-7)$CH_2COOH$-4 (Table **12**, Nos. **92, 94**, **109**, **131**, **137**). $NOPF_6$ was added to

H
H_3C
O
O
H_3C
R
VII

O
O
VIII

X
CN
X
CN
IX

a solution of the acid complex (Nos. 92, 94) in CH_3CN at 0°C. The resulting mixture was stirred for 15 min. $N(C_2H_5)_3$ was added dropwise and the mixture was stirred for an additional 15 min. The reaction mixture was poured into water to yield the free lactone (Formula VII: R = H or CH_3) [46]. A similar reaction with Nos. 131, 137 yielded the lactone shown in Formula VII (R = CH_3 or $CH_2CH{=}CH_2$) [47]. Reaction of No. 109 with iodine in CH_3CN followed by hydrolysis yielded the lactone shown in Formula VIII [47].

$C_5H_5Mo(CO)_2C_7H_{10}R$-4 (Table **12**, Nos. **106, 111, 112, 118**). The ^1H NMR signals of these compounds have been assigned with help of decoupling experiments [47].

No. 118 (R = CN-exo) was used as precursor for a reactive carbanion complex by decyanation with LiC_4H_9-t (Method VIII) or deprotonation with $LiN(C_3H_7\text{-}i)_2$ (Method IX) in THF. Treating of No. 118 with Br_2 gave the free cycloheptenes shown in Formula IX (X = Br). Reaction with $NOPF_6$ followed by addition of H_2O also afforded a mixture of free cycloheptenes (Formula IX, X = OH) [46].

$C_5H_5Mo(CO)_2C_7H_7$ (Table **12**, No. **144**). Variable temperature ^1H NMR measurements in acetone-d_6 between −60 and +30°C indicate fast rotation of the C_7H_7 ligand [5]. Selected ^1H NMR spectra (δ values) are listed in the following table. Spin-decoupling experiments in CF_2Cl_2/toluene-d_8 at −107°C resulted in J(4, 5 = 6, 7) = 4.9, J(1, 7 = 3, 4) = 7.0, J(1, 2 = 2, 3) = 6.6 Hz for the cycloheptatrienyl ligand. At higher temperatures only one singlet was found. Figures of the spectra were given [9]. The ^1H NMR spectrum in $CFCl_3/CS_2$ (2:1) at −100°C showed resonances at δ = 2.10 (t, H-2; J(H, H) = 6.7 Hz), 4.27 (t, H-1, 3; J(H, H) = 6.7 Hz), 5.00 (d, H-5, 6 overlapping with C_5H_5), 6.00 (unresolved t, H-4, 7) ppm; at +25°C a singlet was obtained for the C_7H_7 ligand. The complex crystallized below −110°C from $CFCl_3/CS_2$ (2:1) during NMR experiments [7].

solvent	T (°C)	C_5H_5	H-4, 7	H-5, 6	H-1, 3	H-2	C_7H_7 (singlet)
$CF_2Cl_2/C_6D_5CD_3$ (2:1)	−107	4.07	6.20	5.19	3.76	1.13	
$CF_2Cl_2/C_6H_5CD_3$ (2:1)	−17	4.61					4.66
$CF_2Cl_2/C_6D_5CD_3$ (1:1)	−102	4.04	6.47	5.46	3.84	1.20	
$CF_2Cl_2/C_6D_5CD_3$ (1:1)	+10	4.45					4.69
$CHFCl_2/C_6D_5CD_3$ (1:1)	−118	5.09	6.36	5.36	4.37	2.00	
$CHFCl_2/C_6D_5CD_3$ (1:1)	+9	5.02					4.84
$CHFCl_2$	−118	5.36	6.50	5.40	4.59	2.32	
$CHFCl_2$	+10	5.16					5.02

The principal pathway of rearrangement is a 1,2-shift of the metal about the C_7H_7 ring. The activation energy of 12.5 ± 0.6 kcal/mol and log A of 16.0 ± 0.6 should be regarded with some suspicion because of the possibility of systematic error and the rather high value of log A. Using absolute rate theory, the free energy of activation varies between 8.8 and 9.7 kcal/mol [9, 10].

The IR spectrum in cyclohexane solution exhibits 4 ν(CO) absorptions, indicating the presence of two conformers. They correspond to exo and endo arrangement of the C_7H_7 ring with respect to the C_5H_5 ligand (see General Remarks on p. 251). Assuming an equal molar absorptivity an equilibrium ratio of about 0.35 between the two conformers can be derived which does not vary significantly between −10 and +60°C [9].

Single crystals were obtained by recrystallization from pentane at −15°C. No. 144 crystallizes in the hexagonal space group $P6_3/m-C_{6h}^2$ (No. 176) with the unit cell parameters a = 13.320(3), c = 11.658(3) Å; Z = 6 molecules per unit cell, D_{calc} = 1.714, and D_{meas} =

References on pp. 303/4

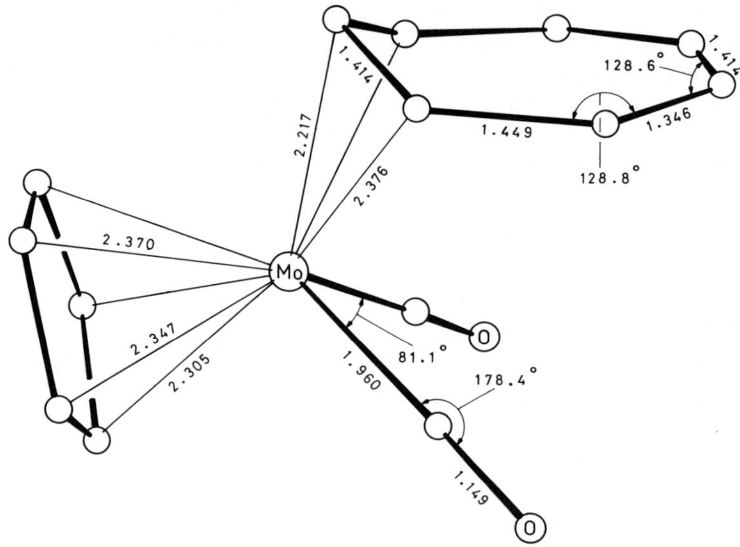

Fig. 53. The molecular structure of $C_5H_5Mo(CO)_2C_7H_7$ [11].

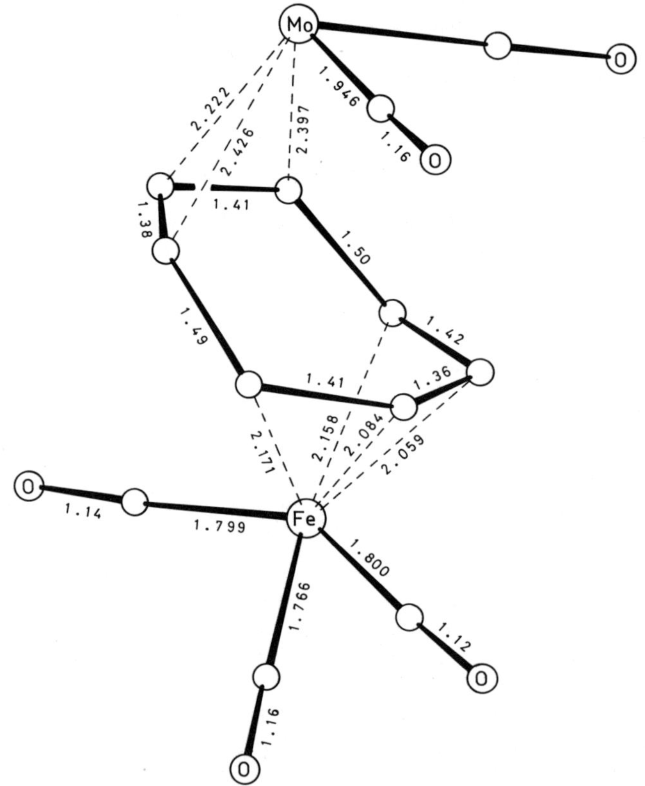

Fig. 54. The molecular structure of $C_5H_5Mo(CO)_2C_7H_7Fe(CO)_3$; the C_5H_5 ring is omitted [11].

References on pp. 303/4

1.69(2) g/cm^3 (by flotation in aqueous ZnI_2). The molecular structure with the main bond distances and angles is shown in **Fig. 53** [11] and was mentioned in [14].

Attempts to hydrogenate the coordinated cycloheptatrienyl ligand at atmospheric pressure using Pd or Rh on charcoal failed. Treatment with $(CN)_2C{=}C(CN)_2$ in C_6H_6 gave a yellow precipitate that was not obtained pure. Irradiation with $Fe(CO)_5$ did not give the heteronuclear complex $C_5H_5Mo(CO)_2C_7H_7Fe(CO)_3$ (No. 145) [2]. This complex was obtained by irradiation in ether/$Fe(CO)_5$ (2:1) for four days or with $Fe_2(CO)_9$ in heptane at 80°C for 24 h [8].

$C_5H_5Mo(CO)_2C_7H_7Fe(CO)_3$ (Table **12**, No. **145**). The ^{1}H NMR spectra are temperature-dependent. Figures of the spectra were given between −71 and +50°C [8]. ^{1}H NMR spectra (δ values) are given in the following table. The activation parameters for the 1,2-shift are: E_A = 13 ± 1 kcal/mol, log A = 12.7 ± 0.8, $\Delta S^{\ddagger}_{273} = -2\,cal \cdot mol^{-1} \cdot K^{-1}$, $\Delta S^{\ddagger}_{251} = +2\,cal \cdot mol^{-1} \cdot K^{-1}$ [8].

proton	$CDCl_3$	$CDCl_3$/toluene-d_8 (4:1)	pattern
H-2	4.98	4.82	distorted triplet
H-1, 3	4.50	4.37	triplet
H-4, 7	3.83	3.77	complex
H-5, 6	5.40	5.37	doublet of doublets

No. 145 crystallizes from ethanol in the orthorhombic space group $Pna2_1-C^9_{2v}$ (No. 33) with the unit cell parameters a = 11.838(6), b = 6.773(3), c = 20.918(8) Å; Z = 4 molecules per unit cell, D_{calc} = 1.77, and D_{meas} = 1.80 (1) g/cm^3 (by flotation in aqueous BaI_2). The molecular structure with the main bond distances and angles is shown in **Fig. 54** [11] and was mentioned in [14].

References:

[1] King, R. B.; Bisnette, M. B. (Tetrahedron Letters **18** [1963] 1137/41).
[2] King, R. B.; Bisnette, M. B. (Inorg. Chem. **3** [1964] 785/9).
[3] King, R. B. (Inorg. Chem. **5** [1966] 2242/3).
[4] King, R. B.; Fronzaglia, A. (J. Am. Chem. Soc. **88** [1966] 709/12).
[5] King, R. B. (J. Organometal. Chem. **8** [1967] 129/37).
[6] Hayter, R. G. (J. Organometal. Chem. **13** [1968] P 1/P 3).
[7] Bennett, M. A.; Bramley, R.; Watt, R. (J. Am. Chem. Soc. **91** [1969] 3089/91).
[8] Cotton, F. A.; Reich, C. R. (J. Am. Chem. Soc. **91** [1969] 847/53).
[9] Faller, J. W. (Inorg. Chem. **8** [1969] 767/71).
[10] Anderson, A. S. (Diss. Yale Univ. 1971; Diss. Abstr. Intern. B **22** [1972] 2551).
[11] Cotton, F. A.; De Boer, B. G.; La Prade, M. D. (23rd IUPAC Congr., Boston, Mass., 1971, Vol. VI, p. 1).
[12] Benfield, F. W. S.; Forder, R. A.; Green, M. L. H.; Moser, G. A.; Prout, K. (J. Chem. Soc. Chem. Commun. **1973** 759/60).
[13] Wong, K. L.; Brintzinger, H. H. (J. Am. Chem. Soc. **97** [1975] 5143/6).
[14] Bennett, M. J.; Pratt, J. L.; Simpson, K. A.; Lishigman, L. K.; Takats, J. (J. Am. Chem. Soc. **98** [1976] 4810/7).
[15] Davidson, J. L.; Green, M.; Nyathi, J. Z.; Scott, C.; Stone, F. G. A.; Welch, A. J.; Woodward, P. (J. Chem. Soc. Chem. Commun. **1976** 714/5).
[16] Green, M. L. H.; Luong-Thi, T.; Moser, G. A.; Packer, I.; Pettit, F.; Roe, D. M. (J. Chem. Soc. Dalton Trans. **1976** 1988/92).

[17] Bottrill, M.; Green, M. (J. Chem. Soc. Dalton Trans. **1977** 2365/71).
[18] Atwood, J. L.; Rogers, R. D.; Hunter, W. E.; Bernal, I.; Brunner, H.; Lukas, R.; Schwarz, W. (J. Chem. Soc. Chem. Commun. **1978** 451/2).
[19] Green, M.; Nyathi, J. Z.; Scott, C.; Stone, F. G. A.; Welch, A. J.; Woodward, P. (J. Chem. Soc. Dalton Trans. **1978** 1067/80).
[20] Adams, M. A.; Folting, K.; Huffman, J. C.; Caulton, K. G. (J. Organometal. Chem. **164** [1979] C 29/C 32).

[21] Bottrill, M.; Green, M. (J. Chem. Soc. Dalton Trans. **1979** 820/5).
[22] Mills III, W. C.; Wrighton, M. S. (J. Am. Chem. Soc. **101** [1980] 5830/2).
[23] Faller, J. W.; Shvo, Y.; Chao, K.; Murray, H. H. (J. Organometal. Chem. **226** [1982] 251/75).
[24] Griffiths, M.; Knox, S. A. R.; Stansfield, R. F. D.; Stone, F. G. A.; Winter, M. J.; Woodward, P. (J. Chem. Soc. Dalton Trans. **1982** 159/65).
[25] Davidson, J. L. (J. Chem. Soc. Dalton Trans. **1983** 1667/70).
[26] Faller, J. W.; Murray, H. H.; White, D. L.; Chao, K. H. (Organometallics **2** [1983] 400/9).
[27] Adams, H.; Bailey, N. A.; Gauntlett, J. T.; Winter, M. J. (J. Chem. Soc. Chem. Commun. **1984** 1360/1).
[28] Faller, J. W.; Chao, K.-H. (Organometallics **3** [1984] 927/32).
[29] Faller, J. W.; Chao, K.-H.; Murray, H. H. (Organometallics **3** [1984] 1231/40).
[30] Hughes, R. P.; Reisch, J. W.; Rheingold, A. L. (Organometallics **3** [1984] 1761/3).

[31] Pearson, A. J.; Khan, M. N. I. (J. Am. Chem. Soc. **106** [1984] 1872/3).
[32] Pearson, A. J.; Khan, M. N. I. (Tetrahedron Letters **25** [1984] 3507/10).
[33] Green, M.; Greenfield, S.; Kersting, M. (J. Chem. Soc. Chem. Commun. **1985** 18/20).
[34] Hughes, R. P.; Kläui, W.; Reisch, J. W.; Müller, A. (Organometallics **4** [1985] 1761/6).
[35] Hughes, P. R.; Reisch, J. W.; Rheingold, A. L. (Organometallics **4** [1985] 1754/61).
[36] Pearson, A. J.; Khan, M. N. I.; Clardy, J. C.; Cun-heng, H. (J. Am. Chem. Soc. **107** [1985] 2748/57).
[37] Vanarsdale, W. E.; Winter, R. E. K.; Kochi, J. K. (J. Organometal. Chem. **296** [1985] 31/54).
[38] Faller, J. W.; Whitmore, B. C. (Organometallics **5** [1986] 752/5).
[39] Kreiter, C. G.; Kögler, J.; Nist, K. (J. Organometal. Chem. **310** [1986] 35/46).
[40] Pearson, A. J.; Khetani, V. D. (J. Chem. Soc. Chem. Commun. **1986** 1772/4).

[41] Green, M.; Greenfield, S.; Grimshire, M. J.; Kersting, M.; Orpen, A. G.; Rodrigues, R. A. (J. Chem. Soc. Chem. Commun. **1987** 97/9).
[42] Müller, H.-J.; Beck, W. (J. Organometal. Chem. **330** [1987] C 13/C 16).
[43] Kreiter, C. G.; Wendt, G.; Kaub, J. (J. Organometal. Chem. **352** [1988] 307/19).
[44] Müller, H.-J. (Diss. Univ. München 1988, pp. 1/150).
[45] Pearson, A. J.; Khetani, V. D. (J. Org. Chem. **53** [1988] 3395/6).
[46] Pearson, A. J.; Khetani, V. D. (J. Am. Chem. Soc. **111** [1989] 6778/89).
[47] Pearson, A. J.; Khan, M. N. I. (J. Org. Chem. **50** [1985] 5276/85).
[48] Brunner, H.; Lukas, R. (J. Organometal. Chem. **90** [1975] C 25/C 27).
[49] Wang, Tein-Fu; Lee, Ming-Chao (Proc. 29th Intern. Conf. Coord. Chem., Lausanne, Switz., 1992, Abstr. Papers P 249).

1.5.1.3.2.4 Compounds with One ^4L Ligand

The complexes listed in Table 13 are compounds of the type $[^5LMo(CO)_2{}^4L]^+$. The ^4L ligands are buta-1,3-diene derivatives, trimethylenemethane derivatives, and cyclic di- or polyenes with weakly coordinating counterions like BF_4, PF_6, or CF_3SO_3. The compounds were obtained in many cases by one of the following methods. Further information on preparation is given in the table.

Method I: $^5LMo(CO)_2{}^3L$ (^3L = allyl derivative) was treated with $[C(C_6H_5)_3]X$ in CH_2Cl_2. The products were in most cases precipitated by the addition of ether [2, 4, 11].

Method II: A solution of $C_5H_5Mo(CO)_2{}^3L$ (^3L = allyl or derivative), $(C_2H_5)_2O \cdot HBF_4$, and a fivefold excess of the diene in ether was stirred for 24 h [6, 10, 15].

Method III: $[^5LMo(CO)_2(NCCH_3)_2]X$ was allowed to react with an excess of the buta-1,3-diene or cyclic diene ligand in a polar solvent [1, 10]. In the preparation of the trimethylenemethane cation, an excess of a methylenecyclopropane derivative in CH_2Cl_2 was used as the ^4L precursor [3, 5, 16, 17].

Method IV: $^5LMo(CO)_2{}^3L$ (^3L = allyl derivative) was treated with CF_3SO_3H in CH_2Cl_2 at -78 °C (for the trans-coordinated diene) to ambient temperature [20].

Method V: $(^5LMo(CO)_3)_2$ was treated with $AgBF_4$ in the presence of an excess of the ^4L ligand in CH_2Cl_2 for 24 h [1, 10, 18, 19] or the ^4L precursor methylenecyclopropane for 4 h [5].

Method VI: A solution of $^5LMo(CO)_2(CR)_3C{=}CH_2$ (a coordinated methylenecyclobutenyl with R = CH_3, C_6H_5) in ether was cooled to 0 °C and treated with $(C_2H_5)_2O \cdot HBF_4$. The product precipitated [7].

With the exception of Nos. 7 and 13, the buta-1,3-diene derivatives are coordinated in the cis form. Isomerization of the trans-coordinated butadiene derivatives occurs in solution above -40 °C [20]. The endo and exo isomers are shown in Formula I. The numbering used for the ^1H NMR spectra and formulas is shown in Formula II.

The preferred conformation (see "Further information" of No. 17) of the trimethylenemethane fragment is shown in Formula III. The numbering used for the NMR spectra and formulas (the number indicates the R position) is shown in Formula IV. The numbering for the cyclohexa-1,3-dienes is shown in Formula V and that for the cyclohepta-1,3-dienes in Formula VI. C_9H_7 means indenyl and the numbering is given in Formula VII.

The reaction of $[^5LMo(CO)_2{}^4L]^+$ (^4L = buta-1,3-diene derivative) with hydride reagents (e.g., $Na[H_3BCN]$, $NaBH_4$) in THF leads to the formation of the corresponding neutral

References on pp. 319/20

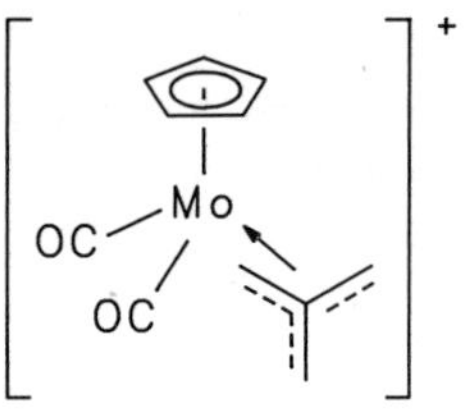

III 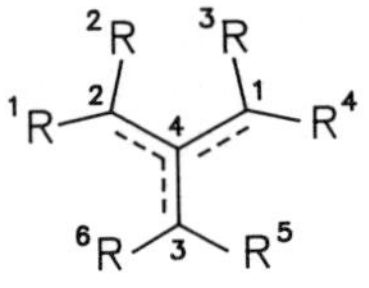 IV 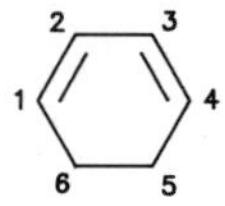 V 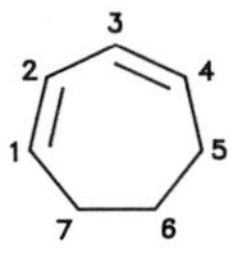 VI 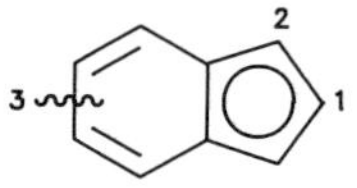 VII

complexes, $^5LMo(CO)_2{}^3L$ (^3L = allyl derivative) [1, 2]. Reaction with $Na[M(CO)_5]$ (M = Mn, Re) in THF at low temperatures results in addition of the $(CO)_5M$ unit to the end of the butadiene giving the allyl complex, $^5LMo(CO)_2C_3H_{4-n}R_nCH_{2-m}R'_mM(CO)_5$ [16, 17].

Addition of nucleophiles (e.g., H^-, OH^-, or SR^-) to the coordinated trimethylenemethane ligand leads to the formation of neutral allyl complexes, $^5LMo(CO)_2{}^3L$ [5]. The addition of nucleophiles also takes place with coordinated cyclodienes. Various reactions of this type are given in "Further information".

Table 13
Compounds with One ^4L Ligand.
An asterisk indicates further information at the end of the table.
For explanations, abbreviations, and units see p. X.

No.	compound	method of preparation (yield) properties and remarks
^4L = derivative of buta-1,3-diene		
*1	$[C_5H_5Mo(CO)_2C_4H_6]BF_4$	I, ^3L = $C_3H_4CH_3$-anti,1 [2]; II (30%); II with irradiation (40%) [15] exo to endo ratio 0.21 in acetone-d_6 at −60°C [2] ^1H NMR (acetone-d_6, −70°C): endo isomer: 2.28 (d, H-1, 6; J(H, H) = 9.6), 3.07 (d, H-2, 5; J(H, H) = 7.4), 6.13 (s, C_5H_5), 6.45 (s, H-3, 4); exo isomer: 2.09 (d of d, H-1, 6; J(H, H) = 8.02, 2.0), 3.36 (d of d, H-2, 5; J(H, H) = 7.4, 2.0), 6.13 (s, C_5H_5), 6.17 (s, H-3, 4); (acetone-d_6, +50°C): 2.23 (d, H-1, 6; J(H, H) = 9.2), 3.18 (d, H-2, 5; J(H, H) = 7.4), 6.08 (s, C_5H_5), 6.39 (m, H-3, 4) [2]; similar data in [15]; $\Delta G^{\ddagger}$ = 14.1 kcal/mol for the endo/exo equilibration [2] IR (nitromethane): 1985, 2005, 2055 (ν(CO)) [2]; (acetone): 2015, 2061 (ν(CO)) [15]
2	$[C_9H_7Mo(CO)_2C_4H_6]BF_4$	III, in CH_2Cl_2 for 24 h (62%) [1] yellow crystals [1] ^1H NMR (CD_3NO_2, 32°C): 0.70 (br s, H-1, 6), 3.40 (br s, H-2, 5), 5.50 (br s, H-3, 4), 5.9 (t, C_9H_7, H-1; J(H, H) = 3.0), 6.36 (d, C_9H_7, H-2), 7.1 to 7.6 (m, C_9H_7, H-3); did not show any appreciable

References on pp. 319/20

Table 13 (continued)

No.	compound	method of preparation (yield) properties and remarks
		sharpening of the diene part on cooling to −90 °C in acetone-d_6, suggesting that both isomers are present even at low temperatures [1] IR (Nujol): 2020, 2070 (ν(CO)) [1] reaction with $NaBH_4$ in THF yields $C_9H_7Mo(CO)_2C_3H_4CH_3$-anti,1 [1]
3	$[(CH_3)_5C_5Mo(CO)_2C_4H_6]BF_4$	V [10] reaction with $LiCu(CH{=}CHC_5H_{11})_2$ in THF at −78 °C yields $(CH_3)_5C_5Mo(CO)_2C_3H_4$-$CH{=}CHC_5H_{11}$ [10]
*4	$[C_5H_5Mo(CO)_2C_4H_5CH_3\text{-}2]BF_4$	I, ^3L = $C_3H_3(CH_3)_2$-1,3 [2] ^1H NMR (acetone-d_6, −50 °C): endo isomer: 1.77 (d, CH_3; J(H, H) = 5.9), 2.20 (d, H-1; J(H, H) = 10.7), 3.41 (m, H-6), 2.98 (d, H-5; J(H, H) = 7.8), 6.05 (s, C_5H_5), 6.22 and 6.34 (m, H-3, 4); (acetone-d_6, +60 °C): 1.81 (d, CH_3; J(H, H) = 6.2), 2.55 (br, H-5), 3.25 (H-1, 6), 5.90 (C_5H_5), 6.15 and 6.25 (m, H-3, 4) [2]
5	$[C_9H_7Mo(CO)_2C_4H_5CH_3\text{-}2]BF_4$	III, in CH_2Cl_2 for 24 h (50%); V [1] yellow crystals ^1H NMR (CD_3NO_2, 32 °C): −0.4 and 1.2 (br m, H-1, 6), 1.8 (br m, CH_3), 3.4 (br m, H-5), 4.5 (br m, H-3, 4), 5.6 to 6.3 (br m, C_9H_7, H-1, 2), 6.3 to 7.4 (m, C_9H_7, H-3) [1] IR (Nujol): 2010, 2070 (ν(CO)) [1] reaction with excess $NaBH_4$ in THF at −78 °C affords $C_9H_7Mo(CO)_2C_3H_3(CH_3)_2$-anti,1,3; with ca. 2 equivalents of 1-morpholinocyclopent-1-ene in CH_2Cl_2, $C_9H_7Mo(CO)_2C_3H_3(CH_3\text{-}1)$-CHC-(O)$(CH_2)_3$-cyclo is formed [1]
6	$[(CH_3)_5C_5Mo(CO)_2C_4H_5CH_3\text{-}2]O_3SCF_3$	IV, ^3L = $C_3H_4CH{=}CH_2$-anti; also obtained by the isomerization of the corresponding trans isomer (No. 7) in solution above −40 °C [20] exo to endo ratio 7:3 [20] ^1H NMR (CD_2Cl_2): endo isomer: 1.10 (d, H-6; J(H-5, 6) = 10.4), 1.68 (d, CH_3-2), 2.02 (s, $(CH_3)_5C_5$), 2.65 (d, H-5), 5.82 (d of d, H-3; J(H-1, 3) = 10.0, J(H-4, 3) = 5.2), 6.05 (d of d of d, H-4; J(H-5, 4) = 7.5) [20]
7	$[(CH_3)_5C_5Mo(CO)_2C_4H_5CH_3\text{-}2]O_3SCF_3$	IV, ^3L = $C_3H_4CH{=}CH_2$-syn (high); trans isomer [20] ^1H NMR (CD_2Cl_2, 212 K): exo isomer: 1.90 (s, $(CH_3)_5C_5$), 1.94 (d, CH_3-2), 3.17 (m, H-4), 3.19

References on pp. 319/20

Table 13 (continued)

No.	compound	method of preparation (yield) properties and remarks
7 (continued)		(m, H-5), 3.48 (d of d, H-3; J(H-2, 3) = 12.6, J(H-4, 3) = 6.7), 3.79 (d, H-6), 4.24 (d of q, H-1; J(H-3, 1) = 12.6); endo isomer: 1.90 (d, CH_3-2), 1.93 (s, $(CH_3)_5C_5$), 3.24 (m, H-6), 3.34 (d of d, H-3; J(H-1, 3) = 12.0, J(H-4, 3) = 7.5), 3.65 (d of t, H-4; J(H-3, 4) = J(H-5, 4) = 7.0, J(H-6, 4) = 12.9), 4.24 (d, H-5; J(H-4, 5) = 6.8), 4.99 (d of q, H-1) [20] ^{13}C NMR (CD_2Cl_2, 228 K): exo isomer: 10.3 (q, $(\mathbf{C}H_3)_5C_5$; J(H, C) = 129.0), 18.4 (q, CH_3-2; J(H, C) = 129.6), 68.8 (t, CH_2; J(H, C) = 165.3), 87.7 (d, CH; J(H, C) = 171.3), 94.6 (d, CH; J(H, C) = 170.9), 101.6 (d, $\mathbf{C}HCH_3$; J(H, C) = 152.8), 108.7 (s, $\mathbf{C}_5(CH_3)_5$), 119.0 (q, CF_3; J(F, C) = 317.8), 225.6 and 233.2 (s, CO) [20] rearranges in solution above −40°C to the cis derivative (No. 6) [20]
8	$[C_5H_5Mo(CO)_2C_4H_5CH_3\text{-}3]BF_4$	I, ^{3}L = $C_3H_3(CH_3)_2$-1,1 [2]; II (35%) [15] ^{1}H NMR (acetone-d_6): 2.00 (d of d, H-6; J(H-5, 6) = 1.3, J(H-4, 6) = 9.7), 2.23 (d of d, H-1; J(H-6, 1) = J(H-4, 1) = 2.5), 2.56 (s, CH_3), 3.06 (d of d, H-5; J(H-6, 5) = 1.8, J(H-4, 5) = 7.5), 3.12 (d of d, H-2; J(H-5, 2) = J(H-4, 2) = 1.7), 6.03 (s, C_5H_5), 6.29 (d of d of d of d, H-4) [15]; similar data in [2] IR (acetone): 2008, 2059 (ν(CO)) [15] reduction with $Na[H_3BCN]$ in THF yields $C_5H_5Mo(CO)_2C_3H_3(CH_3)_2$-1,1; reaction with $Na[Re(CO)_5]$ in THF at −78°C affords exo-$C_5H_5Mo(CO)_2C_3H_3(CH_3\text{-anti},1)CH_2Re(CO)_5$-1 [17]
9	$[C_9H_7Mo(CO)_2C_4H_5CH_3\text{-}3]BF_4$	V (50%) [1] yellow crystals [1] ^{1}H NMR (CD_3NO_2, 32°C): 0.0 (br m, H-2, 5), 2.3 (s, CH_3), 3.4 (br m, H-6), 4.50 (br m, H-4), 5.8 (t, C_9H_7, H-1; J(H, H) = 3.0), 6.2 (d, C_9H_7, H-2), 7.5 (s, C_9H_7, H-3) [1] IR (Nujol): 2010, 2070 (ν(CO)) [1] reduction with $NaBH_4$ in THF affords $C_9H_7Mo(CO)_2C_3H_3(CH_3)_2$-1,1; while with 1-morpholinocyclopent-1-ene in CH_2Cl_2, $C_9H_7Mo(CO)_2C_3H_3(CH_3)CH_2CHC(O)(CH_2)_3$ is formed [1]
10	$[C_9H_7Mo(CO)_2C_4H_5CH{=}CHC_5H_{11}\text{-}2]BF_4$	I, ^{3}L = $C_3H_4CH_2CH{=}CHC_5H_{11}$; a mixture of (E, E)- and (E, Z)-1,3,5-triene cations [10]

References on pp. 319/20

Table 13 (continued)

No.	compound	method of preparation (yield) properties and remarks
11	$[(CH_3)_5C_5Mo(CO)_2C_4H_5CH{=}CHC_5H_{11}\text{-}2]BF_4$	I, $^3L = C_3H_4CH_2CH{=}CHC_5H_{11}$ [10] treatment with $(CH_3)_3NO$ in CH_3CN solution at 25°C affords good yields of free (E, E)-undeca-1,3,5-triene [10]
12	$[(CH_3)_5C_5Mo(CO)_2C_4H_5OH\text{-}2]O_3SCF_3$	IV, $^3L = C_3H_4C(O)H$-anti,1 (high); also obtained by the rearrangement of No. 13 in solution above −40°C [20] exo to endo ratio 7:3 [20] ^1H NMR (CD_2Cl_2): exo isomer: 1.41 (d of d, H-6; J(H-4, 6) = 11.7), 1.94 (s, CH_3), 2.20 (d of d, H-5; J(H-4, 5) = 8.5), 4.11 (d of d of d, H-4; J(H-3, 4) = 6.3), 4.30 (d of d, H-3), 5.67 (d, H-1; J(H-3, 1) = 9.3); endo isomer: 1.26 (d of d, H-6; J(H-4, 6) = 12.5), 1.98 (s, CH_3), 2.27 (d of d, H-5; J(H-4, 5) = 8.3), 4.24 (d of d of d, H-4; J(H-3, 4) = 6.3), 4.72 (d of d, H-3; J(H-1, 3) = 9.2), H-1 not stated [20] IR (CH_2Cl_2): 1933, 2000 (ν(CO)) [20] is reversibly deprotonated with $N(C_2H_5)_3$ in CH_2Cl_2 to the precursor allyl complex [20]
13	$[(CH_3)_5C_5Mo(CO)_2C_4H_5OH\text{-}2]O_3SCF_3$	IV, $^3L = C_3H_4C(O)H$-syn,1 (high); trans isomer [20] exo to endo ratio 7:3 [20] ^1H NMR (CD_2Cl_2, 240 K): exo isomer: 1.82 (s, CH_3), 2.72 (d of d, H-5; J(H-4, 5) = 6.9), 3.07 (d of d, H-3; J(H-1, 3) = 10.8, J(H-4, 3) = 7.7), 3.22 (d of d of d, H-4; J(H-6, 4) = 11.4), 3.35 (d of d, H-6), 6.79 (d, H-1); endo isomer: 1.88 (s, CH_3), 2.53 (d of d, H-3; J(H-3, 1) = 10.1, J(H-4, 3) = 7.4), 2.56 (d, H-6; J(H-4, 6) = 11.5), 3.65 (d, H-5; J(H-4, 5) = 6.4), 4.13 (d of d of d, H-4), 7.61 (d, H-1) [20] rearranges in solution above −40°C to yield the cis isomer No. 12 [20] treatment with $N(C_2H_5)_3$ at −78°C affords the precursor complex [20]
14	$[C_5H_5Mo(CO)_2C_4H_4(CH_3)_2\text{-}3,4]BF_4$	I, $^3L = C_3H_2(CH_3)_3$-1,1,2 [2] ^1H NMR (acetone-d_6): 2.03 (s, H-1, 6), 3.09 (d, H-2, 5; J(H, H) = 1.5), 5.98 (s, C_5H_5) [2] reaction with $Na[Re(CO)_5]$ in THF at −78°C affords exo-$C_5H_5Mo(CO)_2C_3H_2((CH_3)_2$-anti,1,2)$CH_2Re(CO)_5$-1 [17]

References on pp. 319/20

Table 13 (continued)

No.	compound	method of preparation (yield) properties and remarks
15	$[C_9H_7Mo(CO)_2C_4H_4(CH_3)_2\text{-}3,4]BF_4$	III, in CH_2Cl_2 for 24 h (60%); V [1] yellow crystals [1] ^{1}H NMR (CD_3NO_2): −0.62 (s, H-1, 6), 2.39 (s, CH_3), 3.60 (s, H-2, 5), 5.70 (t, C_9H_7, H-1; J(H, H) = 3.0), 6.19 (d, C_9H_7, H-2), 7.1 to 7.5 (m, C_9H_7, H-3) [1] IR (Nujol): 2000, 2060 (ν(CO)) [1] reduction with excess $NaBH_4$ in THF affords $C_9H_7Mo(CO)_2C_3H_2(CH_3)_3\text{-}1,1,2$ [1]
16	$[C_5H_5Mo(CO)_2C_4H_3(C(O)OCH_3\text{-}3){=}CHCH_3]BF_4$	II, at −40°C, $^3L = C_3H_3(C(O)OCH_3\text{-}2){=}CHCH_3$-syn,1 (60%) [6] yellow crystals [6] ^{1}H NMR (acetone-d_6): 5.7 and 5.8 (C_5H_5) [6] unstable, decomposes fast [6] reaction with $NaOCH_3$ affords $C_5H_5Mo(CO)_2C_3H_3({=}CHCH_3\text{-}1)(C(O)OCH_3$-syn,3)$CH_2C(O)CH_3$-anti,3 [6]
^{4}L = derivatives of trimethylenemethane		
*17	$[C_5H_5Mo(CO)_2C(CH_2)_3]BF_4$	III, for 16 h (70%) [5, 16, 17] cream-colored crystals (from CH_2Cl_2/ether) [5] ^{1}H NMR (CD_3NO_2, −90°C): 3.23 (d, H-5, 6; J(H-2, 3, 5, 6) = 5.0), 3.49 (d, H-2, 3), 3.64 (s, H-1, 4), 5.88 (s, C_5H_5); (CD_3NO_2, +25°C): 3.36 (s, CH_2), 5.88 (s, C_5H_5); T_c = −60°C and $\Delta G^{\neq}$ = 9.9 ± 1.0 kcal/mol for the $C(CH_2)_3$ rotation [5] ^{13}C NMR (CD_3NO_2): 65.3 (CH_2), 91.7 (C_5H_5), 115.3 (**C**$(CH_2)_3$), 215.8 (CO) [5] IR (Nujol): 2020, 2065 (ν(CO)) [5] reaction with $Na[Re(CO)_5]$ in THF at −78°C affords exo-$C_5H_5Mo(CO)_2C_3H_4CH_2Re(CO)_5\text{-}2$ [16, 17]
*18	$[(CH_3)_5C_5Mo(CO)_2C(CH_2)_3]BF_4$	V, followed by column chromatography on alumina with CH_2Cl_2 and treatment of the resulting $(CH_3)_5C_5Mo(CO)_2C_3H_4CH_2OH\text{-}2$ with aqueous HBF_4 in propionic anhydride (58%) [3, 5] air-stable, cream-colored crystals (by addition of ether) [5] ^{1}H NMR (CD_3NO_2): 2.1 (s, CH_3), 2.8 (s, CH_2) [3]; ($CDCl_3$): 2.31 (s, CH_3), 2.90 (s, CH_2); T_c < −70°C for the $C(CH_2)_3$ rotation, but the rotation is not completely frozen out even at −90°C [5]

References on pp. 319/20

Table 13 (continued)

No.	compound	method of preparation (yield) properties and remarks
		^{13}C NMR (CD_3NO_2): 11.6 (CH_3), 69.0 (CH_2), 106.6 (**C**$_5(CH_3)_5$), 119.0 (**C**$(CH_2)_3$) [3]; ($CDCl_3$): 12.0 (CH_3), 68.0 (CH_2), 105.0 (**C**$_5(CH_3)_5$), 117.0 (**C**$(CH_2)_3$), 218.0 (CO) [5] IR (Nujol): 2004, 2056 (ν(CO)) [5]; (CH_2Cl_2): 2012, 2055 (ν(CO)) [3]
19	$[C_5H_5Mo(CO)_2C(CH_2)_2C(CH_3)_2\text{-}5,6]BF_4$	III, with dimethylmethylenecyclopropane (70%) [5] cream-colored crystals [5] ^{1}H NMR (acetone-d_6): 2.12 (s, CH_3), 3.09 (br s, H-1, 4), 3.42 (br s, H-2, 3), 5.94 (s, C_5H_5) ^{13}C NMR (acetone-d_6): 24.5 (CH_3), 54.8 (C-1, 2), 92.5 (C_5H_5), 109.1 (C-4), 125.3 (C-3), 218.0 (CO) [5] IR (CH_2Cl_2): 2005, 2054 (**ν**(CO)) [5] reduction with $NaBH_4$ affords $C_5H_5Mo(CO)_2C_3H_4CH(CH_3)_2\text{-}2$ [5]
20	$[C_5H_5Mo(CO)_2C(CH_2)(CH)_2(CH_3)_2\text{-}1,4]BF_4$	III, cis-2,3-dimethylmethylenecyclopropane (70%) [5] ^{1}H NMR (CD_3NO_2): 1.88 (d, CH_3; J(H-2, 3, CH_3) = 3.0), 3.95 (d, H-5, 6; J(H-2, 3, H-5, 6) = 3.0), 4.0 (m, H-2, 3), 5.8 (s, C_5H_5) [5] ^{13}C NMR (CD_3NO_2): 16.3 (CH_3), 61.7 (C-3), 83.5 (C-1, 2), 92.1 (C_5H_5), 112.0 (C-4), 217.3 (CO) [5] IR (CH_2Cl_2): 2006, 2055 (ν(CO)) [5] no conversion into No. 22 in refluxing nitromethane [5]
21	$[(CH_3)_5C_5Mo(CO)_2C(CH_2)(CH)_2(CH_3)_2\text{-}1,4]BF_4$	V, with cis-2,3-dimethylmethylenecyclopropane; for workup see No. 18 (60%) [3, 5] yellow crystals [3] ^{1}H NMR (CD_3NO_2): 1.8 (d, CH_3; J(H-5, 6, CH_3) = 9.0), 2.1 (s, $(CH_3)_5C_5$), 2.55 to 2.8 (m, H-5, 6), 3.75 (d, H-2, 3; J(H-5, 6, H-2, 3) = 3.0) [3, 5] ^{13}C NMR (CD_3NO_2): 11.0 ((**C**$H_3)_5C_5$), 14.9 (CH_3), 61.9 (C-3), 87.9 (C-1, 2), 105.8 (**C**$_5(CH_3)_5$), 115.6 (C-4) [3]; (CD_2Cl_2): 11.0 ((**C**$H_3)_5C_5$), 14.9 (CH_3), 61.3 (C-3), 87.8 (C-1, 2), 104.6 (**C**$_5(CH_3)_5$), 114.4 (C-4), 219.4 (CO) [5] IR (CH_2Cl_2): 2000, 2045 (ν(CO)) [3,5] no conversion into No. 23 in refluxing CH_3NO_2 [5] reaction with $NaBH_4$ (X = H) or chromatography on basic Al_2O_3 (X = OH) affords complexes of the type $(CH_3)_5C_5Mo(CO)_2C_3H_2(CH_3)_2CH_2X$ [5]

References on pp. 319/20

Table 13 (continued)

No.	compound	method of preparation (yield) properties and remarks
22	$[C_5H_5Mo(CO)_2C(CH_2)(CH)_2(CH_3)_2\text{-}1,3]BF_4$	III, with trans-2,3-dimethylmethylenecyclopropane (70%) [5] ^1H NMR (acetone-d_6): 1.85 (d, CH_3-1; J(H-2, CH_3) = 7.4), 2.01 (d, CH_3-3; J(H-4, CH_3) = 6.6), 2.97 (d of d, H-5; J(H-6, 5) = 1.6), 3.58 (d, H-6), 4.26 (q, H-4), 5.13 (d of q, H-2; J(H-5, 2) = 3.5), 5.96 (s, C_5H_5) [5] ^{13}C NMR (CD_3NO_2): 16.0 (CH_3-1), 16.7 (CH_3-3), 55.5 (C-1), 82.9 (C-2), 90.0 (C-3), 92.5 (C_5H_5), 112.0 (C-4), 218.6 (CO) [5] IR (CH_2Cl_2): 1997, 2049 (ν(CO)) [5] no isomerization to No. 20 in refluxing CH_3NO_2 [5]
23	$[(CH_3)_5C_5Mo(CO)_2C(CH_2)(CH)_2(CH_3)_2\text{-}1,3]BF_4$	V, with trans-2,3-dimethylmethylenecyclopropane; for workup see No. 18 (60%) [3, 5] yellow crystals [3] ^1H NMR (CD_2Cl_2): 1.61 (d of d, H-5; J(H-6, 5) = 1.1), 1.91 (d, CH_3-3; J(H-4, CH_3) = 7.0), 2.0 (s, $(CH_3)_5C_5$), 2.72 (q, H-4), 3.6 (d, H-6), 4.53 (d of q, H-2; J(CH_3, H-2) = J(H-6, 2) = 7.0) [5] ^{13}C NMR (CD_2Cl_2): 12.5 ($(\mathbf{C}H_3)_5C_5$), 15.1 (CH_3-1), 16.7 (CH_3-3), 62.1 (C-1), 87.3 (C-2), 89.0 (C-3), 106 ($\mathbf{C}_5(CH_3)_5$), 115.8 (C-4), 221.7 (CO) [5] IR (CH_2Cl_2): 1996, 2040 (ν(CO)) [5] no conversion into No. 21 in refluxing CH_3NO_2 [5] reaction with $NaBH_4$ (X = H) or chromatography on basic Al_2O_3 (X = OH) affords complexes of the type $(CH_3)_5C_5Mo(CO)_2C_3H_3(CH_3\text{-}1)\text{-}C(X)HCH_3\text{-}2$ [5]
^4L = four-membered cyclic diene		
24	$[(CH_3)_5C_5Mo(CO)_2C_4(CH_3)_4]BF_4$	VI [7] ^1H NMR ($CDCl_3$): 2.04 (s, $(CH_3)_4C_4$), 2.17 (s, $(CH_3)_5C_5$) [7] ^{13}C NMR ($CDCl_3$): 8.2 ($(\mathbf{C}H_3)_4C$), 10.2 ($(\mathbf{C}H_3)_5C_5$), 84.1 ($\mathbf{C}_4(CH_3)_4$), 106.0 ($\mathbf{C}_5(CH_3)_5$), 220.2 (CO) [7] IR (CH_2Cl_2): 1985, 2037 (ν(CO)) [7]
25	$[C_5H_5Mo(CO)_2C_3(C_6H_5)_3CCH_3]BF_4$	VI (91%) [7] yellow, air-stable solid (from CH_2Cl_2/ether) [7] ^1H NMR ($CDCl_3$): 2.81 (s, CH_3), 5.65 (s, C_5H_5), 7.18 to 7.43 (m, C_6H_5)

References on pp. 319/20

Table 13 (continued)

No.	compound	method of preparation (yield) properties and remarks
		^{13}C NMR ($CDCl_3$): 13.5 (CH_3), 80.8 (**C**CH_3), 89.4 and 95.4 (**C**C_6H_5), 97.3 (C_5H_5), 127.8, 129.0, 129.3, 130.1, and 131.6 (C_6H_5), 214.8 (CO) [7] IR (CH_2Cl_2): 2010, 2054 (ν(CO)) [7] reaction with $[N(P(C_6H_5)_3)_2]Cl$ in refluxing THF for 12 h affords $C_5H_5Mo(CO)(Cl)C_3(CH_3)_3CC_6H_5$ [7]
^4L = six-membered cyclic diene		
*26	$[C_5H_5Mo(CO)_2C_6H_8]BF_4$	I, ^3L = cyclo-C_6H_9, at 0°C for 1 h (nearly quantitative yield) [4]; II (34%) [15] yellow powder (from CH_2Cl_2/ether) [4] or (CH_3NO_2/ether) [15] ^1H NMR (acetone-d_6): 2.05 and 2.30 (m, H-5, 6; J(A, B) = 11.4), 5.05 (m, H-1, 4), 6.01 (s, C_5H_5), 6.11 (m, H-2, 3) [15] IR (acetone): 2013, 2059 (ν(CO)) [15]
*27	$[C_5H_5Mo(CO)_2C_6H_8]PF_6$	I, ^3L = cyclo-C_6H_9, at 0°C for 1 h (98%) [4] yellow powder (from CH_2Cl_2/ether), m.p. > 300°C (dec.) [4] ^1H NMR (acetone-d_6): 1.44 and 2.31 (d, H-5, 6; J(H, H) = 13.3), 5.05 (m, H-2, 3), 6.05 (s, C_5H_5), 6.15 (m, H-1, 4) [4] IR (CH_2Cl_2): 1962, 2017 (ν(CO)) [4]
28	$[C_9H_7Mo(CO)_2C_6H_8]BF_4$	III (50%); V [1,10] yellow crystals [1] ^1H NMR (CD_3NO_2, 32°C): 1.8 to 2.2 (m, H-5, 6), 4.70 (s, H-1 to 4), 6.1 (t, C_9H_7, H-1; J(H, H) = 3.0), 6.29 (d, C_9H_7, H-2), 7.4 to 7.7 (m, C_9H_7, H-3) [1] IR (Nujol): 2000, 2060 (ν(CO)) [1] reduction with excess $NaBH_4$ in THF affords $C_9H_7Mo(CO)_2C_6H_9$ [1]
*29	$[C_5H_5Mo(CO)_2C_6H_7CH_3\text{-}5]PF_6$	I [9], ^3L = cyclo-$C_6H_8CH_3$-4, at 0°C for 75 min (86%) [4] yellow solid, m.p. > 300°C (dec.) [4] ^1H NMR (acetone-d_6): 1.11 (d, CH_3), 1.94 and 2.09 (d, H-6), 3.38 (q, H-5), 4.82 and 4.87 (m, H-2, 3), 6.04 (s, C_5H_5), 6.09, 6.22 (m, H-1, 4) [4] IR (CH_2Cl_2): 1963, 2018 (ν(CO)) [4]
*30	$[C_9H_7Mo(CO)_2C_6H_7CH_3\text{-}1]BF_4$	I, ^3L = exo-$C_6H_8CH_3$-1 or $C_6H_7(CH_3\text{-}4)OCH_3$-4; III, in CH_2Cl_2 for 12 h [10]
*31	$[C_9H_7Mo(CO)_2C_6H_7CH_3\text{-}2]BF_4$	III, in CH_2Cl_2 for 12 h (high) [10]

References on pp. 319/20

Table 13 (continued)

No.	compound	method of preparation (yield) properties and remarks
*32	$[C_9H_7Mo(CO)_2C_6H_7C(O)OCH_3\text{-}1]BF_4$	III, in CH_2Cl_2 for 12 h (high) [10]
33	$[C_9H_7Mo(CO)_2C_6H_7OH\text{-}1]BF_4$	II, ^3L = cyclohex-1-en-3-yl-4-one, in CH_2Cl_2 at 0°C [14]
34	$[C_9H_7Mo(CO)_2C_6H_7OH\text{-}1]O_3SCF_3$	IV, ^3L = cyclohex-1-en-3-yl-4-one, in CH_2Cl_2 at 0°C; also formed by the decomposition of No. 36 [14]
*35	$[C_9H_7Mo(CO)_2C_6H_7OCH_3\text{-}1]BF_4$	III, in CH_2Cl_2 for 12 h (high) [10, 14]; also formed in the reaction of $C_9H_7Mo(CO)_2{}^3L$ (^3L = cyclohex-1-en-3-yl-4-one) with $CF_3SO_3CH_3$ in CH_2Cl_2 at 0°C [14]
36	$[C_9H_7Mo(CO)_2C_6H_7OSi(CH_3)_3\text{-}1]O_3SCF_3$	III, in CH_2Cl_2; also formed in the reaction of $C_9H_7Mo(CO)_2{}^3L$ (^3L = cyclohex-1-en-3-yl-4-one) with $CF_3SO_3Si(CH_3)_3$ [14] unstable compound, is rapidly desilylated to No. 34 [14]
*37	$[C_9H_7Mo(CO)_2C_6H_6(CH_3)_2\text{-}1,4]BF_4$	V (52%) [18, 19] yellow, needle-shaped crystals, m.p. 125 to 126°C (from CH_2Cl_2/ether) [19] ^1H NMR (CD_3NO_2): 1.84 (s, CH_3), 1.7 to 2.0 (m, CH_2), 4.3 (s, CH=), 5.94 (t, C_9H_7, H-1), 6.37 (d, C_9H_7, H-2), 7.5 to 7.8 (m, C_9H_7, H-3) [19] IR (CH_3CN): 1940, 2010 (ν(CO)) [19]
38	$[C_9H_7Mo(CO)_2C_6H_6(CH_3)_2\text{-}2,5]BF_4$	II, ^3L = $C_6H_7(CH_3\text{-}1)R\text{-}4$, R = CH_3, in CH_2Cl_2 at −78°C in the presence of 2,3-dichloro-5,6-dicyano-1,4-benzoquinone (high) [10]
39	$[C_9H_7Mo(CO)_2C_6H_6(CH_3\text{-}2)C_2H_5\text{-}5]BF_4$	like No. 38 (R = C_2H_5) [10]
^4L = seven-membered cyclic diene		
*40	$[C_5H_5Mo(CO)_2C_7H_{10}]PF_6$	I, ^3L = C_7H_{11}, at 0°C for 1 h (95%) [11] yellow crystalline solid (from CH_2Cl_2/ether), m.p. > 300°C (dec.) [11] ^1H NMR (CD_3CN): 1.15 (m, H-6, exo), 1.35 (m, H-6, endo), 2.07 (m, H-5, 7, exo), 2.28 (m, H-5, 7, endo), 4.81 (d of t, H-1, 4; J(H, H) = 7.32, 3.47), 5.64 (s, C_5H_5), 5.78 (d of d, H-2, 3; J(H, H) = 7.34, 3.47) [11] IR (CH_2Cl_2): 1960, 2016 (ν(CO)) [11]
*41	$[C_5H_5Mo(CO)_2C_7H_9CH_3\text{-}5]PF_6$	I, ^3L = $C_7H_{10}CH_3\text{-}4$, at 0°C for 1 h (90%) [11] yellow, crystalline solid (from CH_2Cl_2 or CH_3CN/ether), m.p. 170°C (dec.) [11]

References on pp. 319/20

Table 13 (continued)

No.	compound	method of preparation (yield) properties and remarks
		^{1}H NMR (CD_3CN): 1.07 (d, CH_3-exo; J(H, H) = 6.9), 1.20 (m, H-6, exo), 1.55 (m, H-6, endo), 2.22 (m, H-5, endo and H-7, exo), 2.55 (d of q, H-7, endo; ^{2}J(H, H) = 18.12, ^{4}J(H, H) = 4.94), 4.81 (m, H-1, 4), 5.75 (s, C_5H_5), 5.88 (d of d, H-2, 3; J(H, H) = 7.57, 3.55) [11] IR (CH_2Cl_2): 850, 1960, 2016 (ν(CO)) [11]
*42	$[C_5H_5Mo(CO)_2C_7H_9CH_2CH{=}CH_2\text{-}5]PF_6$	I, ^{3}L = $C_7H_{10}CH_2CH{=}CH_2$-4, at 0°C for 1 h (90%) [11] crystalline solid, m.p. 140 to 141°C (from CH_2Cl_2/ether) [11] ^{1}H NMR ($CDCl_3$): 1.28 (m, H-6, endo, exo and H-7, exo), 2.12 (q, **CH_2**CH=, 1H; ^{2}J(H, H) = J(H-5, CH_2) = J(CH=, CH_2) = 13.4), 2.24 (m, H-7, endo and **CH_2**CH=, 1H), 2.63 (m, H-5, endo), 4.73 (br d, H1 or H4; J(H, H) = 9.2), 4.94 (m, H-4 or H-1), 5.10 (d, CH=CH_2, 1H; J(H, H) = 18.7), 5.12 (d, CH=CH_2, 1H; J(H, H) = 10), 5.73 (m, CH=CH_2, 1H), 5.80 (s, C_5H_5), 6.02 (m, H-2, 3) [11] IR (CH_2Cl_2): 1955, 2016 (ν(CO)) [11]
*43	$[C_5H_5Mo(CO)_2C_7H_9(C_6H_4OCH_3\text{-}4)\text{-}5]PF_6$	I, ^{3}L = $C_7H_{10}(C_6H_4OCH_3$-4)-4, at 0°C for 40 min (90%) [11] yellow crystals (from CH_2Cl_2/ether), m.p. > 155°C (dec.) [11] ^{1}H NMR ($CDCl_3$): 1.26 (m, H-6, exo), 1.55 (m, H-7, exo), 2.31 (m, H-6, endo), 2.75 (m, H-7, endo), 3.31 (d of d, H-5, endo; J(H, H) = 9.95, 4.69), 3.75 (s, CH_3), 4.58 (br d, H-4; J(H, H) = 9.38), 4.94 (m, H-1), 5.78 (s, C_5H_5), 6.07 (m, H-2, 3), 6.83 (d, C_6H_4, 2H; J(H, H) = 7.92), 7.07 (d, C_6H_4, 2H) [11] IR (CH_2Cl_2): 1960, 2021 (ν(CO)) [11]
^{4}L = eight-membered ring		
44	$[C_5H_5Mo(CO)_2C_8H_{12}]BF_4$ (C_8H_{12} = cycloocta-1,3-diene)	II (15%) [15] ^{1}H NMR (acetone-d_6): 2.82 (m, CH_2), 5.08 (m, CH=), 6.03 (s, C_5H_5) [15] IR (acetone): 1996, 2049 (ν(CO)) [15]
45	$[C_9H_7Mo(CO)_2C_8H_8]BF_4$	III (31%); V [1] red crystals [1] IR (Nujol): 1965, 2040 (ν(CO)) [1] tends to decompose in solution [1]

References on pp. 319/20

*Further information:

[$C_5H_5Mo(CO)_2C_4H_6$-cis]BF_4 (Table **13**, No. **1**). Reduction with Na[H_3BCN] in THF at −78°C affords $C_5H_5Mo(CO)_2C_3H_4CH_3$-anti,1 stereospecifically. Alkylation with isobutyraldehyde pyrrolidone enamine in CH_3CN at 0°C gives $C_5H_5Mo(CO)_2C_3H_4CH_2C(CH_3)_2C(O)H$-anti,1 [2]. Reaction with Na[M(CO)$_5$] (M = Mn, Re) in THF at −78°C for 45 min affords exo-$C_5H_5Mo(CO)_2C_3H_4CH_2M(CO)_5$-syn,1 [16, 17].

[$C_5H_5Mo(CO)_2C_4H_5CH_3$-2]BF_4 (Table **13**, No. **4**). Reduction with Na[H_3BCN] in THF at −78°C gives a mixture of $C_5H_5Mo(CO)_2{}^3L$ complexes, the major products being those in which ^3L is $C_3H_3(CH_3$-anti,1)CH_3-syn or $C_3H_3C_2H_5$-anti,1 [2]. Treatment with Na[Re(CO)$_5$] in THF at −78°C for 2 h affords exo-$C_5H_5Mo(CO)_2C_3H_4(CH_3$-3, anti)$CH_2Re(CO)_5$-syn,1 [17].

[$C_5H_5Mo(CO)_2C(CH_2)_3$]BF_4 (Table **13**, No. **17**). Extended Hückel molecular orbital calculations suggest that the observed syn stereochemistry (established for No. 18 by X-ray crystal analysis) is electronically preferred and the barrier of rotation of the $C(CH_2)_3$ ligand relative to the $C_5H_5Mo(CO)_2$ fragment is high. The calculated barrier to rigid rotation of flat $C(CH_2)_3$ is 48.4 and for bent $C(CH_2)_3$ 48.5 kcal/mol. The calculated energy difference is 45.5 (flat $C(CH_2)_3$) and 46.7 (bent $C(CH_2)_3$) kcal/mol between syn (Formula III) and anti ($C(CH_2)_3$ rotated about 60°) conformers [5].

[$(CH_3)_5C_5Mo(CO)_2C(CH_2)_3$]$BF_4$ (Table **13**, No. **18**) crystallizes in the orthorhombic space group Pbca−D_{2h}^{15} (No. 61) with the unit cell parameters a = 12.822(2), b = 12.311(3), c = 22.660(4); Z = 8 ion pairs per unit cell, D_{meas} (by flotation) = 1.5, and D_{calc} = 1.58 g/cm^3. The molecular structure with the main bond distances and angles is shown in **Fig. 55**. A figure of the disordered BF_4 anion and the content of the unit cell were also given [5].

Reaction with nucleophiles affords the allyl complexes, $(CH_3)_5C_5Mo(CO)_2C_3H_4CH_2X$-2. X = H is obtained in the reaction with $NaBH_4$ in THF, X = CH_3 in the reaction with $LiCu(CH_3)_2$ in

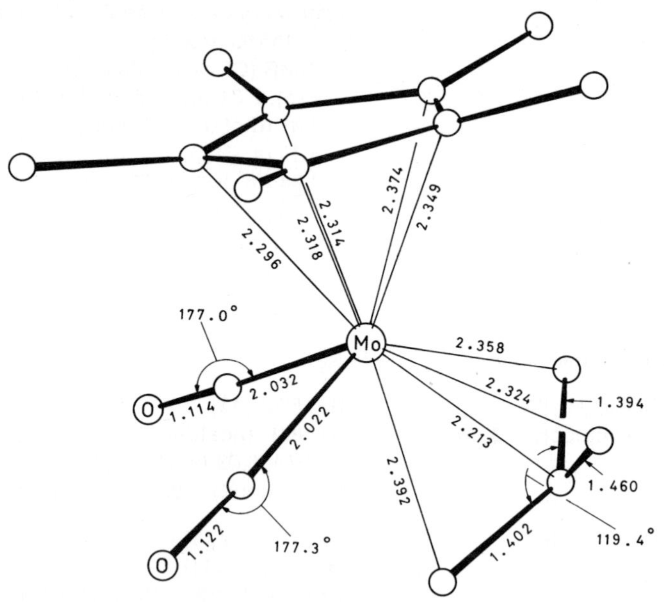

Fig. 55. The molecular structure of [$(CH_3)_5C_5Mo(CO)_2C(CH_2)_3$]$^+$ [5].

ether, X = OH on treatment with basic Al_2O_3 in CH_2Cl_2, and X = SC_6H_5 in the reaction with $NaSC_6H_5$ in THF [5].

[^{5}LMo(CO)$_2$C$_6$H$_{8-n}$R$_n$]X (Table **13**, Nos. **26**, **27**, **29** to **32**, **35**, **37**) were allowed to react with a variety of nucleophiles mostly in THF to give $C_5H_5Mo(CO)_2{}^3L$ with a cyclic allyl ligand, as described in the previous section. The addition of the nucleophile occurs at the 4-position (or 1-position in substituted derivatives) of the cyclohexa-1,3-diene ligand. The ^{3}L fragments obtained and the conditions used are given in the following table, if not stated THF is used as solvent.

No.	nucleophile (conditions)	^{3}L ligand obtained
26	2-methyl-1-propenyl-1-pyrrolidine (in CH_3CN at 0°C)	cyclo-$C_6H_8C(CH_3)_2C(O)H$ [4]
26	$Na[Re(CO)_5]$ (at −78°C)	cyclo-$C_6H_8Re(CO)_5$-4 [17]
27	$Na[D_3BCN]$ (at 0°C)	cyclo-C_6H_8D-4 [4]
27	CH_3MgBr (at 0°C)	cyclo-$C_6H_8CH_3$-4 [4]
27	$Na[C(R)(C(O)OCH_3)Y]$ (R = H, CH_3, Y = $C(O)OCH_3$; R = H, Y = $SO_2C_6H_5$, $C(O)CH_3$)	cyclo-$C_6H_8C(R)(C(O)OCH_3)Y$-4 [8, 9, 12]
27	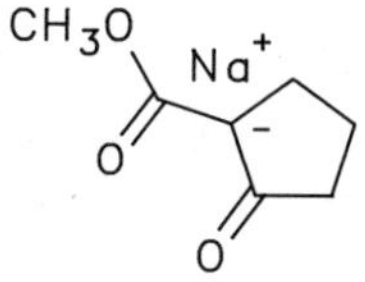	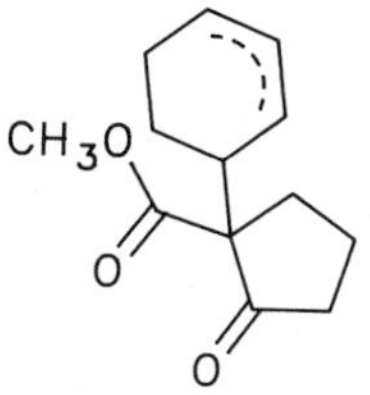 [8, 12]
27	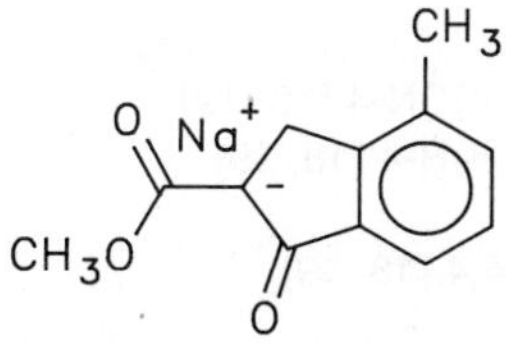	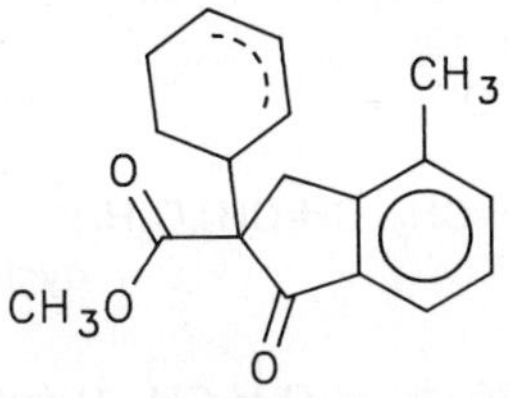 [8, 12]
27	NaCN (in CH_3CN/H_2O)	cyclo-C_6H_8CN-4 [13]
27	4-$CH_3OC_6H_4MgBr$	cyclo-$C_6H_8C_6H_4OCH_3$-4 [8, 12]
29	CH_3MgBr (at 0°C)	cyclo-$C_6H_7(CH_3)_2$-4,6 [4, 9]
29	$Na[CH(C(O)OCH_3)_2]$	cyclo-$C_6H_7(CH_3$-4$)CH(C(O)OCH_3)_2$-6 [8, 9, 12]
29	$Na[CH(SO_2C_6H_5)C(O)OCH_3]$	cyclo-$C_6H_7(CH_3$-4$)CH(SO_2C_6H_5)C(O)OCH_3$-6 [9, 12]
29	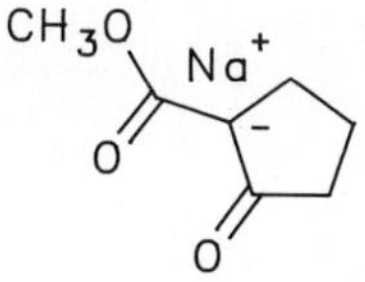	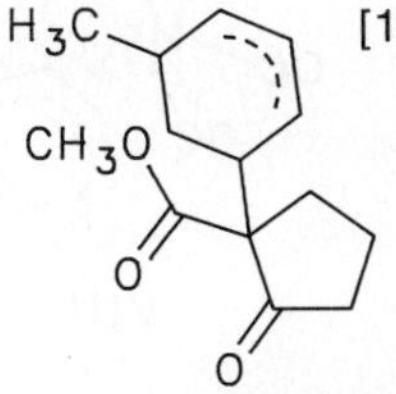 [12]
29	NaCN (in CH_3CN/H_2O)	cyclo-$C_6H_7(CH_3$-4$)CN$-6 [13]
30	$Na[(s\text{-}C_4H_9)_3BH]$	cyclo-$C_6H_8CH_3$-1 [10]
30	$LiCu(CH_3)_2$	cyclo-$C_6H_7(CH_3)_2$-1,4 [10]
30	C_2H_5MgBr	cyclo-$C_6H_7(CH_3$-1$)C_2H_5$-4 [10]

References on pp. 319/20

No.	nucleophile (conditions)	^{3}L ligand obtained
30	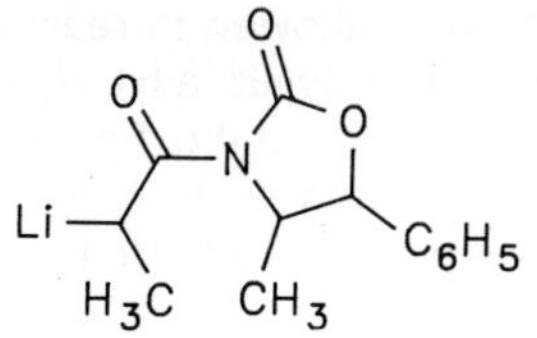 [structure: Li–CH(CH$_3$)–C(O)–N(oxazolidinone, 4-CH$_3$, 5-C$_6$H$_5$)]	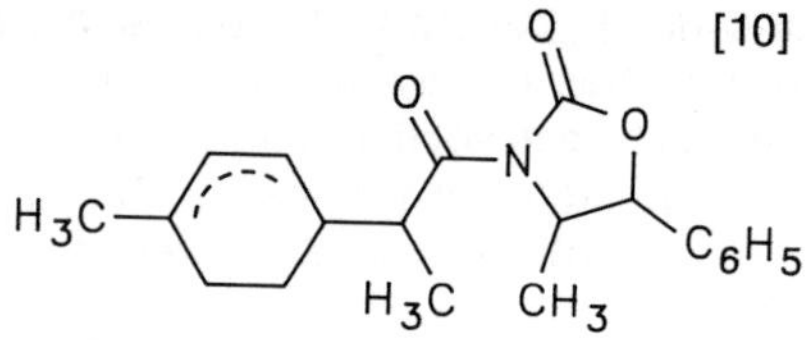 [structure: H$_3$C–cyclohexadienyl–CH(CH$_3$)–C(O)–N(oxazolidinone, 4-CH$_3$, 5-C$_6$H$_5$)] [10]
31	$K[(s\text{-}C_4H_9)_3BH]$	cyclo-$C_6H_8CH_3$-3 [10]
31	$LiCu(CH_3)_2$	cyclo-$C_6H_7(CH_3)_2$-3,4 [10]
32	$K[(s\text{-}C_4H_9)_3BH]$	cyclo-$C_6H_8CH_3$-3 [10]
32	$LiCu(CH_3)_2$	cyclo-$C_6H_7(CH_3)_2$-3,4 [10]
35	$K[(s\text{-}C_4H_9)_3BH]$	cyclo-$C_6H_8OCH_3$-4 [10, 14]
35	RMgX ($R = CH_3$, $X = I$; $R = C_6H_5$, $X = Br$)	cyclo-C_6H_7(R-4)OCH_3-4 [10, 14]
37	$LiOC(OCH_3){=}CHR$ ($R = H, CH_3$)	cyclo-$C_6H_6((CH_3)_2$-1,4)C(R)HC(O)OCH_3-4 [18, 19]
37	$LiOC(OSi(CH_3)_3)C{=}CHCH_3$	cyclo-$C_6H_6((CH_3)_2$-1,4)C(CH_3)HC(O)OH [18, 19]
37	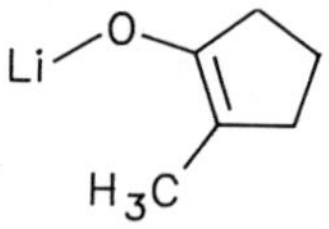 [structure: 1-(LiO)-2-methylcyclopentene]	[structure: 2-methyl-2-(1,4-dimethylcyclohexadienyl-1)cyclopentanone; H$_3$C, H$_3$C, CH$_3$] [18, 19] 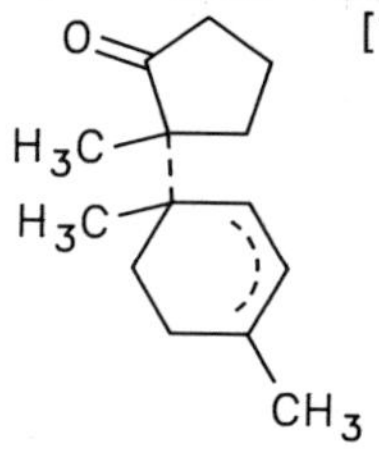
37	$Na[C(R)(C(O)OCH_3)Y]$ ($R = H$, $Y = C(O)OCH_3$; $R = H, CH_3$, $Y = SO_2C_6H_5$) [18, 19]	cyclo-$C_6H_6((CH_3)_2$-1,4)C(R)(C(O)OCH_3)Y-4
37	NaCN (in CH_3CN/H_2O)	cyclo-$C_6H_6((CH_3)_2$-1,4)CN-4 [18, 19]
37	RMgBr ($R = CH_3$, $CH_2CH{=}CH_2$, $CH{=}CH_2$, C_6H_5)	cyclo-$C_6H_6((CH_3)_2$-1,4)R-4 [18, 19]
37	$LiCu(CH_3)_2$	cyclo-$C_6H_6(CH_3)_3$-1,4,4 [18, 19]

Deprotonation of No. 30 ($^4L = C_6H_7CH_3$-1) with $LiN(C_3H_7\text{-}i)_2$ affords $C_9H_7Mo(CO)_2C_6H_6CH_3$ (Formula VIII); treatment of this product with $(C_2H_5)_2O \cdot HBF_4$ regenerates No. 30 [10].

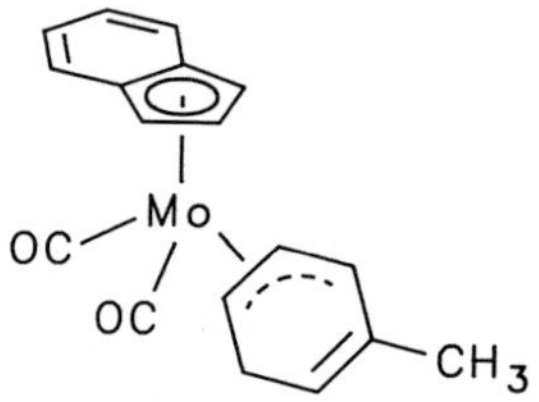

[structure: indenyl–Mo(CO)$_2$–(methylcyclohexadiene); labels: Mo, OC, OC, CH$_3$]

VIII

[C$_5$H$_5$Mo(CO)$_2$C$_7$H$_9$R-5]PF$_6$ (Table **13**, Nos. **40** to **43**) react with a variety of nucleophiles mostly in THF with formation of $C_5H_5Mo(CO)_2{}^3L$ (^3L = cycloheptenyl derivative). The addition of the nucleophile occurs at the 1- or 4-position of the diene. These reactions are given in the following table [11].

No.	nucleophile (conditions)	^{3}L ligand obtained
40	CH_3MgBr (excess at 0°C)	$C_7H_{10}CH_3$-4
40	RMgBr (R = $CH_2CH{=}CH_2$, $C_6H_4CH_3$-4)	$C_7H_{10}R$-4
40	$Na[CH(C(O)OCH_3)Y]$ (Y = $C(O)OCH_3$, $SO_2C_6H_5$,$C(O)CH_3$)	$C_7H_{10}CH(C(O)OCH_3)Y$-4
40	O, Na^+, CH_3O, O	O, CH_3O, O
40	NaCN (in CH_3CN/H_2O)	$C_7H_{10}CN$-4
41	CH_3MgBr (excess at 0°C)	$C_7H_9(CH_3)_2$-4,7 and $C_7H_9(CH_3)_2$-4,5 in a 2:1 ratio
41	4-$CH_3OC_6H_4MgBr$ (excess in CH_2Cl_2/THF)	$C_7H_9(CH_3$-7)($C_6H_4OCH_3$-4)-4 and $C_7H_9(CH_3$-5)-($C_6H_4OCH_3$-4)-4 in a 5:1 ratio
41	$Na[CH(C(O)OCH_3)_2]$	$C_7H_9(CH_3$-7)$CH(C(O)OCH_3)_2$-4 and $C_7H_9(CH_3$-5)-$CH(C(O)OCH_3)_2$-4 in a 2:1 ratio
41	$Na[CH(SO_2C_6H_5)C(O)OCH_3]$	$C_7H_9(CH_3$-7)$CH(SO_2C_6H_5)C(O)OCH_3$-4
41	O, Na^+, CH_3O, O	H_3C, O, CH_3O, O
41	NaCN (in CH_3CN/H_2O 5:1)	$C_7H_9(CH_3$-7)CN-4 and $C_7H_9(CH_3$-5)CN-4 in a 3:2 ratio
42	$Na[CH(C(O)OCH_3)_2]$	$C_7H_9(CH_2CH{=}CH_2$-7)$CH(C(O)OCH_3)_2$-4
42	$Na[CH(SO_2C_6H_5)C(O)OCH_3]$	$C_7H_9(CH_2CH{=}CH_2$-7)$CH(SO_2C_6H_5)C(O)OCH_3$-4
43	CH_3MgBr (excess in THF/CH_2Cl_2)	$C_7H_9((C_6H_4OCH_3$-4)-7)CH_3-4
43	$Na[CH(C(O)OCH_3)_2]$	$C_7H_9((C_6H_4OCH_3$-4)-7)$CH(C(O)OCH_3)_2$-4

References:

[1] Bottrill, M.; Green, M. (J. Chem. Soc. Dalton Trans. **1977** 2365/71).
[2] Faller, J. W.; Rosan, A. M. (J. Am. Chem. Soc. **99** [1977] 4858/9).
[3] Barnes, S. G.; Green, M. (J. Chem. Soc. Chem. Commun. **1980** 267/8).
[4] Faller, J. W.; Murray, H. H.; White, D. L.; Chao, K. H. (Organometallics **2** [1983] 400/9).
[5] Allen, S. R.; Barnes, S. G.; Green, M.; Moran, G.; Trollope, L.; Murrall, N. W.; Welch, A. J.; Sharaiha, D. M. (J. Chem. Soc. Dalton Trans. **1984** 1157/69).
[6] Giulieri, F.; Benaim, J. (J. Organometal. Chem. **276** [1984] 367/76).
[7] Hughes, R. P.; Reisch, J. W.; Rheingold, A. L. (Organometallics **3** [1984] 1761/3).
[8] Pearson, A. J.; Khan, M. N. I. (Tetrahedron Letters **25** [1984] 3507/10).
[9] Pearson, A. J.; Khan, M. N. I. (J. Am. Chem. Soc. **106** [1984] 1872/3).
[10] Green, M.; Greenfield, S.; Kersting, M. (J. Chem. Soc. Chem. Commun. **1985** 18/20).

[11] Pearson, A. J.; Khan, M. N. I. (J. Org. Chem. **50** [1985] 5276/85).
[12] Pearson, A. J.; Khan, M. N. I.; Clardy, J. C.; Henry, H. C. (J. Am. Chem. Soc. **107** [1985] 2748/57).
[13] Pearson, A. J.; Khetani, V. D. (J. Chem. Soc. Chem. Commun. **1986** 1772/4).
[14] Green, M.; Greenfield, S.; Grimshire, M. J.; Kersting, M.; Orpen, A. G.; Rodrigues, R. A. (J. Chem. Soc. Chem. Commun. **1987** 97/9).
[15] Gusev, O. V.; Krivykh, V. V.; Petrovskii, P. V.; Rybinskaya, M. I. (Izv. Akad. Nauk SSSR Ser. Khim. **1987** 1635/7; Bull. Acad. Sci. USSR Div. Chem. Sci. **1987** 1532/4).
[16] Müller, H.-J.; Beck, W. (J. Organometal. Chem. **330** [1987] C 13/C 16).
[17] Müller, H.-J. (Diss. Univ. München 1988, pp. 1/150).
[18] Pearson, A. J.; Khetani, V. D. (J. Org. Chem. **53** [1988] 3395/6).
[19] Pearson, A. J.; Khetani, V. D. (J. Am. Chem. Soc. **111** [1989] 6778/89).
[20] Benyunes, S. A.; Green, M.; Grimshire, M. J. (Organometallics **8** [1989] 2268/70).

Empirical Formula Index

In the following index the compounds are listed in the order of increasing carbon content. Empirical formulas of ionic compounds are given in brackets; ions as well as components of solvates and adducts are separated by a period.

Page references are printed in ordinary types, table numbers in bold face, and compound numbers in the tables in italics.

Ligand Formula Index

Ligands (except CO) are used in the following Ligand Formula Index to locate a compound in the order of increasing carbon content of the respectively ligand. The number of identical ligands in a compound and the nature of bonding are not taken into consideration, so that several compounds may be listed at one position, if need be. On the other hand, compounds having two or more different types of carbon-containing ligands occur at two or more positions, respectively. The following examples illustrate the arrangement.

trans-$C_5H_5Mo(CO)_2(P(C_6H_5)_3)C_5F_4N$-4

C_5F_4N	C_5H_5	$C_{18}H_{15}P$	—	CO	96, **6**, *76*
C_5H_5	C_5F_4N	$C_{18}H_{15}P$	—	CO	96, **6**, *76*
$C_{18}H_{15}P$	C_5F_4N	C_5H_5	—	CO	96, **6**, *76*

Page references are printed in ordinary types, table numbers in boldface, and compound numbers in the tables in italics.

Physical Constants and Conversion Factors

Avogadro constant N_A (or L) = 6.02214×10^{23} mol^{-1}
Faraday constant F = 9.64853×10^{4} C/mol
molar gas constant R = 8.31451 J·mol^{-1}·K^{-1}
molar volume (ideal gas) V_m = 2.24141×10^{1} L/mol
(273.15 K, 101325 Pa)

Planck constant h = 6.62608×10^{-34} J·s
elementary charge e = 1.60218×10^{-19} C
electron mass m_e = 9.10939×10^{-31} kg
proton mass m_p = 1.67262×10^{-27} kg

1 kg = 2.205 pounds
1 m = 3.937×10^{1} inches = 3.281 feet
1 m^3 = 2.642×10^{2} gallons (U.S.)
1 m^3 = 2.200×10^{2} gallons (Imperial)

Force	N	dyn	kp
1 N	1	10^{5}	1.019716×10^{-1}
1 dyn	10^{-5}	1	1.019716×10^{-6}
1 kp	9.80665	9.80665×10^{5}	1

Pressure	Pa	bar	kp/m^2	at	atm	Torr	lb/in^2
1 Pa = 1 N/m^2	1	10^{-5}	1.019716×10^{-1}	1.019716×10^{-5}	9.86923×10^{-6}	7.50062×10^{-3}	1.450378×10^{-4}
1 bar = 10^6 dyn/cm^2	10^{5}	1	1.019716×10^{4}	1.019716	9.86923×10^{-1}	7.50062×10^{2}	1.450378×10^{1}
1 kp/m^2 = 1 mm H_2O	9.80665	9.80665×10^{-5}	1	10^{-4}	9.67841×10^{-5}	7.35559×10^{-2}	1.422335×10^{-3}
1 at (technical)	9.80665×10^{4}	9.80665×10^{-1}	10^{4}	1	9.67841×10^{-1}	7.35559×10^{2}	1.422335×10^{1}
1 atm = 760 Torr	1.01325×10^{5}	1.01325	1.033227×10^{4}	1.033227	1	7.60×10^{2}	1.469595×10^{1}
1 Torr = 1 mmHg	1.333224×10^{2}	1.333224×10^{-3}	1.359510×10^{1}	1.359510×10^{-3}	1.315789×10^{-3}	1	1.933678×10^{-2}
1 lb/in^2 = 1 psi	6.89476×10^{3}	6.89476×10^{-2}	7.03069×10^{2}	7.03069×10^{-2}	6.80460×10^{-2}	5.17149×10^{1}	1

Work, Energy, Heat	J	kW·h	kcal	Btu	eV
1 J = 1 W·s = 1 N·m = 10^7 erg	1	2.778×10^{-7}	2.39006×10^{-4}	9.4781×10^{-4}	6.242×10^{18}
1 kW·h	3.6×10^{6}	1	8.604×10^{2}	3.41214×10^{3}	2.247×10^{25}
1 kcal	4.1840×10^{3}	1.1622×10^{-3}	1	3.96566	2.6117×10^{22}
1 Btu (British thermal unit)	1.05506×10^{3}	2.93071×10^{-4}	2.5164×10^{-1}	1	6.5858×10^{21}
1 eV	1.602×10^{-19}	4.450×10^{-26}	3.8289×10^{-23}	1.51840×10^{-22}	1

1 cm^{-1} = 1.239842×10^{-4} eV 1 Hz = 4.135669×10^{-15} eV

1 hartree = 27.2114 eV 1 eV ≙ 23.0578 kcal/mol

Power	kW	hp	$kp\cdot m\cdot s^{-1}$	kcal/s
1 kW = 10^3 J/s	1	1.35962	1.01972×10^{2}	2.39006×10^{-1}
1 hp (horsepower, metric)	7.3550×10^{-1}	1	7.5×10^{1}	1.7579×10^{-1}
1 $kp\cdot m\cdot s^{-1}$	9.80665×10^{-3}	1.333×10^{-2}	1	2.34384×10^{-3}
1 kcal/s	4.1840	5.6886	4.26650×10^{2}	1

References:

Mills, I. (Ed.), International Union of Pure and Applied Chemistry, Quantities, Units and Symbols in Physical Chemistry, Blackwell Scientific Publications, Oxford 1988.

The International System of Units (SI), National Bureau of Standards Spec. Publ. 330 [1972].

Landolt-Börnstein, 6th Ed., Vol. II, Pt. 1, 1971, pp. 1/14.

ISO Standards Handbook 2, Units of Measurement, 2nd Ed., Geneva 1982.

Cohen, E. R., Taylor, B. N., Codata Bulletin No. 63, Pergamon, Oxford 1986.